International Association of Geodesy Symposia

Klaus-Peter Schwarz, Series Editor

Springer-Verlag Berlin Heidelberg GmbH

International Association of Geodesy Symposia

Klaus-Peter Schwarz, Series Editor

Symposium 101: Global and Regional Geodynamics
Symposium 102: Global Positioning System: An Overview
Symposium 103: Gravity, Gradiometry, and Gravimetry
Symposium 104: Sea Surface Topography and the Geoid
Symposium 105: Earth Rotation and Coordinate Reference Frames
Symposium 106: Determination of the Geoid: Present and Future
Symposium 107: Kinematic Systems in Geodesy, Surveying, and Remote Sensing
Symposium 108: Applications of Geodesy to Engineering
Symposium 109: Permanent Satellite Tracking Networks for Geodesy and Geodynamics
Symposium 110: From Mars to Greenland: Charting Gravity with Space and Airborne Instruments
Symposium 111: Recent Geodetic and Gravimetric Research in Latin America
Symposium 112: Geodesy and Physics of the Earth: Geodetic Contributions to Geodynamics
Symposium 113: Gravity and Geoid
Symposium 114: Geodetic Theory Today
Symposium 115: GPS Trends in Precise Terrestrial, Airborne, and Spaceborne Applications
Symposium 116: Global Gravity Field and Its Temporal Variations
Symposium 117: Gravity, Geoid and Marine Geodesy
Symposium 118: Advances in Positioning and Reference Frames

Advances in Positioning and Reference Frames

IAG Scientific Assembly

Rio de Janeiro, Brazil, September 3-9, 1997

Convened and Edited by

Fritz K. Brunner

Springer

Volume Editor
Professor Dr. Fritz K. Brunner
Engineering Surveying and Metrology
Technical University of Graz
Steyrergasse 30
A-8010 Graz
Austria

Series Editor
Professor Dr. Klaus-Peter Schwarz
University of Calgary
Department of Geomatics Engineering
2500 University Drive N. W.
Calgary, Alberta T2N 1N4
Canada

ISBN 978-3-642-08425-6 ISBN 978-3-662-03714-0 (eBook)
DOI 10.1007/978-3-662-03714-0

Library of Congress Cataloging-in-Publication Data
Advances in positioning and reference frames : international symposium no. 118 : Rio de Janeiro, Brazil, September 3-9, 1997 / convened and edited by Fritz K. Brunner.
p. cm. -- (International Association of Geodesy symposia : Symposium no. 118)
Includes bibliographical references.

1. Global Positioning System--Congresses. 2. Geodesy--Congresses.
I. Brunner, F. K. (Fritz K.), 1943- . II. Series: International Association of Geodesy symposia : symposium 118
G109.5.A39 1998 623.89 '3--dc21 98-25812 CIP

Originally published by Springer-Verlag Berlin Heidelberg New York in 1998.

Softcover reprint of the hardcover 1st edition 1998

Typesetting: Camera ready by editor/author
Cover layout: design & production GmbH, Heidelberg

SPIN: 10680111 32/3020 - 5 4 3 2 1 0 - Printed of acid-free paper

FOREWORD

Klaus-Peter Schwarz, IAG President
The University of Calgary

The Scientific Assembly of the International Association of Geodesy (IAG) was held from September 3 to 9, 1997 in Rio de Janeiro, Brazil, in conjunction with the 18th Brazilean Congress of Carthography. This was the first time that one ofthe major IAG meetings took place in Brazil. It provided an opportunity to showcase the progress of geodetic work in South America through campaigns such as SIRGAS. It also provided an opportunity for a large group of international experts to present the state of the art in geodesy and geodynamis and to interact with their hosts on possibilities of future cooperation. For the IAG, it continued a trend, started in Beijing four years ago, to hold major geodetic meetings outside of Europe and North America. The International Geoid School which was held in Sao Paulo following this meeting showed another facet of this growing internationalization of IAG activities and services.

The scientific program of the meeting consisted of three symposia and two special sessions, namely

Symposium1: Advances in Positioning and Reference Frames
Symposium 2: Gravity and Geoid
Symposium 3: Geodynamics
Special Session 1: IAG Services
Special Session 2: Geodesy in Antartica.

Papers presented at the first symposium are published in volume 1 of these proceedings, while papers of symposia 2 and 3 as well as special session 2 are contained in volume 2. Papers presented at special session 1 will be published separately. More details on the individual volumes are given in the prefaces written by the convenors.

The meeting was jointly organized by the IAG and the Brazilean Society of Cartography, Geodesy, Photogrammetry and Remote Sensing. The two local organizing committees worked closely together to economize on some of the organizational aspects and to guarantee a smooth running of two conferences in the same place. Thanks go to our Brazilean colleagues for their hard work and their warm hospitality. On the geodetic side, special thanks go to Professor D. Blitzkov, the national representative of the IAG who chaired the IAG Local Organizing Committee (LOC), and the dedicated group of individuals working working with him. Similarly, Professor W. Torge, past president of the IAG, who provided the liason between the LOC and the IAG Executive deserves a special vote of thanks. Finally, the symposium and session convenors who not only organized the scientific program but also took care of organizing the review rocess and editing these proceedings, are thanked for their outstanding efforts.

PREFACE

Over the past few years we have experienced tremendous progress in GPS positioning. Therefore, it seemed appropriate to organise a special symposium for the review of current developments in precise GPS positioning with special emphasis on the work of the related Special Study Groups of IAG.

"Advances in Positioning and Reference Frames" was one of the symposia held during the Scientific Assembly of IAG in Rio de Janeiro in September 1997. This symposium comprised six sessions: Maintenance and Densification of Reference Frames, GPS Reference Networks, the SIRGAS Project, Current Developments in Precise GPS Positioning, GPS Kinematic Applications, and a very successful Poster Session. During the symposium, 25 invited papers were presented and 70 posters were displayed.

This volume contains the reviewed contributions to the symposium. Its contents reflect the exciting and steadily growing developments of fundamental GPS work as well as novel applications of static and kinematic GPS surveying techniques. The maintenance and the densification of reference frames are treated for the purpose of establishing global and regional GPS networks. The scientific achievements of the South American Geocentric Reference System project (SIRGAS) are discussed. Congratulations to all SIRGAS contributors for their outstanding achievements! Several contributions review the state of the art of GPS analysis techniques, ambiguity resolution methods, as well as GPS antenna and site problems. New applications of kinematic GPS positioning and the quality control issues of real-time GPS positioning are presented.

The individual sessions of the symposium were organised by the conveners Yehuda Bock, Claude Boucher, Luiz Fortes and Chris Rizos, who also carried the main burden of reviewing the manuscripts. Ms. S. Schmuck assisted in the organisation of the papers. I would like to thank everybody who helped with the preparation of the symposium and the proceedings, especially the authors for their excellent contributions.

May 1998

Fritz K. Brunner

CONTENTS

Session 1: Maintenance and Densification of Reference Frames
Convener: Claude Boucher

WGS 84 - Past, Present and Future — 1
J.A. Slater and S. Malys (invited)

ITRF Densification and Continuous Realization by the IGS — 8
G. Blewitt, C. Boucher, P.B.H. Davies, M.B. Heflin, T.A. Herring and J. Kouba (invited)

The Use of the EUREF Permanent GPS Network for the Maintenance of the European Terrestrial Reference Frame — 18
C. Bruyninx, J. Dousa, W. Ehrnsperger, N. Fachbach, J. Johansson, F. Vespe, L. Ferraro, A. Nardi, M. Figurski, T. Springer, G. Weber and A. Wiget (invited)

The European Vertical GPS Reference Network Campaign 1997 - Concept and Status — 27
J. Ihde, J. Adam, W. Gurtner, B.G. Harsson, W. Schlüter and G. Woeppelmann (invited)

From UELN-95 to EVS 2000 - European Activities for a Continental Vertical Datum — 35
W. Augath, F. Brouwer, H. Lang, J. van Mierlo and M. Sacher

Global GPS Networks and the Determination of Earth Rotation Parameters — 43
V. Ashkenazi, T. Moore, L.J. Symons, W. Chen and M.H.O. Aquino

Realization of the ITRF-94 in Thailand and Malaysia by Use of a Combined Network for Geodynamics and National Survey — 49
M. Becker, E. Reinhart, H. Seeger, S. Mingsamon, C. Boonphakdee and S. Abu

The ITRF96 Realization of the International Terrestrial Reference System (Abstract) 57
C. Boucher, Z. Altamini and P. Sillard

Modernizing the Reference Framework for Canada´s Maritime Provinces 59
R.J. Gaudet

Permanent GPS Tracking Network including the Mediterranean Area 66
G. Weber, M. Becker, P. Franke, W. Schlüter and H. Seeger

Session 2: GPS Reference Networks
Convener: Fritz K. Brunner

The Brazilian Network for Continuous Monitoring of GPS (RBMC): Operation and Products 73
L.P.S. Fortes, R.T. Luz, K.D.Pereira, S.M.A. Costa and D. Blitzkow (invited)

Towards an Austrian GPS Reference Network 79
R. Weber (invited)

Real-Time Differential GPS Error Modelling in Regional Reference Station Networks 86
L. Wanninger (invited)

Processing Strategies for Regional GPS Networks 93
M. Rothacher, T.A. Springer, S. Schaer and G. Beutler (invited)

A Continental Wide Area Differential GPS Strategy 101
M.H.O. Aquino, V. Ashkenazi and T. Moore

Possibility of a Dynamic Cadastre for a Dynamic Nation 107
G. Blick and D. Grant

Preliminary Stability Test for the Regional GPS Tracking Stations in Taiwan 114
C.C. Chang and R.G. Chang

The Use of Heuristics in the Design of GPS Networks 120
P. Dare and H. Saleh

Transformations of GPS Coordinates and Heights in Trinidad 125
K. Edwards, J. Barbaste and J. Saleh

Vectors Connecting the Geodetic Points at Metsähovi and Sjökulla 131
J. Jokela

Use of the Finnish Permanent GPS Network (FinnNet) in Regional GPS Campaigns 137
H. Koivula, M. Ollikainen and M. Poutanen

The New Swiss National Height System LHN95 143
U. Marti and D. Schneider

The Non-Fiducial Approach Applied to GPS Networks 149
J.F.G. Monico and E.S. da Fonseca Junior

The Statefix West Australian GPS Network 155
M.P. Stewart, H. Houghton and X. Ding

POLREF-96 - The New Geodetic Reference Frame for Poland 161
J.B. Zielinski, A. Lyszkowicz, L. Jaworski, A. Swiatek, R. Zdunek and S. Gelo

Session 3: SIRGAS Project
Convener: Luiz Paulo S. Fortes

An Overview of the SIRGAS Project (Abstract) 167
L.P.S. Fortes (invited)

The Definition and Realization of the Reference System in the SIRGAS Project 168
M. Hoyer, S. Arciniegas, K. Pereira, H. Fagard, R. Maturana, R. Torchetti, H. Drewes, M. Kumar and G. Seeber (invited)

Time Evolution of the SIRGAS Reference Frame 174
H. Drewes (invited)

Results of the SIRGAS 95 GPS Network Processing at DGFI/I 180
K. Kaniuth, H. Drewes, K. Stuber, H. Tremel and J. Moirano (invited)

The Integration of Brazilian Geodetic Network into SIRGAS 187
S.M.A. Costa, K.D. Pereira and D. Beattie

Adjustment of the New Venezuelan National GPS Network within the SIRGAS Reference Frame 193
H. Drewes, H. Tremel and J.N. Hernández

Realization of a Geocentric Reference System in Argentina in Connection with SIRGAS 199
J. Moirano, C. Brunini, H. Drewes and K. Kaniuth

Some Considerations Related to the New Realization of SAD-69 in Brazil 205
L.C. Oliveira, J.F.G. Monico, M.C. Santos and D. Blitzkow

The IGS Regional Network Associate Analysis Center for South America at DGFI/I 211
W. Seemüller and H. Drewes

The Redefinition of the Geodetic Reference System of Uruguay into SIRGAS Frame 217
W.H. Subiza, R.R. Pérez, F. Barbato and S.M.A. Costa

Session 4: Current Developments in Precise GPS Positioning
Convener: Yehuda Bock

GPS Ambiguity Resolution for Navigation, Rapid Static Surveying, and Regional Networks 223
P.J. de Jonge (invited)

GPS Antenna and Site Effects 229
J.M. Johansson (invited)

Precise GPS Positioning Improvements by Reducing Antenna and Site Dependent Effects 237
G. Seeber, F. Menge, C. Völksen, G. Wübbena and M. Schmitz (invited)

IGS Orbit, Clock and EOP Combined Products: An Update 245
J. Kouba and Y. Mireault (invited)

The Use of GPS for Monitoring of the Ionospheric Disturbances 252
L.W. Baran and I.I. Shagimuratov

An Integrated GPS Monitoring System for Site Investigation of Nuclear Waste Disposal 259
R. Chen and J. Kakkuri

Determination of Stochastic Models of GPS Baselines in GPS Network Adjustment 265
X. Ding, M. Stewart and J. Chao

The Assisi Landslide GPS Network 271
D. Dominici, F. Radicioni, S. Selli and A. Stoppini

Use of a Regional Ionospheric Model in GPS Geodetic Applications 277
M. Gende, C. Brunini and A. Kleusberg

Instantaneous Ambiguity Resolution for Medium-Range GPS Kinematic Positioning Using Multiple Reference Stations 283
S. Han and C. Rizos

Exploiting the SIRGAS Colocations for Determining Elevation Dependent Phase Center Variations of Geodetic GPS Antennas 289
K. Kaniuth

Real-Time Failure Detection and Repair in Ionospheric Delay Estimation Using GPS by Robust and Conventional Kalman Filter State Estimates 295
L.-S. Lin, C. Rizos and S. Mertikas

GPS Levelling Results from Two Test Areas in Finland 301
M. Ollikainen

Quality Control Algorithms for Permanent GPS Receiver Applications 307
C. Rizos, L.-S. Lin, S. Han, L. Troyer and S. Mertikas

Fast Ambiguity Resolution in Network Mode 313
P.J.G. Teunissen, P.J. de Jonge, D. Odijk and C.C.J.M.Tiberius

Session 5: GPS Kinematic Applications
Convener: Chris Rizos

High Precision GPS Kinematic Positioning: Progress and Outlook 319
C. Rizos and S. Han (invited)

Kinematic Positioning Using Adaptive Filters and Multiple DGPS Receiver Configurations 325
A.M. Bruton (invited)

Estimating the Residual Tropospheric Delay for Airborne Differential GPS Positioning 331
J.P. Collins and R.B. Langley (invited)

High-Accuracy Airborne Integrated Mapping System 337
D.A. Grejner-Brzezinska (invited)

Airborne GPS Performance during a Photogrammetric Project 343
P. Baldi, M. Marsella and L. Vittuari

Signal Distortion in High Precision GPS Surveys (Abstract) 349
F.K. Brunner and H. Hartinger

Testing GPS Approaches for Civil Aviation in the Swiss Alps 351
A. Geiger, M. Scaramuzza, M. Cocard, H.-G. Kahle, H. Lang, R. Aebersold, B. Meier and A. Dosé

Static and Kinematic Positioning with GPS for the Construction and Maintenance of High Speed Railway Lines 357
H. Kahmen

Kinematic GPS for Ice Sheet Surveys in Greenland 363
K. Keller, R. Forsberg and C.S. Nielsen

A Reweighted Filtering Algorithm and its Application to Open Pit Deformation Monitoring 369
M. Jia, M. Tsakiri and M. Stewart

Permanent Automatic GPS Deformation Monitoring Systems: A Review of System Architecture and Data Processing Strategies 375
C. Roberts and C. Rizos

Modified GPS-OTF Algorithms for Bridge Monitoring: Application to the Pierre-Laporte Suspension Brigde in Quebec City 381
R. Santerre and L. Lamoureux

The GPS Component of the Project for Digital Mapping of the KARST Aquifer System Near Curitiba, Brazil (Abstract) 387
M.C. Santos, A.J.B. Vieira and H. Firkowski

Kinematic GPS Positioning with Adaptive Kalman Filtering Techniques 389
J. Wang, M. Stewart and M. Tsakiri

WGS 84 - PAST, PRESENT AND FUTURE

James A. Slater and Stephen Malys
National Imagery and Mapping Agency
Bethesda, Maryland 20816-5003, U.S.A.

ABSTRACT

The World Geodetic System (WGS) was conceived as a practical geodetic reference system that maintains consistency with the best scientific terrestrial reference system at the time but also retains some stability. It has evolved from its beginnings as WGS 60 to its present manifestation as WGS 84. A significantly improved Earth Gravitational Model and global geoid were released in 1996, and the realization of the WGS 84 reference frame is consistent with the International Terrestrial Reference Frame (ITRF) 1994 at the 5-cm level. Evaluations of WGS 84 relative to ITRF94 have been made by comparing International GPS Service (IGS) GPS orbits with the National Imagery and Mapping Agency's precise orbits and by computing WGS 84 positions for a globally distributed set of IGS stations with known ITRF94 coordinates. The results indicate that there is no practical difference between ITRF94 and WGS 84 for mapping, charting, navigation and many survey applications.

HISTORICAL PERSPECTIVE AND MOTIVATION

WGS 72 and Before

Almost 40 years have passed since the U.S. Department of Defense (DoD) began developing a World Geodetic System (WGS) in the late 1950s. The impetus for this was the need to inter-relate maps, charts and geodetic information from different local datums, and the development of satellites and intercontinental ballistic missiles. WGS 60 was the first system introduced by the DoD and only used satellite data to derive the ellipsoid flattening. Large amounts of Doppler and optical satellite tracking data and advances in computer technology prompted development of WGS 66 and then WGS 72. The WGS 72 ellipsoid and associated parameters were defined in a manner consistent with the approach used by the International Union of Geodesy and Geophysics (IUGG) in its establishment of Geodetic Reference System 1967 (GRS 67). An equipotential ellipsoid was adopted along with four defining parameters -- semi-major axis, Earth's gravitational constant and angular velocity, and the second degree zonal harmonic coefficient of the geopotential. Datum transformations were derived to convert local coordinates to WGS 72 for stations where Doppler satellite receivers were colocated over local geodetic control points. A geoid and geopotential model were also derived as a part of WGS 72.

Development of World Geodetic System 1984 (1984-1993)

The World Geodetic System 1984 (WGS 84) is a Conventional Terrestrial Reference System which includes in its definition a geocentric coordinate system, a reference ellipsoid, a consistent set of fundamental constants, and an Earth Gravitational Model and associated global geoid. [DMA, 1991] During the last few years, several refinements have been made to the system, but the name "WGS 84" has been retained. As has always been the case, enhancements to the system have become necessary due to demands from users for higher accuracy and precision, new and higher quality sources of data, and improved data processing techniques. An additional factor has been the development and impact of the Global Positioning System (GPS).

Inadequacies in WGS 72, better computational techniques, improved satellite data (Doppler, satellite laser ranging, satellite radar altimetry), new surface gravity data, and concurrent efforts to upgrade other geodetic reference systems in Europe, Australia and North America all contributed to the development of WGS 84. The initial realization of the WGS 84 reference frame was based on a world-wide set of coordinates derived from Transit satellite Doppler data in the NSWC 9Z-2 reference frame. The origin and scale of this frame were adjusted to align it with the Bureau International de l'Heure (BIH) Terrestrial Reference System of 1984. Absolute positions computed in the new reference frame had an accuracy (1-sigma) of ± 1 m in each horizontal component and ± 2 m in height.

In order to maintain consistency with the recommendations of the IUGG, Geodetic Reference System 1980 (GRS 80) was adopted for use in WGS 84. [Moritz, 1980] A geocentric equipotential ellipsoid of revolution was chosen as the reference ellipsoid and four fundamental constants were defined:

Semi-major axis of ellipsoid of revolution (a) = 6378137 m
Earth's gravitational constant including atmosphere (GM) = 3986005×10^8 m^3s^{-2}
Mean angular velocity of the Earth (ω) = 7292115×10^{-11} rad s^{-1}
Normalized second degree zonal harmonic coefficient of the gravitational potential (C_{20})
= $-484.166685 \times 10^{-6}$

The value for C_{20} was computed from the GRS 80 value for geopotential coefficient J_2 (108263×10^{-8}) using the relationship $C_{20} = -J_2/\sqrt{5}$ and truncating the result to 8 significant figures. The permanent tidal deformation was excluded from C_{20} for compatibility with GRS 80. Geometrical constants derived from C_{20} and J_2 have minute differences; therefore, the ellipsoid used with WGS 84 is referred to as the WGS 84 ellipsoid rather than GRS 80. For all practical purposes, the GRS 80 and WGS 84 ellipsoids are identical. The difference in the semi-minor axes is 0.1 mm and the difference in the flattening is less than 1×10^{-10}.

A new Earth Gravitational Model was developed using the latest satellite Doppler, GPS, laser ranging and radar altimetry data along with surface gravity data. The model was a spherical harmonic expansion of the gravitational potential to degree and order 180. A global geoid was derived from this with a mean accuracy of 2-6 m.

WGS 84 was designed to be a practical standard global, geocentric horizontal datum for mapping, charting, geodesy and navigation. Local datums could be mathematically related to WGS 84 and to each other through datum transformations derived from colocated positions. As of 1994, transformation parameters were available for over 115 local and regional datums. In addition, WGS 84 point positions were accurate enough to provide geodetic control for maps and charts up to 1:5,000 scale which satisfied most requirements.

ENHANCEMENTS (1994-1996)

Reference Frame and Fundamental Constants

As of June 1987, the Defense Mapping Agency (DMA) had established five permanent GPS monitor stations in Bahrain, England, Australia, Ecuador and Argentina. These were geographically located to complement the U.S. Air Force's GPS monitor stations in Diego Garcia, the United States (Colorado and Hawaii), Ascension Island, and Kwajalein. WGS 84 coordinates for these stations were computed from Transit satellite Doppler observations. Since the precise and broadcast orbits for GPS were defined in this Transit-based realization of the reference frame, geodetic positions derived from them were also in this frame.

In 1992, the International Earth Rotation Service (IERS) Standards recommended a value of $3986004.418 \times 10^8\ m^3s^{-2}$ for GM. [McCarthy, 1992] Comparisons using this value and the original WGS 84 value indicated that the WGS 84 GM value was introducing a 1.3 m radial error in DMA's and the Air Force's GPS orbits. Based on this information, DMA and the Air Force adopted the new IERS GM value for high accuracy orbit determination work. Users whose work was not affected by this small change in GM continued to use the original GM value.

A number of studies between 1988 and 1994, showed evidence of a systematic ellipsoid height bias between Doppler-derived coordinates and GPS-derived coordinates for the same site. In order to eliminate the height bias and at the same time obtain a self-consistent GPS-realization of the WGS 84 reference frame, DMA with the help of the Naval Surface Warfare Center generated a revised set of station coordinates based solely on GPS observations. [Swift et al., 1994] Data from 24 IGS stations and the 10 DoD stations were used to estimate the DoD station coordinates, by constraining eight of the IGS stations to their ITRF91 coordinates during the processing. The Nuvel NNR-1 plate motion model was used to propagate station coordinates to a common epoch. The resulting WGS 84 station coordinates were aligned with the ITRF at about the 10-cm level. Thus, through the use of GPS and a refined GM value, both the precision and accuracy of the WGS 84 reference frame were improved. The new realization of the reference frame was designated WGS 84(G730), the "G" denoting GPS-derived and "730" denoting the GPS week number (2 January 1994) when the new coordinates were placed into operational use at DMA. The GPS Master Control Station implemented these changes in August 1994. [Malys and Slater, 1994]

DMA added two monitor stations to its tracking network in 1995 -- Beijing and Washington, D.C. During the same period, the tracking station in Australia was moved from Smithfield to Salisbury and the GPS antenna was repositioned at the station in England. In order to maintain a self-consistent set of network coordinates and retain the accuracy achieved in 1994, all the DoD WGS 84 station coordinates were recomputed in 1996 using the same approach that had been employed in 1994. This time GPS data from 18 IGS stations were used, with 11 of them constrained to their ITRF94 coordinates during the data processing. [Cunningham and Curtis, 1996] Excluding the stations that were physically moved, the coordinate adjustments for all the other stations were 11 cm or less in each coordinate axis. This latest realization of the reference frame is designated as WGS 84(G873) corresponding to GPS week number 873 (29 September 1996) when the coordinates were implemented in the National Imagery and Mapping Agency's (NIMA, new name for DMA) operations. These changes were implemented by the Air Force on 29 January 1997. The accuracy of the coordinates is estimated to be better than 5 cm (1-sigma) in each component, reflecting an extremely good alignment with ITRF94.

New Earth Gravitational Model EGM96 and Geoid

An improved Earth Gravitational Model (EGM96) and an associated global geoid were developed through the combined efforts of the National Aeronautics and Space Administration's Goddard Space Flight Center (NASA/GSFC), NIMA and Ohio State University. [Lemoine et al., 1996] Large amounts of new surface gravity data were incorporated into the model including many areas of the world not represented in the earlier model. Many new sources of satellite data were also used including GEOSAT, TOPEX/POSEIDON, Lageos, Lageos-2, Ajisai, Starlette, Stella, GPSMET, and GEOS-1. EGM96 is a spherical harmonic expansion of the gravitational potential, complete to degree and order 360. The WGS 84 EGM96 geoid was generated from the spherical harmonic model and WGS 84 parameters and is available as a 15' x 15' grid of geoid heights. The new geoid has an average uncertainty of $\pm$0.5-1.0 m worldwide compared to $\pm$2-6 m for the original WGS 84 geoid. A comparison of 1,038,240 15' x 15' gridded geoid heights from the EGM96 geoid and the original WGS 84 geoid shows a mean difference of 0.6 m (σ = 1.6 m) and minimum and maximum differences of -13.0 m and 16.8 m, respectively.

Revised Fundamental Parameters

EGM96 has a new estimated dynamic value for the second degree zonal harmonic coefficient. This is inconsistent with the C_{20} value, and thus the flattening, of the original definition of the WGS 84 ellipsoid. Since this inconsistency has no impact on practical applications of WGS 84, a decision was made early in the gravity modeling project to retain the original WGS 84 ellipsoid semi-major axis and flattening values. Furthermore, a derived geometric value for the second degree zonal coefficient is used to define the expression for "normal" gravity for the WGS 84 ellipsoid needed for creating a global geoid. Based on this, the four fundamental parameters for WGS 84 are now defined as follows:

Semi-major axis of ellipsoid of revolution (a) = 6378137.0 m
Earth's gravitational constant including atmosphere (GM) = 3986004.418 x 10^8 m^3s^{-2}
Mean angular velocity of the Earth (ω) = 7292115.0 x 10^{-11} rad s^{-1}
Reciprocal flattening (1/f) = 298.257223563

COMPARISONS WITH ITRF94

Although the realization of the WGS 84 reference frame is coincident with ITRF94 at the 5-cm level, NIMA's precise WGS 84 GPS orbits and absolute WGS 84 geodetic positions computed from a single GPS receiver occupation of a site will exhibit somewhat more variability in practice. This can be attributed to day-to-day variability in the amount and quality of data received from each of the 12 permanent DoD tracking stations, and to processing and modeling differences introduced by different positioning software and data collection protocols.

Evaluation of GPS Orbits

IGS precise orbits (ITRF94) are routinely compared with NIMA's precise WGS 84 orbits for all GPS satellites. For each day, a 7-parameter similarity transformation is computed between the two sets of orbits. The mean parameter values for the aggregate constellation are recorded daily. In order to specifically compare WGS 84(G873) and ITRF94, these daily means have been averaged for the first 165 days of 1997 and are shown along with their standard deviations in table 1. Mean translations ($\Delta X, \Delta Y, \Delta Z$) are less than or equal to 0.2 cm in each component and mean rotations (RX,RY,RZ) are between 0 and 0.6 mas in a component.

Table 1. WGS 84(G873) to ITRF94 Similarity Transformation Based on NIMA and IGS Orbits for 165 Days in 1997

	ΔX (cm)	ΔY (cm)	ΔZ (cm)	RX (mas)	RY (mas)	RZ (mas)	SCALE (ppb)
Mean	0.1	-0.2	0.1	0.0	0.4	0.6	-0.5
StdDev	2.9	2.3	1.4	0.3	0.2	0.4	0.2

Evaluation of Geodetic Positions

NIMA normally computes absolute (single receiver) point positions using software and procedures that are different from those used to realize the WGS 84 reference frame. In order to evaluate WGS 84 positions relative to ITRF94, GPS data from a globally distributed set of 17 IGS stations were used to compute WGS 84 positions using NIMA's normal production software ("GASP") and the Jet Propulsion Laboratory's GIPSY-OASIS II software for comparison. ITRF94 coordinates and velocities were obtained from the IGS for all of the stations. NIMA's WGS 84 precise orbits and satellite clock corrections were held fixed during the computations and Selective Availability corruptions were removed from the data. The station velocities were used to propagate the ITRF94 coordinates from 1993.0 to 1997.0.

Seven non-consecutive days of data between 2 December 1996 and 9 January 1997 were obtained for each site. For the GASP solutions, each 24-hour data set was broken into three 8-hour segments which were processed separately and then averaged to produce one daily solution. For the GIPSY solutions, each 24-hour data set was processed as a single session. Some difficulties were encountered by both programs in processing some of the data sets, resulting in the elimination of some of them in the results presented here. Table 2 shows the deviations of the GASP and GIPSY WGS 84 solutions from the ITRF94 coordinates in latitude, longitude and ellipsoid height for each site. On a global basis the average difference between WGS 84 and ITRF94 coordinates over all sites aggregated together is very small. In general the most disparity is in the height component, with differences up to 72 cm. The GIPSY solutions are more precise than the GASP solutions but still show relatively large unexplained differences with ITRF94 at some sites.

FUTURE EVOLUTION

It should be apparent that a consistent effort has been made over the years to maintain WGS 84's compatibility with the best geodetic reference systems of the time. Advances in technology and methodology, and demand for increased accuracy and precision will continue to drive the development. There is always a need to maintain the reference system at a higher level of fidelity than the applications it supports. Yet, WGS 84 is intended to be a practical system as well, and as such, the number and frequency of changes made to the system must be kept to a minimum. Every change can affect a vast

Table 2. WGS 84 Positions Compared to ITRF94 Positions for Selected IGS Sites
(WGS 84 - ITRF94)

Site	Mean GASP Solutions No. of Daily Solns.	ΔLat (cm)	ΔLong (cm)	ΔEl.Ht. (cm)	GIPSY Solutions No. of Daily Solns.	ΔLat (cm)	ΔLong (cm)	ΔEl.Ht. (cm)
Arequipa	5	8.3	11.5	33.3	5	3.4	4.9	30.4
Fairbanks	7	-2.5	-24.5	2.8	7	-5.6	-21.1	-5.7
Fortaleza	7	-4.9	1.9	45.0	7	-1.7	2.4	53.4
Hartebeesthoek	7	-6.2	-6.1	-0.5	6	-2.9	-3.2	-15.2
Irkutsk	7	-9.8	20.7	13.2	7	-1.9	9.4	-4.1
Kerguelen	7	1.6	6.7	10.2	7	3.3	5.1	16.2
Kokee Park	7	-1.8	-30.1	10.9	6	-5.5	-25.3	20.2
Maspalomas	7	3.7	4.6	-6.3	6	-1.6	5.8	45.9
Ft. Davis	7	-4.5	-14.9	-71.8	6	-8.5	-24.7	-31.0
Ny Alesund	6	-1.9	-2.6	-14.9	5	-0.8	8.8	5.6
Santiago	7	14.5	-36.3	27.1	6	3.9	1.6	-12.8
St. John's	7	-4.6	12.3	-10.2	6	5.4	8.0	-24.6
Taipei	7	-12.3	44.0	29.9	6	-5.5	3.1	3.6
Tidbinbilla	7	4.5	32.8	-20.3	6	7.9	9.6	-1.1
Tsukuba	7	-8.3	18.7	52.2	7	-5.7	3.4	9.4
Wettzell	3	-4.0	-12.8	7.4	6	-4.5	-0.8	2.8
Yellowknife	7	-0.9	-25.4	18.4	7	5.9	-18.7	-6.5
Mean (all sites)		-1.7	0.0	7.4		-0.8	-1.8	5.1
Standard Deviation		6.7	22.4	28.8		4.9	12.3	22.8

number of users. At this time, refinements to the reference frame are at the decimeter level and really do not have an impact on most mapping, navigation and routine surveying.

In the future, NIMA will add stations to its GPS tracking network, which at some point will result in new realizations of the WGS 84 reference frame. As more surface gravity and satellite data become available, it will become possible to refine EGM96, explore higher degree and order gravitational models, and improve the global geoid. Tectonic plate motion is now being accounted for in WGS 84 geodetic positioning through the use of station velocities or a plate motion model. NIMA will continue to develop and revise datum transformations because, despite the availability of GPS and many national efforts to convert to geocentric datums, much of the geospatial information in the world is still referenced to local datums. It is also an opportune time to consider the practical implementation of a global vertical reference system since GPS provides highly accurate ellipsoid height measurements and geoid undulations are now available globally at meter-level accuracies. In summary, WGS 84 will continue to be maintained as a practical geodetic reference system that provides the accuracy, precision and consistency required by the user community.

ACKNOWLEDGMENTS

The authors wish to thank the NIMA Geodesy and Geophysics Department staff responsible for the daily GPS orbit processing and comparisons with the IGS orbits. We especially want to thank Mr. Robert Ramsey, Mr.Todd Sherman and Mr. Terry Timblin for carrying out all of the point positioning data processing.

REFERENCES

Cunningham, J. and Curtis, V.L., *WGS 84 Coordinate Validation and Improvement for the NIMA and Air Force GPS Tracking Stations*, NSWCDD/TR-96/201, Nov. 1996, Dahlgren, Virginia.

Defense Mapping Agency, Technical Report, *Department of Defense World Geodetic System 1984*, DMA TR 8350.2, Second Ed., 1 Sept. 1991.

Lemoine, F.G., Smith, D.E., Smith, R., Kunz, L., Pavlis, E.C., Pavlis, N.K., Klosko, S.M., Chinn, D.S., Torrence, M.H., Williamson, R.G., Cox, C.M., Rachlin, K.E., Wang, Y.M., Kenyon, S.C., Salman, R., Trimmer, R., Rapp, R.H., and Nerem, R.S., *The Development of the NASA/GSFC and NIMA Joint Geopotential Model*, Proc. of the International Symposium of Gravity, Geoid, and Marine Gravity 1996 (GraGeoMar96), Tokyo, Japan, 30 Sep - 5 Oct 1996 (in preparation).

Malys, S. and Slater, J., *Maintenance and Enhancement of the World Geodetic System 1984*, Proc. ION GPS-94, Salt Lake City, Utah, Sept. 1994, pp.17-24.

McCarthy, D.D., Ed., *IERS Standards (1992),* IERS Technical Note 13, July 1992, Observatoire de Paris, France.

Moritz, H., "Geodetic Reference System 1980", Bulletin Geodesique, Vol. 54, No. 3, Paris, France, 1980.

Swift, E.R., Gouldman, M.W., Merrigan, M.J. and Curtis, V.L., *GPS Orbit Estimation and Station Coordinate Improvement Using a 1992 IGS Campaign Data Set*, NSWSCDD/TR-94/267, Oct. 1994, Dahlgren, Viriginia.

ITRF DENSIFICATION AND CONTINUOUS REALIZATION BY THE IGS

G. Blewitt[†], C. Boucher, P. B. H. Davies, M. B. Heflin, T. A. Herring and J. Kouba

[†]Department of Geomatics, University of Newcastle upon Tyne
Newcastle upon Tyne, NE1 7RU, United Kingdom

Abstract

The tools and infrastructure are now in place to realize a global, terrestrial, kinematic reference frame with few millimeter precision, which is both spatially dense, and can be reliably updated on a frequent basis (e.g., monthly). A procedure is outlined by which IERS can continuously realize the ITRF using methodology developed by the International GPS Service (IGS). However, spatial densification not only requires such a processing scheme, but also requires the recruitment of interested groups to participate in this venture.

INTRODUCTION

It is appropriate that Geodesy, a discipline over 2000 years old, is approaching a watershed as we move into the new millenium. Around 200 B.C., Eratosthenes used the difference in direction of the sun's rays between Syene and Alexandria to infer the radius of the Earth, producing an answer only 2% in error [*Torge*, 1980]. (Careful as he was, this surely required a degree of luck). Within the last two decades, the development of space geodesy has allowed us to measure the 12700 km distance across the Earth between any two points with an accuracy of 1 cm, which is 0.0000001%, or one part per billion. This requires no luck, in the sense that the results are repeatable at this level; the various techniques (SLR, VLBI, GPS, and DORIS) also agree at this level.

The real breakthrough in this last decade of the millenium, is that positioning with 1-cm accuracy can now be routinely achieved on a daily basis, by anyone armed with a single dual-frequency GPS receiver, appropriate software, and products from the International GPS Service for Geodynamics (IGS). With this revolution in the making, it is right for us to step back and reconsider our, perhaps, outdated notions of reference frames. The idea of control networks has evolved with space geodesy, but the basic idea has until recently remained as *Bomford* [1980] describes triangulation control networks: "To constitute the main framework on which less precise observations may be based..."

Indeed, this describes the "fiducial concept", the idea prevalent in the 1980's that precise VLBI or SLR measurements could provide the framework within which densification could be achieved using GPS. Although GPS has since become a truely global technique, similar ideas are also reflected in the IGS program to densify the IERS Terrestrial Reference Frame (ITRF). This program is again based on the principles of heirarchical networks, starting with the IGS Global Network, which is then densified by regional networks to produce the IGS Polyhedron (100-200 stations) [*Blewitt et al.*, 1993a, 1995].

This IGS Polyhedron can then act as "active" control points, relative to which users can position themselves. However, it is a misconception, perhaps too commonly held, to think that the reasons for the Polyhedron are analogous to the triangulation network heirarchy described by Bomford. GPS relative positioning is now essentially independent of distance, using the right tools and techniques, so why do we need the Polyhedron?

The real issue is one of reference frame, and the fact that all stations are moving at a level which is very significant compared to the positioning precision. We need the Polyhedron as an active reference frame, so that we can better interpret what our positional coordinates actually mean. As a concrete example, major earthquakes can displace these so-called "stable reference points" on a regional scale [*Blewitt et al.*, 1993b, *Bock et al.*, 1993]. In this example, we must be careful to define the time evolution of the frame such that points well outside the area of co-seismic deformation do not appear to be displaced due to the earthquake. This sounds so obvious, yet it is no longer obvious what to do when unpredictable motions are not strictly regional phenomena, but extend over all scales. This fact has been realized ever since the network design stage of IGS:

> *"We do not enjoy the benefit of potential fields or elastic waves to define a physical averaging of pointwise properties of the Earth, so we must rely on a hierarchy of networks, with sparser regional networks serving as fiducial sites in the survey of denser local networks"* [Minster et al., p.24, *1993*]

We can therefore view the Polyhedron as providing a primary kinematic description of the Earth's shape as it changes. Although the "reference system" can account for predictable changes as prescribed by conventional models, only the "reference frame" can account for true motions. We should therefore emphasize the kinematic aspects of reference frames if we are to keep pace with these technological advances. How do develop and realize this is the purpose of this paper.

IGS DENSIFICATION PROGRAM

After several years of planning [*Mueller and Beutler*, 1992], the International GPS Service for Geodynamics (IGS) was officially established in 1993 by the International Association of Geodesy. Ever since an initial pilot phase beginning June 1992, the IGS has been coordinating the operations and analysis a global network of GPS stations. The IGS officially commenced operations in January 1994, by which time approximately 40 to 50 IGS stations had become operational.

The expanding global network of high precision GPS receivers (Figure 1) was seen to present an opportunity to produce a reference frame which is (i) dense, (ii) of a reasonably homegeneous quality, (iii) of few-millimeter accuracy on a global scale, (iv) readily accessible to GPS users, and (v) ideal for monitoring variations in the Earth's shape, and for providing kinematic boundary conditions for regional and local geodetic studies [*Blewitt et al.* 1993a, 1995]. The challenge was to be able to analyze cohesively the data from an ever increasing number of receivers, such that near-optimal solutions could be produced. Although ideally all data should be analyzed simultaneously to produce a single solution, in practise this is computationally prohibitive

This led to the "distributed processing approach," which, at the algorithm level, partitions the problem into manageable segments [Figure 1], and, at the organizational level, delegates responsibility to analysis centers who would naturally have an interest in the quality of the solutions. Another characteristic of this approach is a level of redundancy, such that a meaningful quality assessment can be made by other, independent groups. Distributed processing was developed as a method which could be carried out as a natural extension to the existing operations of the IGS.

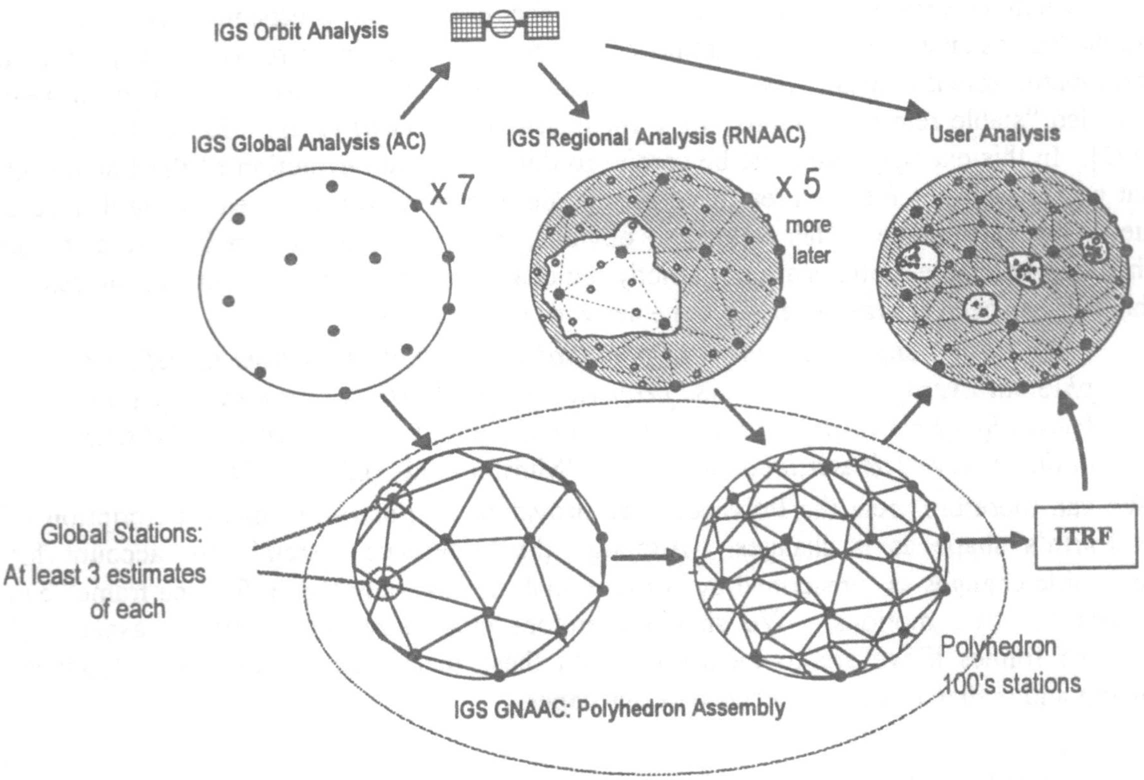

Figure 1: Schematic explanation of the distributed processing approach. Our proposal is for science groups to operate as RNAACs. The GNAACs would then take care of reference frame consistency, and input into ITRF.

Following a planning workshop at JPL in December 1994 [*IGS*, 1995], a pilot program was initiated in September 1995 to test these ideas. Global Network Associate Analysis Centers (GNAACs) were set up at Newcastle University, MIT, and JPL. A format was developed for the exchange of coordinate solutions, covariance matrices, and site information (SINEX format) [*SINEX Working Group*, 1996]. Initially these GNAAC's combined solutions for global network station coordinates provided every week by the seven Analysis Centers, producing a single unified SINEX file. Approximately one year later, Regional Network Analysis Centers (RNAACs) began submitting regional GPS solutions, computed using weekly published IGS orbit solutions. These regional solutions were then assimilated into the unified global solution by the GNAACs, what is known as the "IGS polyhedron solution."

Although currently undergoing final review, the pilot program has been viewed broadly as a success, demonstrating few-millimeter repeatability in weekly solutions for geocentric coordinates of not only the global stations, but also the regional stations. However the actual process of densification (new GPS stations) is still less than adequate in many parts of the globe. For example, tide-gauge benchmark monitoring could help. Additional GPS stations installed at island tide-gauge sites will undoubtedly be greatly welcomed by IGS, especially as oceanic regions of the globe are systematically undersampled (which is the primary reason for the lack of stations in the ocean-rich southern hemisphere). Furthermore, the IGS Densification Program provides a natural way for science groups to participate in IGS. We expand on this point in a later section.

IGS Global Network Associate Analysis Centers: Methods and Results

Blewitt et al. [1995] discuss the following components of the GNAAC activities (previously called "Type Two Analysis" during the planning stages): (i) detection of inter-agency information discrepancies (e.g. in antenna heights); (ii) monitoring of solution consistencies (inter-agency, and with respect to ITRF); (iii) weekly publication of a combined global solution; (iv) weekly publication of an IGS polyhedron solution (global plus regional networks); (v) periodic publication of kinematic solutions (e.g., station height velocity, plate tectonic Euler vectors, etc.), with submission to the International Earth Rotation Service (IERS) with the goal of improving the ITRF.

Global Network Associate Analysis Centers (GNAACs), at Newcastle University (NCL), MIT, and JPL have so far been operating during the pilot phase of the IGS Densification Program. All three GNAACs have in principal been using similar (but not identical) approches and that results are quite comparable. For the purpose of this paper, methods and results from NCL are briefly highlighted

Now two years since the inception of the IGS Densification Pilot Program, the NCL GNAAC is continuously achieving all stated objectives [*Davies and Blewitt*, 1996, 1997]. Taking the most recent submission at the time of writing, coordinate solutions for 132 stations are presented, of which approximately 50% are global stations (defined as being analyzed by at least 3 Analysis Centers), and 50% are regional. A total of 54 regional station solutions derive from 3 RNAACs which cover South America, Europe, and Japan.

We have developed combination procedures [*Davies and Blewitt*, 1996, 1997] which aim to (1) minimize bias from datum assumptions, (2) minimise bias from unrealistic covariance matrices; (3) utilize the inherent redundancy of overlapping networks to remove outliers objectively. The first is achieved by applying a loosening transformation to each input covariance matrix [*Blewitt*, 1997], which can be interpreted as the inverse of reference frame projection [*Blewitt*, 1992]. The second is achieved by variance component estimation [*Rao and Kleffe*, 1988; *Sahin et al.* 1992]. The third is achieved by applying reliability analysis theory [*Kosters and Kok* 1989].

Our weekly, long-term repeatability in station height has a best case value of 3 mm, median of 7 mm, and worst case of 19 mm. This is to be compared with the best Analysis Center solutions (best case 4 mm, median 9 mm, worst case >30 mm). We conclude that GNAAC analysis not only provides a consistent unique solution, but also a more reliable solution (in the statistical sense of the word). The IGS Densification Program methodology should not be viewed as compromising solution quality, but rather as a preferred alternative to unilateral analysis.

In summary, GNAAC results show that the combined solutions produce coordinate time series with a smaller variance than any individual AC network solution, even more so for stations with the worst levels of precision. This indicates that reliable outlier detection, due to redundancy from the multiple solutions, is a major reason for this improvement. Therefore, we can conclude that GNAAC solutions are not only more precise, but more reliable than any individual contributing AC solution.

IERS STRATEGY FOR ITRF

Since its establishment in 1988, the International Earth Rotation Service (IERS) published on an annual basis a new realization of the International Terrestrial Reference System (ITRS). Each year, the IERS Central Bureau collected among contributing analysis centers their solutions for Earth Rotation Parameters together with the associated Terrestrial Reference Frame (TRF) (and Celestial if available) data. These contributions are currently available for several space techniques: SLR, LLR, VLBI, GPS and DORIS.

Such a realization is now widely known under the label ITRF*yy*. Up to *yy*=94, this solution was obtained by a combination of all data submitted to the IERS Central Bureau at the begining of *yy*+1. It was assumed that all individual analysis centers would provide by this was their best and most complete individual solution, in particular including data for the year *yy*.

The succession of these results, from ITRF88 to ITRF94, did actually show an improvement both in accuracy for positions of stations as well as in the geographical coverage of the corresponding network. Nevertheless, the accuracy reached was ranging from a few millimeters to a few centimeters, and a lot of details concerning modeling were raised at this level (in particular relativistic and geodynamical effects).

At that time, IERS decided to establish a primary solution upon specifications of an international working group (WG on ITRF datum). Although this is exactly a task for this WG to recommend the precise definition of what should be a primary ITRF solution, we can agree that the main concept is to provide a consistent, homogeneous set of positions (and time variations) at the subcentimetric level (in precision and reliability) for a well distributed network. This work is presently in progress, in particular by performing research (experimental and theoretical) on the geocenter, which is one of the new topics to be investigated before establishing these general recommendations.

Considering on the other hand the numerous and increasing requirements of the users, mostly outside the restricted IERS community, IERS decided to continue the annual series of ITRF publications for a complete solution along the line of the series up to ITRF94.

Starting for *yy*=96, ITRF*yy* is now specified to be the best complete solution produced on an annual basis by IERS. In particular, any data set expected to bring a useful contribution to this goal should be included, and not only the data from the annual submission to the Central Bureau. Specific work is currently underway for GPS and DORIS contributions, with the possibility of producing updated realizations of ITRFyy, perhaps every month. Both techniques are growing significantly and providing each year new stations and improved present day position quality.

Cooperation with IGS

Close cooperation with IGS is very useful and already under expansion, with mutual benefits:

- since the begining , the IGS organization stimulated individual analysis centers to submit solutions in response to the IERS Central Bureau's annual call for data.
- IGS uses for operational purposes ITRF*yy* solutions. IGS represents a strong user group in favor of these annual solutions. ITRF96 will therefore be very welcome.
- conversely, the GPS contribution for ITRF96 is very strong, including results from IGS densification pilot experiment.
- using these ITRF solutions (or maybe in the future a cleaner primary solution which is under consideration by a ad hoc working group on ITRF datum), IGS could consider by its ITRF densification strategy to provide an operational TRF solution similar to rapid service for EOP.

We propose that routine monthly GNAAC solutions be submitted to the IERS as input to a continuous realization of the ITRF. Not only would this have benefits from the standpoint of reliability, but also non-linear station motion would also be represented by such an approach (e.g., due to possible monument instability, the earthquake cycle, or seasonal loading effects). This approach would also be particulary useful for recently installed stations whose velocities are not yet very well determined.

As an IERS user group, IGS could then use the continuously updated realization of ITRF for its products. Of course, users such as IGS may not wish to update its reference frame every month, however the opportunity will always be there to perform a frame update when needed. For example, IGS recently experienced a slight degradation in the network of available primary stations used to define the frame of the orbits and Earth rotation parameters. Although IGS considered producing its own realization of ITRF to improve the situation, it was decided to wait for the imminent ITRF96. Clearly, a set of sub-annual reference frame realizations would have been useful under these circumstances.

ORGANIZATIONAL STRUCTURE FOR ITRF DENSIFICATION

We can broaden the definition of "ITRF Densification" to mean temporal as well as spatial densification. For temporal densification, we have already described a scheme whereby IGS can contribute to, say, monthly realizations of the ITRF. Of course, other space geodetic techniques (e.g., DORIS).could also be coordinated towards this goal.

For spatial densification, we have described a scheme designed on the principle of distributed processing, which easily allows the incorporation of data from many permanent GPS networks. We haven't explored how RNAAC's can be "recruited" to ensure a good distribution of stations to suit a variety of scientific purposes. Since IGS is a service organization, and not primarily in the business of scientific investigation, we should promote the idea that RNAAC's be connected to some science group, which has its own objectives and agenda. This should ensure that scientific needs are met, while decentralizing the operational burden away from other IGS components.

An organizational structure for this is presented in Figure 2, which reproduces a diagram presented at a recent workshop for monitoring sea level held at JPL in March 1997 [*Blewitt et al.*, 1997]. Simple lines connecting the boxes indicate the organizational hierarchy. Arrows indicate data flow.

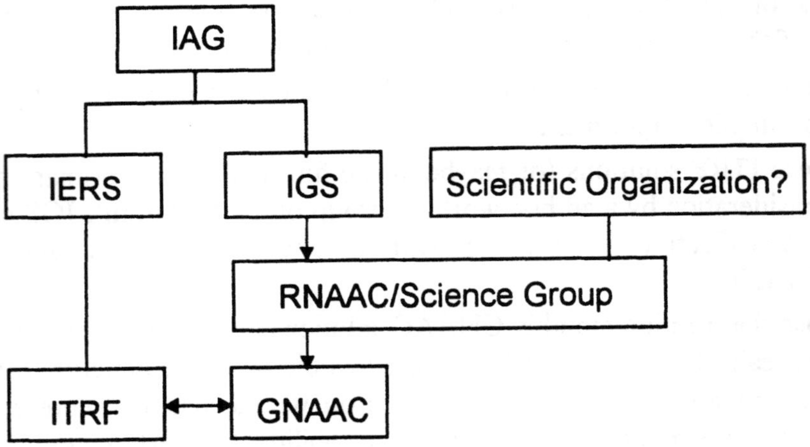

Figure 2: Chart illustrating organizational links and data flow to facilitate the activity of tide-gauge benchmark monitoring (explained in text).

Figure 3 expands this idea specifically for the community interested in monitoring tide gauge benchmarks. We use this here as an example of how Figure 2 might be practically realized . Starting with the bottom right hand side, we have the goal of this organization, which is the production of a database (DB) of the coordinates and velocities of tide gauge benchmarks available at the Permanent Service for Mean Sea Level (PSMSL), which formally reports to the Commission of Mean Sea Level and Tides (CMSLT), under the umbrella of the International Association for the Physical Sciences of the Oceans (IAPSO).

The structure described so far is essentially in place. What remains to be done, is to include GPS data from tide gauge sites into the dataflow, which would be analyzed by new RNAAC's. Figure 3 shows each RNAAC as a part of a science group which falls under the International Association of Geodesy (IAG) through the Special Commission 8 on Sea Level and Ice Sheet Variations (SC8). Special Commission 8's terms of reference look as if they have been written especially for this task, since they not only mention geodetic observing programs to investigate sea level change, but also interdisciplinary communication betwee geodesists, geophysicists, and oceanographers. Science groups are also connected to the CMSLT to make the collaboration with oceanographers explicit, and for the practical necessity for expertise on tide gauge selection. It would be natural for science groups to be regional, given that they act as RNAACs. To complete the loop, the Science Groups access both the tide gauge records and the geodetic records from the PSMSL for scientific interpretation.

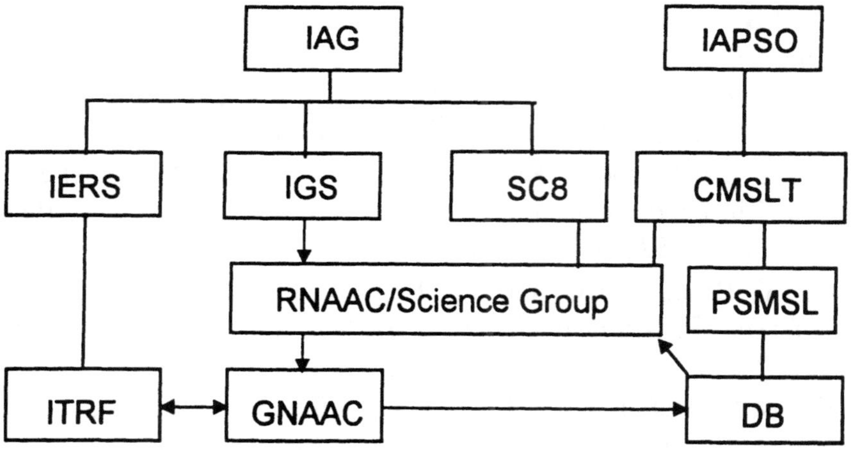

Figure 3: Chart illustrating organizational links and data flow to facilitate the activity of tide-gauge benchmark monitoring.

CONCLUSIONS

A procedure has been outlined by which IERS can continuously realize the ITRF, using IGS methodology which has been tested over the past 2 years during its ITRF Densification Pilot Project. We propose that this project move out of the pilot phase, and be made official. This will involve the resolution of some technical details and coordination between

IGS and IERS, however much of those details have been already resolved with the development of the SINEX format for the exchange of geodetic solutions.

We have also pointed out that spatial densification not only requires a processing scheme, but also requires the recruitment of interested groups to operate the networks and participate in the scheme. We have identified scientific groups as likely candidates for operating IGS Regional Network Associate Analysis Centers. We show, for example, that the IGS, IERS, and the tide gauge benchmark community can be served by RNAAC's serving as part of the IAG Special Commission 8 "Sea Level and Ice Sheets", in collaboration with the IAPSO Commission for Mean Sea Level and Tides.

In conclusion, although there is certainly room for improvement in geodetic techniques and reference frame definition, ultimately the user relies on products which may not necessarily reflect the best currently achievable accuracy. What we have presented here is one way forward to improve the ITRF as a useful, accurate, and reliable product.

REFERENCES

Blewitt, G, "GPS Data Processing Methodology: From Theory to Applications," in *GPS Geodesy*, Eds. A. Kleusberg and P.J.G. Teunissen, Lecture Notes on Earth Sciences, Springer (in press, 1997)

Blewitt, G., P.B.H. Davies, T.Gregorius, R.S. Kawar, and D.U. Sanli, Sustainable geodetic monitoring of the natural environment using the IGS, in *Methods for Monitoring Sea Level, GPS and Tide Gauge Benchmark Monitoring, GPS Altimeter Calibration*, Jet Propulsion Laboratory, Pasadena, Calif. USA (in press 1997)

Blewitt, G., Y. Bock, and J. Kouba, "Constructing the IGS polyhedron by distributed processing," in *Proc. of the IGS Workshop,* ed. by J. Zumberge, IGS Central Bureau, Pasadena, Calif., USA, p. 21-36 (1995)

Blewitt, G., Y. Bock, and G. Gendt, "Regional clusters and distributed processing," in *Proc. of the IGS Analysis Center Workshop,* Ed. by J. Kouba, pp. 62-91, Ottawa, Canada (1993a).

Blewitt, G., M.B. Heflin, K.J. Hurst, D.C. Jefferson, F.H. Webb, and J.F. Zumberge, "Absolute far-field displacements from the 28 June 1992 Landers earthquake sequence," *Nature, Vol. 361,* p. 340-342 (1993b)

Blewitt, G., M.B. Heflin, F.H. Webb, U.J. Lindqwister, and R. P. Malla, "Global coordinates with centimeter accuracy in the International Terrestrial Reference Frame using the Global Positioning System," *Geophysical Research Letters, 19,* pp. 853-856 (1992).

Bock. Y., Agnew, D.C., Fang, P., J.F. Genrich, B.H. Hager, T.A. Herring, K.W. Hudnut, R.W. King, S. Larsen, J-B. Minster, K. Stark, S. Wdowinski and Wyatt, F.K., "Detection of crustal deformation from the Landers earthquake sequence using continuous geodetic measurements," *Nature, 361,* 337-340 (1993)

Bomford, G., *Geodesy*, Clarendon Press, Oxford (1980)

Davies, P.B.H. and G. Blewitt, "Newcastle upon Tyne Global Network Associate Analysis Centre Annual Report 1996," *IGS Annual Report 1996*, p. 237-252, IGS Central Bureau, Jet Propulsion Lab., Pasadena, California (1997)

Davies P.B.H. and G. Blewitt, "Newcastle upon Tyne IGS Global Network Associate Analysis Centre Annual Report 1995," *IGS Annual Report 1995*, p. 189-202, IGS Central Bureau, Jet Propulsion Lab., Pasadena, California (1996)

IGS, *Densification of the IERS Terrestrial Reference Frame through Regional GPS Networks*, Eds. J.F. Zumberge and R. Liu, IGS Central Bureau, Jet Propulsion Laboratory, Pasadena, California (1995)

Kosters, A.J.M. and J.J. Kok, "Statistical testing and quality analyusis of aobservations and transformation parameters in combining 3-dimensional networks," Geodetic Computing Centre (LGR), Faculty of Geodesy, Delft Univ. of Tech., presented at Symp. S102, IAG Congress, 3-12, Aug. 1989, Edinburgh (1989)

Minster J.B., M. Bevis, Y. Bock, C. Boucher, O. Columbo, B. Engen, A.M. Finkelstein, H. Frey, B. Hager, T. Kato, S. Lichten, P. Morgan, W. Prescott, C. Reigber, S. Rekkedal, B. Schutz, H. Tsuji, and V. Velikhov, "Network design considerations for the International GPS Geodynamics Service," p.23-32, in *IAG Symposia 109: Permanent Satellite Tracking Networks for Geodesy and Geodynamics*, Ed. G. L. Mader, Springer Verlag, Berlin (1993)

Mueller, I. and G. Beutler, "The International GPS Service for Geodynamics - Development and current structure," in *Proc. of the 6th Int. Geodetic Symp. on Satellite Positioning,* Columbus, Ohio, pp. 823-835, 1992

SINEX Working Group, "SINEX - Solution Independent EXchange format," Appendix 1 in *IGS 1996 Analysis Centre Workshop Proceedings*, Eds. R.E. Neilan, P.A. Van Scoy, and J.F. Zumberge, pp. 223-276, IGS Central Bureau, Jet Propulsion Laboratory, Pasadena, California (1996)

Rao C.R. and J. Kleffe, *Estimation of Variance Components and Applications*, Elsevier, Amsterdam (1988)

Sahin, M., P.A. Cross, and P.C. Sellers, "Variance component estimation applied to satellite laser ranging," *Bulletin Geodesique, 66,* 284 (1992)

Torge, W., *Geodesy*, Walter de Gruyter & Co., Berlin (1980)

THE USE OF THE EUREF PERMANENT GPS NETWORK FOR THE MAINTENANCE OF THE EUROPEAN TERRESTRIAL REFERENCE FRAME

C. Bruyninx, *Royal Observatory of Belgium, Av. Circulaire 3, 1180 Brussels, Belgium*
J. Dousa, *Geodetic Observatory Pecny, Ondrejov 244, 25165, Czech Republic*
W. Ehrnsperger, *Bay. Akademy für die Intern. Erdmessung, Marstallpl. 8, 80539 München, Germany*
N. Fachbach, *Austrian Academy of Sciences, Lustbühelstr. 46, 8042 Graz, Austria*
J. Johansson, *Onsala Space Observatory, S-43992 Onsala, Sweden*
F. Vespe, L. Ferraro, A. Nardi, *CGS Centro di Geodesia Spaziale "G. Colombo", Contrada Terlecchia, 75100 Matera, Italy*
M. Figurski, *Institute of Geodesy & Geodetic Astronomy WUT, Pl. Politechniki 1, 00-661 Warsaw, Poland*
T. Springer, *Astr. Inst., University of Berne, Sidlerstr. 5, 3012 Bern, Switzerland*
G. Weber, *Inst. for Applied Geodesy, R. Strauss Allee 11, 60598 Frankfurt, Germany*
A. Wiget, *Federal Office of Topography, Seftigenstr. 264, 3084 Wabern, Switzerland*

Abstract

Taking into account the growing number of permanent GPS stations in Europe, the EUREF subcommission decided in 1995 to coordinate the activities related to this network for the maintenance of the European Reference Frame.

The EUREF permanent GPS network, which is the European densification of the IGS network, presently consists of more than 60 GPS stations, covering 23 countries all over Europe. The processing scheme allows for distributed processing : 10 Local Analysis Centers each analyze a part of the EUREF network. One analysis center is responsible for merging the individual subnetwork solutions into one European solution. This solution in submitted to the International GPS Service for Geodynamics (IGS) within the frame of the IGS Densification Project.

Weekly free-network solutions for Europe are available since April 1996. The mean RMS values of the coordinate residuals of the combined solution with respect to the individual solutions is about 2 mm for the north and east components and 6 mm for the height component. The agreement between the solutions of the different analysis centers is of the same quality as the week-to-week repeatabilities of each analysis center.

In may 1997, EUREF submitted its solution to the International Earth Rotation Service. As a result, all EUREF stations will show up in the next realization of the International Terrestrial Reference System.

Introduction

The main task of the EUREF subcommission is the establishment, maintenance and enhancement of a 3-dimensional European Reference Frame.
In order to have station-to-station relations fixed in the European Terrestrial Reference System (ETRS), this ETRS has been defined as fixed to the stable part of the European plate and coinciding with the International Terrestrial Reference System (ITRS) at epoch 1989,0 (Resolution No 1 of the EUREF Symposium in Florence, 1990).

A first step towards the realization of the ETRS was taken in 1989 with the observation of the EUREF89 GPS campaign which covered Western Europe (Seeger et al., 1992). The processing of the EUREF campaigns is done following the rules set up by the EUREF Technical Working Group. In 1993, these standards recommended the use of the IGS sites, which have known ITRF coordinates, as fiducials for the processing of EUREF campaigns. A year later, at the EUREF symposium in Warsaw, the use of the European permanent stations for the maintenance of the European Reference Frame was recognized (Gurtner, 1994) and European agencies were invited to collaborate with EUREF for the implementation of such a network.
In May of 1995 guidelines for a EUREF permanent GPS Network were set up (Gurtner, 1995) and in October of the same year a EUREF network coordinator (C. Bruyninx, Royal Observatory of Belgium) was designated at the EUREF Technical Working Group meeting in Paris in order to coordinate the activities related to the European permanent network.
In March 1996, EUREF responded to the IGS "CALL FOR DENSIFICATION OF THE ITRF THROUGH REGIONAL GPS ANALYSES AS IGS NETWORK ASSOCIATE ANALYSIS CENTER" and proposed to submit one official European solution to the IGS based on the principle of distributed processing.
The EUREF proposal was officially accepted by the IGS in May 96 and the first solutions for the EUREF permanent network were forwarded to the IGS.

Current Status of the EUREF Permanent Network

The GPS Tracking Network

The EUREF network consists of all the European permanent sites included in the IGS Network, plus additional sites accepted by the EUREF Technical Working Group following similar rules as set up by and for the IGS.
The network consists presently of more than 60 stations, covering 23 countries. 40% of the EUREF stations do not belong to the IGS network.

The data from most of the EUREF stations are available at the IGS Regional Data Center at IfAG (Institute for Applied Geodesy). The data of the remaining stations can be retrieved from five local data centers which give access to a particular EUREF subnetwork (Bruyninx, 1997a).

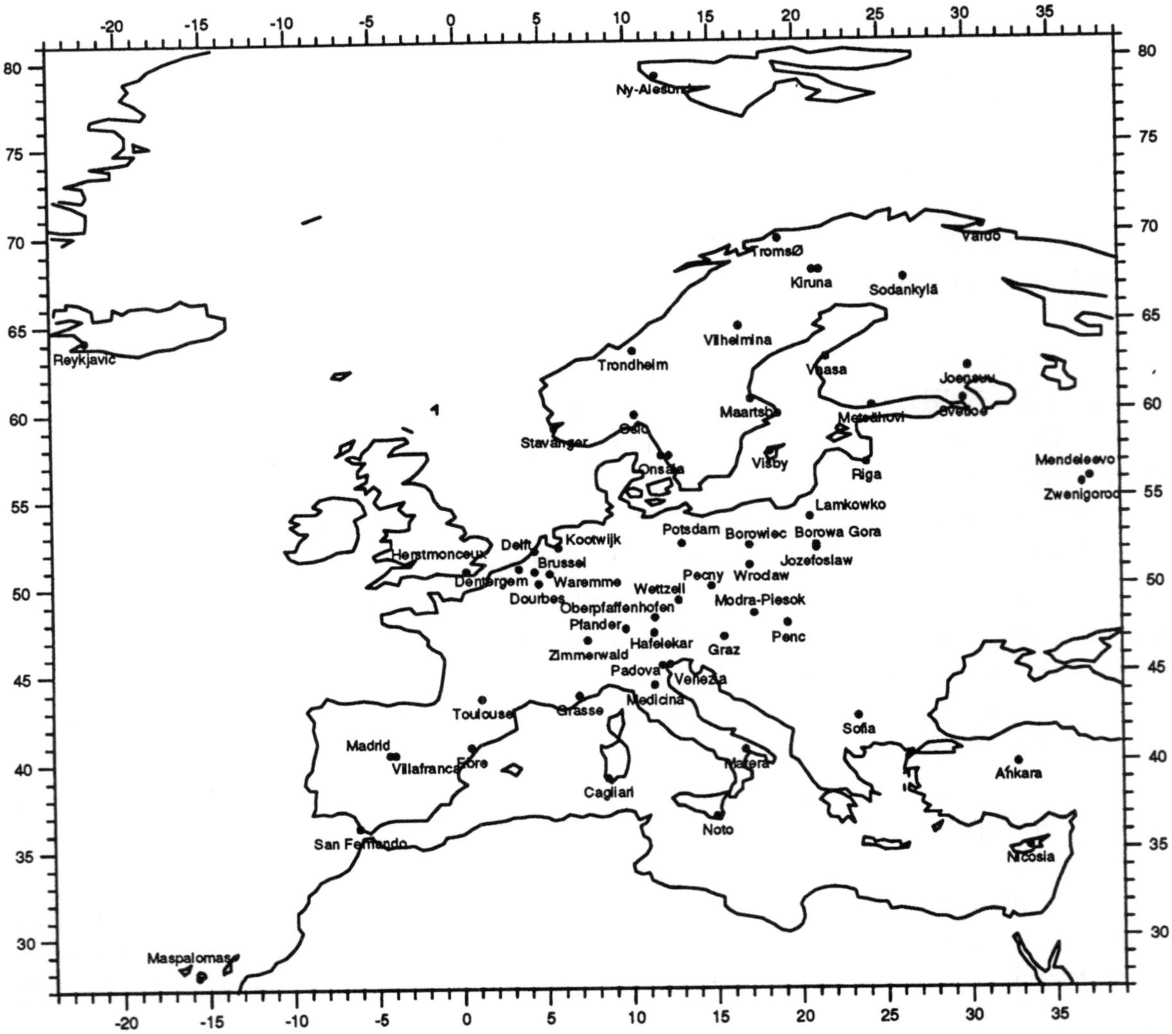

Figure 1: Stations included in the permanent EUREF GPS network (August '97). The EUREF stations Kellyville and Thule (Greenland) are not included in the map.

Data Analysis

When EUREF took, at the end of 1995, the initiative to coordinate the activities related to the European permanent network, it was agreed that the data analysis of the EUREF network would be divided among different European institutions. The main goal of this distributed processing was the dissemination of the knowledge and expertise to routinely process a network of GPS stations for high precision geodesy.
The CODE analysis center agreed to be responsible for merging the individual subnetwork solutions into one European solution.

At present ten analysis centers are involved in the processing of the EUREF network : the Italian Space Agency (Italy), the Bavarian Academy of Sciences (Germany), the Observatory Lustbühel (Austria), the Institute for Applied Geodesy (Germany), the Federal Office of Topography (Switzerland), the Geodetic Observatory of Pecny (Czech Republic), The Nordic Geodetic Commission Analysis Center (Sweden), the Royal Ob-

servatory of Belgium (Belgium), the Warsaw University of Technology (Poland) and the Center for Orbit Determination in Europe (Switzerland).
Taking into account that no specific data analysis recommendations were available to the EUREF analysis centers until recently, the processing strategy used at most of the analysis centers was very similar, although some important differences (for example the elevation cut-off angle) remained.

Since mid 1996 weekly SINEX contributions of the different European subnetworks are combined into one official weekly EUREF solution. This EUREF solution is then submitted to the IGS where the Global Network Associate Analysis Centers combine all global IGS solutions with several regional solutions like EUREF in the frame of the IGS densification pilot project (Zumberge and Liu, 1994).

The EUREF Combined Solution

The Combination Scheme

The results from the EUREF analysis centers are available in the SINEX format and are combined using the Bernese GPS Software program ADDNEQ (Brockmann, 1996) which will generate the weekly EUREF combinations. ADDNEQ is based on the sequential least-squares adjustment. It removes first all constraints from the individual solutions and then applies the covariance rescaling, based on the data sampling used at each individual analysis center.
The combination is statistically correct (equivalent to a common least-squares adjustment using all original GPS observations in one step) assuming that there are no correlations between the observations of each of the sequential solutions. The procedure is not absolutely correct from the statistical point of view because the observations of the anchor sites (sites common to different solutions) are introduced into the combined solutions at least twice.
Two different combinations are made subsequently : one using a free network solution and one using fiducial sites. In both solutions the same stations are used to define the reference frame. In the free network solution the geodetic datum is defined by specifying six no-net constraints (translation and rotations) for the selected reference sites. The free solution is performed since it helps to identify problems with the combination. The fiducial solution gives the official EUREF weekly solution which is available at the anonymous ftp servers at AIUB (EUREF Combination Center, Switzerland), CDDIS (IGS global data center), IfAG (IGS regional data center), and ROB (EUREF Central Bureau, Belgium).

Quality and Optimization of the Data Analysis

The consistency of the solutions from the different EUREF analysis centers is well below the 10 mm level in the horizontal and around 10 mm for the vertical component.

As can be seen in Figure 2, the repeatability of the coordinates from the combined EUREF solution is very good. Only a few stations show slightly higher rms values. The most common causes are site specific errors (like the use of old Rogue receivers in Madrid and Wettzell) or problems with fixing the reference frame, especially for stations at the edge of the network.

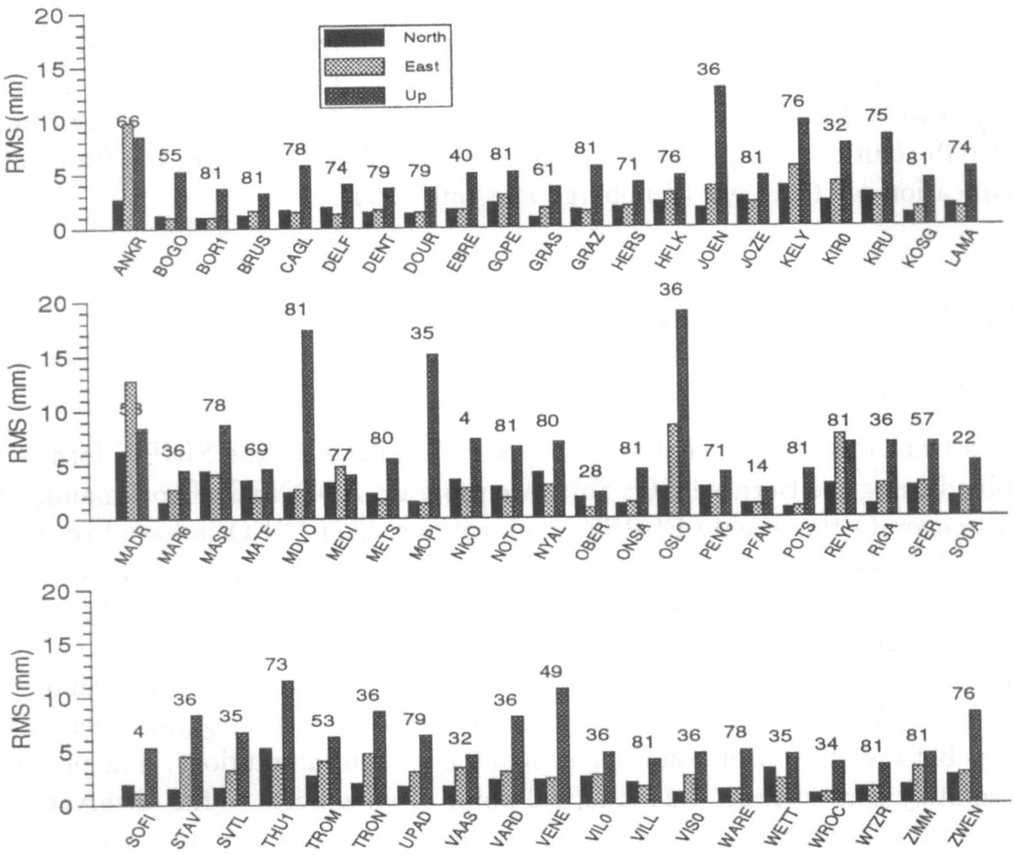

Figure 2: Root mean square errors of the EUREF station coordinates obtained from the weekly EUREF solutions covering the period from January 1996 to August 1997. The numbers written above the north-east-up rms values are the number of weeks that the station was included in the solution.

The data analysis within the EUREF network is presently being reviewed for more optimization :

- All analysis centers follow, since recently, data analysis recommendations which were set up at an "EUREF Analysis Workshop" organized in Brussels on 10-11 may '97 (Bruyninx, 1997);

- The observations from the stations which are processed by several analysis centers are introduced multiple times into the combined solution. The resulting correlations are not taken into account in the combination procedure and they distort the EUREF covariance matrix.
  At the same time, the quality of weekly solution for stations which are processed

by only one analysis center is rather difficult to assess.
Therefore a redistribution of the individual subnetworks has been agreed upon between the network coordinator and the analysis centers. This redistribution aims on one hand to equalize the number of centers analyzing each station and on the other hand to have at least two analysis centers for each station.

- One of the main problems for the correct interpretation of the coordinate time series are undocumented changes of the antennas or their environment at the EUREF stations, e.g. the use of radomes. Examples are shown in Figures 3 and 4.
  EUREF therefore encourages a closer contact between the analysis centers and the station managers through the EUREF Central Bureau. This communication runs two ways :

  1. the station managers who try to document all detected site changes in the station log files ;
  2. a feedback from the analysis center to the station managers (or operational centers) when anomalies are detected during the data analysis.

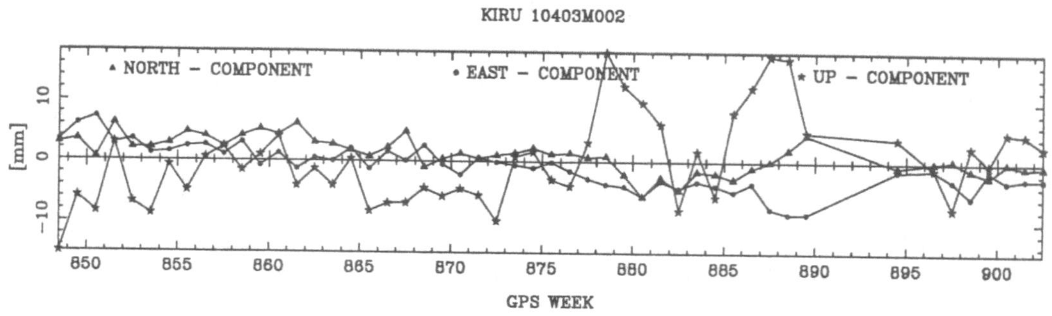

Figure 3: Coordinate time series for Kiruna (from EUREF combined solution). The behavior of the height component can be explained by the fact that there was snow on the antenna in November '96 (week 878-881) and January '97 (week 886-890).

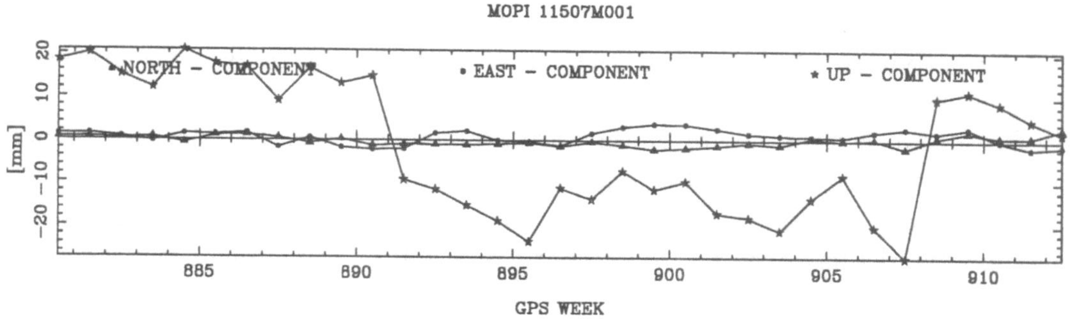

Figure 4: Coordinate time series for Modra Piesok (from EUREF combined solution). A jump of around -2 to -3 cm is detected in the height component at GPS week 891 due to the installation of a radome on the antenna on February 3, 1997. The radome was taken off again on June 2, 1997, before the start of the CERGOP campaign.

EUREF Contribution to the IERS

Early 1997 the International Earth Rotation Service (IERS) decided to produce a new realization of the ITRS, called International Terrestrial Reference Frame (ITRF). It was thought that due to the IGS densification project, the EUREF stations would also show up in the new ITRF results. It turned out, however, that at the moment, due to the "pilot" status of the IGS densification project, there is no guarantee that all the permanent EUREF stations would show up in the new ITRF realization. Therefore EUREF proposed to the IERS to include a separate EUREF contribution in the new ITRF.
In May of 1997, CODE submitted the combined EUREF solution to the IERS (Springer et al, 1997). As a result all EUREF permanent stations will be included in the next ITRS realization.

The EUREF Permanent Network Central Bureau

The EUREF permanent network coordination is performed at the Royal Observatory of Belgium : a fully documented information system (including descriptions of the permanent GPS stations, Local Data Centers, Local Analysis Centers and their subnetworks) is operational and maintained (ftp to `ftpserver.oma.be` or `193.190.249.203`, `cd pub/astro/euref` or the Web site : `http://www.oma.be//KSB-ORB/EUREF/eurefhome.html`).

The consistency between the headers of the RINEX data files and the station description files in the database are weekly checked and if necessary station responsibles are contacted. The product availability is monitored and reports on the combined EUREF solution are made available weekly as feedback to the contributing analysis centers.

Conclusions

The success of the EUREF network has been demonstrated by the fact that since the start of the activities, at the end of 1995, the number of stations, data centers and analysis centers has more than doubled.
Weekly free-network solutions for Europe are available since April 1996. In the beginning of 1997, data analysis guidelines have been set up for the EUREF analysis centers and they are presently followed by all the analysis centers. The distribution of the EUREF subnetworks has been revised recently and will be implemented in the autumn of 1997.
The distributed processing is a success : the consistency between the solutions of the different analysis centers is of the same quality as the week-to-week repeatabilities of each analysis center.

Recognizing the success of the EUREF permanent network, the EUREF Technical Working Group will make, within the next months, the necessary calls for participation to take the activities related to the EUREF network from an experimental stage to an operational service.

Acknowledgments

The authors would like to express their gratitude towards the responsible agencies and representatives at the observation sites, the data centers and analysis centers. Without their labor and commitment this paper would never have been possible.
We would also like to recognize the work of Werner Gurtner who initiated the activities related to the EUREF permanent network. The members of the EUREF Technical Working Group provided support and technical advice.

References

Brockmann, E. (1996), "Combination of Solutions for Geodetic and Geodynamic Applications of the Global Positioning System (GPS)", Ph.D. dissertation, Astronomical Institute, University of Berne, Berne, Switzerland

Bruyninx C., W. Gurtner and A. Muls (1996), "The EUREF Permanent GPS network", Report on the Symposium of the IAG Subcommission for Europe (EUREF) held in Ankara 22-25 May 1996, Veröffentlichungen der Bayerischen Kommission für die Internationale Erdmessung, Vol 57

Bruyninx C. (1997), "Minutes of EUREF Analysis Workshop", Brussels, April 10-11, 1997
(http://www.oma.be//KSB-ORB/EUREF/papers/elacw001/elacw001.html)

Bruyninx C. (1997a), "The EUREF Permanent GPS Network : Activities May '96 - May '97 and Future Plans", Report on the Symposium of the IAG Subcommission for the European Reference Frame (EUREF) held in Sofia 3-7 1997, Veröffentlichungen der Bayerischen Kommission für die Internationale Erdmessung, (in press)

Gurtner W. (1994), "The Use of European Permanent GPS Stations for the Maintenance of the European Reference Frame", Report on the Symposium of the IAG Subcommission for Europe (EUREF) held in Warsaw 8 - 11 June 1994, Veröffentlichungen der Bayerischen Kommission für die Internationale Erdmessung, Vol 54

Gurtner W. (1995), "Guidelines for a Permanent EUREF GPS Network", Report on the Symposium of the IAG Subcommission for the European Reference Frame (EUREF) held in Helsinki 3-6 May 1995, Veröffentlichungen der Bayerischen Kommission für die Internationale Erdmessung, Vol 56

Rothacher, M. , G. Beutler, E. Brockmann, L. Mervart, S. Schaer, T.A. Springer, U. Wild, A. Wiget, H. Seeger and C. Boucher (1996), "Annual Report 1995 of the CODE Analysis Center of the IGS", in IGS 1995 Annual Report, edited by J.F. Zumberge et.al., pp.151-173, IGS Central Bureau, Jet Propulsion Laboratory, Pasadena, California, U.S.A.

Seeger, H., W. Augath, R. Bordley, C. Boucher, B. Engen, W. Gurtner, W. Schlüter, R. Sigl (1992), "Status-report on the EUREF-GPS-Campaign 1989 to the IAG EUREF Subcommission", Veröffentlichungen der Bayerischen Kommission für die Internationale Erdmessung, Vol 52

Springer, T. , W. Gurtner, M. Rothacher, S. Schaer (1997), "EUREF Activities at the CODE Analysis Center", Report on the Symposium of the IAG Subcommission for the European Reference Frame (EUREF) held in Sofia 3-7 1997, Veröffentlichungen der Bayerischen Kommission für die Internationale Erdmessung, (in press)

Zumberge, J.F. and R. Liu (eds) (1994), "Densification of the IERS Terrestrial Reference Frame through regional GPS Networks", Workshop Proceedings, IGS Central Bureau, Jet propulsion Laboratory, Pasadena, California, U.S.A.

THE EUROPEAN VERTICAL GPS REFERENCE NETWORK CAMPAIGN 1997 - CONCEPT AND STATUS

J. Ihde[1), J. Adam[2), W. Gurtner[3), B. G. Harsson[4), W. Schlüter[5), G. Woeppelmann[6)
[1)Bundesamt für Kartographie und Geodäsie*, Außenstelle Leipzig
Karl-Rothe-Straße 10-14, D-04105 Leipzig, Germany
[2)Technical University of Budapest, Department of Geodesy
Mueegyetem rkp. 3.I.61, HU-1111 Budapest, Hungary
[3)Astronomisches Institut der Universität Bern
Sidlerstraße 5, CH-3012 Bern, Switzerland
[4)Statens Kartverk, N-3500 Honefoss, Norway
[5)Bundesamt für Kartographie und Geodäsie*, Fundamentalstation Wettzell
Sackenrieder Straße 25, D-93444 Kötzting, Germany
[6)Institut Géographique National
2. Avenue Pasteur B.P.68, F-94160 Saint-Mandé, France

Abstract

The European Vertical GPS Reference Network (EUVN) is designed to contribute to the unification of different height systems in Europe. The most important practical and scientific aspects are

— contribution to a unique European height datum
— connection of European tide gauge benchmarks as a contribution to monitoring absolute sea level variations
— establishment of fiducial points for the European geoid determination
— preparation of a European Vertical Kinematic Network.

A network of about 195 points distributed over Europe and consisting of 79 EUREF-points, of 53 nodal points of the Levelling Networks of Eastern and Western Europe and 63 tide gauges has been observed in the period of May 21 to 29, 1997 with GPS in order to derive uniform ellipsoidal heights in the frame of ETRS89.

---

*formerly Institut für Angewandte Geodäsie (IfAG)

1. General

The existence and the establishment of geodetic networks such as

— EUREF (European Reference Frame)

— UELN (United European Levelling Network)

— UPLN (United Precise Levelling Network)

lead to the proposal to combine these networks in the height component. Through a GPS campaign such a combination will provide a „European Vertical GPS Reference Network“ (EUVN) for scientific and for practical use.

The realization of the EUVN is based on the Resolution No 3 of the EUREF Symposium in Ankara, 22-25 May 1996.

The authors of the paper are members of the EUVN Working Group, which is instructed by the IAG Subcommission EUREF to prepare and carry out the EUVN project.

2. Objectives

The objectives of the EUVN are

(1) to provide an integrated vertical reference frame for EUREF/ETRS height values at a few centimeter level:

For all EUVN points *P* three-dimensional coordinates in the ETRS89 $(X_p, Y_p, Z_p)_{ETRS}$ and geopotential numbers c_p will be derived. The geopotential number $c_p = W_{o\,UELN} - W_p$ is the difference between the potential of the earth gravity field in the level through the reference tide gauge of the UELN (W_{oUELN}), which best approximates the geoid, and the gravity potential in the EUVN points (W_p).

Finally in the same way the EUVN is a geometrical/physical reference frame. In addition to the geopotential numbers c_p normal heights $h_n = c_p/\bar{\gamma}$ and orthometric heights $h_o = c_p/\bar{g}$ will be provided ($\bar{\gamma}$ is the mean normal gravity between the ellipsoid and the telluroid, $\bar{g}$ is the mean gravity between the geoid and the earth surface).

(2) to connect different European height datums:

In addition to the UELN 73 for West and North Europe and the UPLN 82 for Central and Eastern European countries national height systems exist with different kinds of heights and different zero levels. This project will contribute to a unification of European height systems in the framework of UELN.

(3) to provide fiducial points for the European geoid determination and for future accurate regional geoid computations:

The application of the GPS technique for practical height determination will dramatically be extended if the geoid would be known precisely enough. In order to derive the geoid a European reference geoid is required in the reference System ETRS89 of EUREF and the reference System of UELN. So far there is no precise geoid available for Europe with an accuracy of a few centimeters which fulfils the requirement for the practical applications.

This proposal points out a possibility to evaluate a geoid tailored for the GPS-levelling methods by combining the existing reference network EUREF with the UELN95.

(4) to contribute to the realization of an European vertical datum and to connect different sea levels of European oceans in view of work by PSMSL (Permanent Service Mean Sea Level), also in view of anticipated accelerated sea level rise due to global warming.

At present the zero level for UELN is the tide gauge Amsterdam and for the UPLN is the tide gauge Kronstadt. The level difference is about $h_{Amsterdam} - h_{Kronstadt} = 0{,}15$m. Independent of an uniform height level for the maritime states the knowledge of the level and, under special conditions, of the level changes of the adjacent oceans is vitally important.

(5) to provide contribution to the determination of an absolute world height system (Balasubramania, 1994).

(6) to establish a fundamental network for a further geokinematic height reference system such as UELN 2000 under the special consideration of the Fennoscandian uplift and the uplift in the Carpathian-Balkan region.

(7) to provide data for decoupling the land and sea level components of relative sea level variations, as measured by tide gauges:

As tide gauges provide measurements of sea level relative to a Tide Gauge Benchmark (TGBM), they give access to a local information which generally results from the combination of the sought-after rise in sea level and the vertical movements of the land at the tide gauge site. Therefore, global sea level studies based on tide gauge data require to monitor the vertical crustal velocities at the tide gauge sites with respect to a geocentric reference frame, in order to recover a global geocentric assessment of the sea level variations. In this scope the use of GPS has been recommended by an international group of experts at two occasions (Carter et al., 1989) and (Carter, 1994).

(8) to provide the basis of express the results of the regional European tide gauge GPS surveys in the EUREF reference system (ETRS89):

Following the Carter et al. (1989) recommendations several institutions have carried out regional tide gauge GPS surveys around Europe. At least three common points between two networks are needed for a rigorous coordinates combination process and for the subsequent expression of the regional campaign results in a common geocentric reference frame (Boucher et al., 1994). The design of the EUVN network and its tide gauge component has taken into account this constraint.

3. Design of EUVN

EUVN connects several kinds of heights and will integrate the vertical reference of the European Reference Systems EUREF, UELN (and national height networks), tide gauge sites, as well as the European Geoid.

The EUVN is designed to incorporate the already realized parts of EUREF and UELN as well as the planned European permanent GPS station network.

European Reference Networks for positioning and for heights

EUREF realizes the ETRS89 frame for precise applications of GPS techniques for positioning in Europe with an accuracy about 1 cm. The height components refer to the reference ellipsoid GRS80.

UELN is the „United European Levelling Network" which provides physical heights with respect to the tide gauge of Amsterdam. According to Resolution No. 3 of the EUREF Symposium 1994 in Warsaw, it is the objective to create an uniform height datum for Europe on the basis of UELN73 (Ehrensberger and Kok 1986) which includes the enlargement also to the Central and Eastern European countries.

In the Central and Eastern European countries there already exists an United Precise Levelling Network (UPLN) consisting of first order levelling lines through Bulgaria, East Germany, Czech Republic, Slovakia, Poland, Romania, Russia, Georgia, Estonia, Latvia, Lithuania, Belorussia, Hungary, Ukraine and Moldavia. UPLN was observed in the 1950ies and remeasured in the 1970ies. The readjustment was completed in 1982. It comprises more than 350 nodal points. The lengths of lines vary between 70 km in the western part and 200 km in the eastern part. The reference gauge is Kronstadt/Russia. The adjustment 1982 was carried out using normal heights (UPLN 82).

The goal of the UELN95 is the realization of an unique European height system with an accuracy of better than 1 decimeter. The enlargement to UELN 95 is performed in two parts:

(1) Substitution of data of such network blocks which already are part of UELN 73 but include new measurements.

(2) Step by step adding network blocks of the UPLN.

Tide gauge sites are essential to estimate a possible secular sea level rise. The tide gauge sites will provide all the information for the combination of ellipsoidal heights and physical heights along coast lines. These results will be of great importance for the proposed campaign covering the whole of Europe.

EUVN related observations

Three kinds of observation groups are necessary:

— station measurements for the documentation of situation and the determination of eccentricities between different markers
— GPS campaigns for the determination of the ellipsoidal height of all defined EUVN points
— connecting levellings between the EUVN sites and the UELN nodal points for the determination of the physical height of all defined EUVN points.

4. Schedule of the EUVN97 GPS Campaign

The observation period of the EUVN97 GPS campaign was:

Start time	May 21, 1997, 18:00 UTC
Stop time	May, 29, 1997, 6:00 UTC.

In total the campaign consists of about 195 sites: 66 EUREF- and 13 national permanent sites, 53 UELN and UPLN stations and 63 tide gauges (Figure 1). The EUVN is a joint collaboration of most of the European countries.

The northernmost EUVN point is situated at Ny Alesund (78.9°, 12.0°) on Spitzbergen, the southernmost point is the tide gauge Larnaka (34,9°, 33,6°) on Cyprus. The westernmost point Reykjavik (64,15°, -22,0°) is situated in the North Atlantic, the easternmost point Yozgat (39,8°, 34,8°) in Turkey.

The Baltic Sea Level GPS campaign 1997 and other local GPS activities were performed simultaneously to the EUVN97 GPS campaign. For EUVN and BSL stations and under consideration that at several stations collocations between different receivers were performed more than 200 teams worked in the field.

Three types of receivers were used within the EUVN GPS campaign:

— Turbo Rogue Receivers at permanent sites (IGS standard),
— Trimble SSI or SSE,
— Ashtech Z 12.

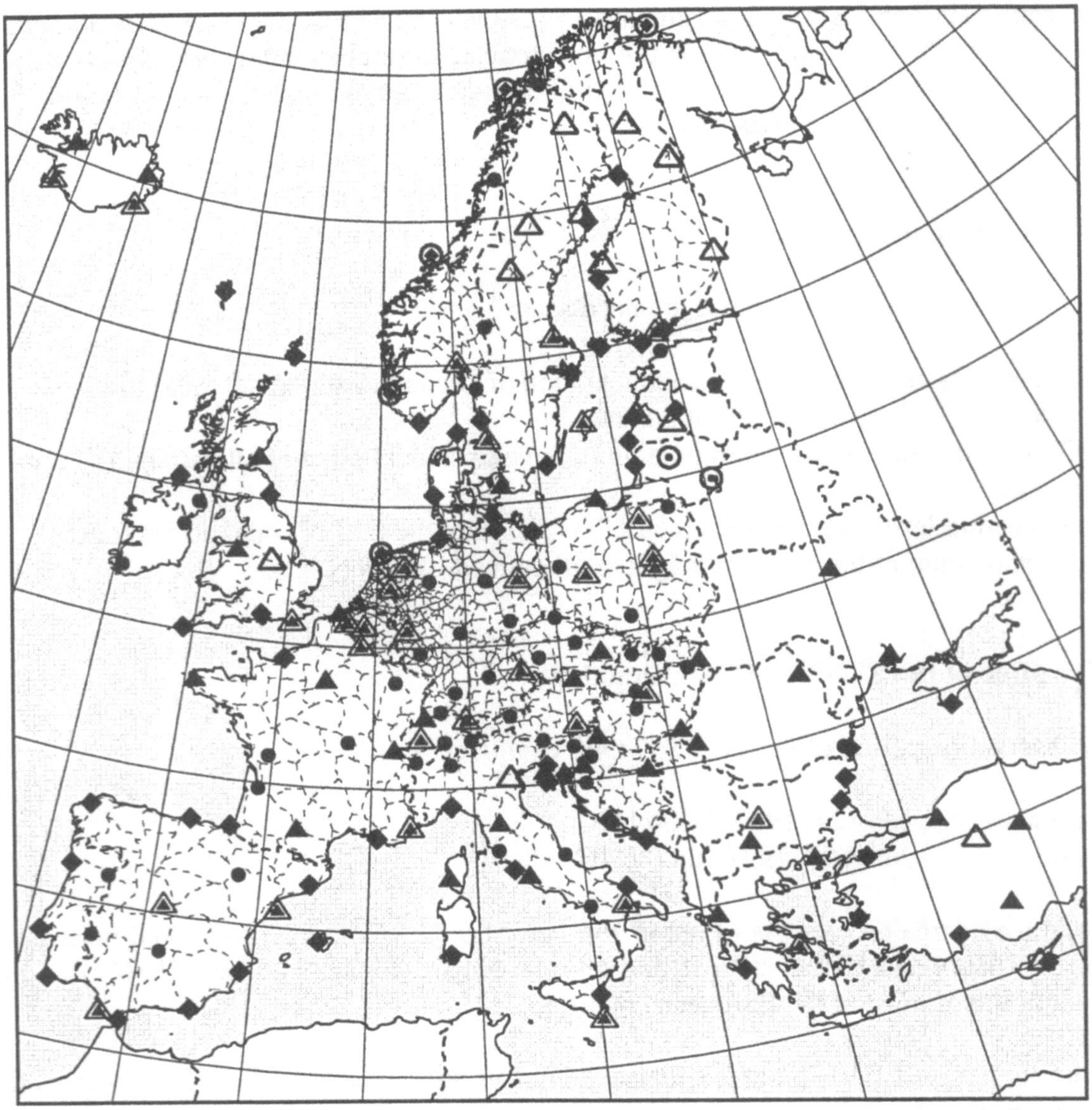

Bundesamt für Kartographie und Geodäsie, January 26, 1998

Figure 1: Distribution of EUVN stations

▲ EUREF sites

◬ GPS permanent stations - EUREF

△ GPS permanent stations

● UELN & UPLN nodal points

⊙ GPS permanent stations - nodal points

◆ Tide gauge sites

◉ GPS permanent stations - tide gauge

UELN lines

Collocation of the receiver and antenna types of some selected observatories is foreseen in order to derive type dependant offsets. For the GPS campaign, the time interval was set to 30 s, the elevation mask is 5°.

5. EUVN GPS Processing

The data preprocessing performed by 9 EUVN Preprocessing Centers (PPC) is mainly a check concerning completeness and consistency of the data and the auxiliary information. The PPCs were requested to prepare complete access information and/or data flow guidelines for the observing agencies before the start of the campaign.

The EUVN Data Center (DC) is established at the Bundesamt für Kartographie und Geodäsie, Aussenstelle Leipzig.

The task of the Analysis Center (AC) is to process the data of a subnetwork. A subdivision of the whole EUVN Network will be done under the aspect of receiver type and regions. 9 European institutions are ready to contribute as Analysis Center.

6. Further Activities — Connection Levellings, Tide Gauge Measurements

To reach the objective of the EUVN as a European Height Reference Network, in the next stage the connection levellings between the GPS points and the UELN/UPLN nodal points must be delivered and the bases for the processing of the tide gauge data which were included into the project must be created.

References

Augath, W.: Proposals for a European Task Force on Vertical Datum within the EUREF-Subcommission. Presented at the EUREF Symposium in Warsaw, June 7-11, 1994, Dt. Geod. Komm. Astronomisch-Geodätische Arbeiten, Heft Nr. 54, S. 171-175.

Balasubramania, N.: Definition a Realization of a Global Vertical Datum. Report No. 427, Dep. of Geodetic Science and Surveying, Ohio State University, Columbus, Ohio 43210-1247, 1994.

Birardi, G.: The Italian North-South GPS Traverse and a proposal for a „first order" Italian geoidal net. Bull. Geod., Berlin 67 (1993)4, pp. 201-209.

Boucher, C., Wöppelmann, G.: Proposal for a European Primary Tide Gauge Network (EPTN), EUREF TWG Meeting, Bad Homburg, Dec. 15-16, 1994.

Brouwer, F. J. J., De Min, E. J.: On the Definition of a European Vertical Datum. Presented at the EUREF Symposium in Warsaw, June 7-11, 1994, Dt. Geod. Komm. Astronomisch-Geodätische Arbeiten, Heft Nr. 54, S. 171-175.

Carter, W. E., Aubrey, D. G., Baker, T. F., Boucher, C., Le Provost, C., Pugh, D. T., Peltier, W. R., Zumberge, M., Rapp, R. H., Schutz, R. E., Emery, K. O. and Enfield, D. B.: Geodetic fixing of Tide Gauge Bench Marks. Woods Hole Oceanographic Institution Technical Report, WHOI-89-31, CRC-89-5, August 1989.

Carter, W. E. (ed.): Report of the Surrey Workshop of the IAPSO Tide Gauge Bench. Fixing Committee held 13-15 December 1993 at the Institute of Oceanographic Sciences Deacon Laboratory, Wormley, UK. NOAA Technical Report NOSOES0006, October 1994.

Ehrnsperger, W., Kok, J. J.: Status and Results of the 1986 Adjustment of the United European Levelling Network - UELN-73, Contributed paper to the Symposium on Height Determination and Recent Crustal Movements in Western Europe, Federal Republic of Germany, Sep. 15-19., 1986.

EUVN Working Group: First circular of the EUVN project, IAG Subcommission for EUROPE (EUREF), Wabern/München, March 1996.

Ihde. J., Schlüter, W.: Proposal for a European Vertical GPS-Reference Network (EUVERN). Presented at the EUREF Symposium, May 3-6, 1995, Kirkkonummi, Finland.

Ihde, J., Schlüter, W., Gurtner, W., Wöppelmann, G., Harsson, B. G., Adam, J.: Concept and Status of the European Vertical GPS Reference Network (EUVN). Presented at the EUREF Symposium in Ankara, May, 1996, Dt. Geod. Komm., Astronomisch-Geodätische Arbeiten, Heft Nr. 57, S. 218-225.

Lang, H., Sacher, M.: Status and Results of the Adjustment and Enlargement of the United European Levelling Network 1995. Presented at the EUREF Symposium, May 3-6, 1995, Kirkkonummi, Finland.

Kakkuri, J.: The Baltic Sea Level Project. Allgemeine Vermessungsnachrichten, Heidelberg, 102 (1995), 8/9 pp. 331-336.

Seeger, H., Adam, J., Augath, W., Boucher, C., Gubler, E., Gurtner, W., Van der Marel, H., Zielinski, J. B. : The new European Reference System EUREF — Status Report 1995. Invited paper at the XXIst General Assembly of IUGG/IAG, Boulder, Colorado, USA, July 2-14, 1995.

Torge, W., Basic, T., Denker, H.: Long Range Geoid Control Through the European GPS Traverse. Dt. Geod. Komm., R. B., Nr. 290, München, 1989.

FROM UELN-95 TO EVS 2000
EUROPEAN ACTIVITIES FOR A CONTINENTAL VERTICAL DATUM

W. Augath[1], F. Brouwer[2], H. Lang[3], J. van Mierlo[4], M. Sacher[3]
[1]Geodetic Institute, Technical University of Dresden, Mommsenstraße 13, D-01062 Dresden, Germany, email: augath@kgise.geo.tu-dresden.de
[2]Rijkswaterstaat, Survey Department, Postbus 5023, NL-2600 GA Delft, The Netherlands
[3]Bundesamt für Kartographie und Geodäsie (former Institut für Angewandte Geodäsie), D-60598 Frankfurt, Germany
[4] Geodetic Institute, University of Karlsruhe, Postfach 6380, D-76131 Karlsruhe, Germany

Abstract

In Commission X (continental networks) a lot of new scientific activities have started. For Europe this task is overtaken since 1989 by the EUREF-subcommission in strong cooperation with the European Survey Agencies represented by the „Comité Européen des Responsables de la Cartographie Officielle (CERCO)". Until now only the European Terrestrial Reference System 1989 (ETRS 89) has already been realized which is identical to the global realization of the International Terrestrial Reference System (ITRS) in Europe at the epoch 1989.0.

The activities for a similar solution for the height component are much more time-consuming. That is why the EUREF-Technical Working Group in view of the different time frames of the main users decided in 1993 to divide the realization into two steps:

Step 1: United European Levelling Net 1995 (UELN-95)

A „quick" solution on the 0,1 m-level as height reference for Europeanwide geoinformation using systems, using existing levelling data sets and a simple static adjustment model.

Step 2: European Vertical System 2000 (EVS 2000)

A realization on the highest possible level (1 cm and better) for National Survey Agencies and scientific requirements. The activities for UELN-95 have to be carried out as usable basis for this step 2.

The paper describes the problems of a change from national to continental datums, the status of the first step UELN-95 and the concept of the realization of EVS 2000.

Concept and Status of UELN-95

The concept of UELN-95 can be summerized in one sentence: „Realization of a vertical datum in Europe as soon as possible for the determination of transformation parameters to the national height datums on the decimeter-level, which can also be used as a first step for a final solution on the highest possible level (EVS 2000)". This causes the following procedure:

- Use of the existing Westeuropean levelling data set UELN 73/86 (geopotential differences between nodal points and connections to Westeuropean tide gauges, published by Ernsperger and Kok (1987) and enlargement to Eastern Europe.
- Use of the Westeuropean datumpoint (tide gauge Amsterdam) and normal heights as common height system.

Block		Number of observations	Sum of redundancies	a posteriori standard deviation [kGal · mm]
Austria	(A)	128	29.020	0.62
Belgium	(B)	54	19.494	1.22
Switzerland	(CH)	13	4.715	1.12
Germany	(D)	745	271.488	0.83
Denmark	(DK)	26	6.813	0.66
Spain	(E)	101	27.255	1.85
France	(F)	175	46.826	2.01
Italy	(I)	97	32.453	1.78
Netherlands	(NL)	951	164.014	1.08
Portugal	(P)	22	5.857	1.77
Great Britain	(GB)	60	15.000	1.72
Norway	(N)	194	70.992	1.67
Finland	(F)	89	20.142	0.76
Sweden	(S)	122	34.865	1.74
Czech Republic	(CZ)	82	28.742	1.10
Hungaria	(HU)	49	12.362	0.54
Slovenia	(SL)	11	1.580	0.51
Poland	(PL)	179	54.863	0.99
Slovakia	(SK)	74	18.518	1.39
	Total	3172	865.000	1.23

Table 1: Accuracy of the national blocks derived from the variance component estimation (UELN-95/10, Status 7/1997)

- Static adjustment model without additional parameters and the use of aposteriori-standard deviations derived from variance component estimations between the national blocks.
- National computing centers and one international center for the data and the common adjustments at Federal Office for Cartography and Geodesy (former IfAG), Germany.

Unfortunately the common levelling data set of the Easteuropean Levelling Net including all former Easteuropean socialist countries was not available. It had to be reconstructed by the Easteuropean Survey Agencies. Missing connections between Eastern and Western Europe were remeasured and new national data sets, remeasured after 1973, were introduced.

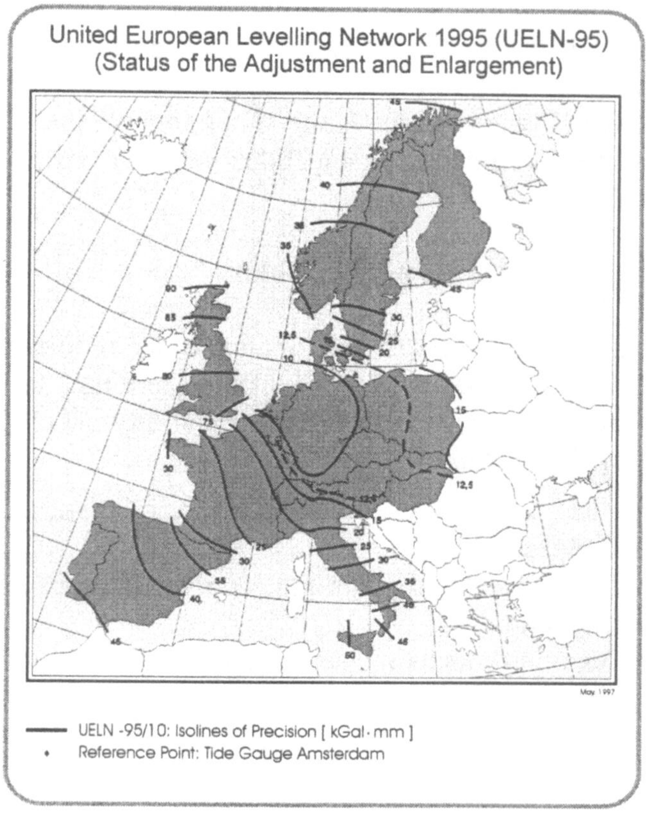

Picture 1: Isolines of standard deviations of the adjusted heights in relation to the datum-point Amsterdam (Status 7/1997)

The adjustments had to be realized step by step:

- UELN-73/94: Repetition of the UELN-73/86-adjustment using weights for the national blocks derived from a variance component estimation.
- UELN-95/1,2: The former block D (Germany) was removed and replaced by the new German First Order Levelling Net „DHHN 92" (East- and Westgermany including new connections).
- UELN-95/3: The former block A (Austria) was replaced by the remeasured Austrian First Order Levelling Net.
- UELN-95/4, 5, 6, 7, 8: Enlargement by the new national levelling networks of the Czech Republic, Hungary, Slovenia, Poland and Slovakia.
- UELN-95/9, 10: The former block NL (Netherlands) was replaced by the Fourth Precise Levelling Net of the Netherlands.

In table 1 the accuracy of the national network blocks derived from variance component estimations are put together. Picture 1 shows the isolines of standard deviations of the adjusted heights (UELN-95/10) in relation to the datum-point Amsterdam (Lang et al. 1997). In the near future the UELN-95-data set will be enlarged by new national blocks of Croatia, Bulgaria and Romania. In Southeastern Europe Yugoslavia is still missing. In Northeastern Europe a connection with the Baltic States, Russia to Scandinavia would be extremely important. In Western Europe some further national block can be replaced (CH, DK) and the new observations between France and the United Kingdom via the Channel-Tunnel will be introduced as well as new hydrostatic connections between Germany and Denmark and another one in Denmark.

Concept of European Vertical System (EVS 2000)

The actual proposal can be described as development of a Continental Vertical System on the one-centimeter-level. It includes components concerning the Vertical Datum, the adjustment model, the network design in the field of reliability and checks against systematic errors as they are put together in table 2.

- **Continental Vertical Datum**
  based on a selected or „mean“ tide gauge and geopotential numbers
- **Kinematic adjustment model**
  using the national levelling epochs and actual kinematic information as repeated GPS/Gravity/SLR/VLBI, Tide gauges
- **Higher reliability**
  by the use of denser levelling networks as done in UELN 73 (national 1st and 2nd order levellings)
- **Check against systematic errors**
  by independent observations (later, e. g. Geoid-traverses on the 1 to 2-cm-level, hydrostatic-levelling-traverses)
- **Connection to a Global Vertical Datum**
  on the 0,01 to 0,1 m-level (e. g. by Improvement of the Global Geoid)

Table 2: Development of a Continental Vertical System on the one-cm-level: EVS 2000

Continental Vertical Datum

Because of the ongoing development of a Global Vertical Datum, the European Vertical Datum should be defined as **continental datum**. The reference surface can be fixed by a selected or a „mean“ tide gauge. Preliminary values of the European sea surface topography and additional special effects will be obtained by the EUVN-GPS-campaign 1997 in combination with the existing UELN 95-results. Until a final decision is made, the existing UELN-definition will be used (tide gauge Amsterdam, geopotential numbers, normal heights).

Kinematic Network

The former UELN-subcommission already proposed the upgrade from the static to a kinematic height network (Remmer 1988). Picture 2 gives an overview over existing levelling epochs in Europe. The kinematic information of the former European levelling epochs which give the relation to the last 50 - 100 years must be completed by actual kinematic information of selected points with repeated height-related observations like GPS/Gravity/SLR/VLBI and tide gauges on the highest possible level. In this field a lot of data are also still existing such as: the ITRS-network, IGS, EUREF permanent sites, sites of regional and local campaigns. For height change determination a special focus for an accurate height connection to the levelling network is necessary as well as a more or less permanent supervision of this connection. Therefore a special network which includes only those stations should be selected (Augath 1990). A first data set could be overtaken from the EUREF-permanent GPS-sides (Picture 3).

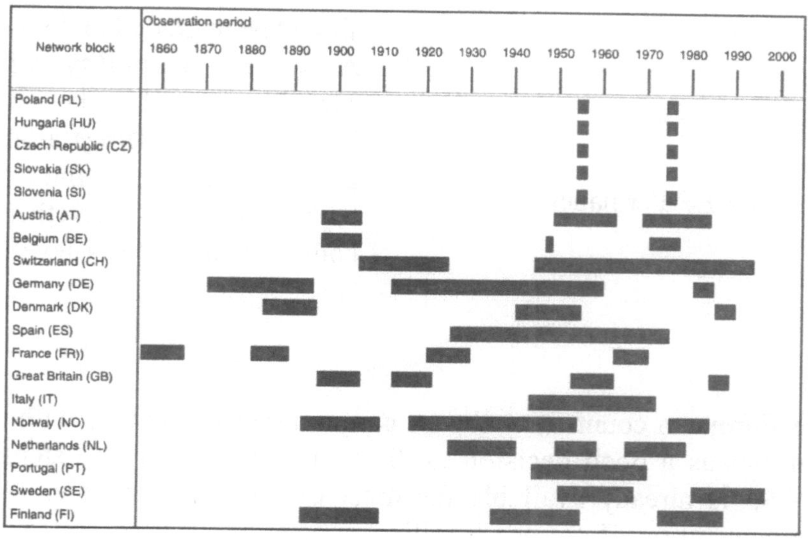

Picture 2: Existing levelling epochs in Europe

Picture 3: The EUREF-Permanent-GPS-side

Reliability

The adjustment of UELN 73/86 used only a reduced number of levelling observations (large loops with 1000 kilometers circuit). The data management of today allows the use of all first (and second) order levelling observations in one adjustment. This will give an upgrade of the reliability.

Check against systematic errors

On the European level independent checks against systematic errors are extremely necessary. The best chance for the next years is given by **3D/Geoid-traverses.** For this purpose the European geoid should be upgraded to the 1-2 cm-level.

A second possibility which is more expensive could be realized with **hydrostatic levellings** along the rivers e. g. with the existing equipment of Rijkswaterstaat (Netherlands) and connections to UELN-levelling lines.

Outlook

Since 1993 the European countries realize a common vertical datum within the EUREF-subcommission. It was a good decision to divide the work into two steps. The „quick" solution UELN-95 is already available for most of the European area. Meanwhile the timeconsuming acquisition of former levelling epochs has started and a first kinematic adjustment project with the levelling data of Switzerland is on the way (Gubler, Wolfram 1997). Nevertheless a lot of work in the field of kinematic modelling and the integration of different height relevant observation data and the upgrade of the quality of the height component derived from space techniques has be done.

References

AUGATH, w. (1990): Proposal for the Establishment of an European Height Supervision Systems (HSS-WE). GPS for Geodynamics, Volume II, pp. 135 - 140, Luxembourg (Walferdange), 1990.

AUGATH, W. (1996): UELN 2000: Possibilities, strategy, concepts. Proposals for a European Task Force on Vertical Datums within the EUREF-Subcommission. Veröffentlichung der Bayr. Kommission für Internat. Erdmessung, No. 54, pp. 170 - 175, Beck-Verlag, Munich, 1994.

BOUCHER, C. (1994): Global Vertical Reference and the IERS Terrestrial Reference System (ITRS): concepts and realizations. Proceedings International Symposium INSMAP 94, Hannover, 1994 and Publication in LAREG, Série Communication, CM6, Paris, 1994.

ERNSPERGER, W., KOK, J. J. (1987): Status and Results of the 1986 Adjustment of the United European Levelling Network - UELN 73. In Pelzer, H. and Niemeier, W. (Eds.): Determination of Heights and Height Changes, pp. 7 - 45, Dümmler Verlag, Bonn, 1987.

GUBLER, E., WOLFRAM, S. (1997): Personel information (unpublished), 1997.

HECK, B., RUMMEL, R. (1989): Strategies for solving the Vertical Datum Problem using terrestrial and satellite geodetic data. IAG-Symposium 104 (Edinburgh): Sea Surface Topography and the Geoid, Springer, 1989.

LANG, H., SACHER, M. (1996): Status and Results of the Adjustment and Enlargement of the United European Levelling Network 1995 (UELN-95), pp. 163 - 169, 1996.

LANG, H., STEINBERG, J. (1993): Zur Entwicklung der Höhennetze auf dem Territorium der neuen Bundesländer. Allgemeine Vermessungsnachrichten, Karlsruhe, 100, pp. 295-309, 1993.

NIEMEIER, W. (1987): Observation Techniques for Height Detemination and their Relation to usual Height Systems. In: Pelzer, H. and Niemeier, W. (Eds.): Determination of Heights and Height Changes, pp. 85 - 108, Dümmler Verlag , Bonn, 1987.

REMMER, O. (1987): The United European Levelling Network - Present State and Future Plans -. In: Pelzer, H. and Niemeier, W. (Eds.): Determination of Heights and Height Changes, pp. 3 - 5, Dümmler Verlag, Bonn, 1987.

RUMMEL, R., REUNISSEN, P. (1988): Height datum definition, Height datum connection and the role of the Geodetic Boundary Value Problem. Bulletin Géodesique, 62, pp. 477 - 498, 1988.

SEEGER, H., AUGATH, W., BORDELEY, R., BOUCHER, C., ENGEN, B., GURTNER, W., SCHLÜTER, W., SIGL, R. (1993): Status Report on the EUREF-GPS-Campaign 1989 to the IAG EUREF Subcommission. Veröffentlichung der Bayr. Kommission für Internat. Erdmessung, No. 52, pp. 26 - 34, Beck-Verlag, Munich, 1993.

TORGE, W. (1990): The Geoid in Europe: Status, Requirement, Prospects. GPS for Geodynamics, Volume II, pp. 119 - 126, Luxembourg (Walferdange), 1990.

WÜBBELMANN, H. (1993): The Adjustment of the German First Order Levelling Net (DHHN 85). In: Arbeitsgemeinschaft der Vermessungsverwaltungen der Länder der Bundesrepublik Deutschland (AdV): Die Wiederholungsmessungen 1980 - 1985 im Deutschen Haupthöhennetz der Bundesrepublik Deutschland, pp. 141 - 154, published by Bayerisches Landesvermessungsamt, Munich, 1993.

GLOBAL GPS NETWORKS AND THE DETERMINATION OF EARTH ROTATION PARAMETERS

V. Ashkenazi, T. Moore, L. J. Symons, W. Chen and M. H. O. Aquino

Institute of Engineering Surveying and Space Geodesy
University of Nottingham
Nottingham
NG7 2RD
United Kingdom

ABSTRACT

Recent developments of the in-house GPS processing software at the Institute of Engineering Surveying and Space Geodesy (IESSG), University of Nottingham, has led to the capability of processing very large or even global GPS networks. This has allowed the investigation of several methods of constraining a network of stations and comparative results of these tests are presented in this paper.

A precise knowledge of the Earth's position and orientation in space is required for a number of astronomical and geophysical purposes, ranging from the tracking of interplanetary spacecraft, to the study of the Earth's interior structure. The new software package enables the determination of Earth Rotation Parameters (ERPs) at regular intervals from the global network of GPS data, thus defining an Earth Rotation Parameter series and improving the final coordinate solution accuracies. Tests performed at the IESSG are discussed and preliminary results are shown.

1 INTRODUCTION

The formation of the International GPS Service for Geodynamics (IGS) has provided the opportunity for processing high precision, global GPS networks within a number of days after data collection. Such easily accessible data can be used within global GPS network adjustments to determine numerous parameters, including the orientation of the Earth.

The movement of the Earth's axis of rotation with respect to its surface, and variations of the rate of rotation of the Earth about this axis, has been predicted for over a century. Techniques used included optical methods such as the timing and locating of solar eclipses, plus occultation's of stars by the moon. The precise knowledge of the Earth's position and orientation in space is required for a multitude of astronomical and geophysical purposes, ranging from the tracking of interplanetary spacecraft, to the study of the Earth's interior structure. These Earth Rotation Parameters may now be determined to a high level of accuracy and resolution, using recent, new space geodetic techniques, such as the Global Positioning System (GPS).

The IESSG has recently developed software which allows the estimation of Earth Rotation Parameters from GPS data. A daily series of Earth Rotation Parameters has been determined and has been compared with the corresponding series computed by IGS Analysis Centres.

2 POLAR MOTION AND UNIVERSAL TIME

The Earth rotates about an axis that has a particular orientation in space. The orientation of this axis changes with respect to time, when viewed against an external non-rotating inertial frame, due to motions known as precession and nutation.

More uncertain is the orientation of the solid Earth with respect to the rotation axis, or alternatively, the location of the intersection of this rotation axis with the Earth's crust. The displacement of this rotation axis from a conveniently defined reference point, the geographic North pole, is termed 'polar motion'. This polar motion is primarily cyclic, due to the dynamic response of the solid Earth to loading caused by the atmosphere, oceans and liquid core.

Polar motion is commonly described using two cartesian components, termed xp and yp, where the xp component may be considered to be the displacement of the Earth's axis of rotation projected on to the IERS Reference Meridian (IRM) and the yp component is the displacement projected on to a meridian 90 degrees west of the IRM.

Not only does the position of the rotation axis vary with respect to the solid Earth, but the rate at which the Earth spins also changes. This results in an excess length of day (Δ LOD) which varies from day to day, and is of the order of 2 milliseconds. The drift in Universal time (UT1), is commonly expressed as UTC-UT1[1] or UT1-TAI [2]. This variation is largely unpredictable and can result in positioning errors of many tens of centimetres at the equator if it is not closely monitored.

To monitor changes in the Earth's orientation in space, two precise stable reference frames must be defined. One must rotate in sympathy with Earth (Earth fixed whose origin is the Earth's mass centre), and one must be independent of the Earth's rotation (Inertial frame), from which changes in orientation can be ascertained.

Similarly, if the Earth fixed (terrestrial) and inertial (celestial) frames are to be used, then transformations between the two frames must be defined. The transformation between inertial and Earth fixed frames involves transformations accounting for polar motion, the earth rotation, nutation and precession and involves several intermediate reference frames. The reader is referred to (**McCarthy, 1992**) for further details.

---

[1] UTC= Co-ordinated Universal Time

[2] TAI= International Atomic Time

3 SOLVING FOR EARTH ROTATION PARAMETERS

If GPS satellite orbits and receiver locations were sufficiently well known, the GPS data may be used to simply monitor changes in the Earth's orientation. When GPS data are processed in practice, satellite orbits, station locations and Earth rotation parameters must all be estimated simultaneously. The estimated orbital parameters are referred to a celestial reference frame, whereas the network of estimated station locations provides the terrestrial reference frame. The relative motion of these two frames allows the Earth orientation parameters to be estimated.

When solving for Earth orientation using GPS techniques, double difference carrier phase data are used to obtain the receiver and satellite positions. Then the initial positions of the receiver and satellite are used to derive observation equations which allow the estimation of changes to the three Earth Rotation Parameter unknowns, namely Δxp, Δ yp and ΔUTC-UT1 in a least squares adjustment.

In fiducial GPS data processing, a certain number of stations are deemed as known 'fiducial' reference sites. Their positions are highly constrained to their coordinate values obtained from previous GPS, VLBI, or laser ranging measurements. Hence coordinates of the free stations in the network may be estimated directly. Since the orientation of the Earth is of interest, these sites should ideally span the globe, with a longitudinal spacing of no more than 90 ° and coverage in both Northern and Southern Hemispheres. Additional sites are required to fill in the terrestrial network redundancy.

In the free network (non-fiducial) approach no station locations are fixed, and only very loose constraints, of the order of 10's to 1000's of metres, are applied. The scale and origin of the network is provided by the satellite force model, and the GPS data. Consequently, only the orientation, provided by the fiducial sites is missing, but the direction of the z-axis is further constrained by the Earth's daily rotation.

In order to make the free (non-fiducial solution) consistent with the scale, origin and orientation of the ITRS, a similarity transformation can be applied. The free (non-fiducial) technique can readily sense changes in the orientation of the Earth over time but cannot establish the absolute orientation of the Earth in space. One disadvantage of this approach is that the time series of Earth orientation variations obtained from 2 or more non-consecutive GPS tests, cannot be combined directly to yield a longer Earth orientation series without some transformation.

4 THE EARTH ROTATION PARAMETER SOFTWARE

An Earth Rotation Parameter estimation routine has been developed at the IESSG and successfully integrated within the existing in-house GPS software suite, known as GAS (GPS Analysis Software).

The routine uses initial *a priori* IERS (International Earth Rotation Service), xp, yp and UTC-UT1 values, taken from the Bulletin B series. The corrections to the approximate values of the three unknown Earth Rotation Parameters, (Δxp, Δyp, ΔUTC-UT1), are then estimated in the least squares process and applied to the initial *a priori* values, thus providing daily estimates of these parameters. The observations input are double difference dual frequency carrier phase GPS data.

The orbit parameters solved for are-

Position, velocity, solar radiation/y-bias for each satellite
xp, yp and UTC-UT1 values per day

The force models used are-

WGS84 gravity model (degree and order 8)
Schwiderski Ocean tide loading model.
Basic solar radiation, spherical satellites.

For the results detailed in this paper, 24 hour two session (arcs) have been processed.

5 TEST DATA AND PROCESSING STRATEGIES

To test the software and strategies, a data set comprising data from 18 IGS tracking stations, evenly distributed across the globe, was collated. The data set covered a 5 day period from 23rd October 1995 to 27th October 1995, (Julian Day 295-300). The selection of these sites was based on the need for a strong network, well represented in the Northern and Southern hemispheres, with baseline lengths in excess of 6000 km, thus providing a truly global GPS network of tracking stations.

Two strategies were tested:

(1) Global network adjustments (no Earth Rotation Parameters estimated)

(2) Global network adjustments estimating and applying Earth Rotation Parameters.

A number of network adjustment strategies were carried out A free (non-fiducial) adjustment was computed with very loose constraints of the order of 10 metres applied to the ITRF stations and 1000 metres to the other stations. A full fiducial network adjustment was carried out, in which constraints of 2cm were applied to the ITRF93 coordinates of 7 stations and all other stations in the adjustment were left free. Fixed station network adjustments were also performed, whereby 1, 4 and 8 stations were fixed rigidly. However, the fixed network strategy results are not presented in this paper.

The aim of this analysis was to assess whether there was any significant improvement, in terms of final coordinate accuracies, when Earth rotation parameters were estimated and applied in the adjustment (**Table 1**). It should be noted that in this paper, accuracy is defined as the difference between the assumed true value of the position of a station (ITRF93 values) and the estimated one. This is expressed by the root mean square (RMS) coordinate differences. For the purpose of this paper, precision is defined by the session to session coordinate repeatability, namely the RMS of the coordinate differences from each session solution from the weighted mean coordinates of all the session solutions.

The daily Earth Rotation Parameters obtained in the tests have also been compared with the corresponding estimates made by four IGS Analysis Centres, over the same 5 day period. RMS values have been quoted together with the mean offset. Both are expressed in terms of milli-arc-seconds (mas) for xp and yp and milli-seconds (ms) for UTC-UT1 (**Table 2**).

It should be noted that these are only preliminary results and more comparisons are to be made with those Earth Rotation Parameters from a greater number of IGS Analysis Centres and the IERS in the near future.

6 ANALYSIS OF RESULTS

Table 1 illustrates that when Earth Rotation Parameters have been estimated and applied, there has been a significant improvement to the final coordinate solution, in terms of accuracy. This is clearly shown in Tests 2 and 4 for example, where X coordinate differences of 2.2 and 1.3cm have been obtained. This complies with the statement, 'that the variation of station coordinates caused by polar motion is recommended to be taken into account, since the polar displacement has a non-zero average over any given time span, and should be considered in global solutions at the cm level '(**McCarthy, 1992).**

It can be seen from **Table 1**, that when the Earth Rotation Parameters are not estimated and applied in the final global network adjustment, the final coordinate solution is degraded in terms of accuracy.

Table 1 RMS Errors of Coordinates with and without the Estimation of Earth Rotation Parameters.

Parameter	Test 1		Test 2		Test 3		Test 4	
	erp off	erp on	erp off	erp on	erp off	erp on	erp off	erp on
RMS (X) cm	4.1	3.8	5.5	3.3	0.35	0.34	5.2	2.9
RMS (Y) cm	6.1	6.0	6.0	6.1	0.34	0.39	2.3	1.8
RMS (Z) cm	7.6	7.4	4.0	4.0	0.56	0.44	2.2	2.7

Test 1=All Free, Test 2=1 station RCM5 constrained 2cm, Test 3= ITRF constrained to 10m others 1000m, Test 4=Fiducial (7 stations constrained to 2cm).

Table 2 indicates that the level of agreement in the pole coordinate values with the external results is on average, better then 0.13 mas xp, 0.10 mas yp and 0.010 ms UTC-UT1 . The results show that they fall within the similar quoted rms accuracies of various IGS bodies, for example, 0.15 mas xp and yp and 0.006 ms UTC-UT1. It should be noted that standard errors of 0.05 mas xp, 0.15 mas yp and 0.03 ms UTC-UT1, have also been achieved. The free (non-fiducial) network adjustment has been used to generate the results shown in Table 2, where 10m constraints were applied to the ITRF stations and all other stations were loosely constrained to 1000m.

Set No.	xp (mas)		yp (mas)		UTC-UT1 (ms)	
	Offset	RMS	Offset	RMS	Offset	RMS
1	0.06	0.16	0.05	0.10	-0.005	0.0149
2	0.05	0.13	0.01	0.16	0.009	0.0112
3	0.07	0.14	0.04	0.10	-0.009	0.0124
4	0.02	0.13	-0.03	0.08	0.007	0.0101

Table 2 Comparison of 1-day Earth Rotation Parameters with External Results.

Set 1=IGS (weighted mean), Set 2=Jet Propulsion Laboratory, Set 3=NOAA, Set 4=ESA/ESOC

7 CONCLUSIONS

A software module has been successfully developed and implemented within the existing software suite. The Earth Rotation Parameters obtained with the Nottingham software agree with those obtained by the four IGS Analysis Centres at a level of 0.13 mas xp, 0.10 mas yp and 0.010 ms UTC-UT1.

A significant improvement occurred in the accuracy of the final station coordinates when the Earth Rotation Parameters were estimated and applied in the global network adjustments. This can be illustrated by an improvement in the X coordinate values of 1.3cm in the fiducial solution and 2.2cm in the 1 station constrained to 2cm solution.

These preliminary test were performed with 1-Day (2 session arcs). It is envisaged that future work will include a 3-day (5 session arc 60 hrs.) smoothed series. In addition, it is anticipated to implement the T30 solar radiation model, (**Fliegel & Gallini, 1996**), within the existing software suite along with various gravity field tests. Similar tests will be carried out to ascertain if any further improvement to the final coordinate accuracies and Earth Rotation Parameter series has occurred.

REFERENCES

Fliegel, H. F. and Gallini, T. E, (1996), *Solar Force Modelling of Block IIR Global Positioning System satellites, Journal of Spacecraft and Rockets, Vol. 33, number 6, pp 863-866.*

McCarthy, D.D., (1992), *IERS Standards (1992), IERS Technical Note 13, Observatoire de Paris, July 1992.*

REALIZATION OF THE ITRF-94 IN THAILAND AND MALAYSIA BY USE OF A COMBINED NETWORK FOR GEODYNAMICS AND NATIONAL SURVEY

M. Becker, E. Reinhart, H. Seeger
Bundesamt für Kartographie und Geodäsie, Frankfurt a. M., Germany
S. Mingsamon, C. Boonphakdee
Royal Thai Survey Department, Bangkok, Thailand
S. Abu
Department of Surveying and Mapping, Kuala Lumpur, Malaysia

ABSTRACT

In 1994 and 1996, two major GPS campaigns for geodynamics, named GEODYSSEA, were conducted in South East Asia. The paper gives an introduction to the GEODYSSEA project and its results. In cooperation with the national survey authorities, this occasion was used to realize the ITRF in several countries of South East Asia. This paper describes the background and results of GEODYSSEA and the additional measurements as well as transformations of the national datums of Thailand and Malaysia to the ITRF-94. By use of the GEODYSSEA data and also by repeated occupations of the first order stations in the respective countries an accuracy of 2 cm in position and about 3 cm in height was obtained for the ITRF-94 coordinates. Depending on the accuracy of the existing local coordinates the transformation parameters between the national datum and ITRF-94 were determined with and accuracy of 5 to 10 cm.

1. INTRODUCTION

The determination of geocentric coordinates in a well defined reference frame is the basis for the establishment and the maintenance of a geodetic infrastructure. It is needed for all activities in GIS, surveying and navigation in all countries. Unified national geodetic datums are required for the integration of regional geodetic and geodynamic GPS campaigns. Up to now, this has posed a problem in the area of South East Asia because there were no permanent GPS receivers of the IGS network deployed, which would have facilitated the connection to the ITRF.

In the framework of the GEODYSSEA project in South East Asia, the survey authorities of Thailand (Royal Thai Survey Department, RTSD) and of Malaysia (Department of Surveying and Mapping, Kuala Lumpur, DSMM) and the Bundesamt für Kartographie und Geodäsie (BKG, formerly IfAG) decided to improve the national survey networks by establishing high precision links to the ITRF.

The objectives are:

- To establish a clearly defined and internationally accepted geodetic reference frame.
- To establish a zero-order GPS Network which would be used as the framework for the national GPS networks.
- To improve the accuracy of the existing points in the national GPS networks by use of the final coordinates from the project as fiducial stations in the GPS network adjustments.
- To establish definitive transformation parameters between the current mapping datum (e.g. *INDIAN 1975*) and the ITRF-94 reference frame.
- To investigate magnitudes and directions of crustal motions in the area to be able to account for the time dependent changes in the coordinates.

2. THE GEODYSSEA PROJECT

The GEODYSSEA Project was carried out under the Contract CI1*-CT93-0337 between the European Commission and a number of ASEAN countries.

The European Partners involved in the geodetic activities are the Bundesamt für Kartographie und Geodäsie, Frankfurt am Main , observations and Analysis; Delft Inst. of Earth Oriented Space Research (DEOS) , analysis; Ecole Normale Superieure(ENS), Paris, analysis; GeoForschungsZentrum Potsdam (GFZ) , observations and analysis, coordinator; Inst. of Physics, Univ. of Bologna, analysis; Royal Observatoire Belgique, Brussels, analysis; Univ. of Newcastle, Dept. of Surveying , analysis.

The ASEAN Partners are BAKOSURTANAL, Cibinong, Indonesia; Dept. of Surveying and Mapping, Kuala Lumpur, Malaysia; Bureau of Energy and Mines, Manila, Philippines; Inst. of Geology, Nat. Center for Nat. Sciences, Hanoi, Vietnam; Royal Thai Survey Department, Bangkok, Thailand; Dept. of Public Works, Brunei.

GEODYSSEA is an acronym for **GEODY**namics of **S**outh and **S**outh-**E**ast **A**sia. The Objectives are the „*Determination of Plate Motions and Crustal Deformations Deduced from Space Geodetic Measurements for the Assessment of Related Natural Hazards in South East Asia*". The geodetic part has as principal goal the measurement of the velocities of the continental plates and the deformations of the Earth's crust by GPS in the region of South-East Asia. The complete GPS network covers the whole area of South East Asia, see Fig. 1.

In both Thailand and Malaysia there are two points which were selected for establishing new permanent marks in bedrock. Field reconnaissance, monumentation and field observations were done by BKG together with the local officers.

Simultaneous GPS observations were conducted in an area of about 4000 by 4000 km on about 40 carefully selected and newly monumented stations. In 1994, 24 hour sessions on days 332 to 336 (Nov. 28th till Dec. 3rd), in 1996 on days 109 to 113 (Apr. 18th to Apr. 23rd) were observed. In addition, data of IGS and AUSLIG sites for the connection to the ITRF were used.

The main geodetic results are given in Tab. 1; for details on the GEODYSSEA solutions see Ambrosius et al. (1997). Fig. 2 shows the ITRF-94 velocities estimated for the network, no vertical velocity components were estimated.

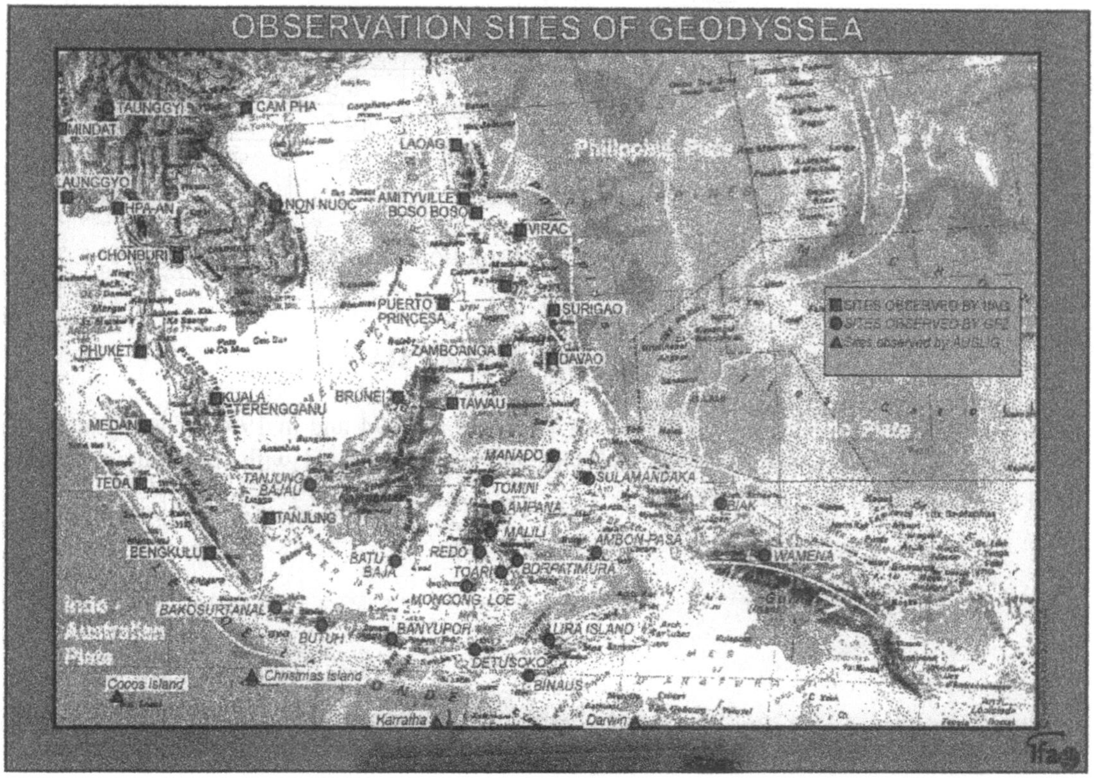

Fig. 1. The GEODYSSEA Network

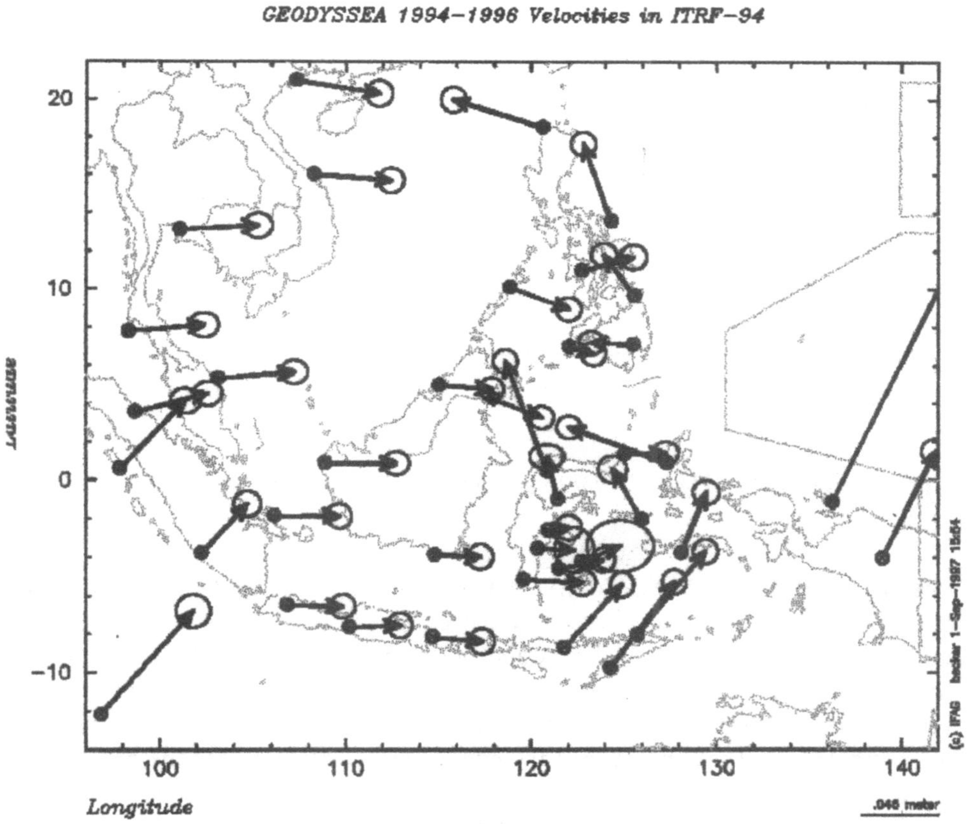

Fig. 2. Velocities of the combined GEODYSSEA solution, ITRF-94 (m/year)

Component	Precision			Accuracy			
	North	East	Up	North	East	Up	
Coordinates 1994	2.3	4.0	7.1	8.2	8.9	4.9	mm
Coordinates 1996	2.8	4.3	7.0	10.0	14.9	11.8	mm
Velocities	3.	5.	not estim.	13.	18.	not estim.	mm / y

Tab. 1. Summary of the GEODYSSEA combined solution accuracy and precision.

3. MEASUREMENTS IN THAILAND AND MALAYSIA

During the observations of the GEODYSSEA project in 1994 and in 1996, existing sites of the first order networks of Malaysia and Thailand were observed with the same accuracy level as specified for GEODYSSEA.

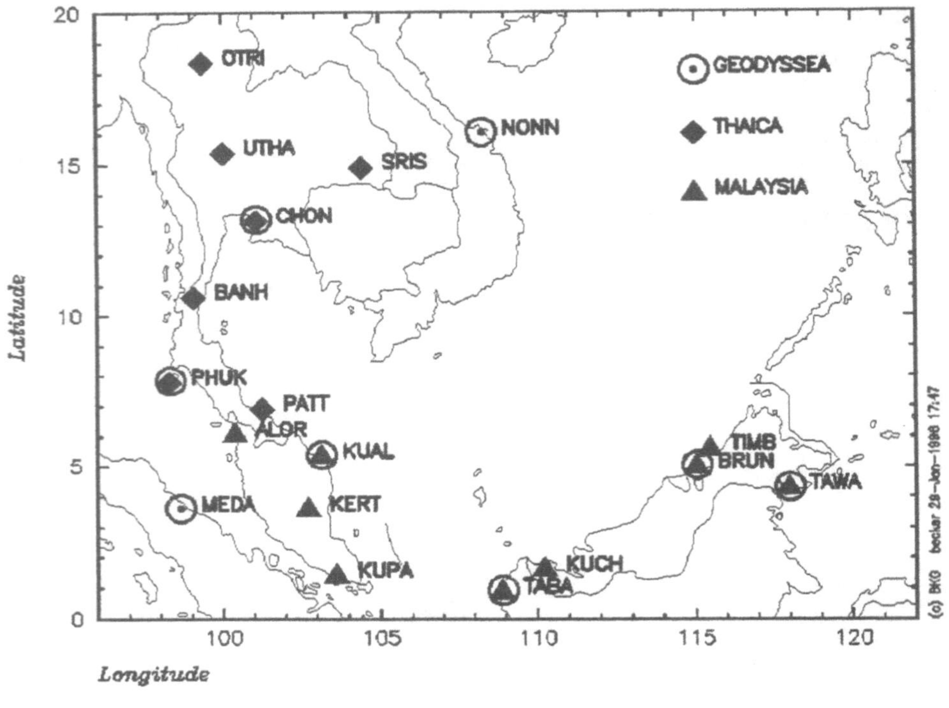

Fig. 3. Points observed in the THAICA94/96 and MALAYSIA94/96 campaigns.

The network designed for the MALAYSIA-94 GPS Campaign included two stations from the existing sites of the first order network of Malaysia. These stations are located separately in two networks, one in Peninsular Malaysia and one in Sabah/Sarawak. In 1996, the same two and five additional points were used, see Fig. 3 and Tab. 2. In Thailand five additional stations were observed in THAICA-94 and three of these points were repeated in THAICA-96, see Fig. 3 and Tab. 2. These points and the connecting points used for the baseline definition are given in Tab.2 and Fig. 3. Simultaneous GPS data were recorded at all of the stations for 24 hour observation sessions. Trimble 4000SSE GPS receivers with Trimble

4000SST L1/L2 GEO GPS antennas were used for the observations. In THAICA-94, TRIMBLE SST with squaring technique had to be used on 4 points.

No	Full Name	ID	Location	Station Description	Days 1994	Days 1996
1.	Bukit Timbalai	TIMB	Malaysia	Malay GPS Network,		
				Doppler 30393 station	332-336	109-113
2.	Bukit Kertau	KERT	Malaysia	Malay GPS Network,		
				Doppler 30398 station	332-336	109-113
3.	Alor Star	ALOR	Malaysia	Malay GPS Network,		
			South Base	Doppler 30404 station	-	109-113
4.	Kuching	KUCH	Malaysia	Malay GPS Network		
				West Base	-	109-113
5.	Tanjung Kupang	KUPA	Malaysia	Malay GPS Network,		
				Doppler 30401 station	-	109-113
6.	Tawau	TAWA	Malaysia	ITRF Reference		
				GEODYSSEA Station	332-336	109-113
7.	Kuala Terenganu	KUAL	Malaysia	ITRF Reference		
				GEODYSSEA Station	332-336	109-113
8.	Tanjung Bajau	TABA	Indonesia	ITRF Reference		
				GEODYSSEA Station	332-336	109-113
9.	Brunei	BRUN	Brunei	ITRF Reference		
				GEODYSSEA Station	332-336	109-113
1.	Chonburi	CHON	Thailand	ITRF Reference	324-326	
				GEODYSSEA Station	332-336	109-113
2.	Phuket	PHUK	Thailand	ITRF Reference	324-326	
				GEODYSSEA Station	332-336	109-113
3.	Uthaithani	UTHA	Thailand	Thai Nat. GPS Net	324-326	
				Datum Point, Doppler,	334-336	109-113
4.	Chomphon	BANH	Thailand	Thai Nat. GPS Net	324-326	109-113
5.	Srisakhet	SRIS	Thailand	Thai Nat. GPS Net	324-326	109-113
6.	Lampang	OTRI	Thailand	Thai Nat- GPS Net	324-326	-
7.	Pattani	PATT	Thailand	Thai Nat. GPS Net	324-326	-

Tab. 2. Points of the MALAYSIA and THAICA campaigns and duration of observations

4. PROCESSING STRATEGY AND RESULTS

Processing and analysis of all campaigns have been carried out at BKG in cooperation with DSMM and RTSD with the Bernese Software, version 4.0. The processing strategy adopted followed the specifications of the EUREF Technical Working Group given in (Gurtner, 1993).

The strategy is summarized as:

- Use of IGS combined orbits and ERP.
- Processing using 24 hours sessions of 15 / 30 sec. data for the 1994 / 1996 campaign.
- Elevation cut-off angle 15 degrees for both campaigns.
- Phase center offsets according to Table IGS-01.
- Zenith delay parameters estimated once per 2-hours.
- Ionosphere-free L3 float solution for the 1994 campaign.
- Ambiguity resolution using the QIF (Quasi Ionospheric Free) strategy in 1996.

- Daily solutions computed for storage of normal equations and the analysis of precision.
- Final coordinates from the combination of normal equations of GEODYSSEA and of all sessions of the Thai/Malay campaigns, constrained to the GEODYSSEA final coordinates in ITRF-94, epoch 1996.3 (Ambrosius et.al,1997).

Detailed descriptions of the processing and the results can be found in (Mingsamon et. al, 1997) and (Becker et.al., 1997). The precision obtained for the national networks is comparable to that of the GEODYSSEA solution for all of the four add-on campaigns. As an example, the repeatability of THAICA94 and THAICA96 results is given in Fig. 4. The day to day repeatability is shown in Tab. 3 with the example of the MALAYSIA baselines.

Taking the coordinate precision of the final coordinates of 2 to 8 mm in horizontal position, from 10 to 20 mm in height, and of about $1\text{-}2 \times 10^{-8}$ for the baseline length, and adding the overall GEODYSSEA accuracy in the ITRF-94, we arrive at an approximate absolute coordinate accuracy of the 5 Thai and the 5 Malay stations in the ITRF-94 global reference frame of at least better than 3cm.

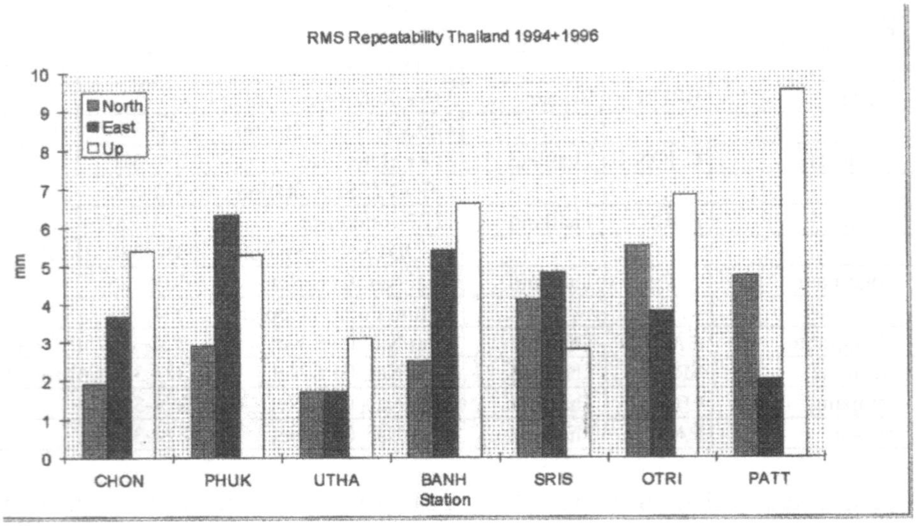

Fig. 4 RMS repeatability of the combined unconstrained 1994 and 1994 THAICA solution.

BASELINE	LENGTH (m)	DLAT (mm)	DLONG (mm)	DHGT (mm)	DLGT (mm)	DLGT/LGT (PPM)
1. BRUN - TIMB	39028.2761	7.9	7.0	23.4	9.3	.238
2. TIMB - TAWA	330198.1849	2.0	5.6	17.5	5.5	.017
3. KUAL - KERT	213007.6309	2.9	7.6	12.6	3.4	.016
4. KERT - KUPA	255537.5578	5.4	1.8	18.3	5.0	.020
5. ALOR - KUAL	318235.0105	3.7	6.4	13.3	7.2	.023
6. KUCH - BRUN	652150.2523	4.2	8.3	12.1	8.5	.013
7. TABA - KUCH	168033.4577	2.3	6.7	18.2	6.8	.041
8. KUPA - TABA	590482.7493	1.7	2.3	17.5	3.2	.005

Tab. 3. Repeatability of the baseline components in the MALAYSIA GPS Campaigns (Dlat, Dlong, Dhgt are the standard deviation of each daily solution.)

The major geodetic results can be summarized as follows:

- GEODYSSEA shows a differential rotational motion of the Sunda Block versus the Eurasian Plate, i.e. the area of Thailand and Malaysia may have a small deviation in their plate velocities with respect to the main portion of the Eurasian plate.
- For the territory of Thailand a uniform velocity field (i.e. that of the Sunda Block) can be used for all sites. Therefore, the two campaigns were combined and time-dependent changes were not considered in the net.
- GEODYSSEA results also indicate a minimum of 12.7 mm/year crustal shortening being absorbed between Peninsular Malaysia and Borneo.
- This leads to time-dependent distortions in the National net of Malaysia. For Peninsular Malaysia and for Sabah/Sarawak different velocities are required.
- Accuracy of ITRF-94, epoch 1996.4 coordinates is estimated to be better than 3 cm.
- These coordinates are now used to compute the Helmert Transformation parameters for the Datum definition of the existing national GPS networks in the ITRF.
- The accuracy of these parameters depends on the quality of the national GPS networks.
- By reason of their high precision, the national networks could be used for studies of intra-plate deformations over the long term.

5. TRANSFORMATION PARAMETERS

The reference frame of the current mapping datum of Thailand is the INDIAN 1975 datum (INDIAN75). At present, the absolute accuracy of the origin point of the first order horizontal network (*Thai National GPS Network*) in the WGS84 reference frame is in the order of 1-2 meters because the WGS84 global reference frame was realized through Doppler observations from the TRANSIT satellite system. The five common stations of the final THAICA results are used to determine the seven parameters of a Helmert Transformation. As can be seen from Tab. 4 the rms of the transformation is about 7 cm. For applications where only a shift is required, an additional 3-parameter transformation is computed.

For Malaysia a datum definition was made in an earlier GPS project and computed by DMA (Report on the Joint Brunei/Malaysia/ Singapore Geodetic GPS Project by 512 STRE November 1992 to June 1993, DSMM, Kuala Lumpur).

Parameters	**Value**	**Std. Dev**	**Value**	**Std.Dev.**
	7-Parameter		Translation Only	
Translation in X	203.884	+- .978 M	205.788	+- .061 M
Translation in Y	832.904	+- .461 M	833.213	+- .061 M
Translation in Z	293.969	+- .468 M	293.720	+- .061 M
Rotation X-AXIS	-.0038	+- .0152"	-	-
Rotation Y-AXIS	.0579	+- .0151"	-	-
Rotation Z-AXIS	.0781	+- .0316"	-	-
Scale Factor	-.017	+- .067 mm/km	-	-
RMS of Transformation		*.0731 M*		*.1365 M*

Tab. 4. Transformation parameters from the coordinates of the Final THAICA solution, ITRF-94,1996.3 and the coordinates of the Thai GPS Net in the INDIAN75 Datum

From Tab. 5 it can be seen that these coordinates are generally in very good agreement with an Helmert transformation rms of 6 cm if we consider a shift of the WGS84 versus the ITRF-94, 1996.3 of about 60 cm in Z-direction and about 20 cm in X-direction.

Parameters	Value of Parameters	Standard Deviations
Translation in X:	0.198	+- .027 M
Translation in Y:	0.034	+- .027 M
Translation in Z:	- 0.588	+- .027 M
RMS of Transformation		+- .0607 M

Tab. 5. Translation of Malaysian coordinates, actual results versus DMA results (WGS-84)

6. SUMMARY AND CONCLUSIONS

GPS campaigns were carried out in order to connect the Thai and Malay national networks to the ITRF global reference frame, to establish the zero order network, to establish a new geodetic surveying and mapping control network, to determine precise transformation parameters, and to improve the accuracy of the existing GPS network. The final coordinates for the 10 new ITRF stations are given in the ITRF-94 (epoch 1996.3) reference frame and have an accuracy which meets the objectives of the GEODYSSEA project. The precision of the final coordinates is at the order of better than 3 cm. The results so far are sufficiently accurate to establish an ITRF-referenced datum for all practical GIS, surveying and navigation purposes. For Thailand, RTSD presently prepares a new adjustment of the national GPS network with about 530 stations using the stations of this study as fiducials. Similar investigations are underway for Malaysia.

A second repetition of the GEODYSSEA GPS campaign is planned which will then yield improved estimates of the velocity field in the area. This will allow for the monitoring of the time dependent changes in the ITRF coordinates. For Malaysia the transition to an active reference system by use of a network of 13 permanently tracking GPS stations and their analysis is planned within the next year. Further works on the complete datum definition will also have to consider the height problem. Therefore, geoid computations will have to be improved in order to allow the complete transformation of all types of terrestrial data.

7. References

Ambrosius, B. et. al. (1997). *The Final Combined Solutions for the GEODYSSEA94/96 Campaigns*. GEODYSSEA Final Report, Brussels, 1997 (in press)

Becker, M., P. Neumaier, E. Reinhart, Dato' Abd. Majid, Samad Abu, C. Boonphakdee (1997) : *Results of the MALAYSIA 94/96 GPS-Campaign Processing at IfAG, GEODYSSEA* Final Report, Brussels, 1997 (in press).

Mingsamon, S., C. Boonphakdee, M. Becker, P. Neumaier, E. Reinhart, H. Seeger, (1997): *Final Results of the THAICA 94/96 GPS Campaigns*. GEODYSSEA Final Report, Brussels, 1997 (in press)

Gurtner, W. (1993). *The Use of IGS Products for Densification of Regional Local Networks*. Report on the EUREF Symposium, Budapest, May 1993.

THE ITRF96 REALIZATION OF THE INTERNATIONAL TERRESTRIAL REFERENCE SYSTEM

Claude Boucher, Zuheir Altamimi, Patrick Sillard
Institut Géographique National
ENSG/LAREG
6-8 Avenue Blaise Pascal
Cite Descartes, Champs-sur-Marne
77455 Marne-la-Vallee, FRANCE
E-mail: boucher@ensg.ign.fr, altamimi@ensg.ign.fr, sillard@ensg.ign.fr

Introduction

As part of the realizations of the International Terrestrial Reference System (ITRS), a new complete solution (called ITRF-C1) has been proposed to be the ITRF conventional frame for several years, and superseding ITRF94.

The analysis of the data collected in 1996 to obtain this solution revealed some inadequacy to insure the robustness and the quality required for ITRF-C1. Moreover, this fact led to the need of more refined specifications. The IERS Working Group on the ITRF Datum was established to develop specifications for ITRF and in particular ITRF-C1, for which a final conclusion has not yet been reached.

In the mean time, the current ITRF in use, namely the ITRF94, based on data up to the end of 1994 requires updates and improvements, considering in particular that in two years, the observations extend the time interval for determining improved velocities, particularly from GPS and DORIS.

Consequently an ITRF96 solution was decided, combining individual solutions coming from those received in 1997 as well as some past data.

Input Data

The solutions provided by the IERS analysis centers and selected for the ITRF96 analysis are 4 VLBI , 2 SLR , 8 GPS and 3 DORIS solutions.

As an improvement of the use of the local ties, all the eccentricities of colocated sites were converted into a complete set of positions for each site, provided in SINEX format. Each SINEX file reflects correlations between the cartesian components of the points within each site.

Data analysis

The current strategy adopted for Terrestrial Reference Frame comparison/combination analysis is twofold: simultaneous combination of positions and velocities using full variance/covariance matrices; rigorous weighting scheme based on the analysis and estimation of the variance components using Helmert method.

The data analysis performed in view of the ITRF96 establishment are mainly: comparison of the individual solutions with ITRF94, combination of the solutions within each technique and a global combination of all the solutions together with the local ties of colocated stations.

Among those selected for the ITRF96 combination, each individual solution was compared to the ITRF94 in order in one hand to estimate the transformation parameters of the system attached to the solution with respect to ITRF94 and, on the other hand, the level of agreement with the ITRF94 values.

In order to assess the relative quality of the individual solutions, independently from the influence of local ties, a combination within each technique was also performed. Matrix Scaling Factors have been rigorously estimated during the combined adjustment of the solutions.

The ITRF96 global combination is achieved with the following properties:

- It includes the 17 selected space geodetic solutions provided by the IERS analysis centers and 70 constructed SINEX files containing positions and covariances, derived from local ties.

Figure 1: ITRF96 Sites

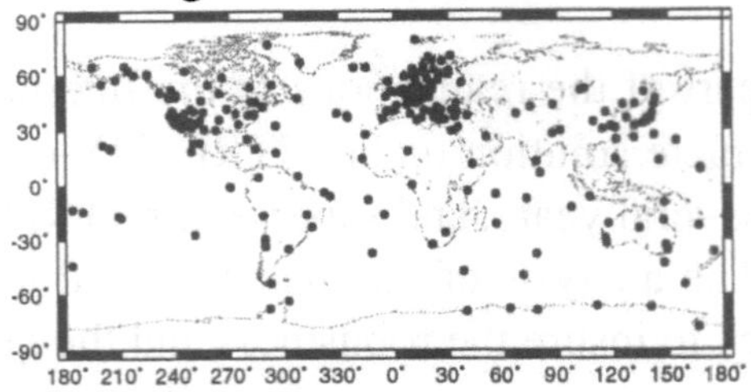

- The reference frame definition (origin, scale, orientation and time evolution) of the combination is achieved in such a way that ITRF96 is in the same system as the ITRF94.
- Velocities are constrained to be the same for all points within each site.

Figure 1 shows the coverage of the 290 sites of the ITRF96. The position formal errors at epoch 1993.0 plotted in Figure 2 demonstrate an improvement of the ITRF96 with respect to ITRF94.

The final results of the ITRF96 will be distributed in fall 1997 by the Terrestrial Reference Frame Section of the IERS Central Bureau.

Figure 2: Position formal errors

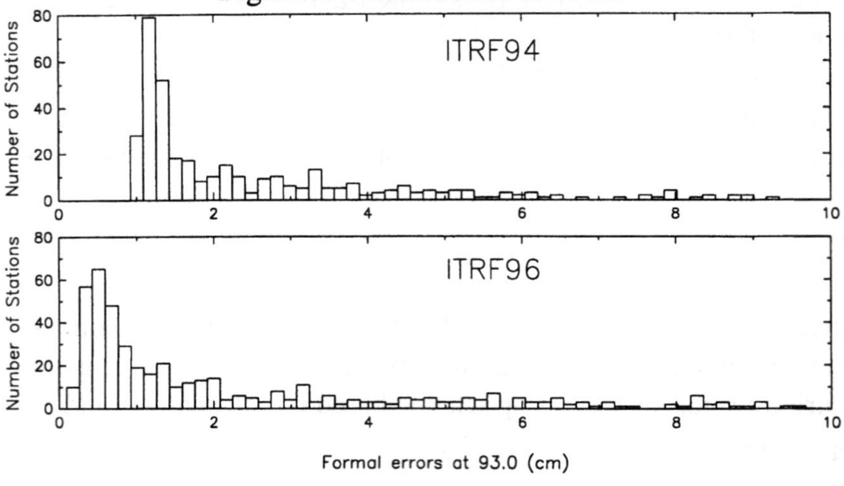

MODERNIZING THE REFERENCE FRAMEWORK FOR CANADA'S MARITIME PROVINCES

Roger J. Gaudet
New Brunswick Geographic Information Corporation
985 College Hill Road
Fredericton, New Brunswick, Canada E3B 5H1

Abstract

The three Maritime Provinces have been managing a geographical reference framework for the past 40 years. During that period a large number of spatial data sets have been created related to this reference framework. There is no doubt that the existing reference framework "Average Terrestrial System of 1977 (ATS77)" served its purpose well. However, without a framework in place which is consistent with the Global Positioning System (GPS), the technology of the next decade, it is likely that many future positioning applications will never be tied to a consistent reference framework. The result will be incompatibility between data sets collected by different individuals and agencies, and will result in additional costs to the users of the data.

The three Maritime Provinces co-sponsored a study of the best way to prepare for the reference framework technology of the future. This study's recommendations imply profound changes for the way the provincial reference framework could be handled. This presentation looks at the history of the provincial framework, the establishment of a GPS based reference framework tied to the Canadian Space Reference System, the migration from the existing framework to the new framework, the human resources and cost implications associated with these actions and the impact on the user community.

History

The primary geodetic network in the Maritime Provinces was established between 1908 and 1927 as part of the North American primary network. Until 1976 the adjustments of the primary geodetic network in the Maritime Provinces was done on the NAD27 datum. In October 1977, a national readjustment of the Canadian framework to a geocentric datum was done by the federal government and became the foundation for the Maritime Provinces' datum identified as the Average Terrestrial System 1977 (ATS77). The primary geodetic network in the Maritime Provinces was also part of the federal government's national readjustment by the Geodetic Survey Division (GSD) of Natural

Resources Canada to a geocentric datum in July 1989 which was defined as the North American Datum 1983 (NAD83). However, the Maritime Provinces never adopted NAD83 because the difference in coordinates between the ATS77 and NAD83 was on the average 4.5 metres and the user community did not feel that the cost of going through another redefinition at the time was justified.

Impetus for the densification of control came as late as 1959 from the provincial government of New Brunswick with the intent of providing a survey coordinate system, originally for cadastral purposes and later for integrated multipurpose use [Roberts 1960, 1966]. The feasibility of undertaking control densification has been brought about by the availability of the tellurometer and the geodimeter [Konecny & Roberts 1963].

The program which was cautiously started in 1959 saw 40,887 permanent survey monuments established in the Maritime Provinces during the next 18 years. These monuments were distributed throughout the Maritime Provinces in an interconnected network of traverses, with spacing of one marker every two kilometres in rural areas and one marker every 300 to 500 metres in urban areas. Monuments were placed along public rights-of-ways. The coordinate values for the secondary networks were obtained by utilizing a method of nonrigorous condition equations and computer programs developed by A.W. McLaughlin and others at the New Brunswick Department of Natural Resources. The adjustment of the secondary network was based on fixing the values of the primary geodetic framework on the 1927 datum. The nominal relative accuracy of a second order survey monument was 1:15 000.

Between 1977 and 1979 the readjustment of the maritime network was done by the Land Registration and Information Service based on the ATS77 framework. This resulted in a homogeneous set of positions across the Maritimes. The nominal relative accuracy of a second order survey monument was 1:20 000.

In 1977 a maintenance program was implemented to replace and densify the secondary networks and over 15,000 new survey monuments were added to the network between 1977 and 1997.

In 1992 the Maritime Provinces recognized the need to review the effectiveness of the existing survey control framework. They hired A. C. Hamilton and Associates to do a review of the existing survey control framework and to recommend a strategy for the provision of an appropriate survey control framework which would serve the region over a long term period, i.e., 20 years or more.

The Report of the Task Force on Control Surveys in the Maritime Provinces (Hamilton & Doig, 1993) recommended profound changes in the way the provincial reference framework could be handled. The control survey framework has been a sound investment for the provinces, however, it is clear that GPS is now sufficiently "operational" to be a cost-effective alternative in providing survey control.

The user community recognized that conversion to the new NAD83 is inevitable, but it reminded the Task Force quite emphatically that conversion to a new reference framework is a costly and time-consuming exercise for them and that whatever is adopted should be designed to serve for many decades. The user community also requested that one definitive set of transformation vectors be prepared so that everyone would be able to make the conversion to the new framework in the same way. Regrettably, a closer examination of the options for a new reference framework revealed that it is possible to design a new framework that will have a definitive set of transformation vectors or to design a new framework that will last for several decades. It is not possible to do both.

If the provinces opted to upgrade the existing network with a reasonable number, say 200 GPS observation points, do a readjustment, and do the conversion to the NAD83 datum, a definitive set of transformation vectors could be derived and the new framework could be ready for adoption in two or three years. As soon as this scenario is adopted, it is predictable that there would be complaints about its accuracy from those who will be using GPS for positioning their surveys. These complaints would increase with time and would continue until a new GPS-based reference framework was finally established and adopted. On the other hand, if the provinces established a GPS-based High Precision Network in the next couple of years, they would have a framework expected to last for several decades but it will require several years to prepare a definitive set of transformation vectors for it.

The Task Force recommended the latter option. Specifically, it recommended that the provinces make a request to the GSD of Energy, Mines and Resources Canada, for the establishment of a High Precision Network (HPN) of some 10 to 12 points as soon as possible. A program to establish an additional 20 to 30 regional High Precision points should be initiated either concurrent with the GSD's program or at a later time.

Maritime Provinces High Precision Network

In 1993, GSD established, as part of the Canadian Space Reference System (CSRS), an array of 12 points in the Maritimes at a nominal spacing of 200 kilometres. In addition to the 12 federal Canadian Base Network (CBN) points, the Maritime Provinces agreed to simultaneously establish another 32 points within the region providing a network with a nominal spacing of 70 kilometres. These 44 points were defined as the Maritime Provinces High Precision Network (MPHPN) and will provide the users of the provincial framework an effective support system for today's technologies. Cooperation between the federal and provincial governments in the establishment of these points has ensured that the network will not only be highly accurate but also highly accessible and completely consistent.

As part of the cooperative federal-provincial project to establish the CBN, the Maritime Provinces agreed to perform site reconnaissance for the federal government in parallel with reconnaissance for their own stations. In choosing the sites for both the federal and provincial CBN stations, the Maritime Provinces followed specifications prepared in 1993 by the GSD. This document outlines seven criteria to be considered in selecting a site for an ACP or an HPN point: obstructions and interference, future land development and permanence, ground stability, accessibility, security, power and communications.

The next aspect considered was the stability of the monumentation. Concrete pillars were built for all CBN points as per the GSD's specifications, including an evaluation of the ground stability at the site. A mixture of monumentation was used for provincial sites; in New Brunswick all sites are monumented with concrete pillars as per the federal specifications, in Nova Scotia and Prince Edward Island both pillars and conventional monument flush to the ground were utilized.

The observation scenarios for the GPS field collection were planned jointly by the GSD and Maritime personnel to optimize observing and moving times for both parties. All personnel participating in the federal and provincial field operations were provided with an intensive one week training program, covering basic GPS theory, field procedures, downloading procedures and data backup procedures.

The GSD occupied all federal CBN points for a minimum of three 24 hour sessions at a data collection rate of 30 seconds. The Active Control Point (ACP) in Saint John's, Newfoundland and Algonquin, Ontario were collecting data at the same rate during the period of the campaign. Turbo-Rogue receivers were used on federal points. The provincial stations were occupied for a minimum of three sessions of eight hours, at a data collection rate of 15 seconds. In order to take advantage of the 24 hour federal sessions, two eight hour provincial sessions were scheduled within each federal session. ASHTECH Z-12 receivers were used for the observations at provincial sites.

As part of the same cooperative arrangement with GSD the GPS data reduction, adjustment and analysis for the Maritime network was done by their staff. Bernese GPS processing software was utilized in the data reduction task and GHOST for network adjustment and analysis software. The objective of the Maritime Provinces was to establish a regional high precision network with an average relative (line) accuracy of 1 ppm for the network. This objective was actually exceeded, as the average 3-D ppm values, based on the 3-D error ellipses (at 95%) of 475 lines, from the minimum constraint adjustment was .252 ppm.

To assist in the management of the MPHPN a set of guidelines was developed. Guidelines for Managing a GPS Based Control System in the Maritime Provinces, Version 1.0 (Geoplan Consultants Inc. in association with Gillis Survey Systems Inc. 1996) addresses the geodetic datum to be used in supporting the control system, accuracy standards, GPS observation procedures, GPS data processing and adjustments, data storage, and the use of validation networks. This document is available on the Web at http://gauss.gge.unb.ca/gps.guidelines.html for your viewing.

Migration from Old to New Framework

During the past two years the Maritime Provinces have been densifying the MPHPN (Figure 1) in order to develop a set of transformation tools for the conversion of huge databases of geographic data collected during the past 20 years. The future densification effort will also benefit the land surveying community in having closer access to the network. The densification project was done according to the guidelines manual previously outlined and used existing provincial control stations to determine the relationship between the two frameworks. The nominal spacing is 30 kilometres with an average relative (line) accuracy of 1 ppm for the network.

As part of the conversion process from NAD27 to NAD83, GSD in cooperation with the provincial government developed the NTv2 software and a grid shift file for converting geographic data between the two datums. The governments of the Maritime Provinces have adopted the NTv2 software as the standard for converting all GIS databases and are presently developing with GSD a regional grid shift file to convert between ATS77 and NAD83 as realized through the CSRS. This will allow for consistency and reliability in converting existing large databases of geographic data from the present ATS network to the new MPHPN. Some of the GPS and GIS software

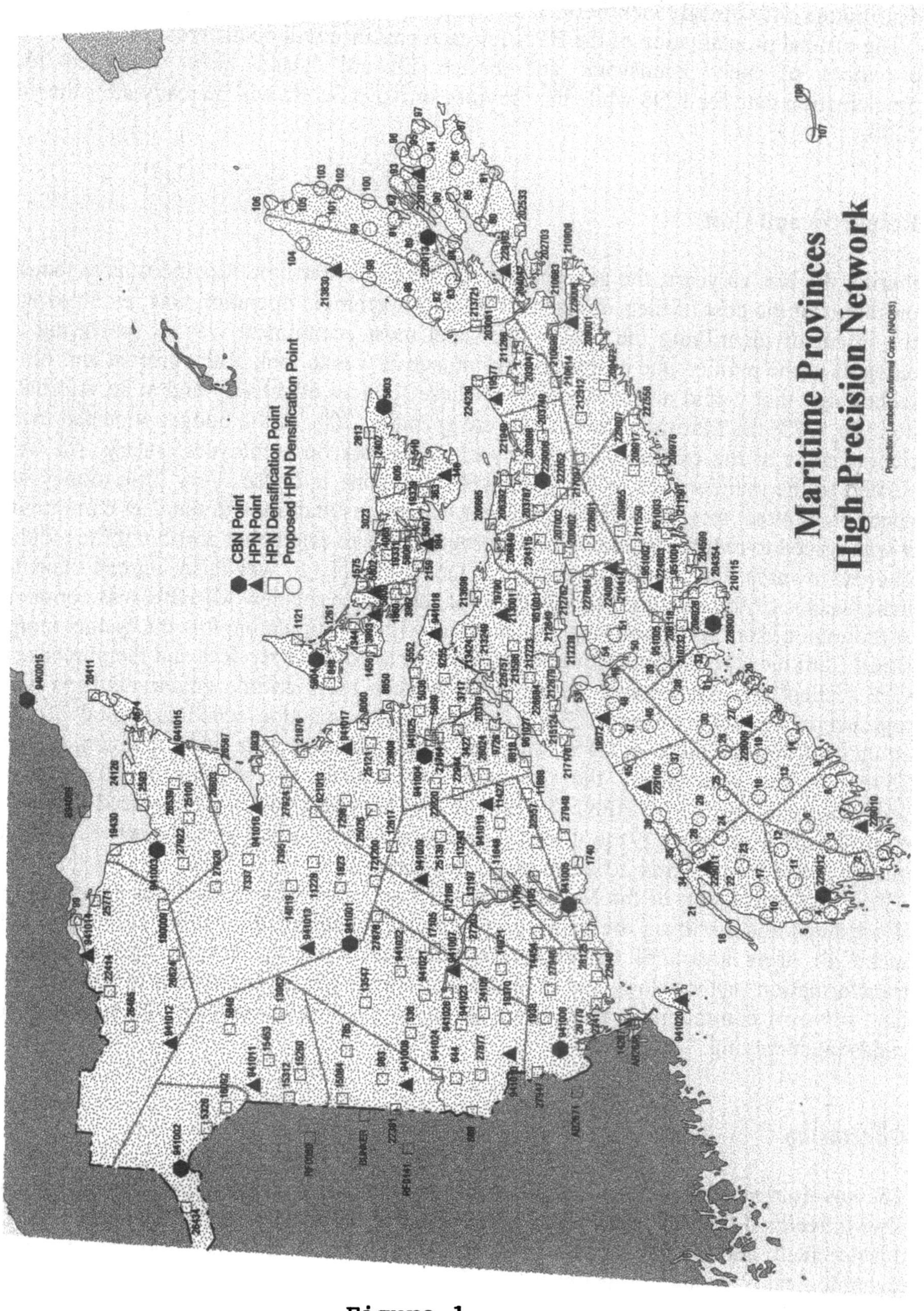
Maritime Provinces
High Precision Network
CBN Point
HPN Point
HPN Densification Point
Proposed HPN Densification Point

Figure 1

distributors have already incorporated the NTv2 software as part of their products.

The official proclamation of the HPN for each province will be different. At present the provinces of New Brunswick and Prince Edward Island have scheduled their proclamation date for 1998 while the province of Nova Scotia will be ready after the year 2000.

Resources and Cost

During the last 10 years, the governments of the three Maritime Provinces have focused on reducing the cost of their operations. Every government operation was reviewed with the intent of identifying the best method and most economical way of delivering the services to the public. One of the primary objective was to look at the present and future technology that would allow us to deliver these services effectively and at an affordable cost. In 1973, the Maritime Provinces were spending 70% of the budget allocated to the maintenance of the existing survey control framework on employees' salary. All work related to the maintenance of the framework was done in-house. Very little money was spent on contract services and new projects. After consultation with the user community, it was agreed to reduce our maintenance program and re-orient that portion of the reduced budget towards the implementation of the MPHPN. It was also agreed that the maintenance of the existing framework would stop once the new MPHPN was complete. A program of staff reduction was implemented with the stipulation that the budget money saved from the staff cutback would be transferred to contract services and the purchase of GPS technology. Agreements were signed between the maritime governments to work together in the acquisition of GPS expertise, implementation schedule, rental of GPS equipment and purchase of GPS software. During the past four years a saving of over $500,000 was realized by the maritime governments by working together in the implementation of the MPHPN. The province of New Brunswick has reduced its survey control staff from seven (7) to three (3) person year during that period and will contract out the services to the private sector as the need arises.

The implementation of the MPHPN has been accepted well by the user communities. Workshops and seminars are being prepared and scheduled during the coming year to make all users aware of the new reference framework, to develop the appropriate transformation methodology and procedures in moving from one datum to another. The land surveyor community is presently working in the development of GPS standards for cadastral surveying.

Conclusion

The governments of the Maritime Provinces and the user communities in the region have always realized the importance of a survey control framework as the foundation for the development and maintenance of spacial data sets. In this fast moving world of technological innovation, it is imperative that the infrastructure be in place to support the modern technology which will allow us to manage our resources and deliver programs more effectively and economically.

The Maritime Provinces have moved again in the forefront by delivering a control survey framework which will serve the citizens of the provinces well into the next century.

References

Geoplan Consultants Inc. in association with Gillis Survey Systems Inc. - Guidelines for Managing a GPS Based Control System in the Maritime Provinces, Version 1.0, March 1996.

Hamilton, A C. and Doig J. F. - Report of the Task Force on Control Surveys in the Maritime Provinces, prepared for the New Brunswick Geographic Information Corporation, Fredericton, New Brunswick, March 1993.

Konecny, G. and Roberts, W. F. - Electronic Surveys in New Brunswick, Journal of the Surveying and Mapping Division, American Society of Civil Engineers, Washington, D.C., October 1963.

Roberts, W. F. - The Need for a Coordinate System of Survey Control and Title Registration in New Brunswick, The Canadian Surveyor, Vol. XV, No. 5, Ottawa 1960. Integrated Surveys in New Brunswick, The Canadian Surveyor, Vol. XX, No. 2, Ottawa, 1966.

PERMANENT GPS TRACKING NETWORK INCLUDING THE MEDITERRANEAN AREA

G. Weber, M. Becker, P. Franke, W. Schlüter, H. Seeger
Bundesamt für Kartographie und Geodäsie (BKG), Frankfurt a. M., Germany

ABSTRACT

In 1995, a proposal was developed within WEGENER to install and operate a network of Permanent GPS Stations around the Mediterranean area and adjacent areas. Established with the intention of determining the velocity of tectonic plate motions, the network has to cover the region between the Azores in the west and the Caspian Sea in the east. At the BKG, a part of this network is processed on a daily basis. Loosely constrained solutions are generated and analyzed to investigate geodynamic effects and their influence on geodetic reference systems in the Mediterranean area.

The time series of coordinates available today cover a period of more than two years. Regional velocities are compared with NUVEL model velocities. The time series are analyzed for short and medium scale variations. Details of the time series analysis are given. Model calculations are presented and their significance for GPS data processing is discussed. The relationship between the obtained results and tidal deformation, and other possible effects, is being studied.

THE EUROPEAN BACKGROUND AND WEGENER

The most precise applications of GPS today make use of Permanent GPS Stations. Apart from the global framework of the International GPS Service for Geodynamics (IGS), additional Permanent GPS Arrays have been established over the last few years mainly through international cooperations. The objectives are to derive e.g.

- accurate GPS satellite ephemeris and earth rotation parameters,
- site coordinates/velocities and information,
- ionospheric and tropospheric delay information.

The accuracy of the resulting products is suitable to support the realization and continuous improvement of the International Terrestrial Reference Frame (ITRF) and the monitoring of changes of the solid and fluid earth.

Based on intensive global cooperation between various national and international authorities, a well coordinated world wide usage of Permanent GPS Arrays has been established in connection with the activities of the IGS. Starting on a national level, data from local or regional Permanent GPS Arrays are observed and processed mainly for the establishment and monitoring of three dimensional Reference Systems. In Europe, results are combined within the IAG Sub-Commission for the European Reference Frame (EUREF)

The BKG introduces it's southern European GPS Permanent activities as part of the WEGENER project.

BKG's CURRENT CONTRIBUTION TO „WEGENER-PERMANENT“

Like other IGS Regional Analysis Centers, the BKG routinely processes a Permanent GPS Array [Weber et. al. 1996]. As of today, a selection of 29 stations are included in a daily analysis, covering a substantial part of Europe and the Mediterranean (see Fig. 1). GPS observations of these permanent stations are uploaded to Frankfurt on a daily basis.

The data analysis is carried out using the Bernese GPS Software Version 4.0, following a processing strategy optimised to determine site movements from a continuous quasi on-line monitoring. The estimated parameters are station coordinates and tropospheric zenith delays only. Ambiguities are resolved using a baseline oriented Quasi-Ionosphere-Free (QIF) strategy.

The activities started in January 1995 with ten IGS stations. Additional stations have been introduced during the following months. The first stations of the German Permanent GPS Network (DREF-Permanent) were included from March 1997. A total number of about 10,000 station coordinate values have been determined. Currently, daily solutions for two and a half years are available. About 5 per cent of daily solutions are missing, mainly due to on-site observation problems.

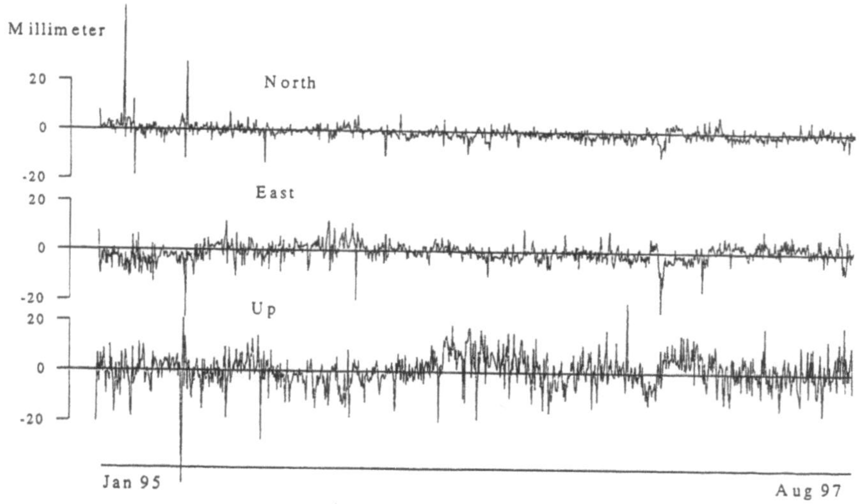

Fig. 2. Time Series of Coordinate Residuals, Jan 95 to Aug 97, Station Wettzell

Solutions are carried out as „loosely constrained“ through the introduction of a priori coordinates with unique standard deviations of ± 1 m. The applied HELMERT transformations (daily solutions vers. a priori solution) lead to time series of residual coordinate components. As an example, these time series are shown in Fig. 2 and 3 for the Permanent GPS Stations Wettzell and Ankara. The daily repeatability varies between ± 3 mm and ± 8 mm for horizontal components and ± 8 mm to ± 20 mm for height components.

to derive a continental wide combined solution [Bruyninx et. al. 1997]. EUREF transfers its European solution to the IGS, which is responsible for deriving a global solution. Finally, these GPS-based results are combined by the International Earth Rotation Service (IERS) with data from various other techniques like Satellite Laser Ranging or Very Long Baseline Interferometry for monitoring the earth rotation, defining the ITRF, etc.

The Bundesamt für Geodäsie und Kartographie (BKG) currently contributes to global networks through the operation of Permanent GPS Stations in Wettzell (Germany), O'Higgins (Antarctica), Lhasa (Tibet), Reykjavik and Hofen (Iceland), Ankara (Turkey), Sofia (Bulgaria) and Nicosia (Cyprus). All data are made available through the IGS Regional Data Center (RDC) at the BKG in Frankfurt.

For regional and local applications the existing networks have to be densified through additional Permanent GPS Stations. Regional or local GPS networks, connected to the global system, provide more detailed information on local kinematics and sea level fluctuations as well as atmospheric and ionospheric influences on observations. In Europe, they will also contribute to the objectives of projects such as WEGENER (Working Group of European Geo-Scientists for the Establishment of Networks for European Research) and SELF (Sea Level Fluctuation in the Mediterranean) or help to satisfy national Differential GPS (DGPS) requirements.

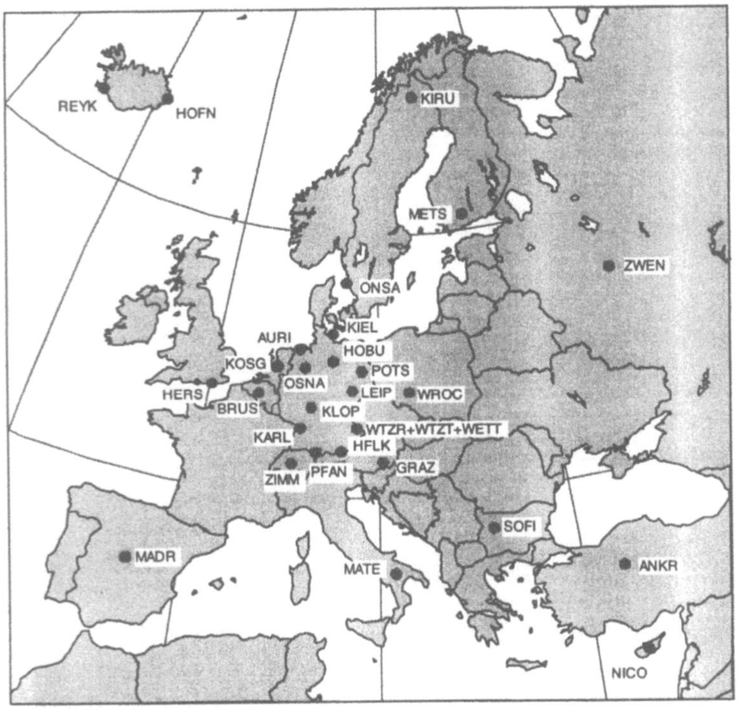

Fig. 1. European GPS Permanent Stations daily processed by BKG

During the WEGENER Conference held in Bologna in October 1995, it was agreed to prepare proposals for a permanent GPS tracking network spanning the area from the Azores along both sides of the Mediterranean Sea into the area of the Caspian Sea [Schlüter 1996].

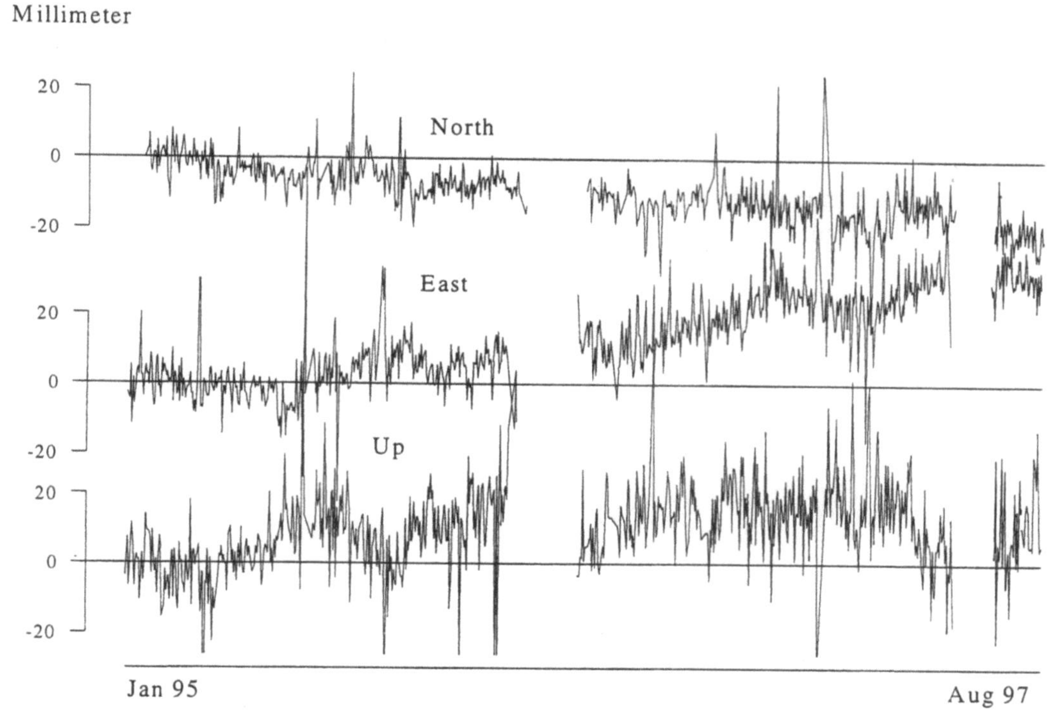

Fig. 3. Time Series of Coordinate Residuals, Jan 95 to Aug 97, Station Ankara

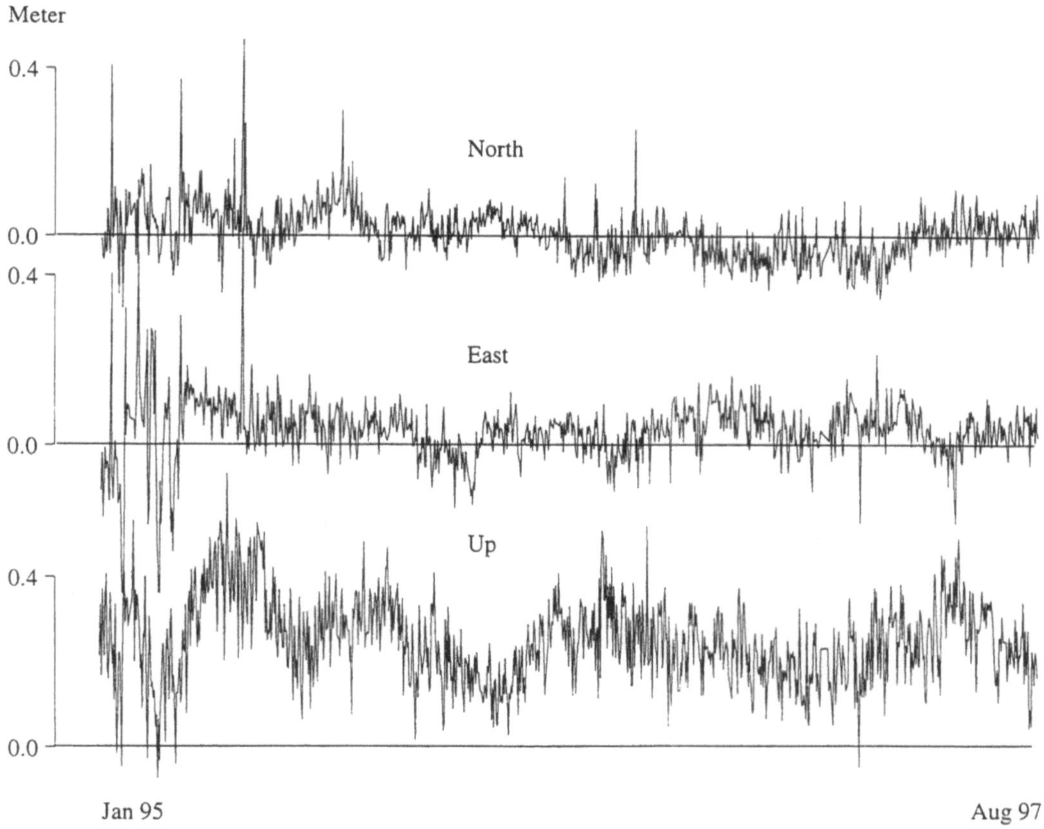

Fig. 4. Time Series of Helmert Transformation Parameters (daily vers. a priori)

The time series of translation parameters as derived from daily applied HELMERT transformations (see Fig. 4) indicate the overall movement of the barycenter of the regional network. The linear approximation of daily station coordinate residuals, added to the linear approximation of the barrycenters translation, permits the estimation of station velocities as shown in Tab. 1 and Fig. 5. Due to the limited regional extension of the network and the intentionally unconstrained procedure, it is not possible to derive full absolute site velocity vectors (see NUVEL model in Fig. 5).

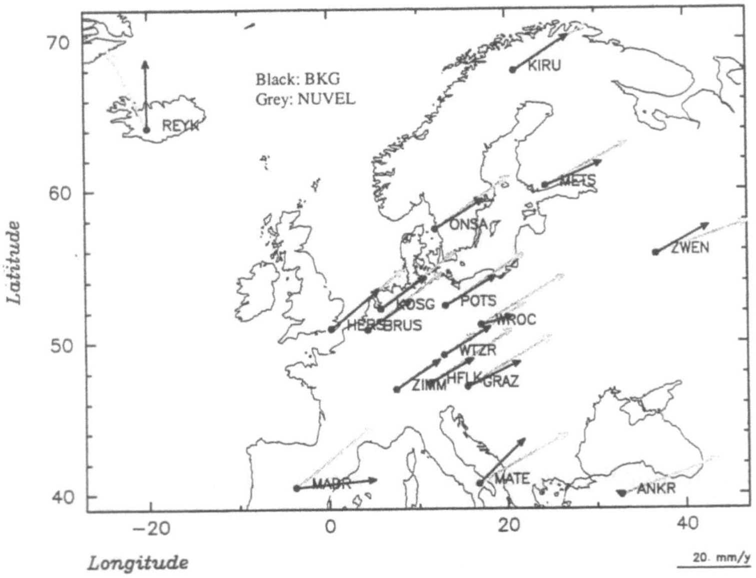

Fig. 5. Site Velocities from BKG Permanent GPS Observations vers. NUVEL

Tab. 1. Relative Site Velocities and Velocity of Barycenter

Site	Days	Rms Fit [mm]	North vel [mm/y]	Rms Fit [mm]	East vel [mm/y]
ANKR	639	± 0.5	7.2	± 0.9	14.6
BRUS	917	± 0.3	0.1	± 0.3	1.7
GRAZ	903	± 0.3	1.4	± 0.4	-0.8
HERS	895	± 0.4	-2.5	± 0.7	0.9
KOSG	891	± 0.2	-0.6	± 0.3	1.4
MADR	858	± 0.9	5.5	± 1.2	-6.7
MATE	907	± 0.5	-3.8	± 0.6	1.2
METS	872	± 0.4	1.6	± 0.4	-1.6
ONSA	905	± 0.3	0.1	± 0.3	0.4
POTS	898	± 0.3	0.1	± 0.3	-0.6
REYK	619	± 0.6	-10.7	± 0.7	13.3
WTZR	870	± 0.2	0.6	± 0.4	0.9
ZWEN	767	± 0.6	0.4	± 0.7	-1.0
HFLK	738	± 0.4	1.4	± 0.5	2.1
KIRU	899	± 0.6	-1.5	± 0.6	-1.0
ZIMM	905	± 0.3	0.2	± 0.4	1.6
Barycenter	---	± 26.3	-23.6	± 26.2	37.7

Fig. 6 shows a Fourier Spectrum for the time series of HELMERT transformation parameters. An interesting effect is the amplitude of about 5 mm for all stations at a wave length of 13.6 days. The reason for this effect might be a missing ocean loading model in the Bernese GPS Software, because the M_f wave group has exactly the same wave length. The Software is currently being modified to account for this effect.

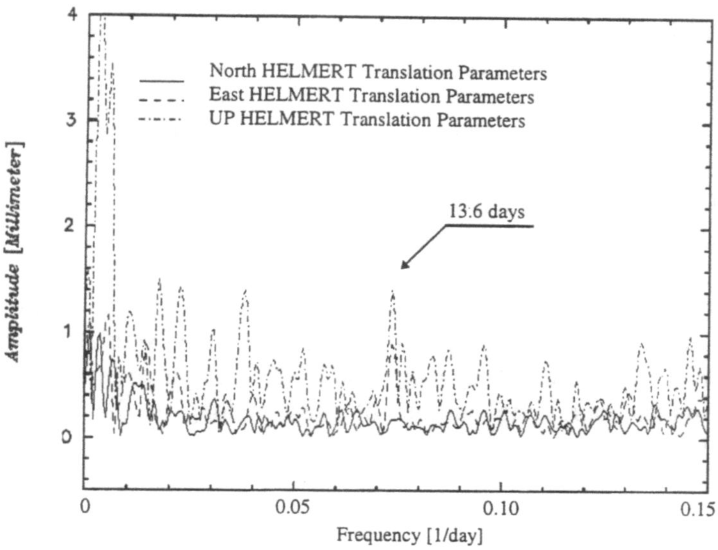

Fig. 6. Fourier Spectrum from Time Series of daily Helmert Transformation Parameters

A part of the daily solutions is the estimation of tropospheric zenith delays every two hours. Results from these estimations (see time series for selected stations in Fig. 7) should be strongly correlated with meteorological observation conditions. The BKG is intends to investigate these correlations for further model improvements and/or a possible forecast of meteorological phenomena.

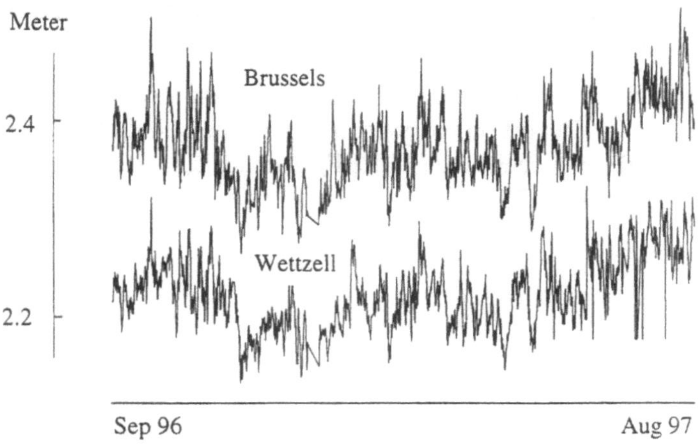

Fig. 7. Time Series of estimated Tropospheric Zenith Delays, Step Width 2 hours

CONCLUSIONS

Developments in the field of data reduction and analysis software have proceeded so far that automatic data handling and processing is reality today. Daily data collected from a set of stations can be processed completely automatic, generating daily solutions. The internal accuracy achievable for a daily solution is better than ± 0.5 cm for a horizontal position and ± 1.5 cm for height components. The precision and stability of these results encourage the inclusion of additional sites for projects such as WEGENER.

The key to the success of permanent networks is the availability of communication links between observation sites, operation centers, data centers and analysis centers. A high speed daily data transfer is required. Developments in telecommunication will allow the inclusion of more sites into the network at moderate additional cost.

In order to support the WEGENER objectives, additional stations are required in the area of the Azores, on the North-African plate e.g. in Marokko, Tunisia, on the Arabian plate (Dubai) and in the area east of the Caspian Sea (Ashgabat). The BGK is prepared to support these activities through operating an increasing number of sites as long as the on-site infrastructure will guarantee daily data transfer to the analysis center.

4. REFERENCES

Bruyninx, C., J. Dousa, W. Ehrnsperger, N. Fachbach, G. Stangl, J. Johansson, C. Ferraro, M. Fermi, A. Nardi, C. Sciarretta, F. Vespe, M. Figurski, M. Piraszewski, J. Rogowski, T.A. Springer, G. Beutler, M. Rothacher, S. Schaer, G. Weber and A. Wiget (1997): *The EUREF Associate Analysis Center.* In: International GPS Service for Geodynamics, Annual Report 1996, Pasadena.

Schlüter, W., H. Seeger, G. Weber (1996): *Proposal for a WEGENER Permanent GPS Tracking Network.* 7th General Assembly of the WEGENER Project, Vila Nova de Gaia, Porto, Portugal, 3-4 June 1996.

Weber, G., W. Schlüter, H. Seeger and P. Franke (1996): *IfAG Contribution to a GPS Permanent Network in Europe.* In: Report on the Symposium of the IAG Sub-Commission for the European Reference Frame (EUREF), Ankara, 22 - 25 May, 1996, BEK, Astr. Geod. Arbeiten, Heft 75.

THE BRAZILIAN NETWORK FOR CONTINUOUS MONITORING OF GPS (RBMC): OPERATION AND PRODUCTS

Luiz Paulo Souto Fortes
Roberto Teixeira Luz
Katia Duarte Pereira
Sonia Maria Alves Costa
Departamento de Geodésia
Fundação Instituto Brasileiro de Geografia e Estatística - IBGE
Av. Brasil 15671, Parada de Lucas, Rio de Janeiro, RJ, 21241-051
Fax: 55-21-481 2747, Brazil

Denizar Blitzkow
Departamento de Engenharia de Transportes
Escola Politécnica da Universidade de São Paulo - EPUSP
Caixa Postal 61548, São Paulo, SP, 05424-970
Fax: 55-11-818 5716, Brazil

ABSTRACT

The Brazilian Network for Continuous Monitoring of GPS (RBMC) is an active geodetic reference network. In this paper its automatic operation and products are described during the first nine months of functioning. It consists of nine permanent GPS stations established in Brazil, in cooperation with many groups. Two of the nine stations (Brasilia and Fortaleza) also belong to the International GPS Service for Geodynamics (IGS), while the remaining 7 are integrated into this service by contributing the corresponding data to the SIR IGS Regional Network Associate Analysis Center (RNAAC) located at DGFI (Deutsches Geodaetisches Forschungsinstitut), Germany. The future activities include the densification of the network in the Amazon region (six stations) and in Northeastern and Southern Brazil (one station in each area).

INTRODUCTION

The Brazilian Network for Continuous Monitoring of GPS (RBMC) is one of the first active geodetic reference networks in South America. IBGE (Fundação Instituto Brasileiro de Geografia e Estatística), as the institution responsible for geodetic activities in Brazil, presented the first proposal for the project in 1991. Since then, while IBGE was looking for funds to establish the network, the project has experienced some changes, in order to keep up with the state-of-art of GPS technology. A description of the status of the project at different times can be found in FORTES (1993), FORTES (1996a) and FORTES (1996b).

RBMC now consists of nine permanent GPS stations, two of which are IGS stations: Brasília (BRAZ) and Fortaleza (FORT), established in cooperation with JPL (Jet Propulsion Laboratory - NASA) and INPE (Instituto Nacional de Pesquisas Espaciais)/USNGS (United States National Geodetic Survey), respectively. The remaining seven stations have been established by IBGE in cooperation with EPUSP (Escola Politécnica da Universidade de São Paulo) and other Brazilian institutions, and sponsored by FNMA (Fundo Nacional do Meio Ambiente). Table 1 lists the names and the four-digit codes of the stations, their approximate coordinates and the date of commencement of operations. Those seven stations will be referred to in this paper as "IGS-SIR" ones, as IBGE is contributing to IGS densification, sending the corresponding data to the SIR IGS Regional Network Associate Analysis Center (RNAAC), located at DGFI (Deutsches Geodaetisches Forschungsinstitut), Germany.

It is important to emphasize that, as all RBMC stations belong to the SIRGAS Reference Network (IBGE, 1997), all GPS positioning in Brazil using them as reference will be automatically integrated into that new system.

Figure 1 shows the current RBMC configuration, with two IGS stations, the seven IGS-SIR ones, and future stations.

TABLE 1: Current RBMC stations

Station	4-digit code	Latitude	Longitude	Beginning of Operation
Fortaleza (IGS)	FORT	-03°53'	-38°26'	13.May.1993
Brasília (IGS)	BRAZ	-15°57'	-47°53'	3.Mar.1995
Curitiba	PARA	-25°27'	-49°14'	13.Dec.1996
Presidente Prudente	UEPP	-22°07'	-51°25'	18.Dec.1996
Bom Jesus da Lapa	BOMJ	-13°15'	-43°25'	18.Feb.1997
Manaus	MANA	-03°07'	-60°03'	28.Apr.1997
Viçosa	VICO	-20°46'	-42°52'	22.May.1997
Cuiabá	CUIB	-15°33'	-56°04'	18.Jun.1997
Imperatriz	IMPZ	-05°30'	-47°30'	18.Jul.1997

STATION DESCRIPTION

Each station is materialized by a centering forced device, specially designed by IBGE for RBMC, which is set up in a very stable structure of the corresponding building. In addition, the following were the criteria for station selection: no obstructions >10 degrees above the horizon; no multipath surfaces close to the GPS antenna; no electromagnetic sources in the range of the GPS frequencies (1.2 to 1.6 Ghz); a restrict public access area; availability of an office close to the antenna location with continuous power supply and telephone facility for data communication.

The IGS-SIR stations are equipped with TRIMBLE SSI GPS receivers, each with 9 L1 and L2 full wavelength channels, able to also measure pseudo-range on L1 and L2; immunity to radio interference; choke ring antenna; 10 Mb of memory; and backup batteries for 16 hours of operation. In addition, there is a PC computer which is used for local and remote operation of the station (see next item) and an uninterrupted power supply (UPS). Detailed specifications of the station can be found in FORTES (1997). Figure 2 shows a diagram of the station components and the connections between them.

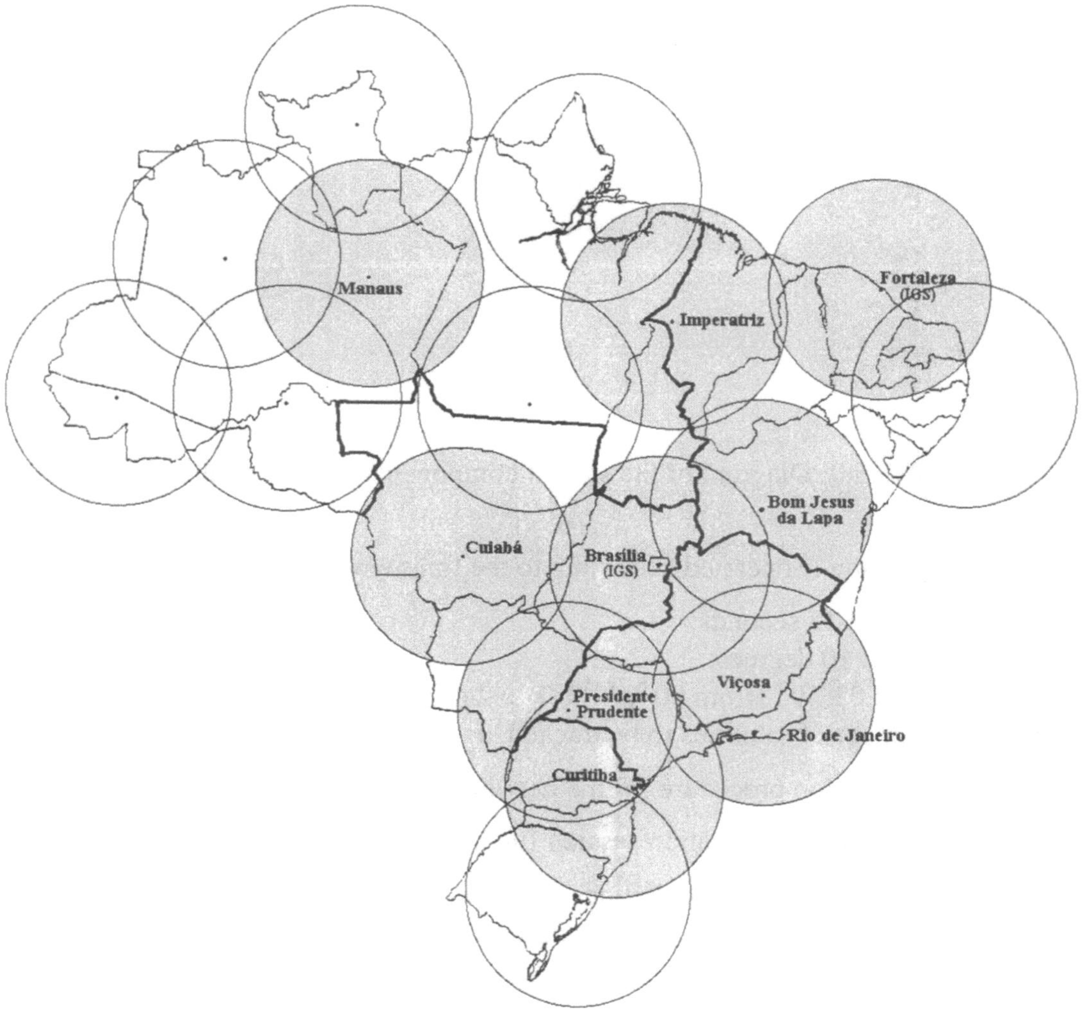

FIGURE 1: RBMC stations. Center of a gray circle represents a RBMC station. Center of a transparent circle represents a station proposed for future densification. Each circle has a 500 km radius

OPERATION

The IGS-SIR stations of RBMC are operated in an automatic way, based on procedures developed at IBGE, while the IGS stations (BRAZ and FORT) follow the procedures established by JPL and USGS, respectively.

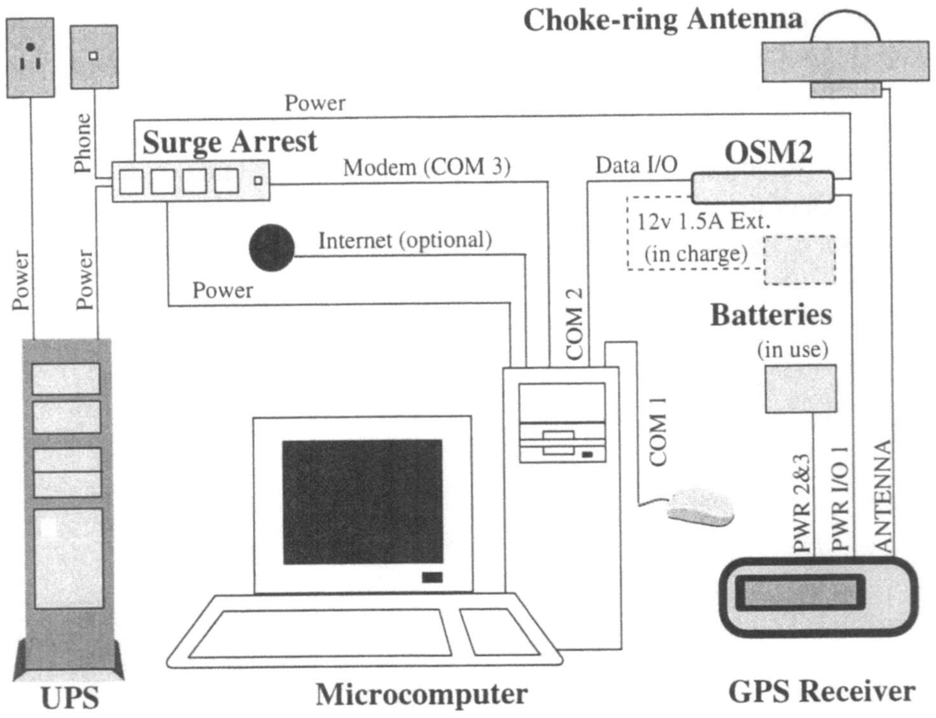

FIGURE 2: Diagram of the station components and the connections

Each IGS-SIR station is operated according to the following tracking parameters:

- Observation rate: 15 seconds;
- Elevation mask: 10 degrees;
- Session length: 23^h59^m (from 00^h01^m UT to 24^h00^m UT);
- Minimum number of satellites to be tracked: 1.

The automatic operation procedure for the network stations is:

1. Closing the previous observation session file and opening the new one at 00^h00^m UT;
2. Executing a batch file on the station microcomputer at 05^h00^m UT (02^h00^m Rio de Janeiro time) to accomplish the following tasks:
   - to download data from the receiver memory to the computer hard disk, erasing older files (in order to make room for new ones);
   - to transform it from binary to RINEX2 format;
   - to accomplish preliminary quality check, using the QC program (GURTNER, 1994);
   - to compress the files using PKZIP program;

3. Downloading the ZIP file, created by the previous procedure, from the station computer to the RBMC control center, located at IBGE Department of Geodesy's headquarters, in Rio de Janeiro, at 06^h00^m UT (03^h00^m Rio de Janeiro time). This is carried out using the communication software pcANYWHERE32 for Windows95 combined with

telephone lines or Internet access, when available. The computer at the control center connects to each station, sequentially, and commands the file transfer.

With the above procedure it is expected that all data collected during a day is available at the control center in the next morning, *before* working hours. The operation also includes the remote control of the stations, from the control center, in order to allow full access to the GPS receiver.

Concerning the IGS stations, the RINEX observation files of BRAZ and FORT are regularly downloaded from JPL and CDDIS (Crustal Dynamics Data Information System) databases, respectively, and made compatible with the others in terms of name and compression utility used.

PRODUCTS

The daily **xxxxddd1.zip** file of each IGS-SIR station is composed of the RINEX files **xxxxddd1.yyo**, and **xxxxddd1.yyn**, and the QC summary file **xxxxddd1.yys**, where **xxxx** is the four-digit identification of the station; **ddd** is the day of the year; and **yy** is the year.

Once in the control center, each ZIP file is stored on an Internet server for future use, according to the directory structure: **/usr/rbmc/yyyy/xxxx**. For instance, the observations for Manaus on day 217, 1997, are located at **/usr/rbmc/1997/mana/mana2171.zip**. In addition, they are also copied to compact discs for backup purposes.

PROBLEMS

The RBMC operation has been very successful, despite its recent start. The only exception occurs at the Imperatriz station (the last one set up), which is experiencing serious communication problems. For the remaining six IGS-SIR stations, Table 2 shows the percentage of data loss. The proposal is to provide data collected at each station during all days of the year and the expectation is to get closer to 0% of data loss after the first year of operation.

Another problem is related to the automatic file transfer from the stations to the control center. In some cases, it fails due to the poor quality of the telephone lines in Brazil. However, this does not imply in data loss, as the file stored in the local computer and in the receiver memory (in the later case, for some days only) is then downloaded manually to the control center during the working hours.

PERSPECTIVES AND CONCLUDING REMARKS

The next step of the project will involve the densification of the network in the Amazon region (6 stations), and in the Northeast and South regions (one station in each) (see Figure 1).

It is also planned to install barometers and thermometers at each station, in order to support meteorological applications, as suggested by Duan et al. (1996).

It can be inferred from what was stated on this paper that Brazil has now a modern active geodetic reference network based on GPS.

TABLE 2: Percentage of data loss for six IGS-SIR stations of the RBMC network

	PARA	UEPP	BOMJ	MANA	VICO	CUIB	TOTAL
number of operating days (by 27.Aug.97)	257	252	188	120	97	69	983
number of days lost	0	9	17	9	0	3	38
% of data lost	0	4	9	8	0	4	4

REFERENCES

DUAN, J. et al.. GPS Meteorology: Direct Estimation of the Absolute Value of Precipitable Water. Journal of Applied Meteorology. 1996, vol.35, Nr.6, pp 830-838.

FORTES, L.P.S.. Brazilian Network for Continuous Monitoring of the Global Positioning System - RBMC. International Association of Geodesy Symposia 109 on "Permanent Satellite Tracking Networks for Geodesy and Geodynamics", Spring Verlag, Berlin, pp 95-101, 1993.

FORTES, L.P.S.. Status of the Brazilian Network for Continuous Monitoring of GPS (RBMC). International Association of Geodesy Symposia 115: GPS Trends in Precise Terrestrial, Airborne, and Spaceborne Applications. Spring-Verlag, Berlin, pp 85-88, 1996a.

FORTES, L.P.S.. Establishment of the Brazilian Network for Continuous Monitoring of GPS (RBMC). End-User Applications Volume of the Trimble Surveying & Mapping Users Conference 1996 Proceedings. Trimble Survey & Mapping Division, Sunnyvale, pp 69-72, 1996b.

FORTES, L.P.S.. Operacionalização da Rede Brasileira de Monitoramento Contínuo do Sistema GPS (RBMC). Master of Science's Thesis. Instituto Militar de Engenharia, Rio de Janeiro, 1997.

GURTNER, W.. IGS Data Quality Checks. IGS Electronic Mail #817, 1994.

IBGE. SIRGAS Final Report, Working Groups I and II. Rio de Janeiro, 1997.

TOWARDS AN AUSTRIAN GPS REFERENCE NETWORK

**R.Weber (Department of Theoretical Geodesy, TU-Vienna,
Gusshausstrasse 27-29,A-1040 Vienna, Austria)**

Abstract

During the past years several activities were launched in Austria to establish a state-of-the-art GPS-network which could serve as a basis for geodynamic applications. Recently, in June 1996, an association of engineering consultants initiated a GPS-campaign which covered the Austrian territory with 330 sites with a mean point-to-point distance of about 25 km. The establishment of such an accurate network (+/- 0.4 cm horizontal, +/- 0.8cm height) will be advantageous to all kinds of surveying activities in the near future and provides a considerable contribution to the densification of the ITRF94 reference frame.

The Institute of Theoretical Geodesy and Geophysics (ITGG) in Vienna is one of the two analyses centers, which were appointed to process the GPS-network data. This paper summarizes the computational efforts at ITGG and puts special emphasis on the algorithms used for reference frame fixing in the analyses. Additionally, problems of modelling phase center variations (seven different antenna types were used in the network) are discussed.

Introduction

Starting in 1990 a densification of the Austrian part of the EUREF-network was realized in several subcampaigns. Due to improvements in receiver technology as well as in orbit and troposphere modelling a substantial increase in accuracy could especially be achieved in 1994 and 1995 (Erker et al., 1997). At that time the network (AGREF= Austrian Geodynamic Reference Frame) comprised of about 80 stations with an accuracy of $\sigma_{xy} = \pm 1\,cm$ in horizontal position and $\sigma_h = \pm 2\,cm$ in height respectively.

In 1996 an association of engineering consultants initiated a further densification in order to decrease the mean point-to-point distance to about 25 km. The resulting network called AREF (=Austrian Reference Frame, see Figure 1) now consists of more than 330 sites altogether including the former AGREF points. The accompanying field campaign was organized in cooperation with the Federal Office of Metrology and Surveying. The Institute of Theoretical Geodesy and Geophysics (University of Technology Vienna ITGG) as well as the Department of Satellite Geodesy (Institute for Space Research, IWFSG) were appointed to serve as Analyses Centers.

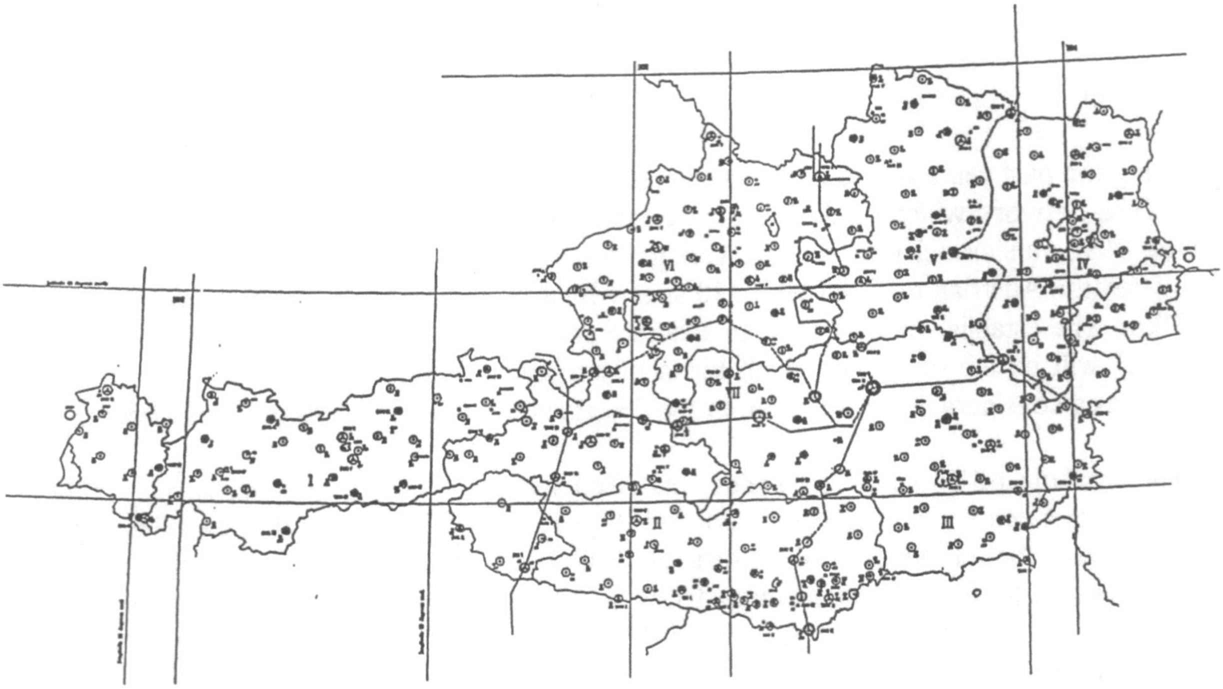

Figure 1: AREF- overall view

The campaign objectives as well as the data-processing guidelines can be summarized as listed below:

a) AREF should represent a precise and homogeneous state of the art reference network with an accuracy better than $\sigma = \pm 1cm$ in position and $\sigma = \pm 2cm$ in height.

b) The field campaign should be limited to 14 days and can be divided in seven 24-hours sessions.

c) The data analysis should deal with not more than 3 hierarchical levels:

- Level A: Global and regional IGS-sites (8 sites, see Figure 2)
- Level B: Campaign-specific reference sites capable of permanent tracking over a period of 14 days (13 sites, see Figure 2)
- Level C: The remaining points in the network (about 50 sites / session)

d) The analysis should be carried out by means of the most recent version (V4.0) of the Bernese GPS Software (Rothacher, Mervart, 1996)

e) Orbits (Zumberge et al., 1996) and estimates of the tropospheric delay at the IGS-sites provided by IGS and CODE respectively have to be used.

Out of a great variety of aspects concerning the data processing, particularly
1) the choice of an appropriate reference frame and
2) the modelling of antenna phase variations
are emphasized in the subsequent paragraphs.

Choice of Reference Frame

In order to be consistent with the IGS precise ephemerides and ERP information, site coordinates were computed within the ITRF93 reference frame at epoch t_s=1996.45. Only two weeks after the campaign, in July 1996, IGS carried out the transition to ITRF94. Thus it was decided, that all results should be converted to this new reference frame at epoch t_s. Coordinates in ITRF94 might be derived by means of equation (1)

$$\vec{X}_{94}(t_s) = \vec{X}_{93}(t_s) + \vec{T}_{93,94}(t_s) + R_{93,94}(m, \omega_x, \omega_y, \omega_z, t_s)\vec{X}_{93}(t_s) \quad (1)$$

$$\vec{T}_{93,94}(t_s) = (-0.094, +0.066, +0.060) \quad [m]$$

$$R_{93,94}(t_s): \omega_x = -2.21 \quad \omega_y = -4.83 \quad \omega_z = +1.95 \quad [mas]$$

$$m = 0.0015 \quad [ppm]$$

where $T_{93,94}(t_s)$ and $R_{93,94}(t_s)$ denote a shift vector and a rotation matrix (similarity transformation) whose components are previously determined based on ITRF sites available in both frames (Graz, Wettzell, Zimmerwald, Padova). The conversion to epoch t_o=1993.0 can be managed by applying velocity field information consistent with the geophysical model NNR-NUVEL1A to all stations, whose individual velocities in ITRF94 are not yet known.

Since customers are not interested to deal with continuously changing coordinates the influence of the ITRF velocity field can be avoided by converting the whole GPS-network to the commonly adopted ETRS89 system at epoch t_E=1989.0. This step would imply the subsequent transformations (Boucher,Altamimi,1995).

$$\vec{X}_E(t_s) = \vec{X}_{93}(t_s) + \vec{T}_{93,E}(t_s) + R_{93,E}(m, \dot{\omega}_x, \dot{\omega}_y, \dot{\omega}_z, t_s)\vec{X}_{93}(t_s)(t_s - t_E) \quad (2.a)$$

$$\vec{T}_{93,E}(t_s) = (+0.019, -0.053, -0.021) \quad [m]$$

$$R_{93,E}(t_s): \dot{\omega}_x = +0.32 \quad \dot{\omega}_y = +0.78 \quad \dot{\omega}_z = -0.67 \quad [mas / year]$$

$$m = 0.0 \quad [ppm]$$

Under the assumption $V_E = 0$ (stable part of the Eurasian plate) we may conclude that

$$\vec{X}_E(t_E) = \vec{X}_E(t_s) \qquad (2.b)$$

It has not been decided yet to which frame the AGREF/AREF -coordinate set should be referred. Of course, ETRF89 would be one opportunity. On the other hand, also less rigorous solutions in ITRF94 or ITRF96 are conceivable. For example, station coordinates may always be determined with respect to a set of ITRFxx-coordinates (AREF-network), 'frozen' at a distinct epoch. The rotation of the stable part of the Eurasian plate can be calculated for any epoch and therefore applied if necessary.

Antenna Phase Center Corrections

The large number of GPS receivers used in the AREF-survey inevitably implies a wide variety of equipment. Tracking data stemming from 5 different receiver types and 7 different antenna types (in alphabetical order: Ashtech L1/L2, Dorne Margolin T, Geotracer 2200 L1/L2, Leica External, Leica Internal, Micropulse Choke-Ring, Trimble Geodetic) had to be processed. Thus especially for the height determination the correct modelling of antenna phase center variations played an important role. Besides the PHAS_IGS.01 file (Rothacher et al., 1995), whose use is recommended by IGS and which contains basically a table of phase-center offsets and elevation dependent coefficients, a further correction table, based on results of antenna calibrations carried out prior to AREF was available. Eventually it was decided to use mainly the coefficients given in the IGS-file (corrections with respect to the Dorne Margolin antenna), because this procedure allows for an understandable interpretation and comparison of the results. Differences to the IGS numbers were only taken into account because of a misleading interpretation of the ARP-definition of the Geotracer 2200 L1/L2 antenna and due to the impact on observations carried out with the Leica external antenna at elevation angles above 15 degree.

Data Processing and Combination

Figure 2 highlights the tie of the campaign specific reference points to the regional IGS-sites. Heavy constraints were put on the coordinates of the IGS-sites according to their standard deviations which were reported in the weekly EUREF-combinations of the regional network. Most of this campaign specific reference stations were continously tracking over 14 days. Thus the standard deviation of their coordinate estimates tended to become small numbers (<3 mm), as expected.

Finally the huge number of regular sites were tied to the reference network. Regular stations were connected to at least 6 stations of the superior frame, including at least 3 stations of the IGS-network. This procedure ensures a convenient tie to the ITRF, even in the case that some of the fiducial stations shall be lost during the next years.

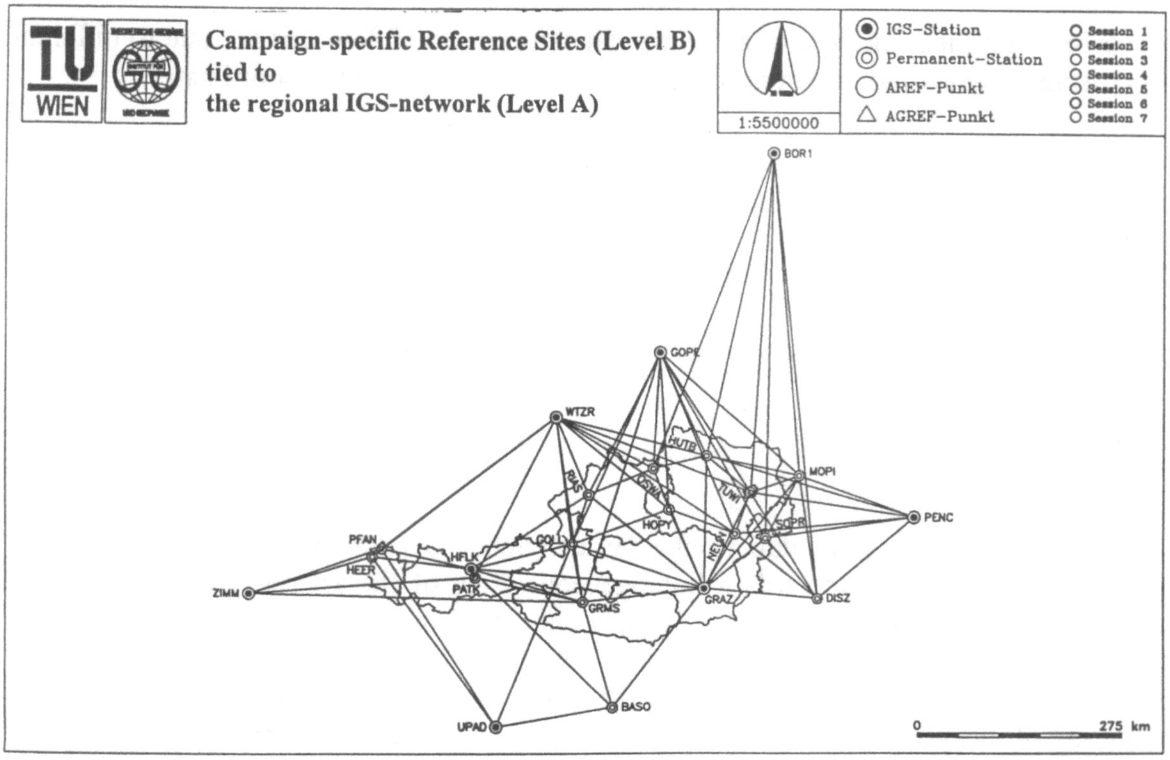

Figure 2: AREF - Reference Sites

The combination of the network was performed at the Normal Equation (NEQ) level. Each of the 'Session-NEQ-files' usually comprises coordinate and troposphere unknowns of about 50 regular points. The combination of this session-NEQs leads to a normal equation system for the whole network and of course a representation in SINEX format which allows for a comparison with the calculations of other Analyses Centers.

Table 1 shows the standard deviation per session of the combined solution, obtained from the results of both Analyses Centers.

Session	1	2	3	4	5	6	7
North	4	1.5	2	1.5	1	2	1.5
East	3	3	2.5	2.5	2	2	2
Up	10	6.5	7.5	5	5	4.5	5.5

Table 1: stdv. in mm

Basically the standard deviation of the horizontal components varies from 1 to 3 mm respectively from 5 to 8 mm in the height. These numbers are quite small and encouraging but one has to keep in mind that both solutions were based on the same data and processed with the same type of software. Therefore these differences are due to different processing schemes and do not necessary reflect the accuracy of the coordinates.

A more reliable estimate of accuracy is based on stations which were occupied several times during the campaign with different equipment. This group comprises about 40 stations and the standard deviations of the horizontal and height components of the combined solution are really very useful numbers to validate the quality of the network.

$$\text{North: } \pm 3.1\,mm \qquad \text{East: } \pm 2.1\,mm \qquad \text{Up: } \pm 7.6\,mm$$

Future Aspects

In terms of accuracy, and due to the large number of stations incorporated, AREF represents a major step towards a homogeneous state of the art reference network. In addition, the combination with AGREF data should allow for the derivation of velocity estimates with respect to the ITRF94 at about 80 sites. To preserve the benefits of AREF, the network has to be maintained during the next years. This goal can either be achieved by means of analysing data obtained from several planned densification campaigns which are tied to AREF, or by re-observation of the whole network in a few years.

Further activities concentrate on the establishment of a network of GPS permanent stations serving real-time applications. Actually there are two systems on trial. The first one distributes RTCM 2.1 messages via Data Radio Channel over the north-eastern part of Austria (Döller et al., 1997). The second system has been installed in a province in the south-eastern part of Austria. It comprises 4 active reference stations with a mean point to point distance of about 50km which transmit correction signals at 75 MHz. These stations were connected via data lines to reduce distance dependent errors within the enclosed area (Seeber, Böder, 1997). Which of these systems, or maybe a combination of both, will finally become a nationwide service cannot be foreseen at the moment. But we are looking forward to support precise real-time applications over the whole Austrian territory still before the end of this century.

References

Beutler G, Mueller I., Neilan R., Weber R. (1994): IGS - Der Internationale GPS Dienst für Geodynamik; Zeitschrift f. Vermessungsw., Vol. 5/94, pp. 221-232 , Stuttgart.

Boucher C., Altamimi Z. (1995): Specifications for Reference Frame Fixing in the Analyses of a EUREF GPS campaign; Bayr.Komm., Vol. 56, pp 265-269, Munich.

Boucher C., Altamimi Z., Duhem L. (1994): Results and Analysis of the ITRF93; IERS Technical Note 18, Central Bureau of IERS, Observatoire de Paris.

Boucher C., Altamimi Z., Feissel M., Sillard P. (1996): Results and Analysis of the ITRF94; IERS Technical Note 20, Central Bureau of IERS, Observatoire de Paris.

Döller H., Erker E., Weber R.(1997): AREF - Das GPS-Grundnetz von Österreich als Basis der modernen telematischen Geodäsie; Paper presented at the Intergeo 97, Karlsruhe.

Pesec P., Sünkel H., Erker E., Imrek E., Stangl G. (1997): AGREF - Das Österreichische Geodynamische Bezugssystem, Inst. F. Weltraumforschung, Graz.

Rothacher M., Mervart L. (1996): Bernese GPS Software, Version 4.0; Astronomical Institute, University of Berne.

Seeber G., Böder V. (1997): New Developments in the Improvement of GPS Kinematic Applications; this Volume.

Seeger H. (1997): The Current Status and Perspectives of EUREF, Reports on Geodesy, Vol. 3/97, pp. 31-76, Warsaw.

Weber R., Walter G., Klotz St. (1995): GPS-relevante Koordinatensysteme und deren Bezug zum Österreichischen Festpunktfeld;
Österr. Zeitschrift f. Vermessung u. Geoinformation, Vol. 4/95, pp.190-200, Vienna.

Zumberge J., Urban M., Liu R., Neilan R. (1996): International GPS Service for Geodynamics, 1995 Annual Report;IGS Central Bureau, JPL, Pasadena.

REAL-TIME DIFFERENTIAL GPS ERROR MODELLING IN REGIONAL REFERENCE STATION NETWORKS

Lambert Wanninger
Geodetic Institute, TU Dresden
D-01062 Dresden, Germany

Abstract

Observations of several GPS reference receivers in a regional network enable two kinds of error modelling and reduction. The spatially correlated errors (orbit, ionosphere, troposphere) are modelled epoch-by-epoch and satellite-by-satellite by 2-D linear interpolation. Furthermore, multipath effects are mitigated by averaging correspondent observations of several receivers. As a result, observations of a virtual reference station are created. It is assumed to be located at the rover's approximate position and its observations are used in the precise baseline positioning of the rover. This approach has been tested with several months of observations including data disturbed by ionospheric irregularities, large tropospheric gradients and large orbit errors. It considerably improved ambiguity resolution and reduced coordinate errors in the positioning of rover receivers.

Introduction

Precise (cm-accurate) GPS positioning uses carrier phase observations and requires the resolution of the double-differenced carrier phase ambiguities. It is usually performed in a baseline-mode with one receiver at a reference station and one rover receiver at a station whose position is to be determined. The effects of ionospheric refraction, orbit errors, and tropospheric refraction on the double-differenced observables grow with increasing baseline length and also depend on the baseline direction. Furthermore, station dependent errors like multipath show severest effects for observation sessions of just a few minutes or less. Therefore, it becomes increasingly difficult to resolve the ambiguities the longer the baseline length and the shorter the observation session. Fast and on-the-fly ambiguity resolutions are often limited to baseline lengths of 5 to 15 km, depending on the actual size of the observation errors and they require continuous observations for at least a few minutes.

In order to reduce the observation errors, GPS-measurements of a regional network of reference stations are used to model the baseline length and direction dependent errors (ionosphere, orbit, troposphere). As a consequence the distance dependence of ambiguity resolution and positioning accuracy is greatly reduced. A further improvement of the reference station observations is obtained through multipath mitigation and quality control in the multistation reference network. As a result, we compute the observations of a virtual reference station located at the rover's approximate position.

Observation errors and their reduction

The dominant error sources in differential GPS positioning and the appropriate ways to model and reduce them in a reference station network will be briefly discussed below.

Although the effects of ionospheric refraction can effectively be removed with dual-frequency observations they adversely affect ambiguity resolution. The reduction of the differential ionospheric refraction is the most important step for the improvement of ambiguity resolution. Large-scale and medium-scale features of the differential ionospheric refraction can be modelled using latitude-differences and longitude-differences as parameters of a model consisting of two coefficients (inclined plane). The models are produced for each satellite at each epoch. A minimum number of three reference stations is required (Wanninger, 1995).

The differential effects of broadcast orbit errors also depend on the length and direction of the baseline. They can be corrected with a similar model to that described above (Wanninger, 1996; Wübbena et al., 1996). The differential effects of tropospheric refraction often do not only depend on the length and direction of the baseline but also on altitude differences. With small altitude differences in the reference station network a two-dimensional-modelling should be adequate, otherwise the model has to be extended to the third (height) dimension.

The size of multipath effects on both code and carrier phase observations depends on the station vicinity and the receiver/antenna-system. As long as no detailed information on the multipath occurrence at the various reference stations exists, multipath is considered to behave stochastically and its effect is reduced by averaging the corresponding observations of all reference stations.

Antenna phase centre offsets and variations should be pre-determined by calibration. No advantage is obtained by combining the observations of several reference stations.

Virtual reference station

Our objective is to combine the observations of several reference stations in such a way that an optimum set of observations of a virtual reference station is obtained, which can be used to determine the position of a rover receiver in baseline-mode. Since some of the observation errors depend on the horizontal position differences of the rover receiver in relation to the reference stations, the rover's approximate position has to be known. Absolute GPS-code-positioning is sufficiently accurate for this purpose. The virtual reference station is assumed to be located at the rover's approximate position.

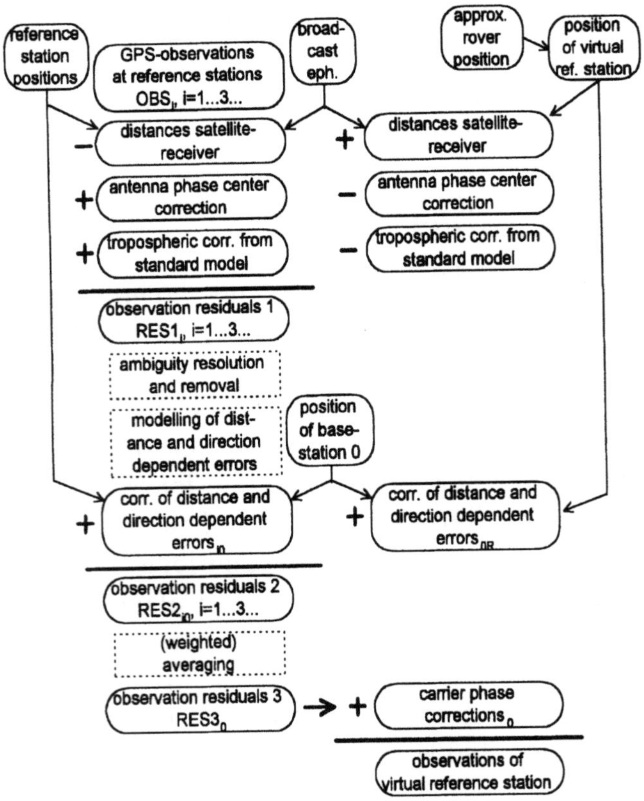

Figure 1: Creation of the observations of a virtual reference station.

Several processing steps have to be performed in order to transform the observations of the network of real reference stations to the observations of the virtual reference station (Fig.1). All processing steps are based on undifferenced carrier phase observables. In the first step the observations are reduced by the satellite-receiver distance computed from the broadcast ephemeris. Furthermore, antenna phase centre corrections and standard tropospheric corrections are applied. The resulting observation residuals are used to resolve and remove the double-differenced carrier phase ambiguities. This step is a prerequisite for all further error modelling and reduction. If ambiguity resolution fails for any observation, it has to be excluded from further data processing. Since the coordinates of the reference stations in the network are precisely known, ambiguity resolution is much simpler to perform than for unknown baselines. On the other hand, distance dependent errors limit the complete ambiguity resolution to maximum distances between reference stations of some 50 km. Most of the difficulties are encountered for low-elevated satellites. From time to time, ambiguity resolution in the post-processing experiences difficulties with short, and typically low, satellite passes. Real-time processing sometimes also fails for the first minutes of data of a rising satellite.

After the successful double-differenced ambiguity removal, two linear combinations of the carrier phase observations are formed. The coefficients of the ionospheric model are calculated from the ionospheric linear combination and the coefficients of the geometric model, which contains the tropospheric and orbit errors, are calculated from the ionosphere-free linear combination.

Both kinds of models are applied to the reference station observations in such a

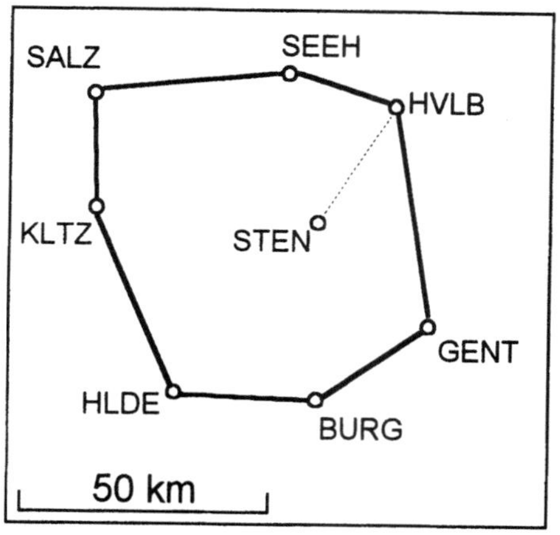

Figure 2: Northern part of the regional GPS reference station network in Sachsen-Anhalt, Germany.

way that the error budget of all reference stations refers to the position of a selected base station, e.g. at the "centre" of the network. Now, simultaneous observations of different reference stations to a satellite should be identical with the exception of multipath, random errors, and the receiver clock error. An averaging process is performed for code and phase on the level of single-differences between satellites, nevertheless undifferenced observation residuals are stored. It reduces both multipath and random errors. In the end, one set of undifferenced observation residuals is obtained, which mainly contains satellite and receiver clock errors and the errors due to ionosphere, orbit and troposphere as they would be experienced at the selected base station.

The observations of the virtual reference station can then be constructed as shown on the right hand side of Figure 1. With the help of the ionospheric and geometric error models the distance dependent errors are shifted from the base station position to the position of the virtual reference station.

The data processing can either be performed at a central computing facility or it can be divided between central facility and users. In the first case, the rover's approximate position has to be known to the central facility. In the second case, the virtual observations are computed for the base station and they are broadcast together with the error models to the user. It is then his task to apply corrections to the (virtual) base station observations in order to obtain the observations of "his" virtual reference station.

Test data set and results

Three months of observations of a regional German GPS-network (Fig.2) with distances between adjacent reference stations of 23 to 40 km have been processed in order to verify the concept of virtual reference stations. The receiver at station STEN has been considered a rover receiver, all other receivers were used as reference receivers. Observations of a virtual reference station at STEN were computed and used to position the receiver at STEN. Ionospheric and geometric error models of the complete data set were evaluated in order to find periods with standard conditions (undisturbed) and disturbed periods of time. Examples of error models are shown in Fig.3. For each epoch and satellite four model coefficients are computed from the reference observati-

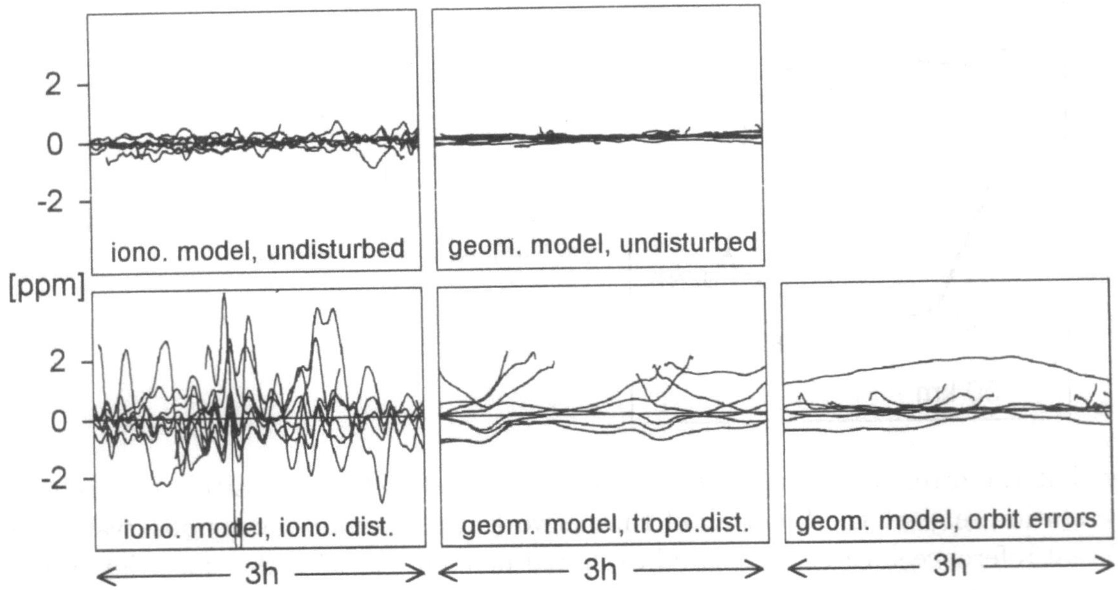

Figure 3: Examples of ionospheric and geometric error models under standard conditions (undisturbed) and under disturbed conditions. Each line represents the time series of model coefficients of a specific satellite. Each example shows either the latitude component or the longitude component of an error model.

ons: latitude and longitude components of the ionospheric and of the geometric model. The numbers are given in units of ppm - parts per million of the coordinate difference for the specific component. The ionospheric model is scaled to the ionospheric effect on L_1-observations.

Most of our data (80-90%) were free of severe disturbances and can be considered as observed under standard conditions. Nevertheless, three kinds of disturbances were detected in the time period from November 1996 to January 1997:

- Medium-scale travelling ionospheric disturbances (MSTIDs), which in mid-latitudes mainly occur during day light hours of winter months (Wanninger, 1995), were found on many days of the November and December data. The error models reveal the characteristic apparent MSTID-periods of 10 to 30 minutes. Maximum rates of change of the error model coefficients can be found in the presence of these disturbances. They reach up to 1.5 ppm/minute.
- Severe tropospheric disturbances due to large humidity and temperature gradients occurred once in the three month period and lasted for more than 24 hours. Observation errors exceeded 1 ppm of the baseline length and mainly affect the height component of the baseline coordinate differences.
- Large broadcast orbit errors of 30 to 50 m were detected for about one satellite pass per week. They produce observation errors of more than 1 ppm of the baseline length.

Under standard conditions single epoch ambiguity resolution succeeded for more than 99% of the observations, and produced incorrect results for just 0.01% of the data. Un-

Table 1: Ambiguity resolution and positioning errors with 5-minute-samples of observations (GPSurvey 2.20).

Baseline		# of samples	Success Rate Ambiguity Resolution	Positioning Errors RMS dN/dE/dH [cm] L_1	L_0
HVLB–STEN (29 km)	Standard Cond.	864	89%	–	1.1/0.7/2.5
	Iono. Disturb.	119	87%	–	1.3/0.8/2.8
	Tropo. Disturb.	106	89%	–	1.6/0.8/7.8
	Large Orbit Err.	126	97%	–	1.3/0.8/2.7
Virt. Ref. Station – STEN	Standard Cond.	864	98%	0.9/0.6/1.9	1.0/0.6/2.4
	Iono. Disturb.	119	96%	1.5/0.9/2.7	1.1/0.7/2.7
	Tropo. Disturb.	106	100%	1.4/0.7/1.7	1.4/0.6/2.4
	Large Orbit Err.	126	99%	0.8/0.7/1.7	0.9/0.6/2.2

der disturbed conditions the success rate decreased to about 95%, with the percentage of incorrect results increasing to about 0.5%.

Table 1 summarizes the results of positioning with 5-minute-blocks of data as performed with the post-processing software GPSurvey 2.20. It compares the baselines from STEN to the closest real reference station HVLB (29 km) to the positioning of STEN with a virtual reference station. The success rate of ambiguity resolution could be improved from about 90% to about 98%. The position estimation of a 29 km baseline is usually performed with the ionospheric-free linear combination L_0. For the positioning with the virtual reference station, however, ionospheric errors have already be considerably reduced, so that an L_1-solution would be more accurate than an L_0-solution because of its lower sensitivity to multipath and random errors. And in fact, the positioning errors are not only smaller using the virtual reference stations as compared to the 29 km baseline, in most cases an L_1-solution is more accurate than an L_0-solution. The only exception that was found was in the presence of ionospheric disturbances. Here, the ionospheric error model removes most of the effects but it does not perform as well as the dual-frequency ionospheric correction.

Conclusions

Combining the observations of several reference stations of a regional GPS-network to create observations of a virtual reference station greatly improves ambiguity resolution and reduces positioning errors. This improvement is mainly achieved by modelling the differential errors of ionospheric refraction, broadcast orbit errors and tropospheric refraction. The redundant data of the reference network are also used for mitigation of multipath and random errors and for a quality control of the observations.

The main prerequisite for the calculation of virtual reference station observations is successful ambiguity resolution for the reference station network. In general, this requirement presents no difficulty for distances between reference stations up to about

50 km, even for data processing in real-time, because the reference station coordinates are well known.

Acknowledgement. The GPS observations were made available by the state survey department of Sachsen-Anhalt, Germany.

References

Wanninger, L. (1995): Improved ambiguity resolution by regional differential modelling of the ionosphere, *Proceedings of ION GPS-95*, pp. 55-62.

Wanninger, L. (1996): Fehlermodellierung in regionalen Referenzstationsnetzen, *41. DVW-Fortbildungsseminar: GPS-Anwendungen und Ergebnisse '96*, Potsdam, Nov. 7-8, 1996, DVW-Schriftenreihe 28/97, pp. 206-218.

Wübbena, G., Bagge, A., Seeber, G., Böder, V., Hankemeier, P. (1996): Reducing distance dependent errors for real-time precise DGPS applications by establishing reference station networks, *Proceedings of ION GPS-96*, pp. 1845-1852.

PROCESSING STRATEGIES FOR REGIONAL GPS NETWORKS

M. Rothacher, T. A. Springer, S. Schaer, G. Beutler
Astronomical Institute, University of Bern
Sidlerstrasse 5, CH–3012 Berne, Switzerland

ABSTRACT

At the Center for Orbit Determination in Europe (CODE), one of the IGS Analysis Centers, we are routinely processing the GPS data of a dense European network. In order to improve the height estimates and, very strongly coupled therewith, the modelling of the troposphere for this network, we decided to include *low-elevation data* into the processing, something that has been done in VLBI analyses for a long time. An appropriate mapping function, the option to weight the observations according to the elevation angle, and the estimation of troposphere gradients were implemented to make best use of the data at low elevations. Problems may arise from phase center variations, multipath, and the troposphere.

143 days have been processed so far using the new processing options and data down to 5 degrees elevation. Comparing the results to a standard 15-degree elevation cut-off solution shows, that the use of low-elevation data together with an elevation-dependent weighting of the observations reduces the scatter in the height estimates by about 20–30 percent. The estimation of tropospheric gradient parameters improves the repeatability in the horizontal site coordinates by a factor of two for sites successfully tracking down to 10 degrees and below.

INTRODUCTION

The *Center for Orbit Determination in Europe (CODE)*, one of seven IGS Analysis Centers (IGS = International GPS Service for Geodynamics), is a collaboration of four European institutions and is located at the University of Berne in Switzerland [*Rothacher et al.*, 1997]. The main task of the CODE Analysis Center consists of the daily generation of global products within the framework of the IGS. In addition, CODE is routinely producing regional GPS solutions for a dense network of European stations, mainly due to the following reasons:

- The European solutions from CODE are combined with other regional European solutions computed by other centers (see [*Bruyninx*, 1997; *Springer et al.*, 1997]) to improve and maintain *EUREF*, the *EUropean REference Frame.*
- Because the European solutions are based on the global products from CODE (GPS orbits, Earth rotation parameters, etc.), these solutions provide a good check for the quality of the daily CODE products.

International Association of Geodesy Symposia, Vol. 118
Brunner (ed.), Advances in Positioning and Reference Frames

- The network is ideal to perform tests with new processing strategies for regional networks.

The *standard* European solution routinely produced at CODE — in addition there are always a few test solutions running — is a full network solutions including the sites shown in Figure 1.

Figure 1: Network of 41 European IGS sites processed on a daily basis by the CODE analysis center.

The correlations between all double-difference observations (sampled at 180 seconds) are correctly modeled [*Beutler et al.*, 1986, 1987]. Tropospheric zenith delay parameters are estimated once per hour for each site using the Saastamoinen mapping function. The elevation cut-off angle is set to 15 degrees. About 85 percent of the ambiguities in the network are resolved to integers using the QIF strategy [*Mervart et al.*, 1994].

NEW SOLUTION STRATEGIES

New processing strategies were developed with the idea, that low-elevation data (from the conventional 15-degree elevation cut-off down to 5 degrees) should be included in the processing. The inclusion of low-elevation data has two major advantages:

- The number of observations increases by about 25 percent when going from 20 degrees to 5 degrees and about 13 percent from 15 to 5 degrees. These numbers are valid for actual data from the European IGS network. When looking at simulated data (full tracking down to 5 degrees) the increase would be almost 40 percent and 28 percent, respectively.
- A better decorrelation between the estimated height and the troposphere delay parameters of the same station: the correlation coefficient changes from -0.96 to about

-0.83 when going from 20 degrees down to 5 degrees elevation cut-off [*Rothacher and Beutler*, 1997].

To make sure that the use of low-elevation data is not going to deteriorate the solution quality, we decided to add three new features to the software (which will be described in more detail below), namely: implementation of an appropriate mapping function, elevation-dependent weighting of the observations, and the option to estimate troposphere gradients.

Niell Mapping Function

The mapping functions developed by [*Niell*, 1993] (one for the dry and one for the wet part of the tropospheric delay) were implemented into our software [*Rothacher and Mervart*, 1996]. The Niell mapping functions were derived from radiosonde data and proved their usefulness for processing VLBI data. To produce the results presented below, only the dry mapping function was used to map the *total* delay.

Elevation-Dependent Weighting

Tropospheric refraction and multipath are responsible for a considerable increase in the scatter of the observations at low elevations. After a few tests with different models for an elevation-dependent weighting of the phase observations, we adopted a weighting function $w(z)$ where the noise increases in the same way with elevation as the tropospheric delay, namely:

$$w(z) = \cos^2(z), \tag{1}$$

where z is the zenith distance of the satellite. Thus an observation at zenith is assumed to have unit weight.

Tropospheric Gradients

When using lower elevation angles, the azimuthal variations of the tropospheric delay will play an increasingly important role. A relatively simple model proposed by [*Davis et al.*, 1993; *MacMillan*, 1995] and generalized here for any mapping function is given by:

$$\Delta r(\alpha, z) = m_f(z) \cdot \Delta r_h + \frac{dm_f(z)}{dz} \cdot \cos\alpha \cdot \Delta r_n + \frac{dm_f(z)}{dz} \cdot \sin\alpha \cdot \Delta r_e \tag{2}$$

where

z,α	...	zenith distance and azimuth angle of the satellite
$\Delta r(\alpha, z)$	...	troposphere delay in the direction (α, z)
Δr_h	...	zenith delay parameter
Δr_n	...	gradient parameter in north–south direction
Δr_e	...	gradient parameter in east–west direction
$m_f(z)$	...	mapping function

Δr_h is the conventional zenith delay and Δr_n and Δr_e are the two gradient parameters of the model. The model corresponds to a tilting of the mapping function by an angle β in the direction α with

$$\beta \approx \frac{\sqrt{\Delta {r_n}^2 + \Delta {r_e}^2}}{\Delta r_h}. \tag{3}$$

New Solution Types

On March 30, 1997 (day 089), four new solution types were implemented into the routine processing of the European network based on the three software modifications discussed above. These solution types are listed in Table 1 together with the standard solution A.

Table 1: Standard solution A and four new solution types.

Solution Type	Mapping Function	Elev.-Dep. Weighting	Elevation Cut-Off	Gradient Estimation
A	Saastamoinen	No	15°	No
B	Niell (dry)	No	15°	No
C	Niell (dry)	Yes	15°	No
D	Niell (dry)	Yes	5°	No
E	Niell (dry)	Yes	5°	Yes

For an easy interpretation of the results, each solution type only differs by one option from the previous one, i.e., the new features are activated one by one. When estimating gradients one set of parameters was set up per day (i.e., one north and east component per station and day).

The results presented in the next section are based on the 143 daily solutions that have been accumulated since the start of the new solution series. Each one-day solution in general contained 30–35 European sites.

RESULTS

To get an idea of the differences between the solution types A to E we discuss two important aspects below: the *systematic effects*, by looking at the parameters of *Helmert transformations* between the solutions, and the *repeatability* of the coordinates over the series of 143 days as a measure of the *quality* of the solutions (there is no ground truth with sufficient accuracy available).

Systematic Effects

Table 2 shows the rms errors and parameters of the Helmert transformations between coordinate solutions generated by combining all 143 daily solutions for each strategy.

Table 2: 7-parameter Helmert transformations between pairs of European solutions generated from 143 days of data using different processing options (see Table 1).

Transformation	RMS of Transf. (mm)			Rotations (mas)			Scale
between	North	East	Up	x-axis	y-axis	z-axis	(ppb)
A and B	0.3	0.3	1.3	0.0	-0.3	0.0	1.9
B and C	0.6	0.5	3.5	0.0	0.0	0.0	1.2
C and D	0.8	0.7	2.6	-0.1	-0.1	0.0	-0.5
D and E	1.0	0.6	1.4	-0.1	0.3	0.0	0.4
A and E	1.5	0.9	2.8	-0.2	-0.1	0.0	2.9

We clearly see that changing the mapping function from Saastamoinen to Niell results in a noticeable scale change of 1.9 ± 0.1 ppb. Weighting the observations also produces a scaling of the network in the same sense (1.2 ppb). All processing changes (A to E) sum up to give a scale change of almost 3 ppb with respect to solution A. The rotations of the network are relatively small.

The rms of the transformation in North, East, and Up directions indicate that the major changes in the height coordinates occur when weighting the observations and by going to low-elevation data, whereas the major change in the horizontal coordinates takes place when estimating troposphere gradients. This will be confirmed in the next section.

Coordinate Repeatabilities

The site Onsala serves as an example here because its receiver performs very well and tracks the satellites down to a few degrees. Due to its location at the shore we may expect quite significant and rapid changes in the behaviour of the troposphere. Figure 2 shows the repeatability in all three components of the site Onsala over the 143 days processed. Let us summarize the most important findings:

- The change of the mapping function did not change the *repeatability* rms error (in contrast to the systematic effects seen in the previous section).
- The biggest improvement in height results from the elevation-dependent weighting of the observations and the inclusion of low-elevation data.
- A small improvement in the horizontal position is obtained by weighting the observations, whereas the data at low elevations degrades the horizontal position.
- The estimation of troposphere gradients has definitely the largest impact on the results by improving the horizontal positions by a factor of two (from 2 to 1 mm) compared to the standard solution A. The small degradation of the height component visible for solution E is due to the reduction of the number of troposphere zenith delays from 24 to 8 per session for this solution (reducing computer resources).

Many pronounced peaks in the North and East coordinate series disappear when estimating gradients. We think it quite improbable that the gradient parameters solely absorb multipath effects and antenna phase center variations. In view of the fact that a permanent site with a good performance collects almost the same amount of data from the same satellite constellation every day (apart from a change in time by 4 minutes a day), phase center and multipath biases in the estimated coordinates should be varying rather slowly with time (periods of weeks to one year).

Figure 3 demonstrates, that improvements similar to those for Onsala may be achieved for most of the sites *tracking down to 10 degrees elevation or below.*

Comparing the standard solution A to the best-performing solution E, we see a clear improvement in the repeatability of both, the vertical and the horizontal components. One has to emphasize in this context that the availability of low-elevation data of good quality is crucial to obtain the type of results shown above. Several sites in Europe are not (yet) delivering data below 15 degrees or are not performing as well (old receiver types, bad environment) and no improvement can be seen for such sites.

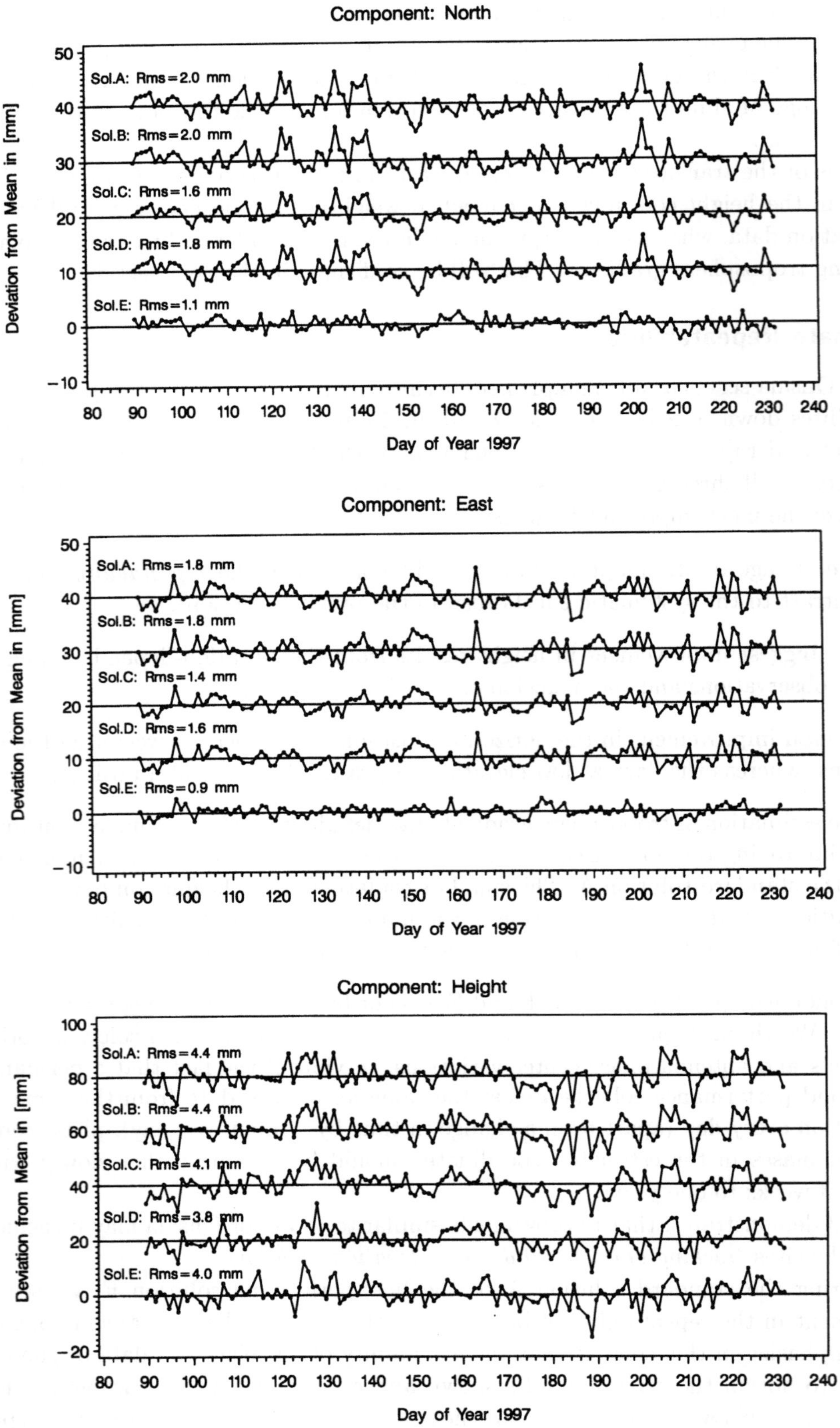

Figure 2: Repeatability of the site coordinates (North, East, and Height) of Onsala for 143 daily solutions using different processing options (see Table 1). The rms error with respect to the mean is given for each solution type.

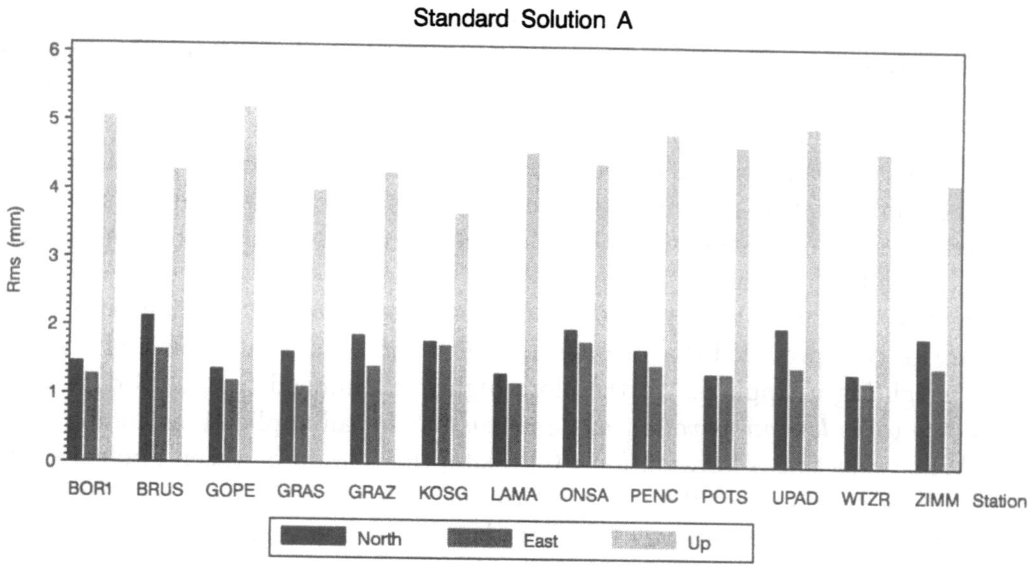

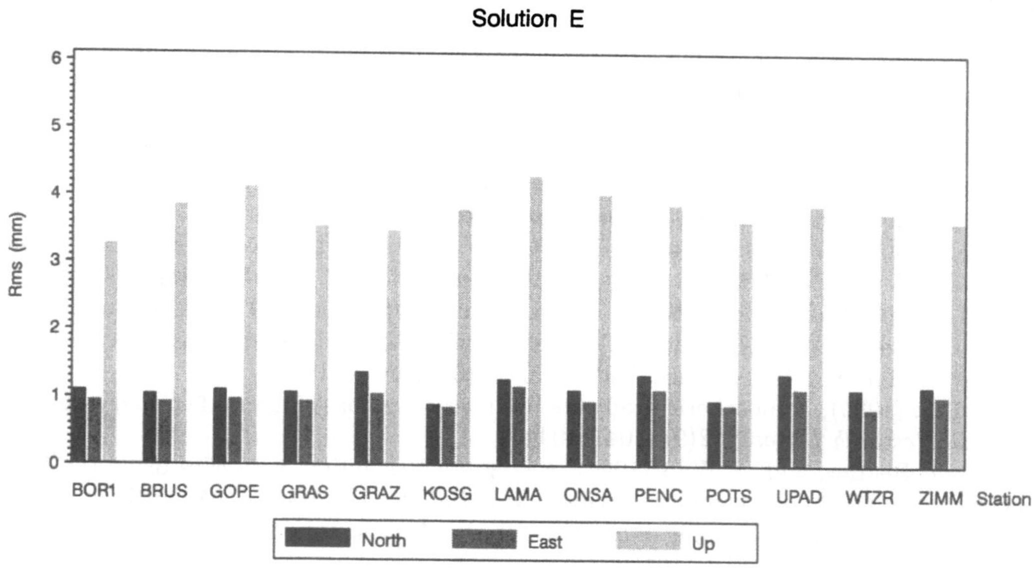

Figure 3: Repeatability of station coordinates in Europe for the standard solution A and the best-performing solution E (see Table 1) over a period of 143 days.

CONCLUSIONS

Several new processing options (inclusion of low-elevation data, a better mapping function, elevation-dependent weighting of the observations, and the estimation of troposphere gradients) to improve the troposphere modelling and thus the station heights were tested using the European IGS network.

From the results we conclude that by including low-elevation data the extremely high correlation between station heights and troposphere zenith delay parameters is considerably reduced. Using elevation-dependent weighting of observations we account for the increased scatter of the data at low elevations and reduce undesirable effects of multipath. The result is a decrease of the daily height rms by about 20–30 percent. Let us mention, that low-elevation data and elevation-dependent weighting of observations are beneficial in many other GPS-related domains like estimation of satellite clocks with code and phase measurements, ionospheric modelling, ambiguity resolution, etc.

The estimation of tropospheric gradients improves the horizontal position estimates by almost a factor of two, provided that enough low-elevation data is available and an elevation-dependent weighting is applied. We therefore highly recommend that GPS data should be collected down to *as low an elevation angle as possible* for all applications aiming at highest precision. When selecting a location for a GPS site, this should be taken into consideration, too. It should be emphasized, that *successful ambiguity resolution* is a pre-requisite to obtain improvements as shown above. Whether the gradients estimated from GPS data may indeed be interpreted as azimuthal variations in the troposphere, remains to be seen, e.g., by comparing the GPS results with meteorological measurements and models.

References

Beutler, G., W. Gurtner, I. Bauersima, and M. Rothacher (1986), Efficient computation of the inverse of the covariance matrix of simultaneous GPS carrier phase difference observations, *Manuscripta Geodaetica, 11*, 249–255.

Beutler, G., I. Bauersima, W. Gurtner, and M. Rothacher (1987), Correlations between simultaneous GPS double difference carrier phase observations in the multistation mode: Implementation considerations and first experiences, *Manuscripta Geodaetica, 12*, 40–44.

Bruyninx, C. et al. (1997), The Use of the EUREF Permanent GPS Network for the Maintenance of the European Terrestrial Reference Frame, in *Proceedings of the Scientific Assembly of the International Association of Geodesy*, Rio de Janeiro, Brazil, September 3–9, (this volume).

Davis, J.L., G. Elgered, A.E. Niell, and C.E.Kuehn (1993), Ground-based measurement of gradients in the "wet" radio refractivity of air, *Radio Sci.*, *28*(6), 1003–1018.

MacMillan, D.S. (1995), Atmospheric gradients from very long baseline interferometry observations, *Geophysical Research Letters, 22*(9), 1041–1044.

Mervart, L., G. Beutler, and U. Wild (1994), Ambiguity Resolution Strategies using the Results of the International GPS Geodynamics Service (IGS), *Bulletin Géodésique, 68*, 29–38.

Niell, A. E. (1993), Global Mapping Functions for the Atmosphere Delay at Radio Wavelengths, *Journal of Geophysical Research, 101*(B2), 3227–3246.

Rothacher, M., and G. Beutler (1997), The Role of GPS in the Study of Global Change, *submitted to Physics and Chemistry of the Earth.*

Rothacher, M., T. A. Springer, S. Schaer, G. Beutler, E. Brockmann, U. Wild, A. Wiget, C. Boucher, S. Botton, and H. Seeger (1997), Annual Report 1996 of the CODE Processing Center of the IGS, in *IGS 1996 Annual Report*, IGS Central Bureau, Jet Propulsion Laboratory, Pasadena, California U.S.A., (in preparation).

Rothacher, Markus, and Leoš Mervart (1996), *The Bernese GPS Software Version 4.0*, Astronomical Institute, University of Berne, September 1996.

Springer, T. A., W. Gurtner, M. Rothacher, and S. Schaer (1997), EUREF Activities at the CODE Analysis Center, in *Proceedings of the International Seminar on GPS in Central Europe*, Penc, Hungary, May 7–9 1997, in press.

A CONTINENTAL WIDE AREA DIFFERENTIAL GPS STRATEGY

M. H. O. Aquino, V. Ashkenazi and T. Moore
Institute of Engineering Surveying and Space Geodesy
The University of Nottingham
Nottingham
NG7 2RD
United Kingdom

ABSTRACT

The paper introduces a new Wide Area Differential (WADGPS) strategy, based on 'Orbit Relaxation'. The technique is simpler to implement and less costly to maintain than the conventional WADGPS approach, based on a full dynamical orbit integration. The scheme is best suited for a single large country, such as Brazil, or a continental region, such as Western Europe. It can also be used as an extension of the American WAAS or the European EGNOS. The orbit improvement is achieved by computing corrections to the elements of the mean Keplerian orbit, thus generating a refined broadcast ephemeris, which is transmitted to the users. The paper describes this new approach, in conjunction with recent developments in modelling error components in WADGPS. Results of tests performed at the Institute of Engineering Surveying and Space Geodesy (IESSG), at the University of Nottingham, with both simulated and real data, are presented and discussed.

1. INTRODUCTION

The provision of WAD (Wide Area Differential) corrections to support a Wide Area Augmentation Service is realised through the operation of a continental or regional WADGPS, which is designed to account for the spatial decorrelation affecting an ordinary DGPS scheme. A key feature of the process is the ability to isolate and model the different error sources individually. One such error component results from the difficulty of accurately coordinating the GPS satellites by the use of their broadcast ephemeris. The use of Orbit Relaxation to improve the satellites orbits in a WADGPS is currently being investigated at the Institute of Engineering Surveying and Space Geodesy (IESSG), of Nottingham University. The operation and maintenance of such a system is simpler and less costly than when the Orbit Integration approach is used. This new solution will work better on a regional basis, as opposed to a global situation, for which the more complex approach is designed.

The orbit relaxation technique has already been addressed in Nottingham with the purpose of improving station coordinates in a fiducial solution of regional networks and short satellites arcs (Aquino, 1990). The objective of the current study is to validate the employment of the resulting orbital corrections in a WADGPS. In particular, a system where both the reference network and the user hardware are based on L1 single frequency receivers is being investigated. Recent developments for the modelling of the ionospheric errors are also incorporated. This however does not prevent the employment of the system when dual frequency receivers are available.

2. IMPLEMENTATION OF ORBIT RELAXATION

The error components in a WADGPS are separated as: orbital errors, atmospheric errors (which can be further separated into a tropospheric and an ionospheric components) and clock related errors. One can employ the orbit relaxation technique to address the orbital errors. It consists of solving for corrections to the broadcast orbits, therefore producing a new "broadcast-like" ephemeris, which can be used to replace the original one.

2.1. COMPUTATION OF WADGPS CORRECTIONS.

Nottingham's current approach for a single frequency based WADGPS breaks down the differential correction into : (a) an independent ephemeris, generated from an orbit integration process; (b) an improved ionospheric model, given by an optimised set of Klobuchar-like parameters; (c) an established model to deal with the troposphere and (d) clock related errors (Ashkenazi et al, 1997).

2.1.1 Orbit Improvement.

L1 pseudoranges are collected from a ground network and relayed to a Master station. A time span of about 72 hours is required to allow for the application of the fiducial technique (Ashkenazi et al, 1989), in a least squares adjustment involving the observed data and the numerically integrated orbit, leading to improved satellites orbits. It was envisaged that the orbit relaxation algorithms could be implemented to replace the orbit integration. The general approach to deal with the error components is unchanged, but for one aspect, regarding the computation of the improved orbits. Corrections for up to six Keplerian elements are included in the double difference observation equations and a two hour data span is used in the least squares solution. A difficulty arises when satellites relevant to the user are not observed at the reference stations, or whose observed arcs are too short to allow the solution to take place. To overcome this problem, a backup approach has to be adopted. One possibility is to rely on the original broadcast ephemeris of those particular satellites, until sufficient data are acquired by the network. A more radical alternative is to ignore them in the user solution. A further option is to use the navigation inversion method for these satellites. An automatic switch between the primary and the chosen backup approaches will be essential for the optimal performance of an operational system.

2.1.2 Ionospheric Modelling.

In the orbit improvement least squares process, ionospheric scale factors are also estimated. The ionospheric delay first approximation is given by the Klobuchar model, whose algorithm requires eight coefficients and corrects for approximately 50 percent of the ionospheric range delay (Klobuchar, 1977). The scale factors accommodate the deficiencies that remain. It is possible then, in a further step, to produce more realistic delays by applying the derived scale factors onto the original delays. Finally, an optimised set of Klobuchar-like coefficients (Ashkenazi et al, 1997) is computed, through the input of the improved delays in a Non-linear Static Estimator (NSE) algorithm (Kee, 1991). This set of improved Klobuchar-like coefficients is transmitted to the user.

2.1.3 Tropospheric Modelling.

Errors due to the troposphere are removed at the Master station by employing the so-called *Magnet Model*, adopted by the Magnavox Magnet processing software (Curley,1988).

Residual unmodelled tropospheric errors are also absorbed by the derived ionospheric scale factors (Chao, 1996). The derived tropospheric delays are not meant to be transmitted to the users, who will be able to derive their own tropospheric delays by employing the same model (i.e. the Magnet model).

2.1.4 Clock Errors.

The process at the Master station ends with the calculation of the satellites clock errors. As all other sources of errors have been properly modelled, it can be assumed that remaining errors are clock related ones. In WADGPS the clock error is not derived from one single reference station, since a network of stations is in place and actually used to generate the corrections. As a consequence, the latter are contaminated by different station receiver clock errors. Hence all reference stations clocks must be related to one master clock, therefore making the transmitted clock terms contain the effect of a unique clock. Clock errors are propagated round the network using satellites which are observed simultaneously by the master clock receiver and one of the other reference station (IESSG and BAe, 1992).

2.2 CORRECTIONS UPDATE.

The system requires a cold start, which used to be of the order of 72 hours for the orbit integration method, and now becomes two hours, with orbit relaxation. Afterwards, these arcs are moved forwards at frequent time intervals and new least square solutions at the Master station are performed, leading to new sets of corrections.

2.3 APPLICATION OF CORRECTIONS BY THE USER.

Once each of the error components is computed and made available to the user, they can then be applied in the following manner: a) replace the broadcast ephemeris with the relaxed ephemeris; b) replace the broadcast Klobuchar parameters by the set of optimised Klobuchar-like coefficients; c) apply the Magnet model for tropospheric correction; d) apply the pseudorange correction transmitted from the Master station, which is represented by the composite clock correction.

3. VALIDATION TESTS

In order to assess the ability of the technique to model the orbital error component, tests using simulated data were performed. A software designed to produce GPS observations free of other error sources was employed. The data was created from arbitrary coordinates of a network of stations, located at six different sites in Brazil, and from real precise satellites ephemeris. The orbit relaxation was performed upon the corresponding (actual) broadcast ephemeris.

A network of five peripheral stations was held fixed, leaving an inner station, Brasília, for user accuracy tests. Several sessions of two hours were then processed, each yielding an improved (relaxed) broadcast ephemeris for all satellites being solved. These orbits were then compared with the precise orbits during that session and over the following one hour. In real time, this one hour extrapolation represents the actual situation. In the simulations, the extrapolated relaxed orbits compare with the precise orbits within 2.5 meters on average. Distances from station Brasília to the reference stations vary between approximately 900 km and 2000 km. Repeated tests showed that the total three-dimensional user error, obtained throughout the one hour extrapolation period, was of the order of 10 cm on average (maximum of 20 cm). It is important to note that two extra satellites, visible from station Brasília, had not been solved for. That is because they were not observed within the session. As a consequence they could not be included in the user solution.

4. TESTS WITH REAL DATA

Initial tests with real field data were performed. The data had been used to evaluate the latest version of the Nottingham WADGPS scheme (Ashkenazi et al, 1997). It was collected from the 3rd to the 5th of May, 1993 at four sites within the Racal Skyfix system, located in Aberdeen, Cadiz, Cyprus and Hammerfest (figure 1). Observations of L1 C/A code and Y1-Y2 cross correlated pseudoranges, along with L1 and L2 phase were recorded by Trimble 4000 SSE receivers at 1 second intervals. The L1 C/A pseudoranges were then used for assessment of the new proposed scheme's performance. Only two preliminary trials can be reported by the time this paper is being written. Cadiz, Cyprus and Hammerfest were adopted as the reference stations. A two hours session, starting at 1 am on the 3rd of May, was chosen for the first trial. The second trial involves a day-time session, from 11 am to 1 pm, on the same day. The strategy outlined in section 2.1 was followed. One ionospheric scale factor per station was estimated for the entire session.

Figure 1 - European Network

4.1 STRATEGIES AND RESULTS / NIGHT TRIAL.

The orbital corrections relate to the Keplerian elements of the broadcast ephemeris referenced to epoch t_0 = 2 am, the center of the session. An attempt to solve for five Keplerian elements of each observed satellite was made (the argument of the perigee was fixed in the solution). Solutions for eight out of 11 satellites observed were achieved. Station Aberdeen played the user's role in the experiment.

The Klobuchar model treats the ionospheric delay as a constant term at the night time period. Since the processed session was chosen to be mostly at night time, almost all the derived ionospheric delays were computed by the night-time constant on its own. However, there was the need to generate a set of improved Klobuchar-like coefficients to be transmitted to the user, who will be navigating one hour ahead and could be observing already during day-time. The following strategy was then devised: a) stations scale factors are derived in the orbital solution least squares and are assumed to cover the deficiencies of the Klobuchar model (in this case represented by the night-time constant); b) in the following step, i.e. the process of computing the clock corrections, each of the three stations scale factor is then applied to the night-time constant; c) in order to accomplish an homogeneous treatment, stations observing partly at day time, have their day-time delays also multiplied by their corresponding scale factors; d) the clock corrections are now in sympathy with the combination of the orbital solutions and the scale factors; e) at the user end a compatible approach has to be employed: a weighted average of the scale factors is derived, using his distance to the reference stations; f) this weighted scale factor, applied to the Klobuchar model (if night-time, to the night-time constant) will then comprise an optimised ionospheric correction, which will play the role of the improved coefficients. The results of employing the procedure are shown in figure 2, where it is compared with a solution based on the precise orbits provided by the Center for Orbit Determination in Europe (CODE), within one hour ahead of the orbit relaxation solution session. Coincidentally, only the eight satellites solved in the orbit relaxation adjustment were visible by the user. The mean error for the relaxed orbits solution is 2.3 m (σ = 1.2 m), whereas for the precise orbits solution it is 2.9 m (σ = 1.2 m).

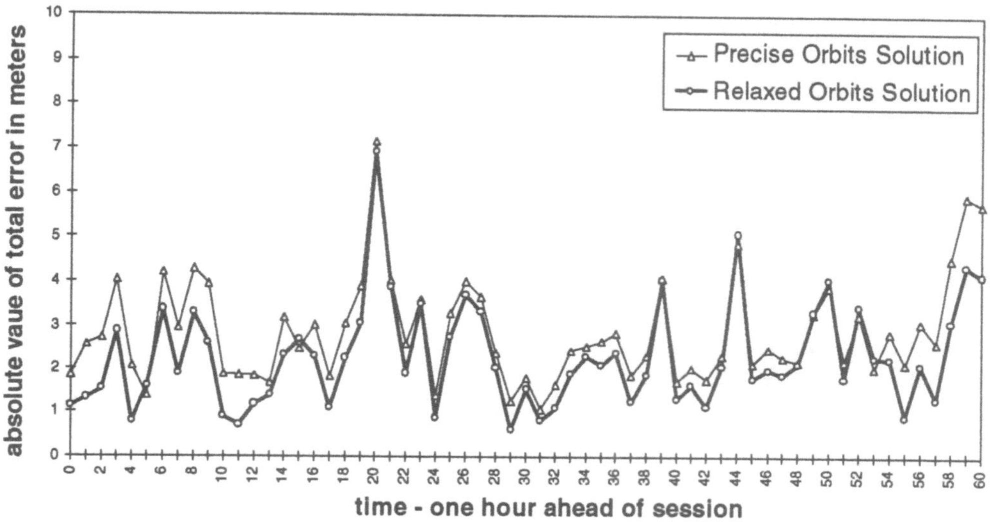

Figure 2 - User Positoning Accuracy : Precise x Relaxed Orbits

4.2 SIMPLIFICATION OF THE METHOD / DAY-TIME TRIAL.

For the day-time session the mean Keplerian elements relate to the reference epoch t_0 = 12 am. Six out of ten satellites present in the session were solved for, along with stations scale factors. The day-time ionospheric delays computed by the Klobuchar model at each station were improved by the application of the corresponding derived scale factors and then used to feed the NSE. The optimised Klobuchar coefficients, the Magnet model and the improved orbits were then used in the process of deriving the clock corrections at the Master station. At the user end each of the computed error component is applied, following what is described in 2.3. The user (at station Aberdeen) mean three-dimensional positioning error, during the one hour just ahead of the session (i.e. from 1 pm to 2 pm), was found to be of 2.0 m (σ = 0.9 m). Two other user solutions were computed for comparison. The first relates to a simplification of the ionospheric delays optimisation process. The use of the NSE to produce optimised Klobuchar coefficients was replaced by the strategy devised for the night time session. The user results are very similar when this strategy is employed to this particular session. The third solution was derived by supposing that IS / CODE precise ephemeris were available to the user, in a similar fashion as for the night trial (Figure 2). An average positioning accuracy of 2.3 m (σ = 0.9 m) was achieved.

5. CONCLUSIONS

A suitable WADGPS scheme is being developed in order to enable the computation of regional WAD corrections. The current WADGPS approach in Nottingham relies on the full dynamic orbit determination process to accomplish a proper accountancy for the orbital error component. This paper introduces an alternative method which has an orbit relaxation technique in its core. The investigations being carried out aim, in particular, at developing a regional WADGPS capable of providing the best possible accuracies based only on a single frequency GPS hardware. This does not imply that its use in the presence of dual frequency receivers is impossible. So far, experiments were undertaken with the use of L1 pseudoranges only. Simulations, performed to validate the orbit relaxation, demonstrated the way it handles the orbital errors can provide final user accuracies within the range of 20 cm, if no other error sources are present. Real data initial experiments gave place to an interesting and challenging situation, when a combination of the ionospheric and the orbital error had to be dealt with. A special concern was dealing with the definition of night-time and day-time when the Klobuchar model is to be used in conjunction with scale factors. A strategy was devised to overcome this problem and results of its implementation yielded user accuracies of the order 2.3 m and 2.0 m, when a night and a day-time session were tested respectively. It should be noted that these figures agree with average results of the latest WADGPS tests performed at Nottingham, based on Orbit Integration.

ACKNOWLEDGEMENTS

The GPS field data used in the real time experiments was provided by Racal Survey Ltd. This research is part of a PhD course being undertaken by M. H. O. Aquino, sponsored by CNPq, Conselho Nacional de Desenvolvimento Científico e Tecnológico, Brasil.

REFERENCES

Aquino, M. H. O. (1990). Improving GPS Positioning Accuracies by Orbit Relaxation. M.Phil Thesis, University of Nottingham, UK.

Ashkenazi, V. , Chao, C. H. J. , Chen, W. , Hill, C. J. and Moore, T. (1997). A New High Precision Wide Area DGPS System. Navigation, Journal of The Royal Institute of Navigation, UK, Vol. 50, No 1.

Ashkenazi, V. , Moore, T. , Whalley, S. and Aquino, M. H. O. (1989). High Precision GPS Positioning by Fiducial Techniques. International Association of Geodesy Symposia, Edinburgh, Scotland.

Chao, C. H. J. (1996). Improved Modelling of High Precision Wide Area Differential GPS. PhD. Thesis, University of Nottingham, UK.

Curley, R. A. (1988). The Use of TI 4100 Receivers and MAGNET Software to Determine Height Differences. MSc. Thesis, University of Nottingham, UK.

IESSG and BAe (1992). Wide Area Differential Corrections R&D Study. Project Report, University of Nottingham and British Aerospace (Space Systems)

Kee, C. , Parkinson B. W. and Axelrad, P. (1991). Wide Area Differential GPS. Journal of The Institute of Navigation, USA,Vol 38. No 2.

Klobuchar, J. A. (1977). Ionospheric Time Delay Corrections for Advanced Satellite Ranging Systems. Propagation Limitations of Navigation and Positioning Systems, AGARD-CP-209.

POSSIBILITY OF A DYNAMIC CADASTRE FOR A DYNAMIC NATION

Graeme Blick and Don Grant
Land Information New Zealand
PO Box 5501
Wellington
New Zealand

ABSTRACT

New Zealand, lying across the Pacific/Australian plate boundary, is subject to ground movements within the country of approximately 5 cm/yr. The existing geodetic control network and datum, New Zealand Geodetic Datum (NZGD) 1949, has been distorted by the effect of this ground movement. Because of these distortions, cadastral boundary coordinates, if held fixed in a coordinate cadastre, would eventually conflict with the position of slowly moving ground marks. However, the use of new technology such as GPS is ideally suited to the development of a coordinate cadastre. A dynamic coordinate cadastre is a possible outcome of recently started projects. A programme is underway to develop a modern control network and new datum (referred to as NZGD 2000) to replace the existing network. This could include up to 30 GPS permanent tracking stations which would monitor the integrity of the new datum and determine rates of ground deformation. A four dimensional (dynamic) datum with coordinates assigned velocities and changing to reflect ground movements is an option. As a separate project, the Department is considering the automation of the survey and land title systems. The automation of the survey system would see the digital capture of land parcel dimensions. Through connection to the geodetic network, geodetic coordinates would be assigned to parcel boundary points resulting in a coordinate cadastre. As the geodetic network moves to reflect ground movements, adjustment for this could be applied to boundary points forming a dynamic coordinate cadastre. In this manner, coordinates of boundary points would encapsulate available evidence of their true ground positions. This would not be a legal coordinate cadastre, but in the absence of other evidence the coordinate could be accepted as evidence to define parcels. To take the model further, it is conceivable that the cadastral surveyor in performing surveys would contribute to the geodetic network and dynamic datum. This possible model is debated.

INTRODUCTION

New technology such as the Global Positioning System (GPS) is ideally suited to take advantage of a coordinate cadastre because of its ability to survey and coordinate points directly in terms of geodetic control. In the near future, it may be possible to routinely obtain centimetre level accuracy real time positioning over wide areas using Wide Area Differential GPS (WADGPS) networks and satellite communication techniques. This, and the possible adoption and integration of other technologies such as inertial systems, is expected to result in the ability to routinely coordinate cadastral boundary points to required accuracy relative to a reference network. Such a system could result in large savings to government in maintenance of the spatial infrastructure and savings to surveyors and their clients in undertaking cadastral surveys. It is conceivable that a highly accurate GPS reference network and the survey methods described above, may enable the number of geodetic survey control points in New Zealand, (over 30,000 in the geodetic database) to be reduced dramatically to a few thousands or hundreds. The future may also see well defined physical features in the environment (eg houses or manhole covers) being used to densify geodetic control for cadastral surveys.

In a dynamic (kinematic) environment such as New Zealand's, several challenges must be addressed and overcome before the development of dynamic coordinate cadastre can be contemplated.

THE CHALLENGES

New Zealand is a Dynamic (Kinematic) Nation

New Zealand is a dynamic (kinematic) nation, lying as it does across the Australian/Pacific tectonic plate boundary. It is subject to ground movements across the country of approximately 5 cm/yr, disregarding the effects of large earthquakes [*Walcott* 1984]. This amounts to approximately 2.5m in the 50 years since the current New Zealand Geodetic Datum 1949 (NZGD49) was observed and established. Added to this are the catastrophic effects of large earthquakes and the more localised effects of volcanic activity, ground creep, and subsidence due to, for example, geothermal extraction.

The earth movements have led to distortions in, and a deterioration in the accuracy of, the current datum. The distortions increase with time causing the coordinates of the trig stations defining NZGD49 to become increasingly incompatible with new survey observations. Bevin and Hall [1995] provide a review of the current datum and distortions within it. These movements also result in movement of cadastral boundary marks, so that boundary coordinates, if they were fixed in a coordinate cadastre, would eventually conflict with the position of the slowly moving ground marks [*Grant* 1995].

Development of a dynamic coordinate cadastre (*a cadastre where coordinates of boundary marks change to reflect their true positions*) will necessitate modelling of earth movements.

Current Spatial Definition of the Cadastre is Provided by Survey Observations

The spatial definition of the New Zealand cadastral system is currently based on physical marks in the ground and survey observations (bearings and distances) between those marks, rather than coordinates. Witness marks placed close to the boundary are used to validate the location of boundary marks or to reinstate missing boundary marks. Because the distance between witness or survey mark and boundary mark is small (tens of meters) the current system has proved to be very robust. It accommodates distortions in the datum and the effects of earth movement because the distortions over small areas are generally insignificant. In the case of catastrophic events, movements cannot be ignored and resurveys are required. In the current system, high relative spatial definition is maintained over short distances. Absolute positional accuracy may be only of decimetre or metre accuracy but the definition of the cadastre is not adversely affected by this.

Such a system may have limitations as we move to new technologies such as GPS with the capability to deliver sufficient accuracy for cadastral survey over much longer distances and the desire to achieve survey economy by operating over these greater distances, eg, WADGPS networks. Distortions in the networks and cadastre then become significant and new ways of accommodating these distortions must be found. Otherwise, new precise observations will need to be distorted to fit the lower accuracy survey network.

CURRENT PROJECTS TO OVERCOME THE CHALLENGES AND HELP DEVELOP THE CONCEPT OF A DYNAMIC COORDINATE CADASTRE

Development of Datum 2000

The New Zealand geodetic control network is the culmination of developments over the past century [*Lee* 1978] . Geodetic triangulation was commenced in 1909 and completed in 1949, forming the 1st Order triangulation of New Zealand. The coordinates of the 1st Order trig stations were held fixed and used to define NZGD49 which is based on the International Spheroid. The current geodetic control network has served the cadastral system well but it now suffers from a number of limitations [*Blick and Rowe* 1997].

Recognising the limitations of the current datum, a programme was commenced by the Department in 1995 to re-observe the 1st order triangulation network (293 stations) which define NZGD49. In addition, a new network was established and surveyed to form the basis for a new datum, tentatively referred to as New Zealand Geodetic Datum 2000 (NZGD 2000). The requirements of the new datum are described by Blick and Linnell [1997].

It is proposed that the new datum will be geocentric and tied to the IERS International Terrestrial Reference System. Three options are being considered, a 'fixed' datum, a 'dynamic' datum [*Grant and Pearse* 1995] or a semi-dynamic datum. As with the current NZGD49 datum, a new 'fixed' datum is not considered appropriate as it too will degrade through time due to the effects of ground deformation and changing user requirements.

Automation of the Survey and Titles Systems

Land Information New Zealand has been charged with the development of a strategy for integrated automation of the survey and title systems in New Zealand [*Grant et al* 1997]. Under this strategy survey and title transactions could be combined into a single generic land transaction with survey and/or title components. This will enable surveyors and solicitors to develop new relationships for creating and submitting transactions in land.

In such an automated system, it is envisaged that cadastral survey data would be submitted in a digital format. The fundamental building block for the survey component of an automated system is a survey accurate coordinate cadastre based on the geodetic network that would allow efficient electronic validation of this new survey data. There will be a requirement for a spatial reference system to underpin the integration of all cadastral survey data into a digital database. The development of a survey accurate coordinate cadastre would see the capture and adjustment of parcel dimensions in terms of the new geodetic datum 2000 network. This would enable all boundary points to be assigned geodetic coordinates.

Development of a GPS Permanent Tracking Network

Under the development of a new geodetic datum (discussed above) is the development of a new geodetic network to support that datum [*Blick and Linnell* 1997]. This new network will include a number of GPS permanent tracking stations to monitor the integrity of the geodetic network. Currently four such stations are operating.

A plan by the New Zealand Institute of Geological and Nuclear Sciences (IGNS) proposes that a number of organisations in New Zealand combine resources to develop a network of approximately 40 geophysical monitoring sites, half of these sites including permanent GPS tracking stations. Partners in such a project may include those with scientific interests, Land Information New Zealand for monitoring the integrity of the new datum 2000, and commercial organisations who may enhance and sell the data. Under such a system the very network that controls the dynamics of the datum may also be used by survey practitioners to define the cadastre (eg by development of a WADGPS network).

IMPLEMENTATION OF A DYNAMIC COORDINATE CADASTRE

Development of a New Datum

In a dynamic country such as New Zealand, any new datum will need to account for the effects of ground movements. Several options can be considered for the implementation of such a datum.

Dynamic Geodetic Datum: A dynamic geodetic datum is one where the actual motion of trig stations is accounted for in some manner. For example, all trig stations could have defined velocities (centimetres per year) which allow their future and past coordinates to be easily calculated. In other words, the datum definition includes a *velocity model.* Such a datum carries within its definition, mechanisms for change. Earth deformation and the increasing accuracy needs of the nation can be accommodated. Changes in coordinates are smooth and reasonably predictable rather than being jerky and unpredictable as with a series of static datums. Such a datum may see coordinates of control points in an automated system changing regularly (weekly, monthly or yearly).

Semi-dynamic Geodetic Datum: It is possible to have a hybrid static / dynamic datum. We can "freeze" the coordinates of a datum at a specified reference epoch just as NZGD49 coordinates were frozen in 1949. However we can develop and maintain a non-zero velocity model to be used with this datum. Therefore, this could be called a "frozen" dynamic datum which is accessed and maintained using dynamic transformations. We can define a new datum with a reference epoch of 1 January 2000 and define a velocity model based on actual observations of earth deformation. If we have observations made in 2005 we can convert the 2000 coordinates of existing stations to 2005, adjust the new observations at that date and convert the coordinates of all new stations back to 2000. By relating the observations to station coordinates at a common epoch, we can ensure that the effects of earth deformation are accounted for. This does not stop the physical network being distorted but it does stop the datum being distorted. We are able to monitor and maintain the changing relationship between the physical network of ground marks and the datum coordinates which are used in calculations.

It is probable that initially a semi-dynamic datum will be developed with a possible future move to a fully dynamic model after the proposed automation of the survey system. A velocity model for New Zealand is being developed by IGNS. This will enable survey observations and coordinates of surveys marks to be managed in a way that reflects the ground motions using either a dynamic or semi-dynamic datum.

Development of a Dynamic Coordinate Cadastre

Connection of the geodetic network to the cadastral system as proposed in the Land Information New Zealand automation project would allow geodetic coordinates to be assigned to all parcel boundaries. It is proposed that this not be a legal coordinate cadastre but that in the absence of other evidence, such as an undisturbed boundary mark, that the coordinate will provide another layer of evidence.

As coordinates of a dynamic control network are allowed to change due to ground deformation, so these movements could be propagated into the coordinate cadastre. The coordinates of boundary points would also move to encapsulate the best available evidence of their true position and accommodate the effects of ground deformation. This would see the development of the dynamic coordinate cadastre.

Keeping Track of the Dynamics

A GPS permanent tracking network could control the datum at the large scale. Other control points will be required to provide origin for surveys to ensure that un-modelled distortions have not crept into the network. While a surveyor may make GPS observations relative to a base station many tens of kilometres from the site of their survey, they could be required to connect to control in the vicinity of the survey to ensure that the modelled coordinates in that area reflect the true position of the ground marks.

Taking this concept further, as surveyors make cadastral surveys and submit data in digital format to the Department under the proposed automated system, the observations used to establish and prove the survey origin (the geodetic component of the survey in a dynamic coordinate cadastre model) could be used to enhance and update the datum. In future well defined physical features in the environment (eg houses) could possibly be used to densify the control network. In this manner, the effects of crustal deformation over small areas could be monitored and used to improve the velocity model. Where large changes were recorded, eg due to the effects of an earthquake, a resurvey of the area might be required to re-definite the datum in the affected area.

Beyond GPS Permanent Tracking Stations and Survey Control Marks

Future survey technology (eg, remote sensing) may soon give the ability to map crustal deformation in detail utilising a few survey marks for calibration and control. The use of Interferometric Synthetic Aperture Radar (InSAR) to interpolate earthquake deformation is an example of what is possible. Utilising such detailed deformation maps along with the ability to easily and precisely fix well defined physical features in the environment could result in reduced need for geodetic control points. Hardware in the environment would take over the current function of dense, lower order control. Each house corner may essentially become a control point. The potential of this, combined with high resolution remote sensing, could provide a built cadastre showing both legal and physical aspects of land ownership.

REFERENCES

Bevin, A.J. and J. Hall. 1995: The review and development of a modern geodetic datum. *New Zealand Survey Quarterly. Issue 1: 14 - 18.*

Blick, G., Rowe, G 1997: Progress towards a new geodetic datum for New Zealand. *Journal of the New Zealand Institute of Surveyors. No. 287, pg 25-29*

Blick, G.H., Linnell, G., 1997: The design of a new geodetic network and datum for New Zealand. *Paper presented to the first Trans Tasman Surveyors Conference 12-18 April, 1997, Newcastle, Australia.*

Grant, D.B., 1995: A dynamic datum for a dynamic cadastre. *The Trans Tasman Surveyor, Vol 40, No 4, pages 22-28*

Grant, D.B., Pearse, M. 1995: Proposal for a dynamic national geodetic datum for New Zealand. *Presented at IUGG General Assembly, 2-14 July Boulder, Colorado, USA*

Grant, D., Dawidowski, T., Winmill, R., 1997: New Zealand land tenure beyond 2000: Full integration and automation of the New Zealand survey and title systems. *Paper presented to the first Trans Tasman Surveyors Conference 12-18 April, 1997, Newcastle, Australia.*

Lee, L.P. 1978: First-order geodetic triangulation of New Zealand 1909-49 1973-74. *Department of Lands and Survey Technical Series No 1.*

Walcott, R.I. 1984: The kinematics of the plate boundary zone through New Zealand: a comparison of short- and long- term deformations. *Geophysical journal of the Royal Astronomical Society 79: 613-633*

PRELIMINARY STABILITY TEST FOR THE REGIONAL GPS TRACKING STATIONS IN TAIWAN

C. C. Chang and R. G. Chang
Department of Surveying and Mapping Engineering
Chung Cheng Institute of Technology
Tahsi, Taoyuan, Taiwan 335, ROC

Abstract

A regional GPS tracking network consisting of eight permanently operating GPS stations in Taiwan and its offshore islands has worked since 1995, in order to establish a high accuracy geodetic network purely based on GPS observations. These GPS stations aim to act as the core stations for establishing a high accuracy three-dimensional reference frame and motivating some other geodetic and geodynamic applications in Taiwan. The GPS measurements of this network are to be expected to an accuracy of a few millimeters. This is to be achieved, through the use of high accuracy GPS technique, by linking this network of stations to several primary stations of the ITRF (IERS Terrestrial Reference Frame). Two one-week GPS data from the preliminary stage of measurements made at these regional GPS tracking stations were collected to compute the network adjusted coordinates and test for their stabilities in three-dimensional components. A test of tropospheric modelling was also carried out to investigate the effectiveness of using associated surface measured meteorological data in high accuracy GPS network adjustment.

Introduction

GPS has been the most recent and most important advance in space geodetic techniques over the last 20 years. Due to its high positioning accuracy and effectiveness of extending the working range from local, regional to continental scale, GPS has now almost entirely surpassed terrestrial methods for high precision geodetic works. Hence, there has been a tremendous interest over the past 5 to 10 years in using GPS to connect some geodetic stations to a global geocentric reference frame for many geodetic applications.

Followed with its economic development and civil applications in Taiwan, a large number of traditional geodetic control points have been found to be seriously damaged or lost. In order to re-establish a high accuracy geodetic network purely based on the GPS observation, a regional GPS tracking network in Taiwan area consisting of eight permanent GPS tracking stations have been set up and operated since 1995.

The growth of continuously operating regional GPS network has been one of the most recent trends in GPS, so this regional GPS tracking network in Taiwan was originally designed to work as the core network for any scale of GPS observations made in this area. These regional GPS tracking stations were also expected to provide their high accuracy three-dimensional coordinates for some high accuracy geodetic, geodynamic, and navigation applications. The coordinate of these tracking stations, therefore, have to be accurately determined and constantly monitored, with respect to a global reference frame.

The network adjustments based on the two one-week GPS observations made in a short time interval were independently carried out with a precise ephemeris calculated by the IGS (International GPS Service for Geodynamics). This aims to measure zero movement of these regional tracking stations between the two one-week GPS measurements, in order to prove the potential of GPS for accurate positioning in Taiwan. The effectiveness of using surface measured meteorological data, opposing to the software-defined standard meteorological data, for tropospheric modelling was also tested with two GPS data sets in the network adjustments.

Data Sets and Data Processing for GPS Network

The data sets used to test the zero movement of eight regional GPS tracking stations were taken from the GPS observation carried out from 30 April to 6 May 1995 and from 11 June to 17 June 1995. The data for each week of GPS observation was taken from seven consecutive days, with twenty-four hour data observed at three IGS stations nearby this area, including USUD in Japan, SHAO in China, and TAIW in Taiwan, and also recorded at eight regional GPS tracking stations located in Taiwan and its offshore islands (see Figure 1). Data from these IGS stations and regional GPS tracking stations were all obtained using Rogue SNR-8 receivers. Two tested data sets were expected to demonstrate insignificant coordinate differences during the short time interval between the two one-week GPS observations.

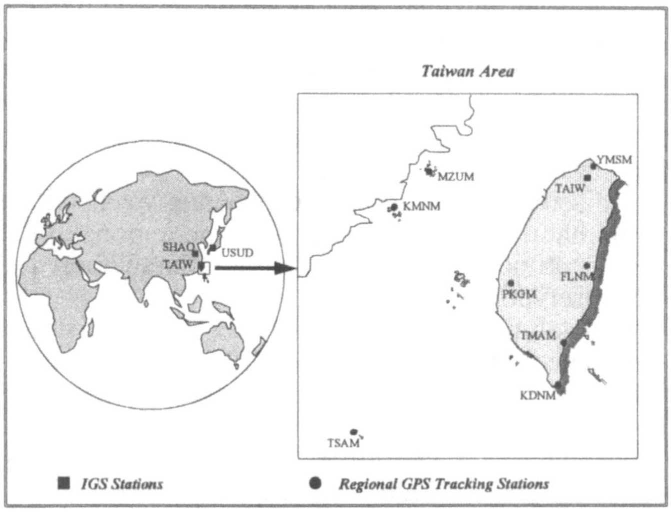

Figure 1 Regional GPS tracking stations in Taiwan area

The GPS network solutions were carried out using the GPS Analysis Software (GAS), developed by the University of Nottingham. The IGS precise ephemeris was used to provide the satellite orbit information in the tests. During the two one-week GPS observations, the IGS precise ephemerides used were for GPS week 799 and 805, respectively. A summary of the options used in the data processing is given in Table 1.

Table 1 Data processing options for the GPS solutions

Reference Frame Definition	
Stations Fixed:	USUD, SHAO, TAIW
Coordinate Source:	ITRF94 GPS Stations *[Boucher et al, 1996]*
Tectonic Plate Model:	ITRF94 Velocity Field *[Boucher et al, 1996]*
Reference Epoch:	1995.35
Models Applied	
Ionospheric Delay:	L1/L2 Frequency Combination
Tropospheric Delay:	MAGNET Model, with a Standard Atmosphere
Antenna Phase Center:	Elevation Angle Dependent Model *[Schupler, 1992]*
Earth Body Tides:	IERS Standards *[McCarthy, 1996]*
Adjusted Parameters	
Data Sessions Defined:	14 x 12 hours / session
Stations:	Position (X,Y,Z) of Regional GPS Tracking Stations
Satellites:	Fixed to IGS Precise Ephemeris
Tropospheric Delay:	Zenithal Scale Factor per Station, per Session (Solved as a First Order Polynomial)
Ambiguities:	Integer Free

Results and Analysis for GPS Network Adjustments

Once GPS solutions have been obtained, e.g. 14 single-session network solutions in each one-week GPS observation, the repeatability tests on successive single-session solutions can be performed to test the precision, based on the root mean square deviation of each single-session solution from the weighted mean.

The indicator of accuracy can be tested through the week-to-week agreement for all the regional GPS tracking stations. The so-called week-to-week agreement, also seen as a test of stability, is the discrepancy between the two one-week GPS solutions, which is based on the assumption that the site movement at all the regional GPS tracking stations is theoretically less than a centimeter level over the one month short time interval between the two one-week GPS observations.

The network adjustment solutions are now detailed. The precision and accuracy for the GPS solutions are estimated by two indicators, namely session-to-session coordinate repeatabilities and week-to-week coordinate agreements.

Session-to-Session Coordinate Repeatabilities. The session-to-session repeatabilities, based on the station coordinates, for each one-week GPS observation, namely GPS Week 799 and Week 805, are summarized in Table 2.

Table 2 RMS coordinate repeatabilities for the GPS solutions

Component	GPS Week 799	GPS Week 805
N (mm)	6	7
E (mm)	13	19
H (mm)	16	26

The session-to-session repeatabilities can show the quality of data collected during the two one-week GPS observations. The horizontal coordinate repeatabilities are consistently better than 10 mm in North component and 20 mm in East component. The height repeatabilities are as expected to be worse than the horizontal coordinate repeatabilities with a level of better than 30 mm for both sets of solutions.

It can also be seen that the Week 805 precisions are generally worse than the Week 799 precisions, by around 1 mm in North component, 6 mm in East component, and 10 mm in height. This level of coordinate uncertainty implies that the GPS observations made at the preliminary stage of operation for the GPS tracking stations in Taiwan were not in a very stable condition.

Week-to-Week Coordinate Agreements. The stability test, i.e. the consistency of these two one-week GPS observations, for the regional GPS tracking stations can be carried out by using the week-to-week coordinate agreements. The three-dimensional coordinate agreements between the two one-week GPS observations are given in Table 3.

Table 3 Week-to-week coordinate agreements for the GPS solutions (values shown are Week 805 minus Week 799)

Station	N(mm)	E(mm)	H(mm)
MZUM	-2	3	39
KMNM	3	9	50
FLNM	4	-4	41
PKGM	-2	1	48
TMAM	2	9	62
RMS	3	6	49

The week-to-week coordinate agreements represent the quality of the data from the two one-week GPS observations, with RMS agreements in horizontal coordinates of 3 mm in North component, 6 mm in East component, and around 50 mm in height. The week-to-week coordinate agreements show a normal level of performance on the horizontal coordinates, based on the network adjusted solutions using precise ephemeris. However, the agreements in height are much worse than expected. It is, therefore, important to test more GPS data sets to investigate the coordinate accuracy of height for the GPS tracking stations in Taiwan.

Surface Measured Meteorological Data Applied to Tropospheric Modelling

In order to improve the GPS accuracy, a standard tropospheric model is normally used in the GPS data processing to provide an estimate of the tropospheric zenith delay at a site. It is believed that the standard atmospheric models provide a broad approximation of expected tropospheric conditions, but they ignore the actual atmospheric conditions. Alternatively, surface measured meteorological data are able to be introduced to practically estimate the tropospheric delay errors. A test of applying such surface measured meteorological data to the tropospheric modelling was carried out for GPS observations made at some GPS tracking stations in Taiwan.

Data Availability. A total of four GPS stations, as a part of the continuously operating GPS tracking stations in Taiwan, were selected to test the effect of the tropospheric delay modelling. These four GPS stations, located at YMSM, PKGM, FLNM, and MZUM, were chosen mainly based on the consideration of their geographic distributions in Taiwan and its offshore islands.

Two GPS data sets used for the tropospheric tests were obtained from 8 January to 10 January 1996, and from 22 July to 24 July 1996, respectively. These two three-day data sets were measured in two different seasons, i.e. winter and summer in the same year, to present the most different atmospheric conditions. Surface measured meteorological data were also collected at the GPS sites, provided at one hour intervals for the measurements of temperature, pressure, and relative humidity.

Test Model and Strategy. The tests of tropospheric modelling for the two GPS data sets, namely Winter 96 and Summer 96, were performed with a 12-hour session of GPS data for all the baselines, where FLNM was fixed to its GPS coordinates defined by a longer term of observations.

The surface measured meteorological data sets were practically applied with the Saastamoinen atmospheric model during the GPS network adjustments. The results were then compared to those obtained by alternatively using the Bernese-defined standard meteorological data.

Results and Analysis. The session-to-session repeatabilities based on the results of using Saastamoinen atmospheric model with different types of meteorological data, i.e. defined and measured data, are shown in Table 4. The so-called season-to-season coordinate differences are also shown in Table 5 for the GPS solutions based on using two types of meteorological data.

Table 4 RMS coordinate repeatabilities for the GPS solutions based on two types of meteorological data

Coordinate Component	Winter 96		Summer 96	
	Defined Data	Measured Data	Defined Data	Measured Data
N (mm)	10	10	13	13
E (mm)	23	23	35	35
H (mm)	38	37	30	30

Table 5 Season-to-season coordinate differences for the GPS solutions based on two types of meteorological data (values shown are Summer 96 minus Winter 96)

Station	Defined Data			Measured Data		
	N (mm)	E (mm)	H (mm)	N (mm)	E (mm)	H (mm)
PKGM	-5	14	-18	-4	16	-18
YMSM	2	25	-2	1	26	-5
MZUM	-1	39	53	-2	39	55
RMS	3	28	32	3	29	34

It is clear that the repeatabilities are not effectively improved in all three coordinate components when the surface measured meteorological data were used to estimate the tropospheric delay, where a level of 1 mm difference is only found in height for the Winter 96 solution. However, it can be seen from Table 5 that the season-to-season coordinate differences are increased by 1 or 2 mm in RMS when the surface measured meteorological data were applied.

As surface measured meteorological data are believed to be easily affected by calibration problems and not completely representative of the atmospheric conditions dominating in the upper troposphere, the software-defined meteorological data is somehow still effective to be applied with a standard atmospheric model for tropospheric modelling.

Conclusions and Suggestions

This paper aims to show the preliminary tests for the GPS data observed at the regional GPS tracking stations in Taiwan. An overall summary of the research results is given as follows:

1. Based on the precise ephemeris network solutions carried out for the GPS tracking stations at their early operating stage, the coordinate repeatabilities were shown to be better than 10 mm in North component, 20 mm in East component, and 30 mm in height. However, a 10 mm difference in height repeatability between two one-week GPS observations was shown to be not consistent for the quality of GPS data collected.
2. The coordinate agreements between two one-week GPS observations made in a very short time interval were found to be around 5 mm in horizontal components. However, the agreements in height were found to be significantly biased by a level of 50 mm.
3. The use of surface measured meteorological data was not able to effectively improve the precision of the GPS network solutions, comparing with those obtained by using the software-defined standard meteorological data. The reason is possibly the often biased surface measured data caused by the calibration errors of the measurement equipment, systematic and random observation errors, radiation effects on the equipment, and ground proximity effects *[Brunner and Tregoning, 1994]*.

Acknowledgments

This work has been carried out as part of the research project, entitled "High Accuracy Positioning for the Regional GPS Tracking Stations" and funded by the National Science Council of ROC (NSC Contract 86-2611-E-014-005-T).

References

Boucher, C, Z Altamimi, M Feissel and P Sillard (1996), *Results and Analysis of the ITRF94*, IERS Technical Note 20, Observatoire de Paris.

Brunner, F K, and P Tregoning (1994), Tropospheric Propagation Effects in GPS Height Results Using Meteorological Observations, *Australian Journal of Geodesy, Photogrammetry, and Surveying*, Vol 60, pp 49-65.

McCarthy, D D (editor) (1996), *IERS Conventions (1996)*, IERS Technical Note 21, Observatoire de Paris.

Schupler, B R (1992), GPS Antenna Pattern Data - 11 February 1992, *Personal Communication from B R Schupler*, Bendix, USA.

THE USE OF HEURISTICS IN THE DESIGN OF GPS NETWORKS

Peter Dare and Hussain Saleh
School of Surveying
University of East London
Longbridge Road
Dagenham,
Essex, RM8 2AS
England

Peter@uel.ac.uk

ABSTRACT

One aspect of GPS network design concerns the logistics of the survey. Others aspects include location of stations, length of time for session observations. Efficient logistics enables a survey to be carried out with a lower cost and/or shorter time. A significant influence on the efficiency of the survey comes from the order in which the sessions are observed (the schedule). Previous work has shown that, given a list of sessions, the optimal schedule can be determined given the cost of moving receivers between points and the list of sessions. The solution was obtained by transforming the problem into a Multiple Travelling Salesman Problem enabling more than one working period to be included in the solution. However, the method was only suitable for relatively small networks.

To enable larger networks to be solved, the use of heuristic techniques within the field of Operational Research have been investigated. Heuristics enable optimal or near-optimal solutions to be found for very large problems with a reasonable computation time - optimality, however, is not guaranteed. This paper will describe one particular heuristic (Simulated Annealing) and show how it can be applied to the logistics design of GPS networks.

GPS NETWORK DESIGN

There are many different aspects of GPS network design to be considered. However, this paper will be concentrating on the order of session observations.

A session is when 2 or more GPS receivers are observing satellites simultaneously. In a survey network a number of consecutive sessions ('schedule') have to be observed. It is usually a surveyor's intuition that decides on the sessions to be observed - there is no automated design process to provide a list of the sessions to be observed to meet a specified criterion. Once the sessions have been decided upon, the receiver schedule (i.e., the order in which the sessions will be measured) has to be determined. If S represents the number of sessions, then the number of possible schedules is given by S! which is clearly a very large number for some networks. This paper addresses the problem of determining the optimal (according to some cost criteria) receiver schedule. This aspect of GPS network design has received little attention in the past despite its importance. Instead, previous design work has concentrated on other areas.

OPTIMAL SOLUTION TO LOGISTICS DESIGN PROBLEM

The methodology described in detail in Dare (1995), shows how, for small networks, the optimal order of sessions can be determined whilst allowing for the inclusion of a base (e.g., hotel, office) and permitting multiple returns to the base so allowing the network to be observed over more than one working period. Optimality is defined as minimising some cost criteria represented by a cost matrix containing the cost to move from each point to all other points. The developed method is based upon an Operational Research (OR) solution to the Travelling Salesman Problem developed by Little et al. (1963). As an example, consider the cost matrix shown in Table 1 derived from data supplied by Craymer (1991). The diagonals are costed as ∞ to prevent these movements.

Table 1. Fundy Park survey cost matrix. Costs in minutes of travel time.

From/To	1	2	3	4	5	6
1	∞	15	70	77	90	80
2	15	∞	77	65	85	83
3	70	77	∞	21	24	30
4	77	65	21	∞	21	29
5	90	85	22	21	∞	21
6	80	83	30	29	21	∞

The survey was to provide control for the positioning of a new road passing through Fundy Park, New Brunswick, Canada - two receivers were to be used to coordinate six points. The observational schedule adopted by Craymer is shown in Table 2.

Table 2. Fundy Park survey - observational schedule as used by Craymer (1991).

	Day 1			Day 2				
Rec. 1	2	2	1	2	3	5	6	6
Rec. 2	3	4	4	1	4	4	5	3

Using the data of Table 1, and the method described in Dare (1995), the optimal solution obtained is in Table 3.

Table 3. Fundy Park survey - optimal schedule using Craymer (1991) data.

	Day 1				Day 2			
Rec. 1	2	1	2	2	3	5	6	6
Rec. 2	3	4	4	1	4	4	3	5

The total cost for the computed schedule is 287 minutes which is 13 minutes less than the observed schedule (300 minutes) where an assumed travel time of 30 minutes was used for travel to and from the base. Comparing the observed schedule with the computed schedule the following differences become apparent.

- During the first working day the order of sessions 1,4 and 2,4 is reversed.
- During the second working day the order of sessions 6,5 and 6,3 is reversed.

Although the computed schedule has the sessions spread more evenly throughout the working days this is partly due to the fact that the planned schedule included an additional session in Day 1 which was abandoned due to the inability to locate the two points.

HEURISTICS

The method described above does provide, for 2 receivers, optimal schedules which can span more than one working period.

Nowadays, however, many groups use 3 or more GPS receivers with large numbers of points. Including additional receivers and more points will enlarge the cost matrix so making heuristic approaches necessary.

A Heuristic, within the OR sense, is a technique to determine near-optimal solutions in a reasonable amount of time. They are used extensively within OR to solve large problems that cannot be solved optimally reasonably quickly (see, for example, Reeves (1993)). Although a heuristic does not guarantee optimality, a good heuristic may provide the optimal solution (although this will not be known) or at least close to it.

The basic strategy for a heuristic as applied to GPS schedule design is as follows:

1. Choose an initial schedule
2. Swap two of the sessions to make the schedule cheaper
3. Repeat step 2 until no improvements can be made.

The problem with the above simplistic approach is that the solution is often a local optimum rather that the global optimum (see Figure 1). To obtain the global optimum, it becomes necessary to 'climb out' of the local optimum to allow a later descent to the global optimum. One method to allow this is known as Simulated Annealing.

Figure 1. Local and global optimum.

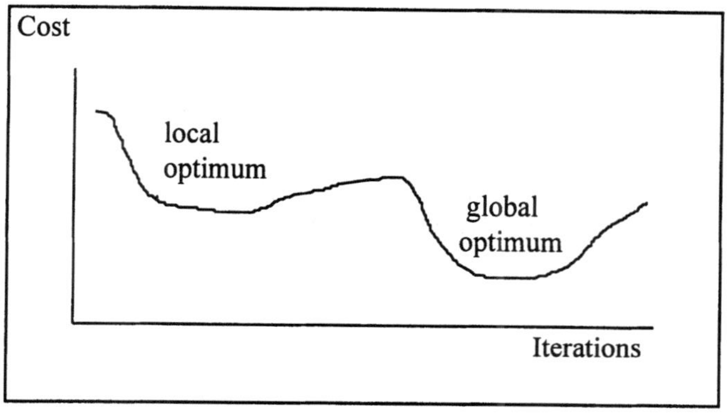

SIMULATED ANNEALING

The cooling of a material in a heat bath is known as annealing. If a solid object is heated to make it melt, the rate of subsequent cooling, and the highest temperature reached, determine the structure of the reformed solid material. Simulated Annealing (SA) is, obviously, a technique to simulate this process. The basic strategy for the SA heuristic as applied to GPS schedule design is as follows:

1. Choose an initial schedule
2. Swap two of the sessions to make the schedule cheaper
3. Repeat step 2 until no improvements can be made.
4. Swap two of the sessions to make the schedule more expensive.
5. Repeat step 4 for a small number of attempts
6. Go back to step 2.
7. Stop when no improvements can be made.

The application of this method to GPS will enable much larger networks (i.e., more points, more receivers) to have optimal or near-optimal schedules designed for them. This clearly will enable surveys to be carried out more efficiently. Further details of the SA method are in Saleh and Dare (1997).

CONCLUSIONS

The optimal schedule for GPS networks can be solved using a modified solution to the Travelling Salesman Problem. The solution allows for the inclusion of an office base with multiple returns to this base when observing over more than one working period.

For larger networks, the use of heuristics is necessary. A good heuristic will provide optimal or near-optimal solutions. A popular heuristic within the Operational Research community is known as Simulated Annealing and the application of this to GPS schedule design is now progressing.

Acknowledgement

This research is sponsored by the Ministry of Higher Education, Syria.

REFERENCES

Craymer, M. (1991). Personal communication.

Dare, P. (1995). Optimal Design of GPS Networks: Operational Procedures. PhD Thesis, University of East London.

Little, J.D., K.G. Murty, D.W. Sweeney and C. Karel. (1963). An Algorithm for the Traveling Salesman Problem. *Operations Research*, **11**, 972-989.

Reeves, C. (1993). (editor) *Modern Heuristic Techniques for Combinatorial Problems*. Blackwell Scientific Publications.

Saleh, H. and P. Dare. (1997).). “A Simulated Annealing Approach for Designing a GPS Surveying Network”. Paper presented at the 2nd Metaheuristic International Conference (MIC’97), Sophia-Antipolis, France, July 21-24.

TRANSFORMATIONS OF GPS COORDINATES AND HEIGHTS IN TRINIDAD

Karla Edwards, Joseph Barbaste and Jarir Saleh
Department of Surveying & Land Information,
The University of the West Indies, St. Augustine, Trinidad & Tobago.

ABSTRACT

GPS/levelling undulations were computed at 20 accurately-levelled stations of a nation-wide GPS network. The prediction accuracy of a multiple regression polynomial fitted through these undulations ,as well as, long wavelength errors of EGM96, OSU91A and CARIB97 over Trinidad are evaluated. Data to be used in the computation of a gravimetric geoid is also summarized.

The 1903 triangulation of Trinidad is readjusted and compared to the 1963 triangulateration in order to decide which control points are reliable for use in horizontal transformation computations and in order to get deformation-free transformation parameters. In addition a multiple regression transformation and the results of a variance-covariance simulation of similarity transformations is presented.

INTRODUCTION

Vertical Coordinate Transformation

A GPS network consisting of 91 baselines (measured with a pair of TI4100 receivers) connecting 33 stations was adjusted in order to facilitate the computation of a GPS / levelling geoid in the country. However, transformations of WGS84 heights to the local height datum require the computation of a precise geoid for the island - a gravimetric geoid. To achieve this, gravity and elevation data in Trinidad, its surrounding sea and the neighbouring part of Venezuela are needed. As can be seen from Fig.3 , there is a large gap of gravity data in the Northern Range of Trinidad. However, the Trinidad and Tobago Lands and Surveys Division is currently pursuing the idea of observing gravity data in that area. They are also currently in the process of occupying about 35 benchmarks with GPS (hence facilitating the computation of geometric undulations at these positions) in order to provide an independent test for the accuracy of the gravimetric geoid.

Horizontal Coordinate Transformations.

The old datum 1903 and the Naparima datum 1963 are still in use in Trinidad, hence two horizontal transformations are necessary, namely:

1. WGS84↔Old 1903
2. WGS84↔Naparima 1963

In order to compute reliable transformation parameters, issues such as the size of the island, the degree of tectonic activity, reliability of the control points, the random and systematic errors propagated from the 1903 and 1963 networks to the transformed coordinates and the number of transformation parameters required must be considered.

To facilitate the computation of horizontal transformation parameters, the Trinidad and Tobago Lands and Surveys Division has been observing GPS baselines between about 120 trig stations that belong to the 1963 and 1903 networks.

PRELIMINARY RESULTS

Vertical Positions:

- ***The GPS Network Adjustment.***

The 33 station network (Fig.1) was adjusted with 1 fixed point (Horn-86). This point was observed and computed by the DMA using precise orbits. Therefore, it is expected to have an absolute accuracy of a few decimetres.

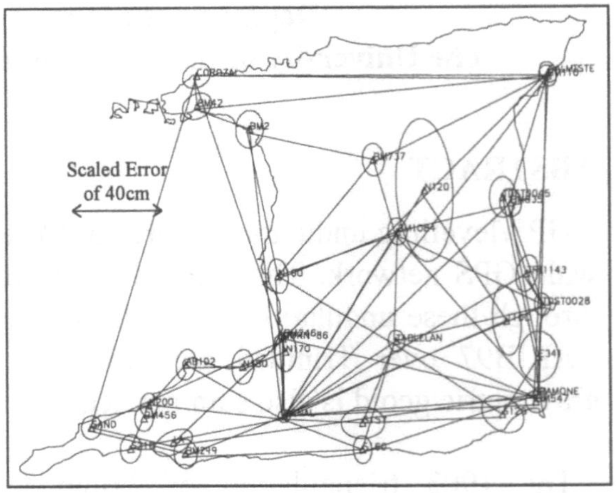

Figure 1 : Baselines of Least Squares Adjustment & Error Ellipses of the Horizontal Positions.

Table 1: Computed Geometric Undulations

Station	Undulation from GPS & Levelling / m	Accuracy / cm
RAMONE	-42.340	11.6
PENAL	-43.948	10.2
110C	-42.677	35.9
BM 1084	-42.841	16.9
BM 110	-42.485	13.7
BM 299	-44.466	19.1
BM 456	-44.497	18.4
BM 547	-42.149	17.1
BM 835	-42.860	16.5
BM 737	-42.486	16.9
COROZAL	-41.425	101.8
S 160	-43.442	17.0
SAND	-45.018	17.0
TABLELAN	-43.544	13.6
TDST0028	-42.814	14.3
TDST0045	-42.645	23.4
TRI 1143	-43.008	19.6
HORN-86	-43.811	5.0
LA FABIAN	-44.814	102.0
PALMISTE	-42.017	100.8
EDAM(UWI)	-42.290	7.8

With the exception of 2 points, the achieved accuracy in the horizontal coordinates is 1cm -12cm and 10 cm - 25 cm in the heights.

- ***A Preliminary Geoid For Trinidad.***

GPS/Levelling geoid undulations were determined using the orthometric heights of 20 benchmarks whose corresponding ellipsoidal heights were derived from the Network Adjustment (Refer to Table 1). These undulations were then interpolated using Multiple Regression Polynomial Fitting (e.g., Draper and Smith, 1977; DMA, 1987) based on the model given by the Defense Mapping Agency (1987). According to the DMA model, 21 stations (an additional benchmark was included which was not originally part of the GPS network) generate a polynomial with no more than 16 unknowns, of which 13 were found to be statistically-significant (using partial F tests).

The resultant polynomial was found to be:

$$N = - 43.551 + 2.283u + 5.659v + 11.214u^2 + 18.562uv + 13.251v^2 - 18.766u2v - 91.492v^3 - 250.20u^3v - 150.05u^2v^2 - 648.85uv^3 + 351.9u^2v^3 + 5607.9u^3v^3$$

where N is geoidal height , $u = \phi - \phi_{mean}$, $v = \lambda - \lambda_{mean}$ and (ϕ_{mean} , λ_{mean}) represents the mean position of 21 benchmarks.

The multiple regression fitting was repeated 21 times, excluding one point from the data each time. The discrepancies between the predicted and known undulations at each excluded point were used to evaluate the prediction error of the model, which when computed was ±11 cm.

Long Wavelength Errors Of EGM96, CARIB97 & OSU91A

Geoid undulations were extracted from CARIB97, EGM96 and OSU91A (Rapp, 1997) at the positions of the 33 stations of the GPS Network. Long wavelength errors were computed using: Long Wavelength Error = $N_{GPS\&Levelling} - N_{GPM}$

The weighted means of these differences were -0.678m, -0.510m and -1.493m respectively.

Relative Geoidal Heights using EGM96

In Figure 2 the difference between the relative geoidal heights, along the baseline AB, of EGM96 and the Preliminary Geoid were defined as:

$dN = \Delta N_{EGM96_{AB}} - \Delta N_{GPS\&levelling_{AB}}$

Figure 2 clearly demonstrates that the relative geoidal height difference between EGM96 and the Preliminary Geoid increases with baseline length. Furthermore, it shows that Mean Sea Level heights, to better than 15 cm, may be obtained at the present time by a combination of relative geoid undulations extracted from EGM96, absolute undulations of a few points on the island and ellipsoidal heights derived from GPS.

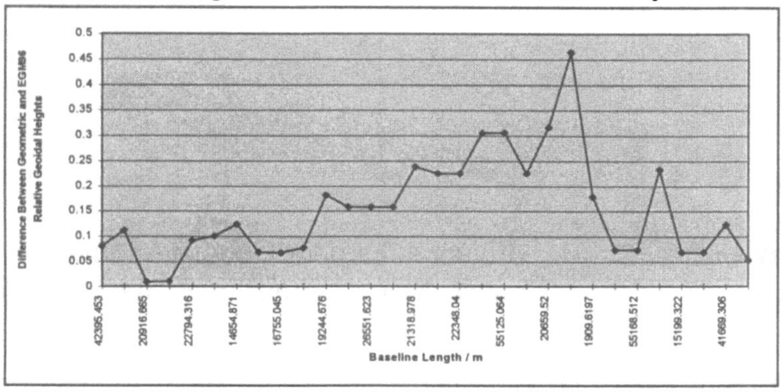

Figure 2: Relative Geoidal Height Difference vs. Distance.

- ***Data for Proposed Gravimetric Geoid***

The Digital Terrain Model (DTM) Of Trinidad

The DTM is needed to compute the terrain corrected gravity anomalies. The DTM of Trinidad was built from 37 topographic maps of scale 1:25000. The height contours were digitized using PC ARC/INFO. However, PC/TIN was used to create a 3D model Triangulated Interpolation Network (TIN) from which a 100m x 100m grid was extracted using Linear Interpolation.

The accuracy of the DTM was evaluated by comparing the levelled heights, of 512 points located along a 20 km 2nd order levelling loop (accuracy ≈1.8cm), with the heights extracted from the DTM using a contour interval of 25ft and 50ft. The accuracy of the DTM at these contour intervals was found to be 1.78m and 2.82m respectively. (Jules-Moore, 1996)

The Gravity Data

LaCoste & Romberg gravimeters were used to collect the terrain gravity observations. The Primary Gravity Base Station used in gravity networks in Trinidad and Tobago is located at Piarco International Airport and refers to IGSN 1971. This data was used to compute the free air anomalies. The distribution of gravity data throughout the country proved to be substantial except in the north where the Northern Range is situated (See Figure 3).

The gravity data positions were homogenized by referencing them to the WGS84 coordinate system while the free air anomalies were referenced to GRS80. Approximately 500 outliers were removed from the gravity anomalies data set. The resulting gravity anomaly contour map is displayed in Figure 4.

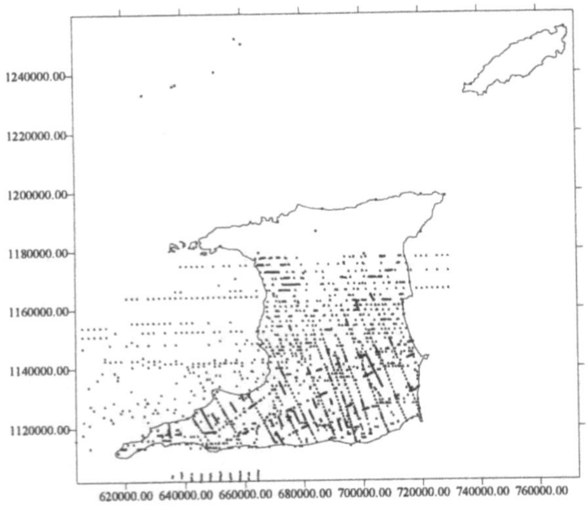

Figure 3: Distribution of Gravity Data in Trinidad

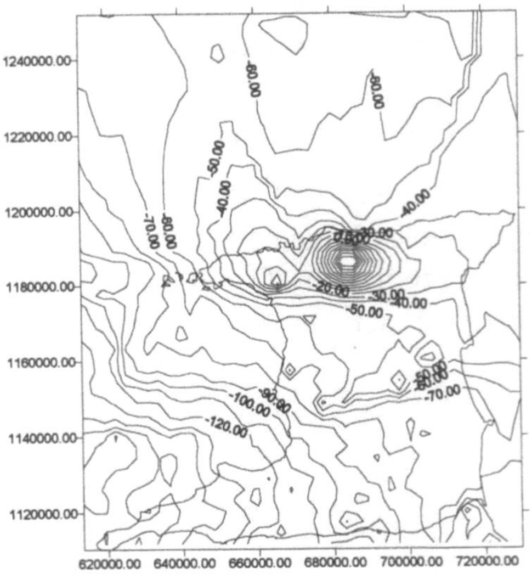

Figure 4: FreeAir Gravity Anomaly Contour Map. Contour Interval of 10 mgal

Horizontal Positions:

- ***Readjustment Of The 1903 Triangulation Of Trinidad***

The 1903 triangulateration network includes about 20000 directions (approximately 3500 directions after averaging) between 700 points. The sets observed from the major 135 stations were unified resulting in about 850 independent directions. 270 triangles were closed with r.m.s. triangular misclosure of 3.1".

The data was reduced to the International Ellipsoid 1924 and then to the UTM projection plane (where the network adjustment took place). The resulting variance component was $\sigma_0^2 \approx 1.4$ and the standard deviations of the adjusted coordinates range from 5 cm - 70 cm.

The adjustment revealed that:

1. stations 20, 131, 167, 274 and 328 (see Figure 5) have statistically-significant local movement (possibly landslides) and should therefore be excluded from any transformations computations.
2. The Los Bajos fault is not active.
3. The El Pilar is *definitely not* a 1.5 cm/year right lateral transform fault! (Demets et al., 1990) However, more investigation is needed to determine the degree of its activity (which is possibly not more than a few mm/year).

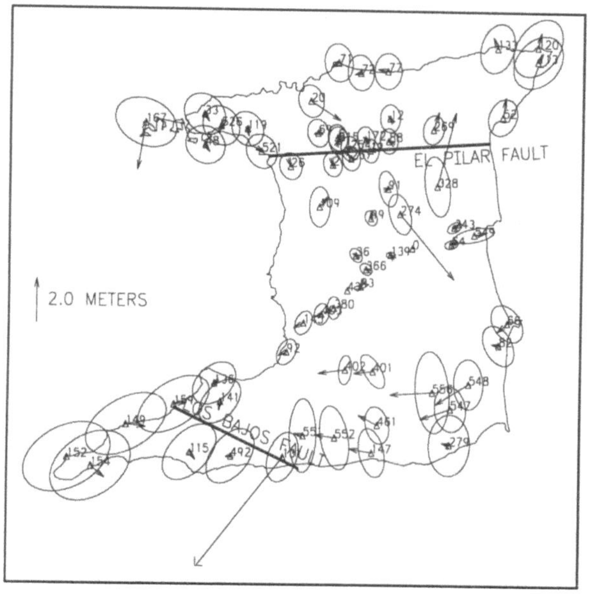

Figure 5: Displacements of 65 Trig.Points from 1903 - 1963 and their 95% Error Ellipses

4. The effect of more than 100 years of oil and gas extraction can be attributed to the movement in the southern part of the island.

- ***Transformations between the 1903 &1963 Local Datums***

The transformation of GPS coordinates to the 1903 datum is done in two steps. First, they are transformed to the 1963 datum and then to the 1903 datum. In order to accurately determine the transformation from 1963 to 1903 datums, Multiple regression polynomial fitting through 105 common triangulation points was used.

For example, the resultant polynomial for transforming the 1903 (Cassini) to 1963 (UTM) was:

$$\Delta E = 260734.563 - 79884.443\,u - 105.028\,v + 0.557\,u^2 + 0.484\,uv - 1.056\,v^2 + 0.316\,u^3 + 0.165\,u^2v^2 - 0.241\,uv^3 - 0.221\,u^3v^2 - 0.117\,u^2v^3 + 0.108\,uv^4 + 0.054\,u^3v^3 + 0.037\,u^4v^3 - 0.003\,u^3v^4 - 0.008\,u^4v^4$$

$$\Delta N = 844538.219 + 105.869\,u - 79884.594\,v + 0.268\,u^2 + 2.075\,uv + 0.349\,v^2 - 0.726\,u^3 + 1.069\,u^2v - 0.282\,uv^2 + 0.017\,v^3 + 0.173\,u^4 - 0.015\,u^3v - 0.859\,u^2v^2 + 0.038\,uv^3 - 0.045\,v^4 - 0.347\,u^4v + 0.745\,u^3v^2 - 0.239\,u^2v^3 + 0.008\,uv^4 - 0.006\,u^4v^2 - 0.086\,u^3v^3 + 0.213\,u^2v^4 + 0.069\,u^4v^3 - 0.118\,u^3v^4 + 0.002\,u^4v^4$$

where E and N are the easting and northing of each station on either datum and $E_{1963}=E_{1903}+\Delta E$, $N_{1963}=N_{1903}+\Delta N$, $u=10^{-5}(E - 419518.371\text{m})$, $v=10^{-5}(N - 306731.054\text{m})$. The estimated prediction error of the polynomial is 30 cm in the easting and northing directions.

- ***Investigation of Similarity Transformations***

A straight forward covariance simulation was done to identify the appropriate similarity transformation (Rapp, 1989) for a country of the size and location of Trinidad. The simulation used an expanding square of points where 9 points are located on its corners, centre and middle of its sides. The assumed accuracies of the local and GPS coordinates are 35cm and 7cm respectively. The square was centred on the datum point of Trinidad's datum while its sides were allowed to increase on the surface of a rotational ellipsoid, from zero to about 40°, with steps of about 200". The covariance matrix of the parameters of various transformation models were estimated at each step. The results may be summarized as follows:

7 Parameters Transformation

The standard deviations of the parameters of a 7-parameter similarity transformation for an island of the size and location of Trinidad would be about 20 meters in each of the three translation parameters, 0.7' in each of the three rotation angles and 2 ppm in the scale factor. Considering that the ΔX translation from WGS84 to the Naparima 1963 datum is about 0.5m and the rotations are of the order of 1", it is clear that most of the computed parameters are statistically insignificant. In order to obtain 7 statistically significant transformation parameters, the country should be about the size of Brazil. However, it was also found that the correlations among the parameters are not a function of the country's size. Furthermore, increasing the number of control points decreases the aforementioned standard deviations by a factor of $\sqrt{(n/9)}$ which is, in the best case, three times. Consequently, 7 parameter transformations should not be used in Trinidad and Tobago or for islands of comparable size and location.

6 Parameters Transformation

A 6-parameter transformation is constituted by three translations, a scale factor, an azimuthal rotation around the vertical of the datum point and a tilt rotation around an arbitrary line of azimuth Az_0 (to account for a certain tilt in the geoid relative to the WGS84 ellipsoid). This investigation revealed that the accuracy of the 3 translations ,which is about 0.1 m each, is not a function of the size of the country at all. The standard deviations of the two rotations are 0.6 and 0.5 second and decreases as a function of $\sqrt{(n/9)}$ where n is the number of points used for the computation of the transformation. If 120 points are used for the computation, the above errors drop to about 0.15". Thus, it is possible to compute 6 physically meaningful, statistically significant transformation parameters, even for a country as small as Trinidad.

ACKNOWLEDGMENTS

The first author wishes to thank the IAG for the travel award to attend IAG97. Thanks to the Campus Committee on Graduate Studies and Research and the Research Fund Committee of U.W.I. who provided additional funding and software.

REFERENCES

Defense Mapping Agency (1987): Department of Defense World Geodetic System 1984: Its Definition and Relationship with Local Geodetic Systems, DMA TR 8350.2, Defense Mapping Agency.

DeMets C, Gordon R. G., Argus D. F., Stein S. (1990): Current plate motions, Geophys. J. Int., 101, pp. 425-478.

Draper N. R. and Smith H (1966): Applied Regression Analysis. John Wiley & Sons, New York.

Jules-Moore, S.P. (1996): An Evaluation of the Accuracy of Elevation Information Generated By Elevation Models, Department of Surveying and Land Information, The University of the West Indies, St. Augustine, Trinidad.

Rapp R. H. (1997): Past and future developments in geopotential modeling, IAG97, Brazil.

Rapp R. H. (1989): Geometric Geodesy, Part II, Department of Geodetic Science and Surveying, The Ohio State University, Columbus, Ohio.

VECTORS CONNECTING THE GEODETIC POINTS AT METSÄHOVI AND SJÖKULLA

Jorma Jokela
Finnish Geodetic Institute
Geodeetinrinne 2, FIN-02430 Masala

Abstract

The Finnish Geodetic Institute has carried out geodetic operations at the Metsähovi research station since the 1970s. During the last ten years a number of new observation sites have been founded for GPS, SLR, VLBI etc. Some of these are located at Sjökulla, 3 km NNW from the original Metsähovi. Questions concerning the positions of the points relative to each other are frequently asked. Both terrestrial and GPS measurements have been used to determine the centering elements and connecting vectors, and measurements have been repeated to detect possible movements of the benchmarks. A brief summary of the observation sites and the local measurements carried out between them is now given.

Observation sites at Metsähovi

Fundamental points.

348 Metsähovi was the main benchmark in the triangulation era. It consists of a central benchmark and three auxiliary benchmarks, all four are bolts in the bedrock.

388 Rogue antenna (METS 10503S011) was installed on the southern edge on the top of a new height stabilized 20-m mast, built for the permanent GPS station in 1992. Since then, continuous observations have been made using Rogue SNR-C and Turbo Rogue SNR-8100 GPS receivers. The reference point of the GPS antenna (now Dorne Margolin B) is now the main GPS benchmark at Metsähovi (Ollikainen et al. 1997).

386 old SLR (METS 10503S001, CDP 7805) and *391 new SLR* (CDP 7806); the co-ordinates refer to the intersection of horizontal and vertical telescope axes in the satellite laser equipment.

Auxiliary markers.

384 Bilby benchmark is a bolt in the bedrock under the Bilby tower. The steel tower, which is older than, and distinct from the GPS tower, was erected when Metsähovi was connected to the first order triangulation network. When the continuous GPS observations started in 1991 (with an Ashtech receiver), the antenna was first placed up on this tower. *385 Azimuth benchmark* was used in astronomical observations in 1978. The astronomical azimuth was determined between 348 and 385. The result, 14° 19' 23.96", is not used in the present work.

It is possible to install three GPS antennas on the top of our GPS mast. The points *387 Ashtech antenna,* located on the north-east edge of the mast, and *389 Trimble antenna,* located on the north-west edge of the mast have been used in some observation campaigns. *390 CIGNet benchmark* (Co-operative International GPS Network) is a bolt in the rock underneath the mast supporting the points 387 – 389.

392, an auxiliary benchmark, a bolt at the corner of the new gravity laboratory building and *393,* another auxiliary point at the corner of the new satellite laser building were used in the last check measurements in 1997.

Observation sites at Sjökulla

Fundamental points.

349 VLBI/GPS (METS 10503M002, CDP 7601) is a steel bolt in the centre of a massive concrete plate (Vermeer and Paunonen 1994). Like all of our bolts, this bolt has a little hole in the top for precise centering. The plate is embedded in and lying on sedimentary deposits. This benchmark was built for the American mobile VLBI system visiting here in 1989. This benchmark was the GPS observation site in the campaign for EUREF 89, too.

380 DORIS (Doppler Orbitography and Radio Positioning Integrated by Satellite) is the oldest benchmark at Sjökulla, dating from 1988. The benchmark is marked during control measurements by a special tool placed in the mounting frame, which is fixed in the antenna base (a massive concrete block) under the DORIS antenna. The bottom plate of the DORIS antenna (10503S013) is 3.057 m higher than the benchmark 380.

Auxiliary markers.

In the check measurements we can monitor the possible movements of the fundamental points. *379* is a steel bolt in the bedrock, surrounded by cultivated land. It was built in 1996 to get a reference marker which is suitable for GPS measurements, too. *381* is a steel bolt in the bedrock in the wood. *382* is a steel bolt in a block of stone in the forest. 381 and 382 were built in 1988 to get stable reference markers for the DORIS antenna.

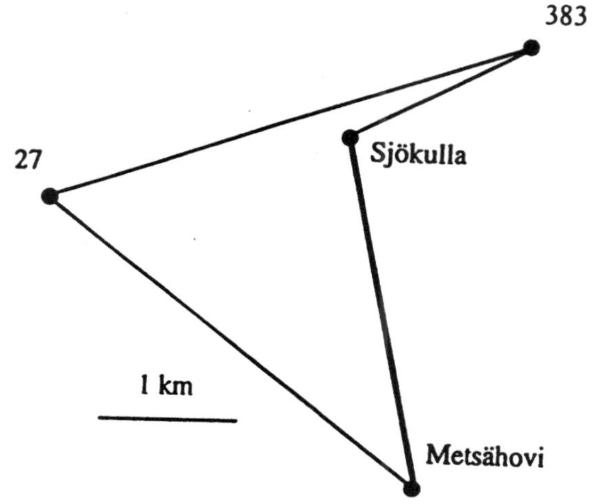

Figure 1 Tie measurements between Metsähovi and Sjökulla

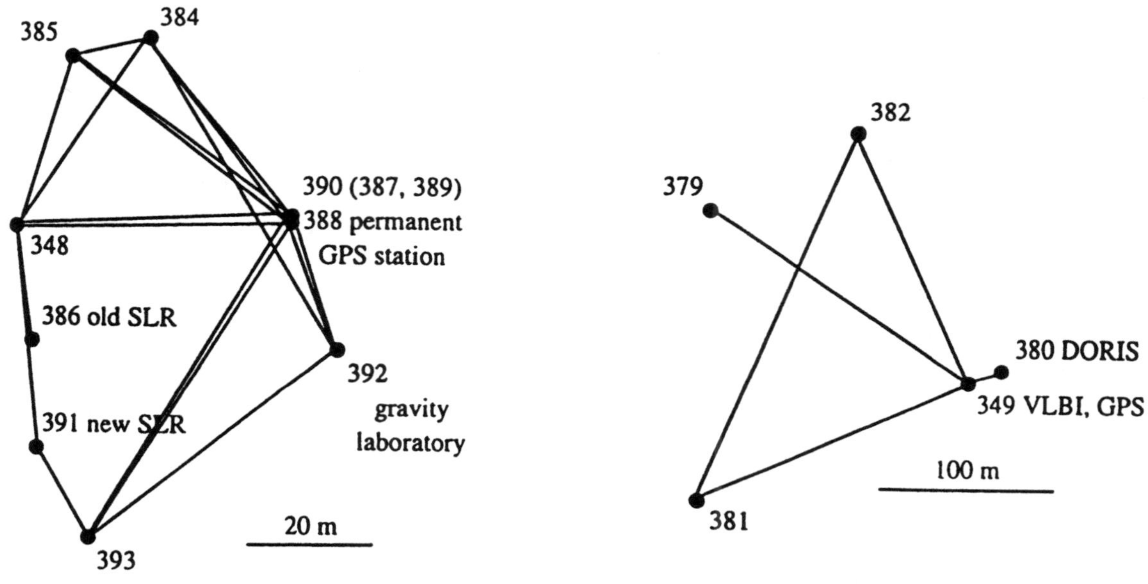

Figure 2 Observation sites at Metsähovi (left) and Sjökulla (right)

Other observation sites

As Metsähovi and Sjökulla are situated in a lowish country, existing triangulation stations on the tops of two hills, Falkberg and Kopumäki, were used to create the measurement contact in the terrestrial tie measurement in 1988. However, coordinates dating from the triangulation were not used in the new computation.

27 Falkberg was selected as a triangulation point in 1919. Measurements for the first order triangulation there were completed in 1926. The central benchmark and the four reference markers are iron bolts built in the bedrock. To see the neighbouring stations Metsähovi and Kopumäki from the forested Falkberg, a new auxiliary benchmark was necessary. *383 Kopumäki* is an old low order triangulation station with an iron bolt in the bedrock as the central benchmark. The dilapidated wooden tower above it was done up good enough for the measurements.

Measurements and computation

A Wild Theomat T2000 and Distomat DI2000, Kern DKM3 theodolite and Mekometer ME3000 and Wild and Kern prisms were used in the initial measurements in 1988-1989. Later on, Wild Theomat T2002 and Distomat DI2002 and Wild prisms have been used. After station adjustments, we have 127 direction observations and 67 distance observations made in 1988 – 1997 in the nets presented in Figures 1 – 2.

The orthometric height of the triangulation benchmark 348 Metsähovi had been determined in 1977 with precise levelling in the Finnish height system N60, $H = 54.013$ m. The levellings between Metsähovi and Sjökulla were started in 1988 to connect all the stations to this height system. Wild N3 and Zeiss Ni002 instruments were used. If we compare the EUREF 89 coordinates, Sjökulla (349) is 34.980 m lower than Metsähovi (388). The difference in orthometric heights determined by precise levelling is 35.062 m. From this we computed the ellipsoidal height differences using the FIN95 geoid (Vermeer 1995), the NKG89 and NKG96 geoids by Forsberg. The results were 34.989 m, 34.969 m and 34.980 m, respectively. The FIN95 geoid was used in the final computation. To avoid discrepancies between ellipsoidal and orthometric heights, the geoidal heights were corrected with -.036 m at Metsähovi and -.027 m at Sjokulla.

GPS observations were made with Mini Rogue, Turbo Rogue and Ashtech geodetic GPS receivers and Dorne Margolin antennas, and processed with Bernese and GPPS softwares. The vector between Metsähovi and Sjökulla was measured under fairly ideal conditions during the DOSE campaign in August 1993. Computation with the Bernese 3.4. software gave (Vermeer and Paunonen 1994):

$$\Delta X = -1\,918.178 \text{ m}, \quad \Delta Y = -1\,548.001 \text{ m}, \quad \Delta Z = +1\,324.664 \text{ m},$$

from Metsähovi to Sjökulla; ±1 mm accuracy was obtained. This vector and the EUREF 89 coordinates of Sjökulla (349 VLBI/GPS),

$$\varphi = 60° \, 14' \, 31.06834'', \quad \lambda = 24° \, 23' \, 03.00378'', \quad h = 59.563 \text{ m},$$

have been used to compute the EUREF 89 coordinates for Metsähovi (388 Rogue antenna),

$$\varphi = 60° \, 13' \, 02.89097'', \quad \lambda = 24° \, 23' \, 43.13725'', \quad h = 94.543 \text{ m}.$$

To make the orientation of the terrestrial nets more reliable, GPS measurements were recently made at every point where it is possible, i.e. at points 348 and 379, too.

Now the EUREF 89 coordinates of benchmarks 349 and 388 have been fixed in the processing of both GPS observations and angle and distance measurements. The first angle and distance observations of the Metsähovi–Sjökulla traverse in 1988-1989 were adjusted in connection with the final adjustment of the Finnish first-order triangulation (Jokela 1994, pp. 119-121). The International 1924 ellipsoid and ED87-FIN coordinates were used then. In consequence of troublesome measurements and unfavourable geometry, no better accuracy than ±1 cm was obtained. Now the terrestrial observations form a loose connection between the two places, but the crucial constraints in the adjustment come through the fixed GPS vector. All the observations are reduced now on the GRS80 ellipsoid. A new version of the original adjustment program described by Jokela et al. 1994 was used.

The computation gives the positions of all observation sites relative to the two fixed points. In the control measurements submillimeter accuracy has been obtained. No horizontal movement has been found. At Sjökulla there are suspicions of a slight subsidence.

Eccentricities expressed in coordinate differences in EUREF 89

The results of the adjustment presented in Table 1 are in EUREF 89 coordinate differences with respect to the Metsähovi GPS station (Rogue antenna, 388). The vector to Sjökulla (VLBI/GPS point, 349) is kept fixed. Errors in terrestrial measurements cause maximum discrepancies of about 3 mm both at Metsähovi and at Sjökulla, and standard errors of about 2 mm. For the whole project such a good accuracy can not be guaranteed, because of the inaccuracies in geoidal heights. More control measurements will be carried out during September 1997. This may cause slight differences in the results. A more exhaustive report of the work presented here will be published by the FGI in the near future.

Table 1
Coordinate differences in EUREF 89. For the point 388,
X = 2 892 571.072, Y = 1 311 843.306, Z = 5 512 633.939.

		ΔX	ΔY	ΔZ
27	Falkberg	– 532.698	– 3 494.985	+ 1 120.716
348	Metsähovi main triang bm	+ 9.785	– 46.090	– 19.140
349	**Sjökulla VLBI/GPS**	**– 1 918.178**	**– 1 548.001**	**+ 1 324.664**
379	Sjökulla DORIS GPS ref bm	– 1 928.900	– 1 716.441	+ 1 380.006
380	Sjökulla DORIS bm	– 1 931.170	– 1 533.411	+ 1 328.240
381	Sjökulla DORIS ref bm I	– 1 794.600	– 1 659.527	+ 1 301.256
382	Sjökulla DORIS ref bm II	– 1 999.033	– 1 655.915	+ 1 403.844
383	Kopumäki	– 3 115.144	– 605.160	+ 1 768.131
384	Metsähovi Bilby bm	– 22.364	– 37.109	– 3.342
385	Metsähovi azimuth bm	–23.126	– 50.596	+ 0.104
386	Metsähovi old SLR	+ 24.474	– 35.526	– 23.276
387	Metsähovi Ashtech antenna	– 0.947	+ 0.201	+ 0.397
388	**Metsähovi Rogue antenna**	**0.000**	**0.000**	**0.000**
389	Metsähovi Trimble antenna	– 0.370	– 1.102	+ 0.440
390	Metsähovi CIGNet bm	– 9.762	– 4.505	– 17.701
391	Metsähovi new SLR	+ 36.219	– 30.251	–35.392
392	Metsähovi grav lab aux bm	+ 4.231	+ 10.259	– 27.521
393	Metsähovi aux p	+ 42.391	– 17.132	– 44.328

References

Jokela J (1994) The 1993 adjustment of the Finnish First-Order Terrestrial Triangulation. Publ. of the FGI 119.

Jokela J, Poutanen M, Konttinen R (1994) The program for the 1993 adjustment of the Finnish First-Order Terrestrial Triangulation. Rep. of the FGI 94:3.

Ollikainen M, Koivula H, Poutanen M, Chen R (1997) Suomen kiinteiden GPS-asemien verkko. Geodeettisen laitoksen tiedote 16.

Vermeer M (1995) Two new geoids determined at the FGI. Rep. of the FGI 95:5.

Vermeer M, Paunonen M (1994) The vector connecting the fundamental points Metsähovi and Sjökulla. In: Gubler E, Hornik H (ed.) Report on the Symposium of the IAG Subcommission for the European Reference Frame (EUREF) held in Warsaw 8 – 11 June 1994. Veröffentlichungen der Bayerischen Kommission für die Internationale Erdmessung 54, pp. 210-215.

USE OF THE FINNISH PERMANENT GPS NETWORK (FINNNET) IN REGIONAL GPS CAMPAIGNS

Hannu Koivula, Matti Ollikainen and **Markku Poutanen**

Finnish Geodetic Institute
Geodeetinrinne 2 , FIN-02430 Masala , Finland
E-mail: FinnNet@fgi.fi

Abstract

The Finnish Geodetic Institute has established a network of 12 permanent GPS-stations in Finland (FinnNet). The planning of the network started during winter 92/93. The first stations were installed in 1994 and the last one was built in the autumn of 1996. The stations are equipped with geodetic Ashtech Z-12 receivers and Dorne Margolin type antennas. All the stations are working fully automatically. The network is used as a 1st order reference frame in Finland for GPS users, for crustal dynamics studies and for providing RTCM corrections in real time applications.

FinnNet is used as a reference frame or as a source of data for numerous national and international GPS campaigns and projects like *DOSE* (Dynamics of the Solid Earth), *BSL* (Baltic Sea Level), *BIFROST* (Baseline Inferences for Fennoscandian Rebound Observations, Sea Level and Tectonics), *EUVN* (European Vertical GPS Reference Network), and *EUREF* permanent GPS Network.

We will introduce FinnNet and its function in Finland, present the first results of the EUREF-FIN campaign and give an overview of the projects FinnNet has taken part in. We will discuss in general how permanent networks can be used in regional GPS measurements and call into question the function and need of large episodic GPS field campaigns in the future.

Introduction

The Fennoscandian Regional Permanent GPS network was established by the Nordic Geodetic Commission in response to an initiative by the directors of the Nordic Mapping Institutes (Kakkuri *et al.* 1995). The primary scientific goals of the project are to use the GPS determinations to estimate the post-glacial deformation of the Earth's crust in Fennoscandia, and to correct the extensive Fennoscandian tide gauge record for the influence of vertical motions (Bifrost Project 1996, Mitrovica *et al.* 1997).

The planning of the Finnish part of the Fennoscandian Regional Permanent GPS network was started in the Finnish Geodetic Institute (FGI) during the winter 92/93. Eleven stations were then chosen for future reconnaissance. The selection of the stations was made according to the following principles (which initially were not very explicit):

- The network should give a good coverage over the whole country in such a way that the maximum land uplift differences and horizontal motions between various parts of the crust can be detected.
- The site should be established on the bedrock.
- The absolute gravity at, or near, the station should be measured.
- The site should be connected to the precise levelling network.

The establishment of the stations

In 1993 steel grid masts of 2.5 m height were erected at five sites. The construction of the steel mast is shown in Fig 3. The height changes due to the thermal expansion of the steel tower, appr. ±0.8 mm during the course of the year, were considered insignificant and will average out anyway during the yearly cycle. The eccentricities of the towers from the reference bolts were measured during Sept., 1993, with rapid static GPS in 1994, and with tachymeter in Sept. 1997.

Three sites, i.e. Olkiluoto, Konginkangas, and Kuhmo were built up in co-operation with the company Posiva Oy. The mount of the antenna is a steel enforced concrete pillar which is more stable than the standard mast. The mounts are a part of high precision local GPS control networks which are periodically remeasured (CHEN and KAKKURI, 1994).

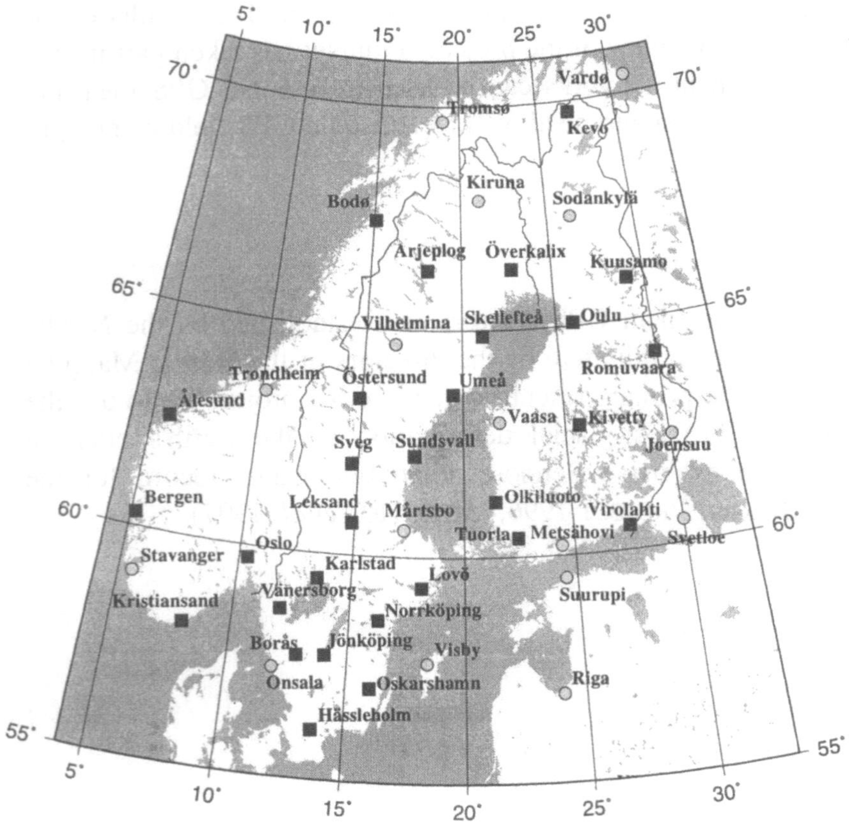

Fig. 1 *Permanent GPS stations in Nordic countries. Circles denote points used in NKG combined solution which is an EUREF Local Network, squares are other permanent stations.*

An existing building was used for the GPS equipment at four stations, i.e. in Tuorla, at the Astronomical Observatory of Turku University, and at Sodankylä Geophysical Observatory. In Oulu a 8.5 m high existing mast has been converted for GPS use and the existing plywood observatory building was also available. At Kevo the antenna mast is 5 m tall due to trees around the point; otherwise it is similar to the "standard" setup. The existing building of Kevo Arctic Research Station is used. Separate instrument cabins were built in Virolahti, Joensuu, Vaasa, Kuusamo, Kivetty, Romuvaara and Olkiluoto.

Receivers, data downloading and current status of the network

Five TurboRogue SNR-8100 GPS receivers were purchased in 1994. The first observations with the new receivers were made during the DOSE'94 Campaign (DOSE = Dynamics Of the Solid Earth), in Aug. 15-26, 1994. The data was collected via telephone line from four sites. i.e. Tuorla, Virolahti, Vaasa, and Sodankylä. Since the campaign two sites, i.e. Virolahti, and Sodankylä have been operating continuously.

Due to frequent malfunctioning of many of the TurboRogues, it was decided to change them to Ashtech Z-12 receivers in 1995. The only exception is the Metsähovi station where a TurboRogue receiver is still being used. The Dorne Margolin type choke ring antennas are retained for compatibility and because choke ring antennas have superior multi-path properties compared to other antenna types (ROCKEN *et al.* 1994).

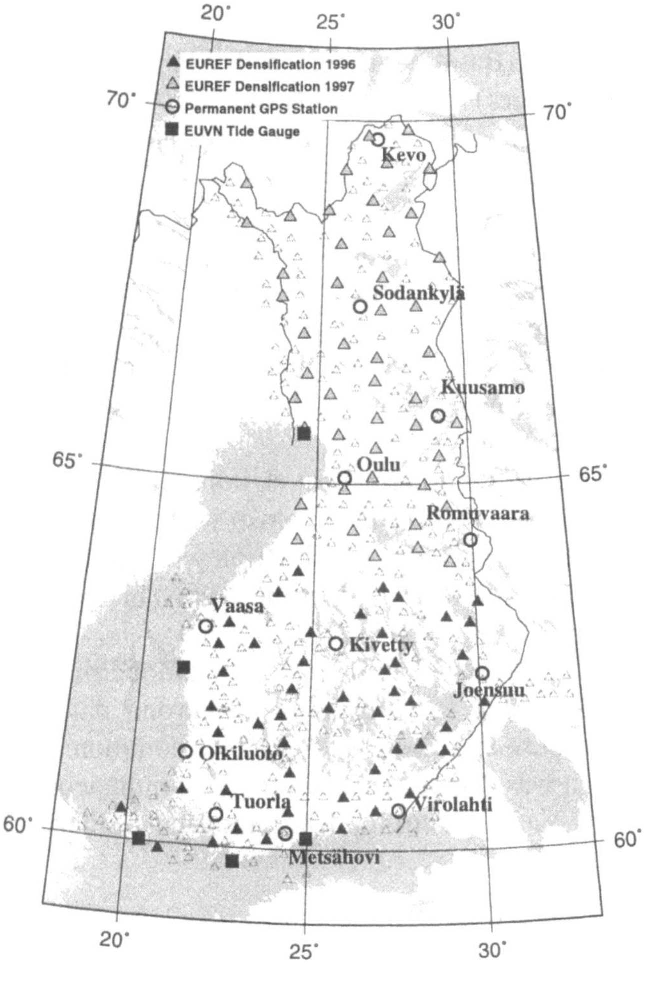

Fig. 2 (*left*) *The Finnish Permanent GPS Network – FinnNet, First order triangulation network, and EUREF densification sites. EUVN Tide gauges are also marked.*

Fig. 3 (*below*) *Construction of the FinnNet antenna mast.*

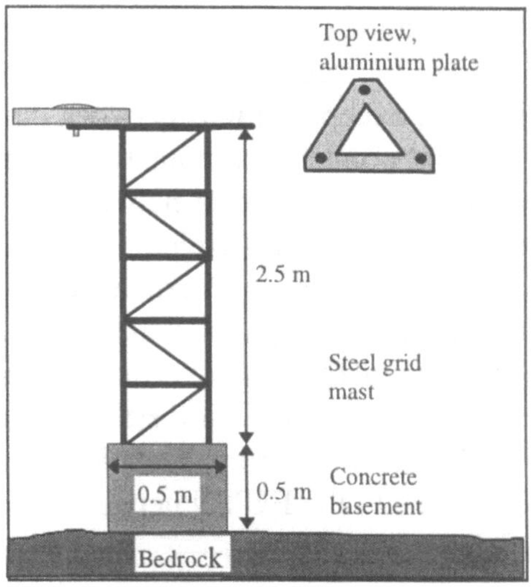

Continuous observations were started in 1994 at Virolahti, Tuorla, Sodankylä, and Olkiluoto, and in addition in 1995 at Vaasa, Joensuu, and Oulu. During 1996 four stations, e.g. Konginkangas, Kuhmo, Kevo and Kuusamo, were taken into operation. Altogether 12 stations are working today.

Data are downloaded automatically every 24 h via an ordinary modem-equipped telephone line. The control is made with an in-house made program running on a Pentium PC under DOS 6.22. For the actual downloading, the Ashtech program Remote is used.

Data are stored on hard disk and a backup is made on another computer. Moreover, data are also sent by ftp to Onsala for BIFROST and NKG EUREF subnetwork computation. CD-R and DAT are used for archiving both original data and RINEX files.

Use of FinnNet data for international GPS campaigns

The base station in the permanent GPS network is Metsähovi Space Geodetic station which has been collaborating e.g. in international SLR observation programs during two decades. Today it belongs also to the IGS network producing GPS data for IGS orbit determinations.

The first observations in FinnNet were made during the DOSE'94 campaign. (see also OLLIKAINEN and POUTANEN 1995). Since then the network has been used in consecutive DOSE campaigns; e.g. for DOSE'94, DOSE'95 and DOSE'96, occupying also several temporary points near the land uplift maximum.

The data have been used also in the BIFROST (Baseline Inferences for Fennoscandian Rebound Observations, Sea Level and Tectonics) project (BIFROST PROJECT 1996, MITROVICA *et al.* 1997). One of the BIFROST goals is to obtain three dimensional vectors from GPS observations for determination of mantle viscosity. First results on land uplift rate already look very promising.

The third BSL (Baltic Sea Level) campaign was organized in May 1997, simultaneously with the EUVN GPS campaign. The Baltic Sea Level Project was initiated already in 1989. The purpose of the project is to interconnect the vertical datums of the countries around the Baltic Sea. For this, a series of GPS campaigns has been arranged, the first one in 1990, the second one in 1993 and the third one in 1997. More than 30 tide gauges are included in the campaign. (POUTANEN 1995)

During the second BSL GPS campaign there were no permanent stations in the Nordic countries, except the IGS network points Metsähovi, Onsala and Tromsø which were used as fiducial points in computation. During the third campaign, the situation was completely changed because a good network of permanent stations existed in the area. This will be used as a background skeleton for computation.

EUVN (European Vertical GPS Reference Network) utilizes also the permanent network. The campaign was arranged in May 1997 for producing backgroud data for unification of European vertical networks. Because there were several sites common with the BSL network, the third BSL GPS campaign was decided to be arranged simultaneously with the EUVN. From FinnNet, Metsähovi, Vaasa, Joensuu, Kuusamo and Sodankylä belong also to the EUVN network.

First results of the EUREF-FIN campaign.

During the years 96 and 97 a GPS campaign called EUREF-FIN was performed where the FGI measured a network of 100 points over Finland. The permanent network is a part of this. Coordinates will be given in the EUREF-FIN reference frame, which is a subset of the original EUREF89 reference frame. The whole network was measured using Ashtech Z-12 receivers equipped with Dorne Margolin type choke ring antennas. The nominal observation time at each point was 48 hours.

The first part of the network was measured in 1996 and the second part of the campaign in Aug.-Sept. 1997. The network was computed with Bernese v.4.0 software. The first solution was computed keeping only Metsähovi fixed. This free network solution is then combined with network of fiducial stations, taken from the EUREF network.

In Fig. 4 we show an example of residuals, obtained during the free network solution. One can see that network shape will have an affect on the results. However, the overall repeatability is on the sub-cm level for FinnNet points and about the same also for EUREF densification points. The whole network is shown in Fig. 2.

Need of large GPS campaigns in the future

Today in Northern Europe, e.g. in Finland, Norway and Sweden, there are approximately 43 continuously working GPS stations. The coordinates of these stations may be computed with centimetre accuracy at any time in whichever global reference frame, thus forming a good local reference frame for any kind of geodetic positioning or geodynamic research.

Using the GPS data recorded at these stations it is possible to determine the position of a GPS receiver with cm-accuracy where-ever at any time in the area of the Nordic

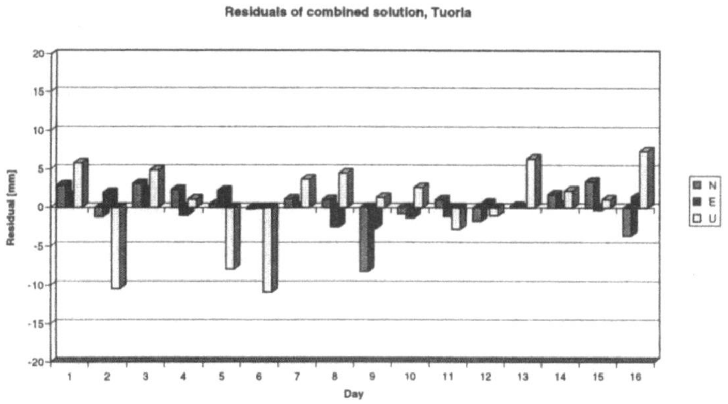

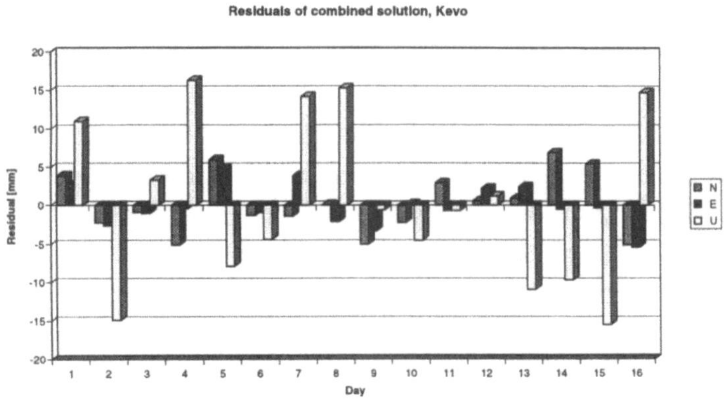

Fig. 4 *Residuals of combined solution in EUREF densification. In this example one can see some dependence of site position in the network. Tuorla is situated in the middle of the measurement area, Kevo is the most distant site, far from other points. In all cases, the N and E component residuals stays well below 0.5 cm, the height component below 1.5 cm.*

Countries. The position may be computed in whichever global reference frame desired.

The question arises: Do we really need to organize a large international GPS observation campaign to achieve the goal which may be obtained using a single GPS receiver together with data recorded at a few permanent GPS stations?

Because the network of permanent stations offer a proper and stable connection to the global reference frames the need for simultaneous observations is diminishing. One has only to assure that when measuring new points,that those are well tied to the old ones. In this way the uniformity of the whole new network is preserved even if measured in small pieces and during a moderate time span. Such a strategy has already been used in EUREF campaigns. The original big European-wide GPS measurement in 1989 was followed by several smaller campaigns.

In the current situation one is tempted to ask if large field campaigns are needed anymore. There could be some special needs for simultaneous observations but in most cases regional measurement with a good tie to the existing reference frame will be sufficient. Also, using data from permanent stations will save one or more GPS receivers during the campaign for measuring the new points. The scheme sounds first trivial one but inertia in changing old habits seems to be quite big. The use of permanent networks in everyone's own measurements could be much more than what it is today.

References

BIFROST PROJECT (BENNET R.A., T.R. CARLSSON, T.M. CARLSSON, G. ELGERED, R.T.K. JALDEHAG, P.O.J. JARLEMARK, J.M. JOHANSSON, B.I. NILSSON, B.O. RÖNNÄNG, H.-G. SCHERNECK, R. CHEN, J. KAKKURI. H. KOIVULA, M. OLLIKAINEN, M. PAUNONEN, M. POUTANEN, M. VERMEER, J.L. DAVIS, P. ELOSEGUI, I.I. SHAPIRO, M. EKMAN, G. HEDLING, B. JONSSON, J.X. MITROVICA and R.N. PYSKLYWEC) (1996). GPS measurements to constrain geodynamic processes in Fennoscandia. *EOS* **77**, No 35, 337.

CHEN, R. and J. KAKKURI (1994). Feasibility study and technical proposal for long-term observations of bedrock stability with GPS. Report YJT-94-02. Nuclear Waste Commission of Finnish Power Companies. Helsinki. 33 pp.

KAKKURI, J., H. KOIVULA, M. OLLIKAINEN, M. PAUNONEN, M. POUTANEN AND M. VERMEER (1995). The Finnish GPS Array FinnNet: Current Status. *Invited paper, IGS Workshop, Potsdam,* 15.-17.5. 1995.

MITROVICA J.X., DAVIS J.L., SCHERNECK H.-G., AND JOHANSSON J.M. (1997). BIFROST Project: GPS measurements to constrain geodynamic processes in Fennoscandia. *European Geophysical Society, XXII General Assembly*, Vienna, Austria 21–25 April, 1997. Annales Geophysicae, **15**, Supplement I, C81.

OLLIKAINEN, M. and M. POUTANEN (1995). GPS-geodesy. In *Geodetic Operations in Finland 1992-1995.* (Ed. J. Kakkuri). FGI. Helsinki.

POUTANEN M (1995). A combined solution of the Second Baltic Sea Level GPS campaign. In *Final Results of the Baltic Sea Level 1993 GPS Campaign* (Ed. J. Kakkuri). Reports of the Finnish Geodetic Institute **95:2**, 115–123.

ROCKEN, C. J. JOHNSON, J. BRAUN, C. MEERTENS, S. PERRY (1994). UNAVCO facility GPS receiver tests. Draft 7/21/94, July 1994.

THE NEW SWISS NATIONAL HEIGHT SYSTEM LHN95

Urs Marti and Dieter Schneider
Federal Office of Topography, Seftigenstr. 264
CH-3084 Wabern, Switzerland

Abstract

The existing old height system of Switzerland (LN02) consists of purely levelled heights without considering gravity measurements. This causes difficulties in the application of modern techniques such as GPS for height determination and in the data exchange with neighboring countries. Therefore, a new height system LHN95, based on geopotential numbers, is being determined at the Federal Office of Topography. LHN95 is treated as a kinematic network with consideration of the Alpine uplift. The corresponding relative vertical movements are determined by repeated observations of the first and second order levelling lines. For LHN95, it was decided to use orthometric heights. This has some advantages when combining levelling with GPS measurements in mountainous regions.

Introduction

The Swiss Federal Office of Topography (SWISSTOPO) is responsible for the establishment, the revision and renewal of national geodetic networks. Between 1989 and 1995 SWISSTOPO established a new geodetic reference system based on a GPS network (LV95) covering the whole country and being well connected to ETRS. Whereas LV95 is a purely geometric network, there is also a growing demand for a new, well-defined height system and an adequate height frame. This paper intends to describe the concept and the procedure for establishing the new height system LHN95 for Switzerland.
The existing National levelling network (LN02) was established at the beginning of this century. It contains basic elements or observations which even go back to the 19th century and is still in use for cadastral and most other surveying and mapping applications. It is based on "observed heights" rather than values corrected with respect to the gravity field. The heights are static, although it is known from systematic remeasurements that there are significant relative height changes in the order of 1 to 2 mm/year caused by tectonic movements of the uppermost crust in the Alps [Gubler et al. 1981].

Concept of the Swiss National Height System LHN95

Project LV95 / LHN95.
In the last few years, SWISSTOPO has been establishing a concept for a new geodetic control (LV95). The backbone of the concept is the new national GPS network (also called

LV95), which is planned to replace the old triangulation (LV03) from the beginning of this century. A particularity of LV95 is that it includes 2 reference systems. The first (CH1903+/LV95) is chosen as close as possible to LV03 and will be used for cadastral surveying and as the base for the national map projections. The second frame (CHTRS95/CHTRF95) is as close as possible to ETRS89 and is used in geodesy for national and international projects [Schneider et al. 1995].
Besides this, a new national height network (LHN95) is part of the new concept. It will be a modern kinematic network computed rigorously in a combined adjustment using all existing levelling and gravimetric data as well as the ellipsoidal heights from the GPS network and the new geoid of Switzerland. The new reference frames are designed to fulfill not only the requirements of cadastral surveying but also those of geodetic control for large engineering projects and international data transfer.

The new geoid of Switzerland.
An important prerequisite for the new height system is an accurate geoid and quasigeoid in order to take advantage of the economic and fast GPS techniques in height determination as well. Therefore, during the last several years the aim of a joint project of the Institute of Geodesy and Photogrammetry (ETH Zurich) and SWISSTOPO was to generate a national geoid and quasigeoid with an accuracy of a few cm over the whole country [Marti, 1997]. About 600 astronomically observed deflections of the vertical and 70 connections between GPS stations and levelled bench marks were used for the computations.
Furthermore, a 25-meter DTM and density models for crustal structures were introduced. After reducing the measurements from the effects of topographic masses, the residuals were interpolated by means of least square collocation. The final geoid was calculated by re-adding the effects of the mass models (remove-restore technique). The result revealed an accuracy of 2 to 3 cm in the flatter areas and of about 5 cm in the Alps.

The new National Height Network LHN95

Why a new height system?
The maintenance of vertical densification networks for cadastral surveying in the Alps using levelling techniques is extremely expensive. Using GPS to determine height would be a much more economic technique for this task. There is a large demand for combining or replacing low order levelling networks with GPS networks in mountainous regions. However, this is not possible with the existing height system LN02, which consists of levelled heights with no consideration of the gravity field and therefore it has no simple relation with the geoid or quasigeoid.
Furthermore, LN02 contains discrepancies of up to 20 cm stemming from errors in the old measurements and adjustment processes and because the effects of recent crustal movements were neglected. Therefore, for high-precision applications the official height systems in general cannot be used.
Another reason for introducing a new height system is the non-compatibility of LN02 with the systems of neighboring countries, making the exchange of data, as needed for several projects, rather difficult. All these considerations lead to the decision to establish the new Swiss National height system LHN95.

Definition of the new vertical datum of LHN95.
The existing Swiss vertical datum (LN02) is based on the height of the reference point "Repère Pierre du Niton (RPN)" in Geneva (373.600 m above sea level) which was introduced in 1902 and derived from a levelling connection to the tide gauge in Marseilles. On the other hand, the European vertical datum (UELN) is based on the tide gauge in Amsterdam. A comparison of the UELN results with a provisional rigorous adjustment of the Swiss first order levelling network revealed an offset of about -13 cm between the two systems (h_{UELN} - h_{LN02}).
Two vertical datums (analogous to the dual definition of the Swiss terrestrial reference system) are defined for LHN95: For both datums, the fundamental point in Zimmerwald is used as reference point. For the local vertical datum definition (CH1903+) the reference value of Zimmerwald is derived from the old (orthometric) height of the RPN, whereas for the global datum (CHTRS95) this value is derived from the geopotential number of a node (Olten) of the UELN.
The basic values of the bench mark heights of LHN95 are geopotential numbers derived from the kinematic network adjustment with the values of the fundamental station in Zimmerwald held fixed. For practical use, these geopotential numbers have to be transformed into either normal or orthometric heights. An orthometric height system was chosen for the official "local" system CH1903+ for cadastral surveying. This is in accordance with the new height system in Austria but in discordance with those of France, Germany and other European countries where normal heights were introduced as the official system. The discussion which lead to this decision is summarized as follows:

- The advantages of normal heights are that they are very easy to calculate and that it is not necessary to make any assumptions on the density distribution in the interior of the earth. Furthermore, the normal heights are in better accordance with levelled heights than orthometric heights.
- The advantages of orthometric heights are only evident in mountainous regions such as the Alps. There the geoid shows a much smoother behavior than the quasigeoid, which is advantageous for the interpolation of geoid undulations. Therefore, it is also easier to derive precise orthometric heights from GPS heights.

Differences between the old (LN02) and new (LHN95) reference frame.
In order to make the transition from the existing levelled heights of LN02 to the orthometric heights of LHN95 the differences between the 2 frames have to be known with a high accuracy. First test computations revealed that these differences reach several decimeters and are strongly correlated with height. Therefore, the first functional approach for the transition from LN02 to LHN95 is a linear function in height. In this way, the residuals can be reduced to less than 5 cm for most bench marks. The residual height differences which are caused by influences such as distortions of the old levelling network, kinematics or the influence of gravity can be modelled by a two-dimensional interpolation. With this model the differences between LN02 and LHN95 can be predicted with an accuracy of 1 to 2 cm, which is enough for low order levelling bench marks and triangulation points.

Combination of levelling networks with GPS.

An important option of LHN95 is the integration of selected GPS-derived heights into the adjustments with the aim of stabilizing the levelling network over long distances. The main problem with this combination until now has been the limited accuracy of the height components of relative GPS positions due to systematic errors caused by tropospheric refraction etc. To contribute a significant improvement of the solution, the accuracy of the height component should be in the order of 1 to 2 cm over a distance of 100 km. But the present solution of the GPS network LV95 shows a relative height accuracy of only 2 to 5 cm, especially in the mountainous part of the country with height differences of up to almost 3000 m between adjacent points. Further improvements are expected from water vapor radiometer measurements and from a better modeling of the troposphere.

Another problem of introducing GPS heights into the adjustment of levelling networks is the fact that orthometric heights or potential differences must be introduced instead of the observed ellipsoidal heights. Therefore, the geoid also has to be known with the same accuracy as the GPS heights.

Kinematic adjustment of LHN95.

The adjustment of all first order levelling data measured during this century is performed in a kinematic way. This means that not only the last measurements are of interest but also the older observations have to be taken into account. Because of the large amount of observations, the adjustment process cannot be performed in a single step but is done in the following sequence:

1. Separate calculation of 'observed' geopotential differences for each levelling line by integrating the product of the observed height differences and the observed or interpolated gravity values.
2. A kinematic adjustment of the 'main points' of the whole network. These are stable markers which have been observed more than once and therefore represent the regional vertical kinematics. This step leads to geopotential numbers and their time derivatives ("velocities") as well as to the variance-covariance information for these main points.
3. The results of step 2 are kept fixed, and the potentials and - where possible - "velocities" of the other stations are now determined by further interpolation.
4. Finally, the orthometric or normal heights are obtained by dividing the geopotential numbers by the mean gravity (or the mean normal gravity) along the plumb line. The mean gravity itself is determined from the observed gravity using the same models for the density distribution of the earth's crust that were used for computing the geoid.

This stepwise determination of heights does not affect the results significantly since the obtained heights differ less than 1 mm from a rigorous solution, as could be confirmed with test calculations. The main problems with this procedure are caused by unstable markers which were observed more than twice and therefore are not consistent with the mathematical model of linear movement in time. And since the degree of freedom of levelling networks usually is very small, the a posteriori variance factor of the least squares solution increases too much if these constraints are not eliminated from the adjustment.

Test computations in the central Alps

A first test of this concept of evaluation was made in an area of the central Alps. This region is most appropriate for testing because of the large variations in the gravity field and the relatively large changes in the vertical velocities ($|dv_z| > 1$ mm/year) of the uppermost crust. The data set used consists of 628 km of levelling lines with a height ranging from 220 m to 2400 m above sea level. All lines were observed at least twice. More than 5000 bench marks are included in the evaluation but only 1200 of those still exist. Out of these, only 780 bench marks were observed more than once, thus allowing the determination of relative vertical velocities.

Until now GPS observations have not been included in the adjustment, however, a comparison of the results between levelled orthometric heights and heights derived from GPS and the new geoid was possible for 15 stations. This comparison showed average differences of ± 2 cm and a maximum value of 7.5 cm for one station.

The main results obtained are potentials, orthometric heights and normal heights in the chosen reference system with their standard deviations as well as vertical velocities (relative to a hypothetical reference station in Aarburg). These results are shown in Fig. 1.

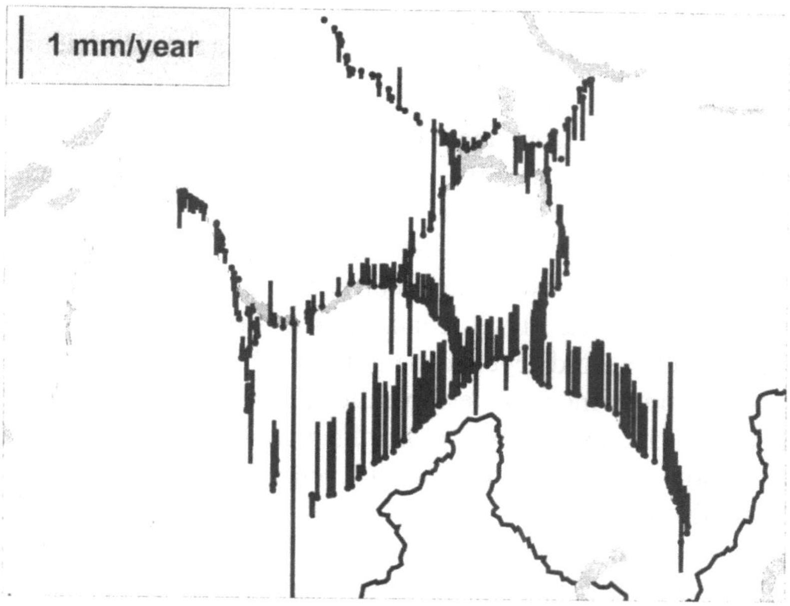

Fig. 1: Vertical movements relative to a marker in Aarburg

The vertical movements show a rather smooth distribution except for unstable markers. In the flatter areas the calculated movements are of no significance. On the other hand, the velocities in the Alpine regions with a maximum of 1.6 mm/year (relative to Aarburg) are highly significant on the 95% level of confidence.

Comparison of obtained results

The results obtained on a part of the test network (the Grimsel line between Balmhof and Gletsch) are now being discussed in more detail. This is a levelling line with a length of 60 km and height differences of up to 1600 meters (Fig. 2). It was observed twice, once in 1920 and again in 1985.

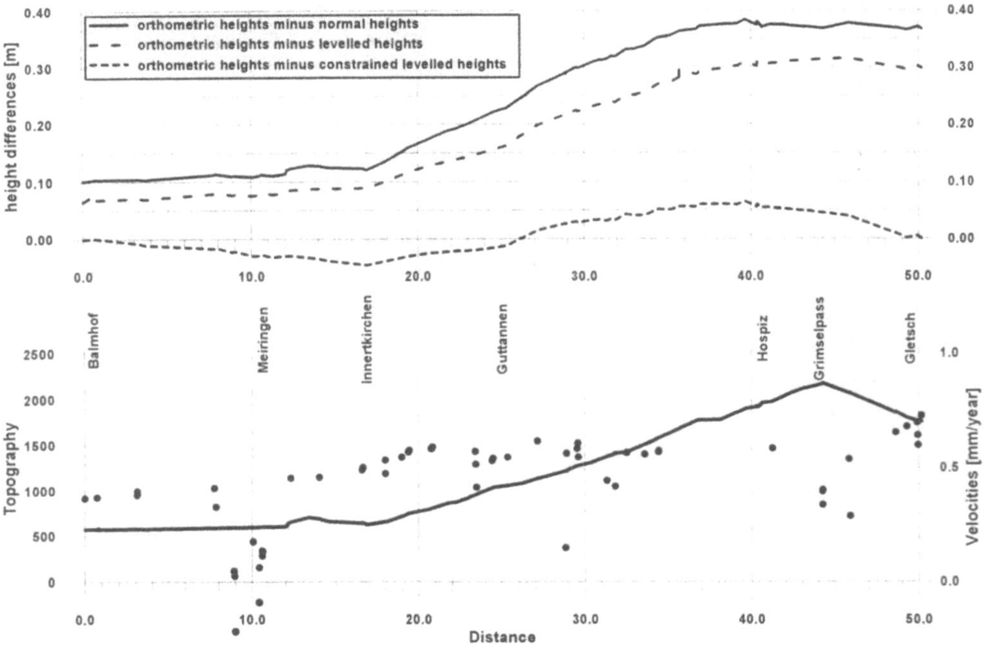

Fig. 2: Comparison of different solutions relative to the orthometric heights

In addition to the topography, the lower part of Fig. 2 depicts the calculated velocities as dots. These velocities increase from about 0.35 mm/year in Balmhof to about 0.7 mm/year in Gletsch. They show a rather smooth behavior except for some unstable points and a whole region near Meiringen which do not follow the general trend. In the top solid curve of Fig. 2, the differences between the determined orthometric and normal heights are obvious. This curve shows a rather high correlation with the topography and reaches maximal values of almost 40 cm. The second, dashed curve shows the differences between orthometric heights and the old levelled heights of LN02. This curve is also correlated with the topography and is more or less parallel to the first curve, which indicates that levelled heights are much closer to normal heights than to orthometric heights. This curve, however, contains not only the systematic differences between orthometric heights and normal heights but also the kinematic modeling and the distortions of the old system. The third curve finally shows the differences between the rigorous orthometric system and a result as it would probably be obtained by a local surveyor, regardless of gravity measurements: The starting and ending stations are kept fixed with their orthometric heights and the resulting discrepancy is distributed linearly. This leads to differences from -4 to +6 cm on this line with its extreme topography.

References

Gubler E., Kahle H.-G., Klingelé E., Müller St. and Olivier R. 1981: Recent Crustal Movements in Switzerland and their Geophysical Interpretation. Tectonophysics Vol.71.

Marti U. 1997: Integrierte Geoidbestimmung in der Schweiz: Diss. ETH Zürich 12015.

Schneider D., Gubler E., Marti U. and Gurtner W. 1995: "Aufbau der neuen Landesvermessung der Schweiz 'LV95': Teil 3: Terrestrische Bezugssysteme und Bezugsrahmen". Berichte aus der L+T Nr. 8, Wabern.

THE NON-FIDUCIAL APPROACH APPLIED TO GPS NETWORKS

João Francisco Galera Monico
FCT/UNESP CEP: 19060-900 Pres. Prudente, SP, Brazil. e-mail: galera@stetnet.com.br

Edvaldo Simões da Fonseca Junior
Denizar Blitzkow
EPUSP-PTR - Caixa Postal 61548 CEP: 05424-970 São Paulo, SP, Brazil

ABSTRACT

The basic fundaments of the fiducial and the non-fiducial approaches are presented. The main advantage of the non-fiducial approach over the fiducial one is the simplicity of computation when the choice of the known stations change. In the traditional fiducial technique all main steps of the processing must be repeated. An example involving the São Paulo State Network, which is composed of 24 stations (plus 3 fiducial stations) is presented. Results obtained from both approaches are quite similar.

INTRODUCTION

The adjustment of a GPS network can either be carried out by using one of the available ephemerides, or by estimating the satellite orbit (state vector) during the processing. At both cases, it is usual to constrain the coordinates of a number of stations to their known values. This approach is referred to as fiducial GPS. In the non-fiducial one, all station coordinates are estimated instead, as well as the satellites state vectors. Such approach has been applied to global GPS experiments, in which fiducial stations are not necessary to provide an origin and scale. This approach may also be suitable for regional networks, but in such a case, instead of estimating the satellites state vectors, one can use an IGS (International GPS Geodynamics Service) precise ephemerides. In order to make the non-fiducial approach solution consistent with that of the fiducial one (both using precise ephemeris), some extra computation should be carried out. The main advantage of the non-fiducial approach over the fiducial one is the simplicity of computation when the choice of the known stations change. In the traditional fiducial technique all main steps of the processing must be repeated, while in the non-fiducial approach only a transformation is required. Therefore, if both approaches provide similar results, it is advisable to make use of the non-fiducial one.

In this paper, the basic fundaments of both approaches will be presented, followed by a description of the software used and an example involving the São Paulo State Network, which is composed of 24 stations (plus 3 fiducial stations). This network was integrated to the ITRF-93 (International Terrestrial Reference Frame 1993) by using the fiducial technique in conjunction with the IGS precise ephemerides (Fonseca, 1996). The results of both approaches will be compared.

THE FIDUCIAL AND NON-FIDUCIAL APPROCHES

The Fiducial Technique

The fiducial GPS technique involves the observation of simultaneous (dual frequency) carrier phase measurements at a network of fiducial stations, whose coordinates are known to a very high order accuracy in a global reference frame, and a number of new stations whose coordinates are required. Full details of the method are given by Ashkenazi et al (1990), and only a brief outline is given here.

The fiducial process involves the computation of a theoretical (integrated) orbit for each of the satellites, based on the forces acting on it. The forces modeled include gravitational attraction (of the moon, sun and the planets, and earth and ocean tides), surface forces (e.g. solar radiation) and other perturbing influences. The theoretical orbit for each satellite is computed by numerical integration of the acceleration vector, starting from a initial state vector (3-d position and velocities) obtained from a previous broadcast or precise ephemeris. These theoretical orbits are improved during the network adjustment. The complete set of carrier phase measurements from all the stations is adjusted by least squares, holding the 3-d coordinates of a minimum of three fiducial stations fixed, and solving for the 3-d coordinates of the new stations, as well as for the corrections to the satellite orbits and other bias parameters (ambiguities, tropospheric scale factors, y-bias, etc.). The resulting adjusted network of ground stations and satellite orbits is positioned, scaled and oriented to the reference frame defined by the adopted coordinates of the fiducial stations held fixed. The high accuracy of the coordinates of the fiducial stations are then transferred to the new stations, via the satellite orbits.

With the establishment of the IGS in 1990 by the International Association of Geodesy (IAG), high quality GPS orbits and Earth rotation parameters of high resolution became available to the scientific community, allowing to expand geographically the current International Terrestrial Reference Frame maintained by the International Earth Rotation Service (IERS) (Mueller, 1993). The precise ephemerides produced by IGS may be used in the fiducial GPS technique, simplifying the approach. Instead of estimating the satellite orbits, one can use the IGS ones and only the coordinates of the new stations have to be estimated. Additionally, IGS can effectively provide fiducial stations coordinates at the epoch of a GPS campaign. For example, such coordinates may be obtained from the most recent realization of ITRF. In the context of this paper, the satellite orbits are not estimated in the GPS fiducial technique. Instead of, the IGS precise ephemerides are used.

The Non - Fiducial Technique

The main difference between the non-fiducial and the fiducial technique is that in the former one, besides the satellites state vectors and other bias, all station coordinates are estimated in the network processing. The non-fiducial technique has been applied to global GPS experiments, in which fiducial stations are not necessary to provide an origin and scale (Heflin *et al*, 1992; Blewitt *et al*, 1992). They are provided by the satellite force model, radio propagation model, and GPS data. This approach may also be suitable for regional networks, but in such a case, instead of estimating the satellites state vectors, it is advisable to use an IGS precise ephemerides. The adjusted network will be referenced to the datum realised by the precise ephemerides.

In order to make the results consistent with those of the fiducial approach, a transformation to the reference frame of the stations used as fiducial has to be carried out. It involves a similarity transformation as follows:

$$\begin{bmatrix} x \\ y \\ z \end{bmatrix}_{N-F} = \begin{bmatrix} X \\ Y \\ Z \end{bmatrix}_{ITRF} + \begin{bmatrix} t_x \\ t_y \\ t_z \end{bmatrix} + \begin{bmatrix} s & -\Theta_z & \Theta_y \\ \Theta_z & s & -\Theta_x \\ -\Theta_y & \Theta_x & s \end{bmatrix} \begin{bmatrix} X \\ Y \\ Z \end{bmatrix}_{ITRF} \qquad (01)$$

where **x, y** and **z** are the Cartesian coordinates derived from the non-fiducial (N-F) network;

X, Y and **Z** are ITRF Cartesian coordinates of the fiducial stations mapped to the time of the campaign;

$\mathbf{t_x}$, $\mathbf{t_y}$ and $\mathbf{t_z}$ represent the offsets in origin;

s represents a difference in scale and

$\mathbf{\Theta_x}$,$\mathbf{\Theta_y}$ and $\mathbf{\Theta_z}$ represent differences in orientation.

In order to apply the transformation given by Equation (01), only a minimum of seven coordinates values (i.e. minimum of three fiducial stations) has to be available. There are also options of estimating less than seven parameters. Usually, only the translation parameters (t_x, t_y and t_z) are estimated.

Once the transformation parameters have been estimated, those stations not included in the transformation may be transformed to the required reference frame. The transformation is the last step involved in the non-fiducial network approach. It means that only this step needs be carried out if the choice of the fiducial stations changes. This is an advantage over the fiducial technique, where all the main steps must be recomputed.

PROCESSING SOFTWARES

GPS Analysis Software (GAS)

The data was processed using GPS Analysis Software (GAS) developed at the University of Nottingham (Stewart *et al*, 1994). GAS process either double-difference carrier phase measurements or pseudo-ranges, formed with different base satellites for each baseline processed. This provides a capability of processing large GPS network, even a global one.

The GPS network processing by GAS can be carried out either with the full fiducial technique (Ashkenazi et al, 1990) or by using the broadcast or the precise ephemerides for the satellites. In all cases, corrections for the effects of the Earth body tide (EBT) and Ocean tide loading (OTL) can be included in the processing. For the former case, the model recommended in IAG standards was used (McCarthy, 1992). The effects of the ionosphere can be greatly reduced by using the ionospheric free observable (Ffoulkes-Jones, 1990). For the tropospheric delay, GAS provides a series of modeling options. For example, one could define zenithal scale factors as unknown parameters. GAS also provides the option of 'fixing' the double-difference carrier phase integer ambiguities to their integer values. This is achieved by one of the two different techniques, namely the ambiguity search technique and the sequential bias fixing (Ffoulkes-Jones, 1990).

The non-fiducial technique can easily be applied by using GAS. At such a case, no fiducial stations are selected and all stations coordinates are estimated in the adjustment. As the model has full rank, a solution will be obtained using conventional algebra.

Cartesian Network Adjustment Program (CARNET)

CARNET enables the integration of several types of conventional observations with GPS and other space geodetic techniques. These observations can be adjusted together, providing a 3-dimensional solution (Lowe, 1994). Several options can be invoked during the network adjustment, e.g. the estimation of transformation parameters or their application once they have been determined.

The Cartesian coordinates or Cartesian coordinate differences estimated in the network adjustment using GAS from several sessions or days can be combined together to produce a final solution using CARNET. The final result is a set of weighted mean values. The main output of CARNET is a list of Cartesian and geodetic coordinates with the respective estimated standard deviations as well as information related to the quality of the processing. CARNET was used in this work to combine all daily GPS solutions.

DATA SET AND PROCESSING STRATEGIES

São Paulo State GPS Network

The data set used refers to the São Paulo State Network, which is composed of 24 stations, plus station Chuá, the origin of the Brazilian Geodetic System. The data was collected during two main campaigns. The first one was carried out during the period of 7 to 18 March 1994 , using 7 dual frequency receivers. Each session was 8 hours long (8:00 a.m. to 17:00 p.m. - local time) with 15 seconds sample rate. A total of 18 stations was occupied. The second campaign took place during the period of 21 November to 3 December of 1994, using 5 (dual frequency) receivers. The duration of each session of this campaign was reduced to 6 hours (8:30 a.m. to 14:30 p.m.), also with 15 seconds sample rate. The number of occupied stations was 18.

In order to connect the network to a global reference frame, the data of 3 IGS stations was transferred via Internet and introduced in the processing. They are the stations KOURU, SANTIGO AND FORTALEZA, which are equipped with IGS standard receivers. Figure 1 illustrates the São Paulo State network with the 3 fiducial stations.

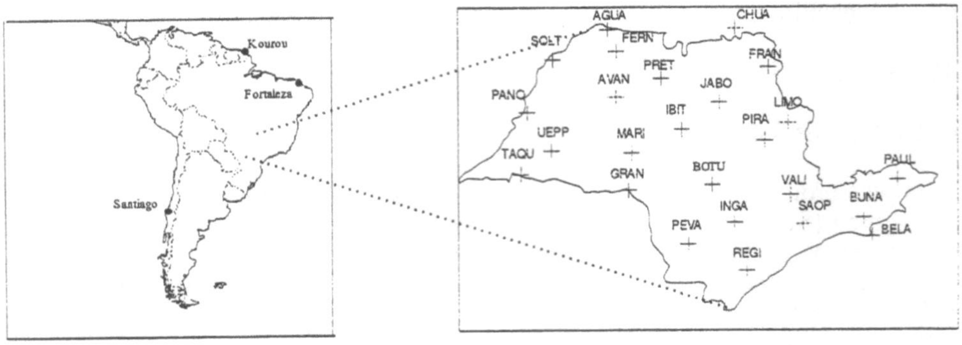

Figure 1: São Paulo State Network with Connection to the IGS Stations

Table 1 summarises the data set used in the processing. Full details can be found in Fonseca (1996).

Table 1: Information of the Data Set Involved in the Processing

	Day of Year (1994)															
Stations	069	070	071	072	073	074	075	076	327	328	329	332	333	334	335	337
CHUA				XX	XX	XX										XX
AGUA				XX	XX	XX										
PRET				XX	XX	XX										
JABO				XX	XX	XX										
PIRA	XX	XX	XX	XX												
LIMO		XX	XX	XX												
FRAN		XX	XX	XX												
FERN						XX	XX	XX								
AVAN						XX	XX	XX								
IBIT						XX	XX	XX					XX	XX		
UEPP								XX	XX	XX	XX	XX	XX	XX	XX	XX
TAQU								XX							XX	
GRAN								XX				XX	XX		XX	
SOLT								XX						XX		
SAOP	XX	XX								XX						XX
BELA	XX	XX							XX							
BUNA	XX	XX							XX							
VALI	XX	XX							XX	XX	XX					
PAUL									XX	XX						
REGI										XX						
INGA											XX	XX				
PEVA											XX	XX				
BOTU												XX	XX			
MARI													XX	XX	XX	
PANO														XX	XX	
FORT(*)		XX	XX	XX	XX	XX		XX	XX	XX	XX	XX	XX	XX	XX	XX
KOUR(*)		XX	XX	XX	XX	XX		XX	XX	XX	XX	XX	XX	XX	XX	XX
SANT(*)		XX	XX	XX	XX	XX		XX								

(*) IGS Stations with 24 hours of data (30 seconds sample rate)

Processing Strategies

Several processing strategies were carried out using the São Paulo State Network data set. Details of each processing can be found in Fonseca, 1996. For the purpose of this paper, the fiducial GPS network processing will be taken into account. In this processing, IGS precise ephemerides were used and the coordinates of the three IGS stations were held fix to the ITRF93. Earth Body Tide corrections based on the IERS standards were applied to the observations (MaCarthy, 1992). The basic observable was the double difference ionospheric free without solution (integer) to the ambiguities and a scale factor was estimated in order to improve the tropospheric refraction model.

For the non-fiducial solution, the same models described above was used, unless of the station coordinates, which were all estimated in the adjustment. In order to make the results consistent, a transformation to the reference frame defined by the three fiducial stations used in the fiducial approach was carried out. Only the 3 translations parameters were estimated.

RESULTS AND DISCUSSIONS

The results obtained are illustrated in terms of repeatabilities, i.e. rms (root mean square) of the differences from the mean. The mean was obtained from all daily solutions, using CARNET software. The repeatability shows the internal consistency of the results, i.e. the precision.

Figure 2 illustrates the repeatabilities of the fiducial and the non-fiducial solutions. In the both cases, the precision in Latitude, Longitude and Height are of the order of 1, 3.5 and 5 cm respectively.

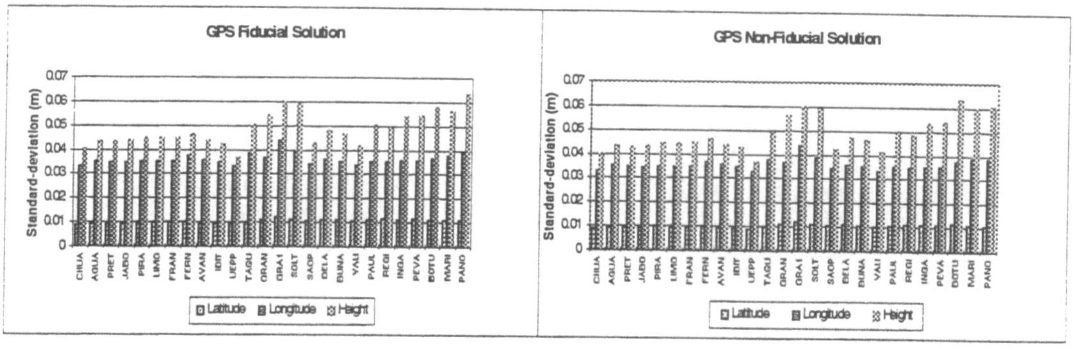

Figure 2: Precision of the GPS Fiducial and Non-Fiducial Solutions

As expected, the Height components provided the worse results, whilst the Latitude components the best ones. From these results one can conclude that both approaches provide similar results in terms of precision. In addition, the stations of São Paulo State network were positioned with resulting precision better than 8 cm (1 sigma).

Figure 3 illustrates the discrepancies between both solutions for the X, Y, Z and h (Height) coordinates components.

The maximum value reached 3.5 cm, in the Y component of station BOTU; almost three times the average discrepancies of the other stations. There is not any clear explanation for such discrepancy at this station. However, this station also provided the worse results in terms of precision. Taking into account the precision of the Y component at both solutions (approximately 5 cm), this discrepancy is within an expected range. On the average, the discrepancies for each of the components are smaller than 1 cm, and the precision smaller than 5 cm.

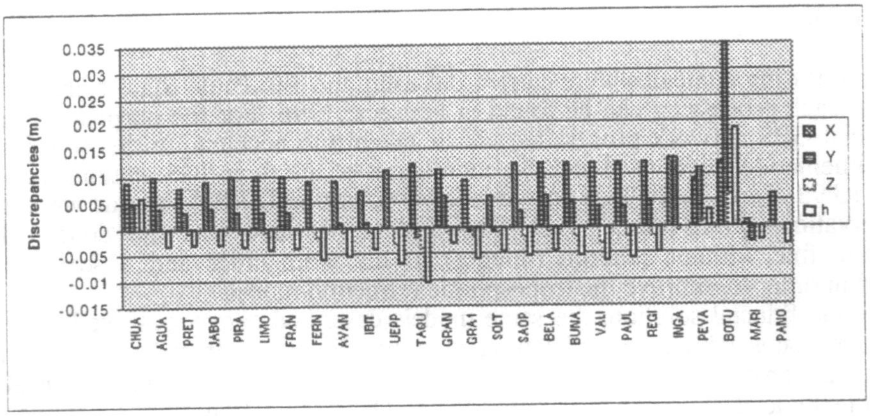

Figure 3: Discrepancies in the X, Y, Z and h components

CONCLUSIONS

The discrepancies between the results of the fiducial and the non-fiducial GPS solutions can be considered statistically insignificant, if one take into account the precision of both solutions, which agree quite well. Therefore, both approaches can be applied in the adjustment of GPS networks. However, the non-fiducial approach provides a more advantageous way of integrating GPS networks. In this method, if one wants to change fiducial stations in order to perform some tests, only a transformation needs be carried out. Therefore, in such cases it is advisable to apply the non-fiducial approach.

REFERENCES

Ashkenazi, V., Moore, T., Ffoulkes-Jones, G.H., Whalley, S. and Aquino M. (1990). High Precision GPS Positioning by Fiducial Technique. In: Bock Y., Leppard N. (eds): Global Positioning System: An Overview, ***Springer, New York, Berlin, Heidelberg, Lond, Paris, Tokyo, Hong Kong, pp 195-202, [Mueller I.I. (ed): IAG Symposia 102].***

Blewitt G., Heflin M. B., Webb F. H., Lindqwister U. J. and Malla R. P. (1992). Global Coordinates with Centimeter Accuracy in the International Terrestrial Reference Frame Using GPS, ***Geophysical Research Letters***, Vol. 19, No.9, pp 853-856.

Ffoukes-Jones G.H. (1990) ***High Precision GPS by Fiducial Techniques***, PhD Thesis, University of Nottingham.

Fonseca E. S. Junior (1996) ***Estudo e Avaliação Metodológica da Rede GPS do Estado de São Paulo***, MSc Thesis, Universidade de São Paulo, Escola Politécnica.

Heflin M., Bertiger W., Blewitt G., Freedman A., Hurst K., Lichten S., Lindqwister V., Vigue Y., Weeb F., Yurck T. and Zumberge J. (1992b) Global Geodesy Using GPS without Fiducial Sites, ***Geophysical Research Letters***, Vol. 19, NO. 2, pp 131-134, January, 24-1992.

Lowe D. P. (1994) ***CARNET User Guide***. Version 3.13. IESSG Publication, University of Nottingham.

McCarthy D. D. (1992) IERS Standards (1992), ***IERS Technical Note 13***, Central Bureau of IERS- Observatoire de Paris

Mueller I. I. (1993) The International GPS Service for Geodynamics: An Introduction. ***Proceedings of the 1993 IGS Workshop***, pp 1-2, Ed. by G.Beutler and E. Brockmann, Univ. of Berne, 1993.

Stewart M. P., Ffoulkes-Jones G. H. and Ochieng W. Y. (1994) ***GPS Analysis Software (GAS) Version 2.2 User Manual***, IESSG Publication, University of Nottingham, UK.

ACKNOWLEDGEMENTS

The first author has financial support of CNPq, the Brazilian Research Council, and FAPESP, the São Paulo State Research Foundation.

THE STATEFIX WEST AUSTRALIAN GPS NETWORK

M P Stewart
School of Spatial Sciences, Curtin University of Technology, Perth, Western Australia

H Houghton
Department of Land Administration, Western Australia

X Ding
Department of Land Surveying and Geoinfomatics, Hong Kong Polytechnic University

ABSTRACT

The STATEFIX project, commissioned and managed by the West Australian Department of Land Administration, represents the densification of the Australian National Network in West Australia to an average spacing of approximately 200km. Some 230 dual frequency baselines (mean baseline length 202km) were observed between March 1996 and November 1996. Data processing and network adjustment were undertaken at Curtin University of Technology. Final results indicate the network is accurate (relative to the ANN) to better than 2cm in the horizontal component (95% confidence), and 6cm in the vertical component (95% confidence). The STATEFIX project has been particularly challenging in terms of the vast area surveyed (over 2,500,000km^2) and the isolated nature of many control points, located in uninhabited desert areas. This paper will outline STATEFIX observation, processing and network adjustment strategies, and describe the precision and accuracy analysis performed for the network.

INTRODUCTION

West Australia is one of the largest yet most sparsely populated states in the world and the establishment a state geodetic control network has been an enormous challenge to surveyors since settlement by Europeans in the early nineteenth century. The STATEFIX project was conceived by the West Australian Department of Land Administration (DOLA) ten years ago as part of an Australia-wide policy to improve geodetic control using satellite-based positioning technology and aid the transition from the existing Australian Geodetic Datum (AGD84) to a geocentric datum, the Geocentric Datum of Australia 1994 ($GDA_{1994.0}$).

STATEFIX uses existing points of the Australian National Network (ANN) as a control framework. The ANN itself represents a densification of the Australian Fiducial Network (AFN) which consists of eight permanent, continuously operating, Rogue GPS receivers on the Australian mainland and Tasmania (Manning and Harvey, 1992). The network was initially observed during the International GPS Service for Geodynamics (IGS) Epoch '92 campaign, July to August, 1992. The AFN, in conjunction with six additional sites beyond the Australian mainland, forms the Australian Regional GPS Network (ARGN). AFN station co-ordinates are based on the ITRF92 at epoch 1994.0, and are estimated to have a precision of a few centimetres (2-4 parts in 10^8).

The Australian National Network (ANN) consists of 78 GPS campaign points spaced at approximately 500 km intervals across Australia. This network was observed between 1992 and 1994; the first GPS observation campaign being conducted during the IGS Epoch '92 campaign and followed by a further nine days of observations during August-September, 1993. It is estimated that the horizontal and vertical precisions of the co-ordinates at the 95 percent confidence level are better than three centimetres and five centimetres respectively (Morgan *et al*, 1996). Figure 1 highlights the location of AFN and AFN stations within Australia.

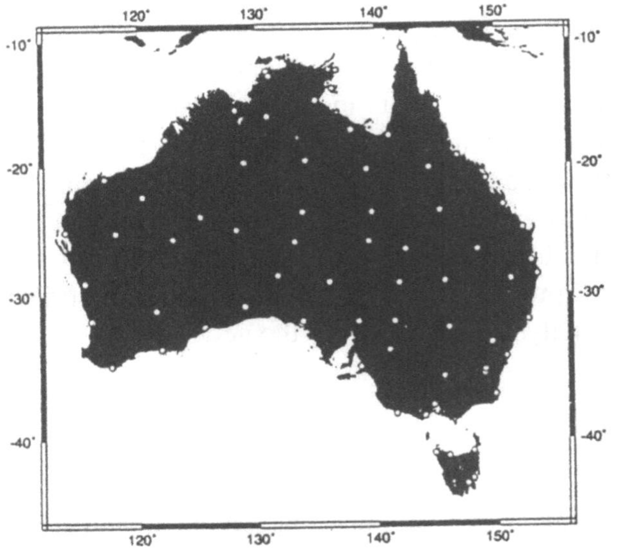

Figure 1 Australian Fiducial Network (AFN) and Australian National Network (ANN)

The STATEFIX project was conceived ten years ago but only executed in 1996. Several reasons were used explain this delay. By 1996 technological advances in GPSreceiver hardware and software, and greater availability of equipment and expertise, were anticipated to ensure the objectives of the project could be easily achieved. The specification set by DOLA (DOLA, 1996) is an absolute accuracy of 5cm (95% confidence) in X, Y, Z and latitude, longitude and height. Furthermore, minimum ionospheric disturbance due to the low in 1996 of the eleven year sunspot cycle was deemed important to ensure success when processing the long baselines required by necessity for this project.

This paper outlines STATEFIX observation, processing and network adjustment strategies, discusses precision and accuracy analysis performed for the network, and offers recommendations for the future densification of regional networks

STATEFIX NETWORK DEFINITION AND DATA COLLECTION

The full STATEFIX network is illustrated in figure 2a. Baseline lengths range from 45km to distances longer than 450km. In total, 228 baselines have been observed to 119 stations. The mean baseline length is 200km.

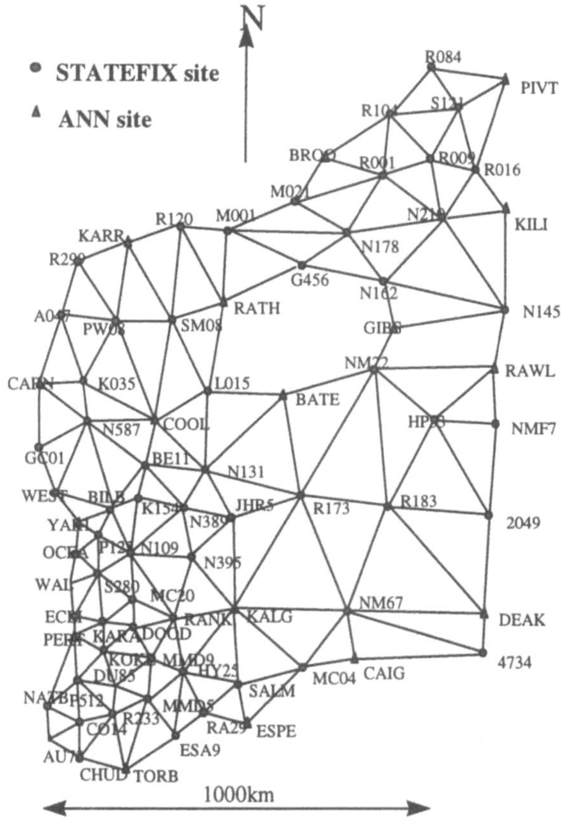

Figure 2a The STATEFIX Network

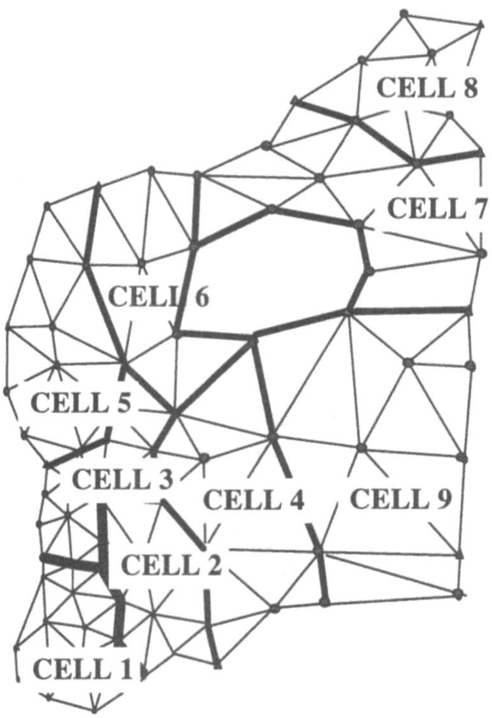

Figure 2b Diagram showing division of STATEFIX Network into 9 cells

For the purposes of observation and processing, the STATEFIX network was divided into 9 individual cells (figure 2b). Because of the size of the STATEFIX network, cells were observed by different private contractors over different time periods in 1996. Baselines of less than 250km in length were observed for a minimum of 12 hours, baselines longer than 250km in length were observed for a minimum of 16 hours. The observation interval was 30 seconds for all baselines. A minimum of three receivers were in use for each cell simultaneously. In total, 198 baselines were observed, with 30 of these being measured twice as they constituted cell boundaries. Several other repeat baselines were observed within individual cells.

All cells, with the exception of cell 3, were observed using Trimble 4000 SSI dual frequency GPS receivers. Cell 3 was observed with Ashtech Z-12 dual frequency GPS receivers. Cells 3 and 5 also incorporated data from the IGS permanent Rogue SNR-8 receiver at Yaragadee (YAR1), whilst cells 5 and 6 incorporated data from a similar receiver (though different antenna) at Karratha (KARR), which is part of the Australian Fiducial Network (AFN).

DATA PROCESSING AND ADJUSTMENT

Baselines from each cell were processed at Curtin University using the in-house SWAG (South West Australian GPS software) GPS processing software suite. The following data processing procedures/models were applied :

a) IGS Precise Orbits:
   IGS precise orbits were held fixed in the processing of all cells.

b) Earth Body Tide Modelling:
   The Wahr two-step Earth body tide model was applied, as per IERS standards (McCarthy 1992) in the processing of all cells.

c) Antenna phase centre modelling:
   The IGS PCV_01 antenna phase centre models (Rothacher and Mader 1996), referenced to the Dorne Margolin T chamber test values of Schupler (1994), were applied.

d) Atmospheric Modelling:
   Processing was undertaken using the ionosphere free linear combination ($\Phi_{IF} = 2.546\Phi_{L1} - 1.984\Phi_{L2}$). Corrections to the zenithal delay tropospheric model parameters in the form of scaling and second order rate of change parameters were estimated as part of the baseline solutions.

e) Ambiguity Resolution:
   Ambiguity resolution was attempted on all baselines using a combination of the wide lane observable and the ionospheric free observable. Ambiguities were successfully resolved on all STATEFIX baselines with the exception of one baseline in cell 8.

f) Length of Data Span:
   The smaller cells, 1, 2 and 3, were processed by dividing the observation time spans into two independent sessions and computing two solutions for each baseline. Cells 4 - 9 were processed using the full 16 hour session.

g) Rigorous Adjustment of Simultaneously Observed Baselines:
   All simultaneously observed baselines were rigorously adjusted taking baseline correlations into account. The resultant baseline vectors and full multi-baseline covariance matrix for each day were then adjusted into the single cell solutions or full STATEFIX network solution.

h) Individual Cell Adjustment
   One ANN point was held fixed in each cell and coordinates of other ANN stations within the cell were computed and compared with their ANN coordinates as an accuracy estimation. Internal cell precision was also studied. Full results and analysis from the individual cell adjustments can be found Stewart *et al* (1997).

i) Full Network Adjustment Strategy:

The full STATEFIX network was adjusted using GEOLAB version 2.4, constraining all ANN station coordinates to 5cm in height and 3cm in plan (95% confidence).

4. RESULTS

i) Relative Errors

Figures 3a and 3b illustrate the relative errors from the STATEFIX network solution at a 95% confidence level. Both the relative horizontal and height errors exhibit a baseline length dependency with error of 0.01 - 0.04 ppm in the horizontal and approximately 0.2ppm in height. Relative horizontal errors are all better than 1cm. Relative height errors are better than 6cm with the exception of baselines connected to site R084 in cell 8, which demonstrate uncertainties of between 7 and 8 cm.

ii) ANN coordinate recovery

In addition to the full solution, a series of further adjustments were conducted which were identical to the full solution except that one ANN point was left free, allowing a direct comparison of STATEFIX coordinates with ANN 'ground truth' coordinates. The results from the coordinate recovery tests are presented in table 1 overleaf. The ANN site coordinates from the additional network adjustments are recovered to better than 2.5cm (±1-2cm) in all components. The largest error is 5.5cm, in the X component at CARN and the overall recovery of DEAK.

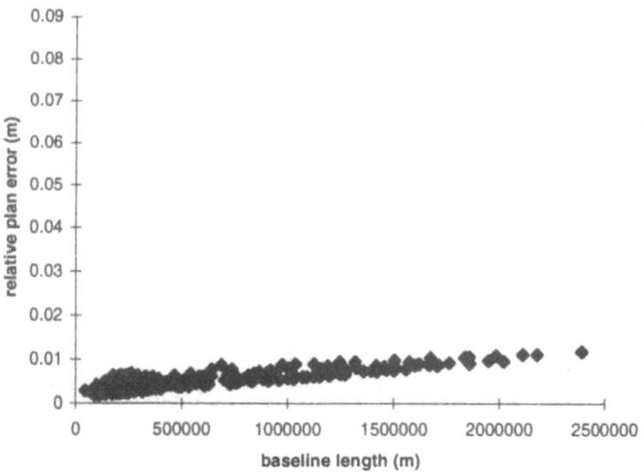

Figure 3a Relative Plan Error for STATEFIX Network Solution

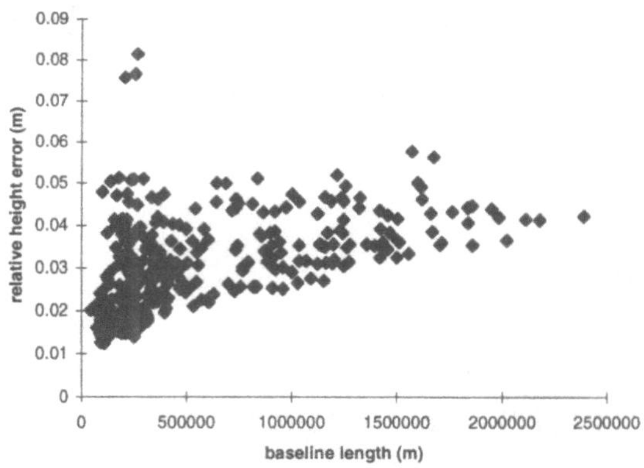

Figure 3b Relative Height Error for STATEFIX Network Solution

SUMMARY

The precision of the adjusted STATEFIX station coordinates within the ANN is better than 1cm in the horizontal and 6cm in height (95% confidence). ANN station recovery tests retrieved ANN coordinates with an accuracy of better than 2.5cm ± 1-2cm. The relative precision of the STATEFIX network is better than 0.04ppm in the horizontal and 0.2ppm in height. Given that the published formal error on the ANN, the reference framework for STATEFIX, is 3cm in the horizontal and 5cm in height (95% confidence), the accuracy of the STATEFIX network can be said to be of a similar order of magnitude to the ANN.

The STATEFIX network has illustrated that given adequate planning, high precision GPS control networks can be established over large, remote regions. The success of the project can be mainly attributed to the existing geodetic control infrastructure offered by the Australian National Network and the long observation spans which ensured adequate data were available to resolve ambiguities, even on long baselines.

Table 1 Recovery of ANN Coordinates from STATEFIX Network Solutions

ANN station	*recovery X (m)*	*recovery Y (m)*	*recovery Z (m)*
BATE	0.005	0.015	0.019
CAIG	0.007	0.037	0.002
CARN	0.055	0.031	0.013
COOL	0.010	0.026	0.007
DEAK	0.056	0.040	0.060
ESPE	0.002	0.020	0.020
KILI	0.010	0.023	0.025
PER2	0.001	0.007	0.013
PIVT	0.012	0.010	0.034
RATH	0.020	0.023	0.005
YAR1	0.025	0.038	0.015
mean	0.018	0.020	0.019
standard deviation	0.019	0.011	0.016

References

McCarthy, D D (ed.), 1996. *IERS Conventions*. IERS Technical Note 21, Central Bureau of IERS - Observatoire de Paris. France.

Morgan P, Y Bock, R Coleman, P Feng, D Garrard, G Johnston, G Luton, B McDowall, M Pearse, C Rizos and R Tiesler, 1996. A Zero Order GPS Network for the Australian Region. *Report ISE TR 96/60*, University of Canberra.183pp

Rothacher, M, and G Mader, 1996. Combination of Antenna Phase Centre Offsets and Variations. Antenna Calibration Set: IGS_01. *IGS Central Bureau (http://igscb.jpl.nasa.gov)*

Schupler, B R, R L Allshouse and T A Clark, 1994. Signal characteristics of GPS user antennas, *Navigation*, 41(3), pp277-295.

Stewart, M P, X Ding, M Tsakiri and W E Featherstone, 1997. 1996 STATEFIX project, final report. *West Australian Department of Land Administration Internal Report.* 47pp (available on request).

POLREF-96
THE NEW GEODETIC REFERENCE FRAME FOR POLAND

J.B. Zieliński, A.Łyszkowicz, L.Jaworski, A.Świątek, R.Zdunek
Space Research Centre, PAS, Department of Planetary Geodesy
00-716 Warsaw, Bartycka 18A, Poland

S.Gelo
Head Office of Geodesy and Cartography
00-926 Warsaw, Wspólna 2, Poland

ABSTRACT. The principles of the new National System of Geodetic Coordinates are the result of analyses carried out by the study group of the Section of Geodetic Networks and the Section of Cartography Geodetic Committee of the Polish Academy of Sciences. As the basis of the new National System of Geodetic Coordinates the European Terrestrial Reference System (ETRF'89) and ellipsoid GRS-80 were adopted. Practical realisation of the new reference system was carried out in the Department of Planetary Geodesy, Space Research Centre using GPS technique. In 1992 precise zero-order network EUREF-POL'92 containing eleven stations was established. In 1994 and 1995 its further densification (356 points) provided POLREF network which is practical realisation of ETRF'89 Reference System in Poland. The paper gives a survey of all works connected with the establishment of the new National Geodetic System in Poland.

OVERVIEW OF CLASSICAL TRIANGULATION NETWORKS IN POLAND

The process of creating the horizontal control networks in Poland has a very long and very complicated history. That history is a reflection of Poland's fate as an independent state. With the reference to historical events, the development of the Polish primary horizontal control net may be divided into the following steps: the period before 1918, the twenty year long inter-war period (1918-1939) and the period after 1945 (Baran,1990).

After 1945 the basic horizontal networks in Poland and in the whole Eastern Europe were related to the Krassovski reference ellipsoid with the initial point in Pulkovo

Observatory. In relation to this geodetic datum two adjustments of the Polish networks were made in 1957-58 and in 1983 (fig.1). National reference system based on Krassovski ellipsoid is called „1942". It has been modified in 1983 as System „JSAG-1983".

Finally the Polish national classical horizontal network consists of the asto-geodetic network having about 55 stations and of the filling network having about 6000 stations. The astro-geodetic network was created on the basis of fragments of different networks which were established on the territory of Poland during over eighty years (1899 - 1981). Densification of the network was carried out in 1955-1970. The results of the adjustment (Gaździcki et al.,1983) show that the average value of a mean relative errors (m_d) of distances are as follows:

- in the case of astro-geodetic network m_d is of the order of 2.5 - 0.65 ppm,
- in the case of the filling network m_d is of the order of 0.45 - 0.31 ppm.

It indicates that network is rather quite good. Further comparisons done twenty five years later showed that the classical network has significant deformations.

Further development of the horizontal geodetic control networks in Poland is connected with the rapid development of satellite observations techniques. In 1983 first experiments with incorporation satellite Doppler technique to strengthen classical network in Poland were carried out. Finally in 1988 the National Doppler Network consisting of ten station was established in Poland. The GPS technique introduced to geodesy in early eighties amounted true revolution in technology of establishing control geodetic networks.

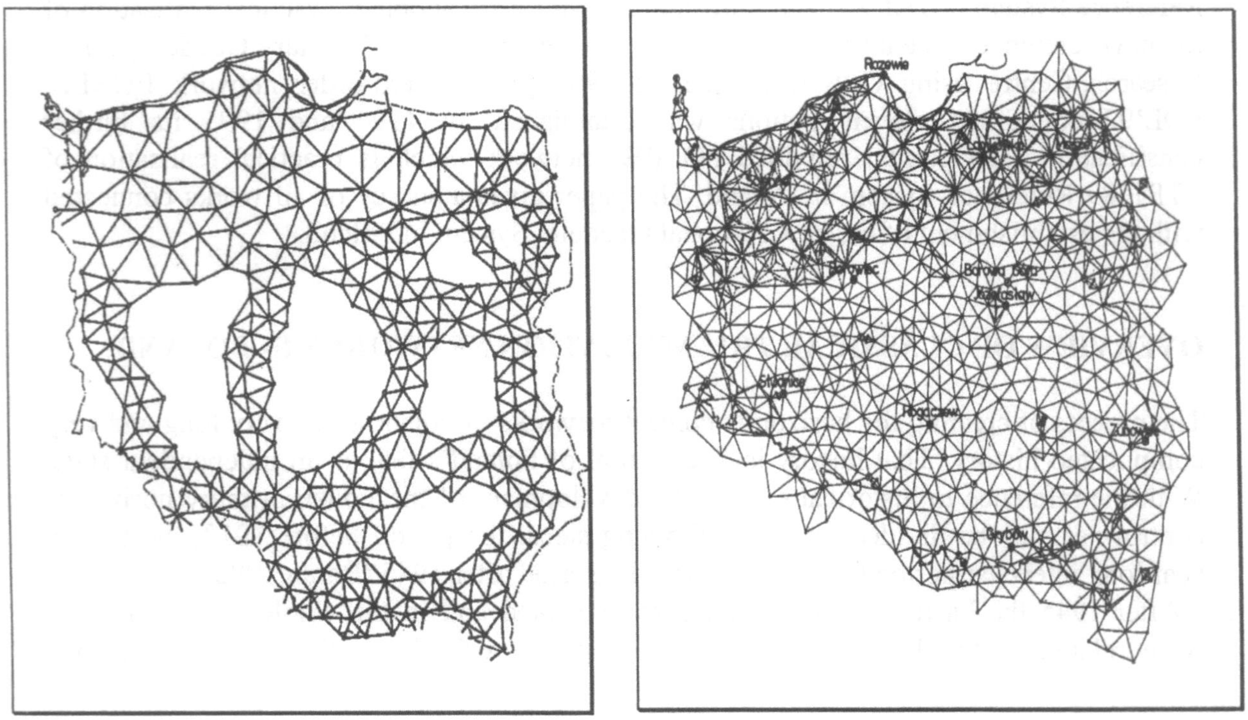

Fig. 1 Polish primary horizontal control network as in 1955 and as in 1983

EUREF-POL'92 GPS OBSERVATION CAMPAIGN AND DATA PROCESSING

Introduction. First concept of the extension of the EUREF network over Poland's territory was presented in 1991 (Baran et al.,1991). It contained eleven stations. Some of them were located in geodetic observatories (Borowiec, Borowa Góra, Józefosław, Lamkówko). Other points of the EUREF-POL network coincided with the National Doppler Network 1986 (Baran,1992). An official presentation of the Polish part of the EUREF network took place in March in Bern, during the EUREF Symposium (Baran, Zieliński,1992). Also in Bern the term and technical conditions of the EUREF-POL'92 campaign were set up.

Observation campaign. The campaign was successfully performed in the period of time from July 4 to 8, 1992 with a one day calibration session in Borowiec. Measurements were made with Trimble 4000 ST receivers, during four days with 10.5 hours of observations each day and one day with four hours session. The observations were carried out with an elevation cut-off angle of 10 degree and a sampling rate of 15 seconds. Among 19 foreign stations there were 15 reference points and 4 newly-positioned points; two of them on German territory (Karlsburg and Kirschberg) and two on the territory of Lithuania (Akmeniskiai and Meskonys). Additionally not planned point Predni Pricka in Czech Republic was included in EUREF-POL'92 network. The Institut für Angewandte Geodäsie (IfAG) had to include 3 additional stations Schönfeld and Kaminke and the station Simeis (Crimea) on the request of Russian colleagues.

Data processing. The data were processed by the team from the Space Research Centre (SRC) and simultaneously by the team from the IfAG in Leipzig. EUREF-POL'92 GPS network was computed with Bernese Software Version 3.4. Strategy of data processing, coordinates of the fixed stations, precision of processing and differences between SRC and IfAG solutions were accounted for in details in (Zieliński et al.,1994).

Final solution. The final solution of EUREF-POL'92 campaign is a combined solution obtained from two sets of coordinates (SRC and IfAG solutions) and their covariance matrices. It was expressed in ETRS89 coordinate system. The rotation to the epoch 1989.0 was computed in ITRF91 with individual velocities as far as available or with velocities of the European plate motion model NNR-NUVEL1. Then the transformation from ITRF91 epoch 1989.0 to ETRF89 was done with a seven parameter Helmert transformation. The final coordinates in ETRF89 at epoch 1989.0 were given in (ibid. table 8). The precision obtained from the daily solution repeatability was: 0.007m for north, 0.012 m for east and 0.022 m for height components.

DENSIFICATION OF THE EUREF-POL'92 NETWORK

Introduction. Just after the establishment of the EUREF-POL'92 network, a project of a new primary geodetic network POLREF was launched. The project assumed that 356 points of the new network should be positioned using GPS technique. Relative accuracy of position should do not exceed 1 - 2 centimetres. Points of the classical astro-geodetic network, where it is possible, should be included to the new network. Points of the POLREF network were monumented to ensure proper conditions for carrying out GPS, levelling and gravimetric observations.

Observation campaign. Network was observed in three campaigns from July 1994 to May 1995. In summer and autumn 1994 Space Research Centre organised two GPS campaigns. As a result of these campaigns 178 points were observed and about 50 % of the territory of Poland was covered with the new network. In spring 1995 the Institute of Geodesy and Cartography carried out the third GPS campaign and observed 178 new points. Common rules were adopted in all three campaigns, namely, length of session 220 minutes and at least two sessions at each POLREF point. GPS measurements were carried out with Turbo Rogue and Wild-Leica 200 receivers. The Turbo Rogue receivers were used for permanent observations at five reference points of EUREF-POL'92 network. The Wild Leica 200 receivers were used at the new points of POLREF network.

Data processing and analysis. Originally, each sub-network was separately adjusted using Trimvec and Bernese software. In the next stage, the east part was solved as one sub-network and the final solution was obtained by linking the eastern and western part with the Bernese Compar routine. Accuracy of the POLREF network meets the demands of the EUREF densification network. The standard deviation of a single observation is 0.39 cm, standard deviation of dB, dL, dh is 0.5 - 1.0 cm, 0.5-1.0 cm, 1.0 - 1.5 cm respectively. The scattering of the errors is rather regular.

TRANSFORMATION PARAMETERS BETWEEN „JSAG-1983" AND EUREF-89

The classical horizontal network in Poland is related to Krassovki ellipsoid and geodetic datum Pulkovo. Any strengthening of the classical network by GPS requires that the possible distorsions in classical network should be investigated beforehand. An obvious way of studying the relative distorsions is to compare the geodetic coordinates of common points in two networks. It requires the knowledge of the transformation parameters between two datum. The are several datum transformation formulas for accomplishing that.

Fig. 2. Residual vectors after 7-parameter transformation of JSAG-1983 to EUREF-89

The most common is 7-parameters transformation which requires the knowledge of the geoid/ellipsoid separations. In 1993 the first gravimetric geoid computed in Poland was fulfilled in the Department of Planetary Geodesy. Since that time the number of available gravity data increased considerably and the last geoid solution by the spherical FFT methods comprises more than 130 000 mean/point gravity anomalies.

Last modification of the Polish National Reference System „JSAG-1983" has been compared with the new satellite network. Because almost all stations of POLREF network are identical with the classical triangulation points thus it was possible to make a comparison of the networks by 7-parameter transformation.". The final coordinates of 339 POLREF points were used to determine the transformation parameters between networks. Classical network was of course two dimensional and for mutual transformation normal heights of the classical network had to be determined. It was done by spirit levelling. The quasigeoidal heights used for converting the normal heights of the triangulation stations to ellipsoidal heights were taken from (Łyszkowicz and Forsberg,1995). The residuals after the transformation of the reference system „JSAG-1983" into EUREF-89 are displayed in fig.2. Distribution of the differences are very regular and only in the north-west part small deformations can be observed. The maximum residuals in this case are 32 cm in horizontal position and 20 cm in height. It means that classical observations have a good quality and could be used in the future.

ACKNOWLEDGEMENTS

The final support provided by the Committee of Scientific Research grant PBZ-08-07 is gratefully acknowledged.

REFERENCES

Baran, W. (1990), History of creating and improvement of triangulation network in Poland, Proceedings of XVIX Congress of International Federation of Surveyors in Helsinki, pp.101.2/3-101/.2/14

Baran, W. (1992), GPS Network for Poland - Status and future plans, Veröffentlichungen der Bayerischen Kommission für die Internationale Erdmesssung der Bayerischen Akademie der Wissenschaften, Astronomisch-Geodätisch Arbeiten, Heft Nr 52, München, pp. 117-123

Baran, W. (1994), New National System of Geodetic coordinates in Poland, Geodezja i Kartografia (Geodesy and Cartography), vol. XLIII, No. 1, pp.51-55

Baran, W., Śledziński, J., Barlik, M., Czarnecki, K., Dobrzycka, M., Gajderowicz, I., Oszczak, St., Pachelski, W., Rogowski, J. Cisak J., (1991), Raport o potrzebie i zasadach założenia podstawowej zintegrowanej sieci geodezyjnej Polski w układzie EUREF, Raport Komitetu Geodezji PAN, Sekcji Sieci Geodezyjnych, Warszawa

Baran, W., Zieliński, J.B. (1992), Design and preparation of the EUREF-POL'92 GPS Campaign in Poland, Veröffentlichungen der Bayerischen Kommission für die

Internationale Erdmesssung der Bayerischen Akademie der Wissenschaften, Astronomisch-Geodätisch Arbeiten, Heft Nr 52, München, pp. 233-235

Baran, W., Zieliński, J.B. (1993), Status Report on participation of Poland in creation of European Reference System _EUREF, Veröffentlichungen der Bayerischen Kommission für die Internationale Erdmesssung der Bayerischen Akademie der Wissenschaften, Astronomisch-Geodätisch Arbeiten, Heft Nr 53, München, pp. 160-163

Baran, W., Zieliński, J.B. (1994), The Use of EUREF-POL Network for Scientific and Practical Purposes, Veröffentlichungen der Bayerischen Kommission für die Internationale Erdmesssung der Bayerischen Akademie der Wissenschaften, Astronomisch-Geodätisch Arbeiten, Heft Nr 54, München, pp. 327-330

Baran, W., Zieliński, J.B. (1995), Realisation of the GPS Primary Network for Poland, - Status Report, Veröffentlichungen der Bayerischen Kommission für die Internationale Erdmesssung der Bayerischen Akademie der Wissenschaften, Astronomisch-Geodätisch Arbeiten, Heft Nr 56, München, pp. 184-186

Baran, W., Zieliński, J.B. (1996): Polish National Report on EUREF Activities in 1995-1996, Veröffentlichungen der Bayerischen Kommission für die Internationale Erdmesssung der Bayerischen Akademie der Wissenschaften, Astronomisch-Geodätisch Arbeiten, Heft Nr 57, München, pp. 288-289

Gaździcki, J., Bokun, J., Deryło-Stępniak, J., Gelo, St. (1983), Present Status of the Primary Horizontal Control Network in Poland (in Polish), Proceedings of the 3rd Geodetic Symposium on Actual Problems of Primary Control Networks, Warsaw

Jaworski, L., Łyszkowicz, A., Gelo, St. (1996), Comparison of the 1-st order national network POLREF with the classic triangulation, Paper presented at the symposium of IAG Subcommission for the European Reference Frame (EUREF) in Ankara, Turkey

Jaworski, L., Gelo, St., Świątek, A., Zdunek, R. Zieliński, J.B. (1996), Adjustment results of the new 1st order national network POLREF, Paper presented at the symposium of IAG Subcommission for the European Reference Frame (EUREF) in Ankara, Turkey

Łyszkowicz, A., Forsberg, R. (1995), Gravimetric Geoid for Poland Area Using Spherical FFT, IAG Bulletin d'Information N.77, IGES Bulletin N.4, Special Issue, Milano, pp.153-161

Zieliński, J.B. , Jaworski, L., Zdunek, R., Engelhardt, G., Seeger, H., Töppe, F. (1993), EUREF-POL 1992 GPS Campaign and data processing, Veröffentlichungen der Bayerischen Kommission für die Internationale Erdmesssung der Bayerischen Akademie der Wissenschaften, Astronomisch-Geodätisch Arbeiten, Heft Nr 53, München, pp. 92-102

Zieliński, J.B., Jaworski, L., Zdunek, R., Engelhardt, G., Seeger, H., Töppe, F., Luthardt, J. (1994), Final Report about EUREF-POL 1992 Campaign, Veröffentlichungen der Bayerischen Kommission für die Internationale Erdmesssung der Bayerischen Akademie der Wissenschaften, Astronomisch-Geodätisch Arbeiten, Heft Nr 54, München, pp. 92-99

Zieliński, J.B., Jaworski, L. (1995), Preliminary Comparison of the Classical and GPS Reference Network in Eastern Poland, GIPSY-OASIS II Course, 17-21 July , Pasadena (USA)

AN OVERVIEW OF THE SIRGAS PROJECT

Luiz Paulo Souto Fortes
President of SIRGAS Committee
Departamento de Geodésia
Fundação Instituto Brasileiro de Geografia e Estatística - IBGE
Av. Brasil 15671, Parada de Lucas, Rio de Janeiro, RJ, Brazil, 21241-051
Fax: 55-21-481 2747; e-mail: fortes@ibge.gov.br

ABSTRACT

A general description of the South American Geocentric Reference System (SIRGAS) Project is presented. Since its establishment, in October 1993, the two working groups ("Reference System" and "Geocentric Datum") of the project have carried out the tasks towards defining a new geocentric reference system and an associated Datum for South America. The achievement of the project objectives were scheduled to be reached on 1997 and the corresponding final reports presented at this IAG Scientific Assembly. This was only possible due to the very high cooperation of many groups, as: the South American countries; IAG, IPGH and NIMA, the sponsors; and many others from North America and Europe. The development of the project was based on a continental GPS network, established during a very successful campaign in 1995. With its conclusion, South America has now an unified reference system, with an accuracy compatible with the up to date positioning techniques, what is of a great benefit to all countries in the continent.

NOTE

This paper is published in "**IBGE. SIRGAS Final Report, Working Groups I and II. Rio de Janeiro, 1997**".

THE DEFINITION AND REALIZATION OF THE REFERENCE SYSTEM IN THE SIRGAS PROJECT

Hoyer M [1].; Arciniegas S.[2]; Pereira K.[3]; Fagard H.[4]; Maturana R.[5]; Torchetti R.[6]; Drewes H.[7]; Kumar M.[8]; Seeber G.[9]

SIRGAS Project Working Group I

ABSTRACT

This paper describes the different stages executed to establish the Reference System to be used in the definition of the geocentric datum in South America, in the frame of the SIRGAS project.

The Working Group I developed the activities to plane, to prepare and to coordinate the GPS measuring campaign, besides the processing organization to be executed by the calculating centers. The analysis and decision about the final solution, represented by the geocentric coordinates of the GPS network, are the last phase of the work.

The GPS observations were carried out from May 26 to June 04, 1995. 58 principal stations were observed in 11 countries. The campaign was possible due to the collaboration of around 30 participant institutions from South, North America and Europe. DGFI (Germany) and NIMA (USA) were the processing centers.

INTRODUCTION

The objective of this paper is to describe the procedure employed by Working Group I of the SIRGAS Project to define the new geodetic reference system for South America and to

[1] Melvin Hoyer ; Universidad del Zulia, Apartado Postal 10311 ; Maracaibo, Venezuela ; Fax 58-61-512197; e-mail: mhoyer@europa.ica.luz.ve

[2] Susana R. Arcienagas; Instituto Geográfico Militar Eldorado, Edificio del IGM; Quito Ecuador

[3] Katia Pereira; IBGE/DEGED; Av. Brasil 15671, Parada Lucas; Río de Janeiro, RJ, Brasil, CEP 21241-051

[4] Herve Fagard; Institute Geographique National SGN; BP 68; 94160 Saint Mande, France

[5] Rodrigo Maturana; Instituto Geográfico Militar; Nueva Santa Isabel 1640; Santiago, Chile

[6] Ricardo A. Torchetti; Instituto Geográfico Militar; Cabildo 381; 1426 Buenos Aires, Argentina

[7] Hermann Drewes; DGFI,ABT.I; Marstallplatz 8; D-80539; Muenchen, Germany

[8] Muneendra Kumar; Defense Mapping Agency/IOG; 4600 Sangamore Road; Bethesda, MD 20816, USA

[9] Gunter Seeber; Institut für Erdmessung, University of Hannover; Schneiderberg 50; D-30167 Hannonver, Germany

establish the corresponding reference frame.

At the first workshop of the project in Asuncion, Paraguay (October 1993), it was agreed that the reference system should coincide with that of the IERS -International Earth Rotation Service - and that the reference frame should be realized by means of the observation of a highly precise network of GPS stations.

The GPS observation campaign was carried out according to the schedule during ten days from May 26 to June 4, 1995. A total of 58 stations were observed. After the collection and preparation of the observation data, which was mainly done at DGFI/I in Munich, DGFI (Deutsches Geodaetisches Forschungsinstitut), IBGE (Instituto Brasileiro de Geografia e Estatistica), and NIMA (National Imagery and Mapping Agency) started with the data processing.

At the workshop in Santiago (August 1996), the preliminary results of these three processing centers were presented and discussed. Important decisions were then made with respect to the pending computations and how they should be completed.

DGFI and NIMA presented their final results in Margarita (Venezuela) in April 1997. At this workshop the procedure for the final, unique solution was defined, and the corresponding computations were done during the workshop.

This paper is dealing with the above mentioned activities and results of the Working Group I. It includes the most important aspects of the GPS observation campaign, some aspects of the processing and the procedure to obtain the definitive results.

The evaluation of the results and the quality of adjusted station coordinates allows the conclusion that the Working Group I accomplished completely the planned objectives. South America has at its disposal one of the most precise continental networks which will serve as a basis for the establishment of a geocentric datum. The immediate task is now to provide the adequate maintenance.

GPS OBSERVATION CAMPAIGN OF THE SIRGAS REFERENCE FRAME

The idea to observe a continental GPS station network and the planification of its configuration started from the beginning of the project at the Asuncion meeting in October 1994.

The assumed initial criterion for the selection of the stations was to include in the network all the existing LASER, VLBI, DORIS, and GPS observatories in South America. In addition one had to look for a homogeneous continental coverage of the net, to guarantee the facile access to the sites and the possibility to perform the GPS measurements. Finally there had to be some overlapping with the official geodetic network of each country.

SIRGAS Working Group I, with the valuable collaboration of the scientific consultants, presented the technical specifications necessary for the GPS measurements in the SIRGAS campaign. These were elaborated taking into account all the aspects which, according the specialist's opinion, had to be included in a project like this.

Before the observations, discussions were carried out and decisions were made with respect to the compatibility and availability of equipment to be used. The idea was to guarantee homogeneity in the technical generation of the instruments and to include as many geodetic GPS receivers as possible from South American institutions. On the other hand, only highly precise receivers should be employed. Four receiver types were selected for this purpose: Ashtech Z12,Leica 200, Rogue/Turbo Rogue, and Trimble SSE (SSI). This selection includes the receivers operating as part of the global IGS network in the region.

A principal difference between receiver types with regard to precise positioning is the relative location of antenna phase centers. The exact location of the phase centers has to be known in the data processing in order to reduce the coordinates to the station's monument marker. If only one type of receiver and antenna is used in a campaign, the relative location of phase centers is identical in all stations and cancels in a relative positioning (with respect to a reference station or in a coordinate difference). If different receiver types are used, the variation of phase center locations between their antennae has to be corrected, i.e. the phase center corrections have to be known for each receiver type antenna.

Since the antenna phase center corrections of the employed receivers were not known in an internationally accepted way (nowadays an IGS recommendation is available), it was decided to co-locate different types of the selected receivers at several sites and to determine their exact three-dimensional distances by local tie measurements. For that reason, nine sites were occupied by two or three different receiver types.

In addition to this, the handling of the observation data was organized. A data center was selected in each country, and two global data centers for the collection of the observations from all countries were established.

The observations were carried out as planned from May 26, 0:00 UT until June 4, 24:00 UT, 1995, i.e., over 10 days in total. Finally, 58 principal stations and 9 eccenters in 11 countries were observed:

Country	No. of stations	Country	No. of stations
Argentina	10	French Guiana	1
Bolivia	6	Paraguay	2
Brazil	11	Peru	4
Chile	8	Uruguay	3
Colombia	5	Venezuela	5
Ecuador	3		
		Total	58

The location of all sites occupied during the SIRGAS campaign is displayed in figure 1

FINAL COORDINATES OF THE SIRGAS 1995.4 REFERENCE FRAME

Independent coordinate solutions of the SIRGAS GPS campaign were computed by DGFI (DGFI, 1997) and by NIMA (NIMA, 1997). DGFI used the Bernese software version 3.4 with some modifications implemented at DGFI, and NIMA used the GIPSY-OASIS II software. The station coordinates computed in the first step by DGFI and by NIMA are

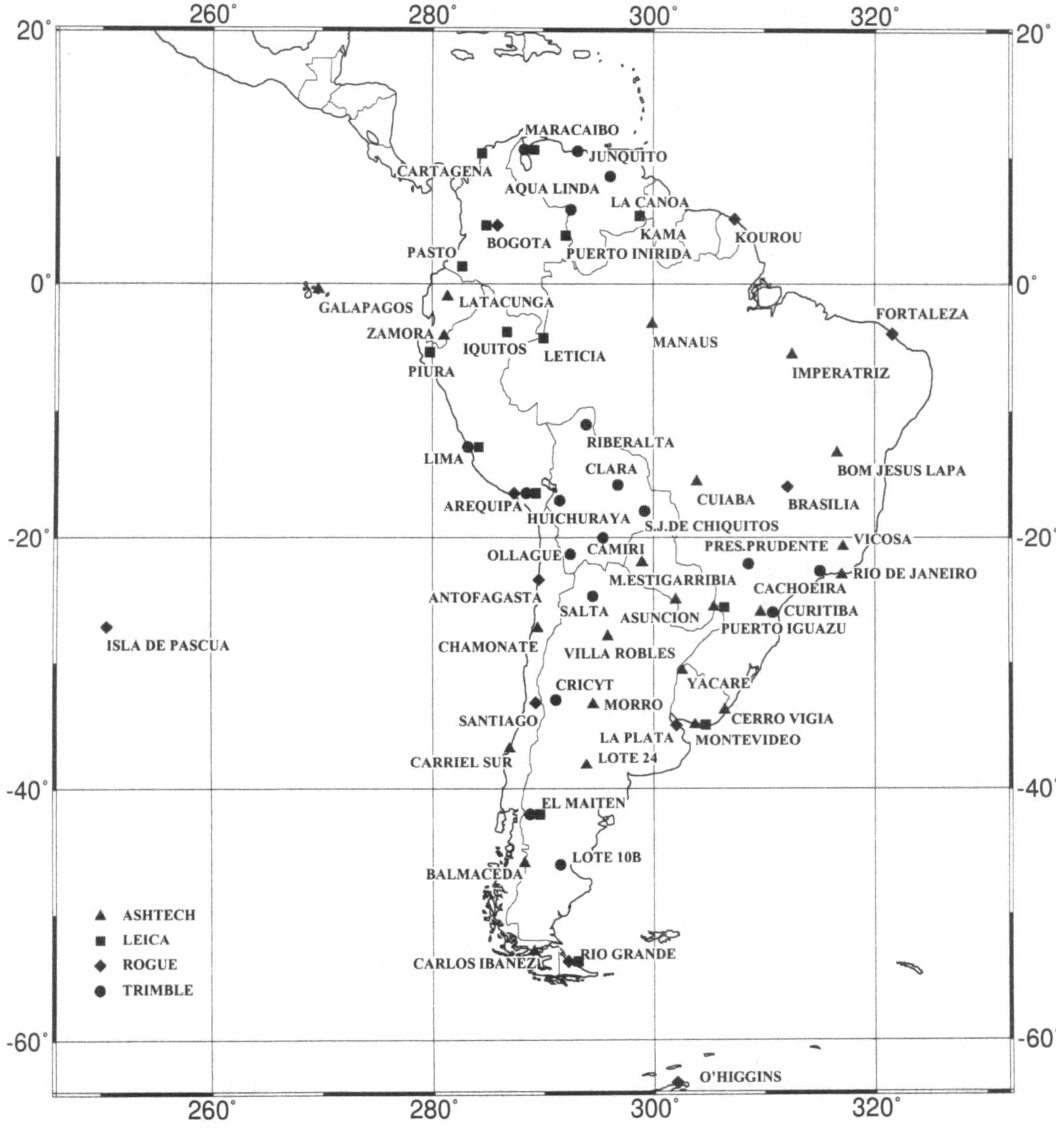

Fig 1: Stations of the SIRGAS reference frame and receiver types of the occupation during the GPS campaign May/June 1995

nearly unconstrained solutions and refer to the IERS (International Earth Rotation Service) Terrestrial Reference Frame 1993 (ITRF93) epoch 1995.4 by fixing the satellite orbits in this frame. DGFI used the International GPS Service (IGS) precise combined orbits and clock offsets while the NIMA results were computed with the JPL orbits and clock parameters.

After applying a seven parameter similarity transformation (Helmert transformation), the maximum difference between the DGFI and NIMA coordinates was 3.5 cm. The rms differences between the solutions were ± 1.0 cm in X, ± 1.4 cm in Y and ± 0.7 cm in Z. DGFI's solution includes a correction for the elevation angle dependent phase center variation of each type of used receiver antennae as proposed by IGS (model IGS_01). This correction is not included in the NIMA solution. The error caused by not including this correction reduces in principal to a constant coordinate (mainly height) offset for each antenna type because of the (nearly) 24 hours permanent observations and consequently full coverage of the satellite passes under all elevation angles. It was therefore decided to combine the two solutions in the following manner:

1. Compute a separate seven parameter similarity transformation between the DGFI solution and each of the five receiver antenna type sub-networks (Ashtech P 700228, Ashtech P 700718, Leica SR299, Trimble 4000ST/TR GP, Dorne Margolin) of the NIMA solution. This will account for the small differences in the antenna corrections as well as the small reference frame differences.

2. Apply the transformation to the five antenna type subsets of the NIMA coordinates. The rms differences between the DGFI coordinates and the transformed NIMA coordinates reduce then to ± 0.7 cm in X, ± 0.9 cm in Y and ± 0.6 cm in Z.

3. Compute the mean of the DGFI coordinates and the transformed NIMA coordinates to produce a combined set of coordinates in the ITRF93.

The final coordinates should be given in the ITRF94 (the most recent realization of the ITRF up to date) according to a previous decision of the SIRGAS Committee. In order to accomplish this, a subset of SIRGAS stations identical with global stations independently determined in the ITRF94 is required. Using these stations, transformation parameters can be computed to convert the SIRGAS coordinates from ITRF93 to ITRF94.

Only four stations (Arequipa, Fortaleza, Kourou and Santiago) on the mainland of South America are included in the official ITRF94 solution of the IERS with station coordinates at epoch 1993.0 and velocities to extrapolate to epoch 1995.4. This was considered too few to derive a good seven parameter transformation. Two nearby SIRGAS stations with IERS-determined ITRF94 coordinates and velocities (Easter Island and O'Higgins) outside the mainland may be added to increase the number of stations identical with the SIRGAS reference frame. Doing so, however, considerable distortions were found in the network.

The International GPS Service (IGS) is computing station coordinate and velocity solutions in the ITRF more frequently than the IERS. At the time of the final SIRGAS computations, two 1996 solutions of the IGS Analysis Centers CODE (Bern) and JPL (Pasadena), respectively, were available, both referring to the ITRF94 and including three additional SIRGAS stations on the South American mainland (Bogota, Brasilia and La Plata). The CODE and JPL solutions are computed with the Bernese and GIPSY software, respectively. Since this software is almost identical with that used by DGFI and NIMA, respectively, it was decided to use the average CODE and JPL station coordinates (referring to ITRF94 and extrapolated to epoch 1995.4) as fiducial coordinates. The maximum component differences between the CODE and JPL coordinates for the nine stations are 3.6 cm in Bogota and Easter Island and 3.8 cm in Santiago.

The final SIRGAS solution was computed by a seven parameter similarity transformation between the mean CODE/JPL coordinates in ITRF94 at epoch 1995.4 and the combined DGFI/NIMA coordinates using the nine IGS stations. The transformation parameters were then applied to the combined DGFI/NIMA coordinates to produce the final SIRGAS coordinates in ITRF94 at epoch 1995.4. The result is given in Table 2.

The final SIRGAS coordinates and their corresponding accuracies are presented in the Final Report of SIRGAS Project (SIRGAS Final Report, 1997). The coordinates of the network have r.m.s. errors with an average of ± 4 mm in each component.

CONCLUSION

A cooperative effort of many institutions and countries permitted the definition of a modern geodetic reference system for South America coinciding with that of the IERS. The reference frame of this system is a highly precise network of 58 GPS stations. The coordinates of the network have r.m.s. errors with an average of ± 4 mm in each component.

REFERENCES

NIMA, Publication GIMGB-97-1603, Fundamental Station Coordinates for the SIRGAS Continental Geodetic Control Networks, 1997.

DGFI, Abt.I, Report on the Processing of the SIRGAS 95 GPS Network at DGFI, 1997.

SIRGAS Final Report, September 1997

TIME EVOLUTION OF THE SIRGAS REFERENCE FRAME

Hermann Drewes
Deutsches Geodätisches Forschungsinstitut, Abt. I (DGFI/I)
Marstallplatz 8, D-80539 München, Germany

ABSTRACT

The geodetic reference system for South America established by the SIRGAS project and adopted by the South American countries provides a set of some 60 stations coordinates over all the continent for the epoch of its observation in May/June 1995 (1995.4). To preserve the consistency with the ITRF, station coordinates have to change in time according to the ITRF kinematic reference frame. The motion of the South American continent cannot completely be modelled by a rigid plate rotation after current models but has to account for extended continuous deformations, in particular along the Andes mountain range. It is proposed to determine individual station velocities of all SIRGAS sites and to derive a continuous velocity field from a combined rigid plate and regional deformation model for the transformation of new stations into the SIRGAS reference frame.

INTRODUCTION

The SIRGAS (SIstema de Referencia Geocéntrico para América del Sur) reference system was adopted by the official representatives of the South American countries in the SIRGAS project during a workshop held in Isla de Margarita (Venezuela) in April 1997. It is defined according to the International Earth Rotation Service (IERS) conventions (McCarthy 1996) using for its realization the IERS Terrestrial Reference Frame (ITRF94, Boucher et al. 1996). The realization of the SIRGAS reference frame is done by a total of 67 sets of station coordinates at 58 sites for the epoch of the GPS observation campaign in May/June 1995 (epoch 1995.4). The materialized stations at the Earth's surface, however, are moving due to recent crustal movements. In the ITRF which is also used for precise satellite orbit parameters, e.g., those of the International GPS Service for Geodynamics (IGS), the station motions are taken into account by providing individual station velocities in addition to the epoch station coordinates.

Therefore, the maintenance of the SIRGAS reference frame has to include - besides the physical maintenance of the monumented sites - the time evolution of the coordinates too, in order to guarantee the consistency between the terrestrial (SIRGAS) and the satellite reference system. For this purpose, station velocities $\underline{V}$ (i.e., coordinate changes dX/dt, dY/dt, dZ/dt) are required for each station. These velocities may either be derived from repeated observations and coordinate determinations, or from crustal deformation models. To include additional stations observed at other than the SIRGAS epoch, we need a continuous velocity model to transform their coordinates to the SIRGAS epoch.

STATION VELOCITIES FROM REPEATED POSITIONING

The stations of the International GPS Service for Geodynamics (IGS) included in the SIRGAS network are equipped with permanently operating GPS receivers. The observed data sets are routinely evaluated providing station coordinates on a weekly basis. The organization of the IGS distinguishes two levels of stations (IGS 1997): IGS global stations and IGS regional stations. The observations of the global stations are processed by seven Analysis Centers (AC) and combined to one common solution by three Global Network Associate Analysis Centers (GNAAC); observations of the regional stations are processed by six Regional Network Associate Analysis Centers (RNAAC). All the sets of data, global and regional, are combined to a so-called polyhedron solution (P-SINEX) by two GNAACs including presently more than 120 stations.

For South America, the RNAAC SIR (i.e. SIRGAS) operated by Deutsches Geodaetisches Forschungsinstitut (DGFI) is processing all the available data from the existing permanently observing stations in this region. These include presently (September 1997) sixteen SIRGAS stations (Seemüller and Drewes 1997). The result is then combined with other solutions by the GNAACs to the global polyhedron. In this way we get weekly coordinates for all the included permanently observing SIRGAS stations in the ITRF and can thus derive station velocities. It is strongly recommended to join this procedure by installing as many permanent GPS receivers as possible at SIRGAS sites. More information may be requested from IGS or the RNAAC SIR.

To derive the velocities of stations not equipped with permanent GPS receivers we need at least one (better more than one) repetition measurement (GPS campaign) at time t_i for coordinate determination. From coordinate changes $\underline{\Delta X}$ $(\Delta X, \Delta Y, \Delta Z) = \underline{X}(t_i) - \underline{X}(t_o)$ divided by the time interval Δt we get then the velocities $\underline{V}$ (V_X, V_Y, V_Z):

$$\underline{V} = \Delta \underline{X} / \Delta t \,. \tag{1}$$

To improve the accuracy of velocity estimations, the time interval between the campaigns should not be too short. Five years seems to be a reasonable time span. It is important to do the adjustment of the repetitive campaigns in the same reference frame as the first SIRGAS campaign, i.e., ITRF94 with its time propagated station coordinates at the observation epoch.

STATION VELOCITIES FROM LITHOSPHERIC PLATE MOTIONS

As long as there has been no repetitive coordinate determination, station velocities cannot be estimated empirically. In order to propagate the coordinate changes with time, approximation models have to be used. A very familiar model is that of plate tectonics. The global crustal deformations are in a first approximation described by the motions of rigid plates (i.e., spherical segments of the globe). These motions can be expressed for each point of a plate by the rotation on a sphere with the geocentric rotation vectors $\mathbf{\Omega}$ (Ω_X, Ω_Y, Ω_Z). The velocity $\underline{V}$ (V_X, V_Y, V_Z) for a station at the Earth's surface with the geocentric coordinates X, Y, Z is then

$$\begin{aligned} V_X &= \Omega_Y \cdot Z - \Omega_Z \cdot Y \\ V_Y &= \Omega_Z \cdot X - \Omega_X \cdot Z \\ V_Z &= \Omega_X \cdot Y - \Omega_Y \cdot X\,. \end{aligned} \tag{2}$$

The IERS adopted the geophysically determined kinematic plate model NNR NUVEL-1A for all the stations for which the velocity has not yet been determined in the ITRF (McCarthy 1996). The South American mainland is covered by two plates in this model: South America and Caribbean plates. The Pacific islands (SIRGAS stations Galapagos and Isla de Pascua) are situated on the Nazca plate and O'Higgins on the Antarctic plate. The rotation parameters of these plates are given in table 1.

Table 1: NNR NUVEL-1A rotation vectors for SIRGAS (from McCarthy, 1996)

Plate Name	Ω_X[rad/10^9a]	Ω_Y[rad/10^9a]	Ω_Z[rad/10^9a]
South America	-1.038	-1.515	-0.870
Caribbean	-0.178	-3.385	1.581
Nazca	-1.532	-8.577	9.609
Antarctica	-0.821	-1.701	3.706

STATION VELOCITIES FROM CONTINUOUS DEFORMATION MODELS

It has to be stated that the hypothesis of rigid plates is only an approximation for the modelling of recent crustal movements, i.e., the station velocities computed by this model may differ from the real station motions. This holds in particular in the plate boundary zones where we have considerable regional deformations, i.e., the plate boundaries cannot be modelled as rigid bodies. Figure 1 shows the South American region with the known plate boundary zones. We clearly identify all the western part of the South American continent along the Andes mountain range as an extended deformation zone. In all this region one cannot use the rotation pole as given for the South American plate in table 1.

A better approximation is a regional rotation vector derived from space geodetic observations (Drewes 1993).

Andes Zone: $\Omega_X = -1.0$ rad/10^9a, $\Omega_Y = -4.0$ rad/10^9a, $\Omega_Z = 1.4$ rad/10^9a.

The difference between computed station velocities in the central Andes using this rotation vector instead of the rotation vector for South America (table 1) is 2 cm/a, i.e., in five years we get a coordinate difference of 10 cm. It is therefore necessary to determine the real deformations from a dense network of stations with observed velocities (see above), and to derive a continuous deformation model for the plate boundary zones. This can be done by a geophysical model, e.g., using the finite element method (Drewes 1993) or by a least squares prediction approach (Drewes 1997).

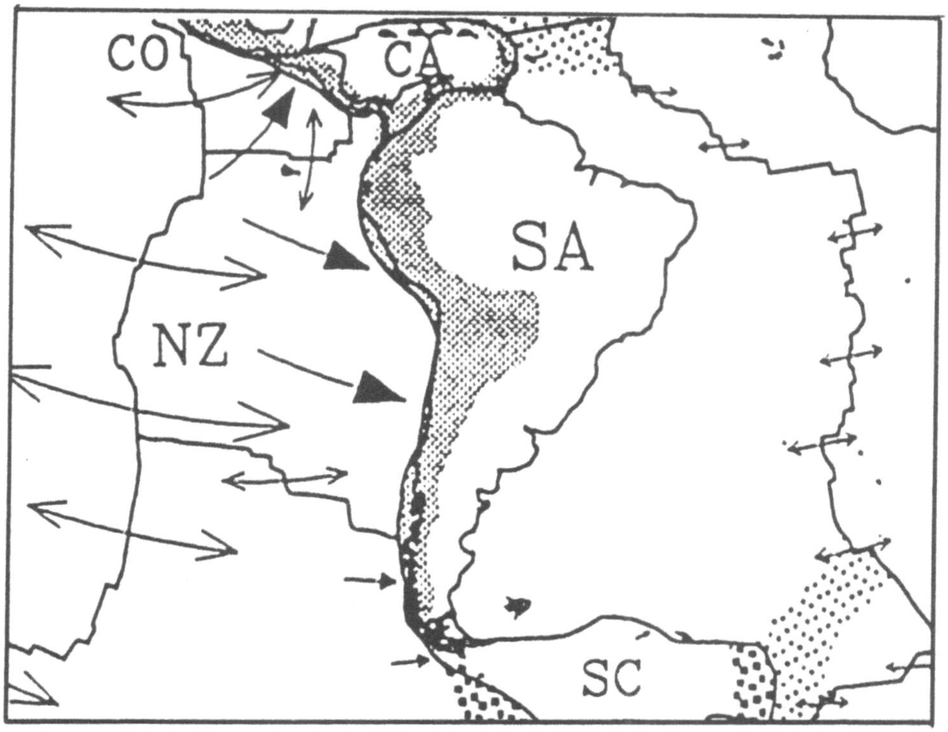

Fig. 1: Lithospheric plates and plate boundary deformation zones (from Gordon 1995)

Besides the continuous deformations we also have to consider discontinuous motions of the Earth's crust, e.g., produced by earthquakes. An example is the July 30, 1995 Antofagasta (Chile) earthquake (two months after the SIRGAS GPS campaign) where horizontal displacements in the order of 70 to 80 cm have been observed (Ruegg et al., 1996, Klotz et al. 1996). The SIRGAS coordinates have to be corrected because of such motions before using them as a reference frame. Such discontinuous motions have also to be reduced when computing a crustal deformation model as described above.

TIME EVOLUTION OF THE SIRGAS REFERENCE FRAME

In the following we shall summarize the requirements for the use of SIRGAS coordinates as fiducial stations in precise geodetic positioning in South America using geodetic space techniques (e.g. GPS). It is assumed that an observation campaign has been performed at a time t_i occupying new stations and SIRGAS stations simultaneously. The result of the data processing ought to be coordinates of the new stations in the SIRGAS reference system, e.g., for national densification of the SIRGAS (= ITRF) reference frame.

The following steps have to be done when using SIRGAS coordinates in the space geodetic (e.g. GPS) data processing:

1. Propagate the SIRGAS coordinates of stations S used as fiducials (connecting points) from epoch 1995.4 (defined epoch for SIRGAS) to the new observation epoch t_i using the station velocities $\underline{V}_S$ derived either from repeated observations or from crustal deformation models:

   $$\underline{X}_S(t_i) = \underline{X}_S(1995.4) + \underline{V}_S \cdot (t_i - 1995.4).$$

2. Perform the coordinate adjustment of new stations N using the observations at epoch t_i in connection with the SIRGAS coordinates $\underline{X}_S(t_i)$.

3. Transform the coordinates of the new stations N from observation epoch t_i back to the SIRGAS reference epoch t_0 = 1995.4 in order to get a homogenous coordinate set in the SIRGAS system:

   $$\underline{X}_N(1995.4) = \underline{X}_N(t_i) - \underline{V}_N(t_i - 1995.4).$$

   As the velocities $\underline{V}_N$ of new stations are normally not known, they have to be derived from crustal deformation models (rigid plate motions and/or continuous deformations) as discussed above.

In particular the last mentioned problem clearly demonstrates the necessity of the inclusion of a continuous crustal deformation model into the SIRGAS reference system. If we don't reduce new coordinates (valid for their observation epoch t_i) to the reference epoch t_0, we will get an inhomogenous set of coordinates referring to different epochs. The difference is 1 to 2 cm per year time interval from the reference epoch 1995.4. To perform the reduction we need a continuous deformation model for the entire continent. A rigid plate kinematic model alone is not adequate for the plate boundary zones. It is strongly recommended to support the geodynamic projects for monitoring crustal deformations in South America and to define an approach (finite element or least squares vector prediction) to compute a continuous velocity field.

REFERENCES

Boucher, C., Z. Altamimi, M. Feissel, P. Sillard: Results and analysis of the ITRF94. IERS Paris, Technical Note No. 20, 1996.

Drewes, H.: A deformation model of the Mediterranean from space geodetic observations and geophysical predictions. Springer, IAG Symposia, No. 112, 373-378, 1993.

Drewes, H.: Sistema de referencia cinemático incluyendo un modelo del movimiento actual de las placas tectónicas. IV Congreso Internacional de Ciencias de la Tierra, p. 230, Santiago de Chile, 1996.

Drewes, H.: Combination of VLBI, SLR and GPS determined station velocities for actual plate kinematic and crustal deformation models. IAG Scientific Meeting, Rio de Janeiro, 1997.

Gordon, R.G.: Present plate motions and plate boundaries. In: T.J. Ahrens (ed.): Global Earth Physics, A Handbook of Physical Constants. AGU Reference shelf 1, 66-87, 1995.

International GPS Service for Geodynamics (IGS): 1996 Annual Report. IGS Central Bureau, JPL Pasadena, 1997.

Klotz, J., J. Reinking, D. Angermann: Die Vermessung der Deformation der Erdober- fläche. Geowissenschaften (14) 389-394, 1996.

McCarthy, D.D. (ed.): IERS Conventions (1996). IERS Paris, Technical Note No. 21, 1996.

Ruegg, J.C., and 14 others: The M=8.1 Antofagasta (N. Chile) earthquake of July 30, 1995: First results from teleseismic and geodetic data. Geoph. Res. Lett. (23), 917-920, 1996.

Seemüller, W., H. Drewes: The IGS regional network associate analysis center for South America at DGFI/I. IAG Scientific Assembly, Rio de Janeiro, 1997.

RESULTS OF THE SIRGAS 95 GPS NETWORK PROCESSING AT DGFI/I.

Klaus Kaniuth, Hermann Drewes, Klaus Stuber, Herbert Tremel
Deutsches Geodätisches Forschungsinstitut, Abt. I (DGFI/I)
Marstallplatz 8, D-80539 München, Germany

Juan Moirano
Facultad de Ciencias Astronómicas y Geofísicas
Universidad La Plata, La Plata, Argentina

ABSTRACT

For establishing an accurate and geocentric reference system for South America, Sistema de Referencia Geocéntrico para América del Sur (SIRGAS), a continental GPS campaign including more than 60 stations has been carried out in May/June 1995. DGFI/I was acting as one of the analysis centers processing the entire data set. The paper describes the applied processing strategy and discusses the achieved results.

INTRODUCTION

The project for the establishment of an accurate geocentric reference system for the South American continent, Sistema de Referencia Geocéntrico para America del Sur (SIRGAS), has been initiated in 1993. The first goal of SIRGAS was the realization of a reference network of some 60 stations equally distributed over the continent by performing a GPS campaign. Besides acting as a data center for the SIRGAS project, the Deutsches Geodätisches Forschungsinstitut, Dept. I (DGFI/I) processed the entire data set acquired during this campaign and computed a unique network solution. This paper discusses the strategy applied for processing the large amount of data, the models and software used, and summarizes the achieved results. The processing has been performed at DGFI/I in collaboration with the University of La Plata, Argentina.

DATA SET

The SIRGAS GPS campaign was successfully carried out from May 26 to June 04, 1995. Following the issued guidelines the majority of stations operated continuously during the ten days period 23 hours per day. Figure 1 displays the sites occupied during the campaign.

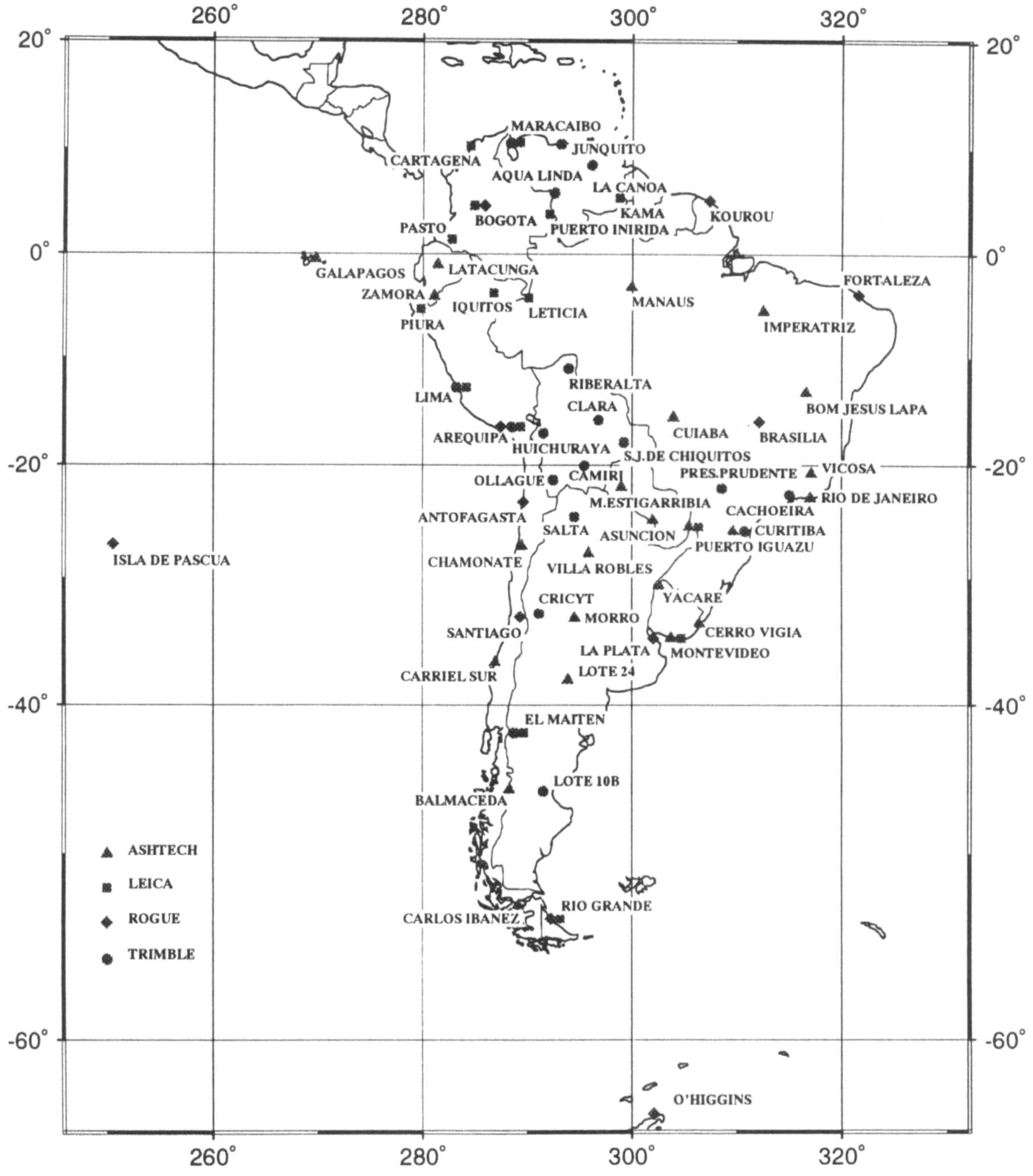

Fig. 1 Location of the sites occupied during the SIRGAS 95 GPS campaign

Figure 1 also indicates that different GPS receiver systems were involved in the SIRGAS campaign. The distribution of receivers reflects the availability of equipment at the participating South American organizations and at the contributing European and North American institutions. In early 1995 the phase center characteristics of the involved GPS antennas were not yet known precisely enough for combining different antenna types when applying techniques differencing between stations. Therefore, co-locations of different receiver/antenna systems were done (figure 1) enabling the combination of receiver-specific sub-networks, provided local ties are available from independent measurements.

PROCESSING STRATEGY

During the SIRGAS campaign extending over ten days and including some sixty stations a total of about $2 \cdot 10^7$ phase observations were collected. Processing such a data set with the Bernese software (Rothacher et al. 1993) which models double differences would require to solve for some $2.5 \cdot 10^4$ parameters. Therefore, we have applied a strategy of distributed processing resulting under the assumption that no correlations are neglected in a solution identical to a one step adjustment. This procedure can be outlined as follows:

- As at the time of the campaign the phase center characteristics of the involved antenna types were not accurately determined, no mixed baselines including different antennas were defined in order to avoid any accuracy degradation. Nevertheless, since the availability of a first set of phase center variation models (Rothacher and Mader 1996) these have been applied.

- Considering also the number of units of each of the receiver types involved in SIRGAS, receiver specific networks including the measurements acquired during a one day session were created.

- Solutions of these daily sub-networks provide a means for assessing the day by day consistency of the results and allow the identification of outliers and processing errors; in addition, local parameters such as ambiguities can be pre-eliminated in these receiver-specific adjustments, and reduced normal equations are then saved for external accumulation with all other subsets.

- These adjustments for setting up the normal equations were performed in the ITRF reference system realized by the International GPS Service for Geodynamics (IGS) combined orbits and compatible Earth orientation parameter series without applying any constraints; consequently adding fiducial information such as local ties between co-located antennas did not require any reprocessing.

- The accumulation and solution of all normal equation systems is done using an external program ACCSOL developed at DGFI/I. Before this final stage no datum realizations or additional fiducial informations are applied.

This approach provides high flexibility, e.g. for analyzing the sensitivity of the network solution to reference frame variations. The preprocessing has been done with the original Bernese software, and the adjustment with a modified version of the Bernese program GPSEST. Among others, the modifications relevant to the SIRGAS project are

- the ability to define constraints for troposphere estimates at co-location sites,
- the availability of advanced troposphere prediction models and mapping functions,
- the capability to store normal equations for the accumulation with ACCSOL.

As already mentioned, the solution is based on the IGS combined orbits and compatible Earth orientation parameter series. A data sampling rate of 30 seconds was used in all adjustments although a number of stations generated observations at a 15 seconds rate. The elevation angle cutoff was set to 15°, because the SIRGAS network includes several sites providing free satellite visibility only above 15°. Considering the observation session length of 23 hours/day and the investigations by Mervart (1995) we have not attempted to resolve the phase ambiguities.

For the a priori calibration of the tropospheric delays the Saastamoinen (1973) zenith delay model and the mapping function proposed by Davis et al. (1985) have been used; residual zenith delays were estimated in the adjustment for 4 hours intervals. Decreasing these intervals to e.g. 2 hours did not result in significantly better daily coordinates repeatabilities.The Bernese software models these zenith delays as a step function which can be constrained and smoothed by assigning standard deviations to the a priori calibration. We have applied ± 1 m for the first absolute parameter and ± 5 cm for the subsequent relative ones; these numbers let the troposphere estimates practically unconstrained.

COMBINATION OF NORMAL EQUATIONS

As described in the previous chapter receiver-specific daily normal equation systems were generated which may be considered to represent unconstrained networks in the IGS orbits' system. The accumulation and solution was done using the program ACCSOL (ACCumulate and SOLve) which provides among others the following options:

- To introduce coordinates of fiducial points as additional observation equations assigning individual weights;

- to set up condition equations constraining certain pairs of troposphere estimates to equality, e.g. at co-location sites;

- to apply local eccentricities information at fiducial stations or for co-located receivers as stochastic information.

The general structure of combining the SIRGAS normal equations in ACCSOL is displayed in figure 2.

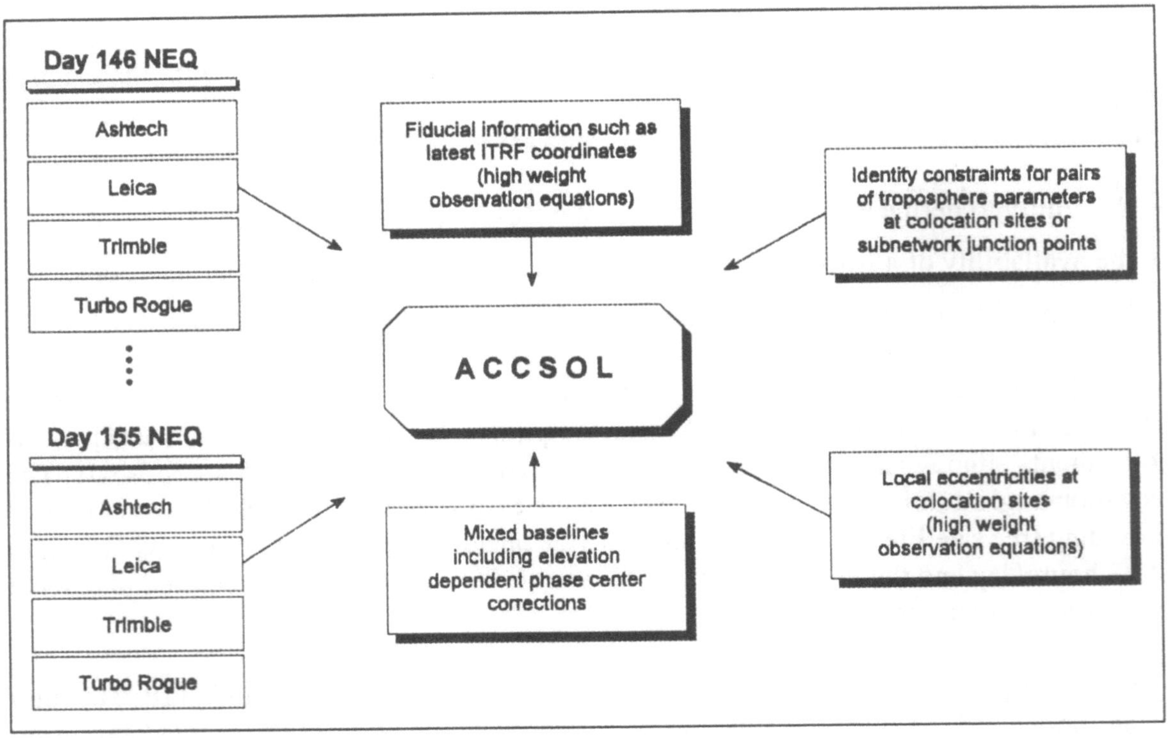

Fig. 2 Combination of SIRGAS normal equations and additional information in ACCSOL

RESULTS

A comparison of the results for the single day receiver-specific network is a means to detect not only data and processing errors but also to assess the achieved accuracies. Therefore, we summarize in table 1 for each receiver type the average root mean square (rms) agreement between the single day adjustments and the ten days solution. The quoted numbers result from similarity transformations which account for the small datum shifts between the unconstrained networks.

Table 1 Consistency of daily solutions: average rms differences in north (N), east (E) and height (H) between single day adjustments and the 10 days solution for each receiver-specific sub-network [mm]

Network	Stations	N	E	H
Ashtech Z-XII	23	3.7	4.8	7.8
Leica 200	15	6.0	7.8	14.3
Trimble SSE	19	5.3	6.9	11.7
Rogue/Turbo Rogue	11	8.6	12.5	14.0

From table 1 it is evident that the Ashtech Z-XII demonstrated the best performance during the SIRGAS campaign. However, the slightly worse consistency for the other receivers does not necessarily reflect a worse tracking performance. Instead the distribution of the stations is obviously of a certain relevance. In case of the Rogue/Turbo Rogue network the processing of the data acquired by the IGS stations close to the equator Bogotá (BOGT), Fortaleza (FORT) and Kourou (KOUR) required particular effort. This holds also for some of the Leica stations in Colombia, Peru and Venezuela which due to lack of sufficient memory operated also at a 30 seconds sampling rate. We suppose these problems to be due to rapid ionospheric delay variations which could not sufficiently be met by those stations in the equatorial region operating at a 30 seconds observation rate.

Another documentation of the agreement of the daily adjustments with the final 10 days solution is presented in figure 3. This figure displays for one baseline per receiver type the departures of the daily baseline length estimates from the mean. The examples shown are selected to be available on each of the 10 days and to represent a wide range of baseline lengths. Again the Ashtech baseline of 1630 km length shows the smallest scatter, but also for the long Leica and Turbo Rogue baselines the daily repeatability is in the order of $3 \cdot 10^{-9}$.

After combining all daily normal equation systems with ACCSOL the unconstrained 10 days solution was transformed to ITRF 94 using seven fiducial points including Richmond (RCM5). We do not present a list of coordinates here, because the final set of SIRGAS coordinates at epoch 1995.4 is a combination of our solution and the results provided by the National Imagery and Mapping Agency (NIMA). The final set of station coordinates is included in (SIRGAS 1997).

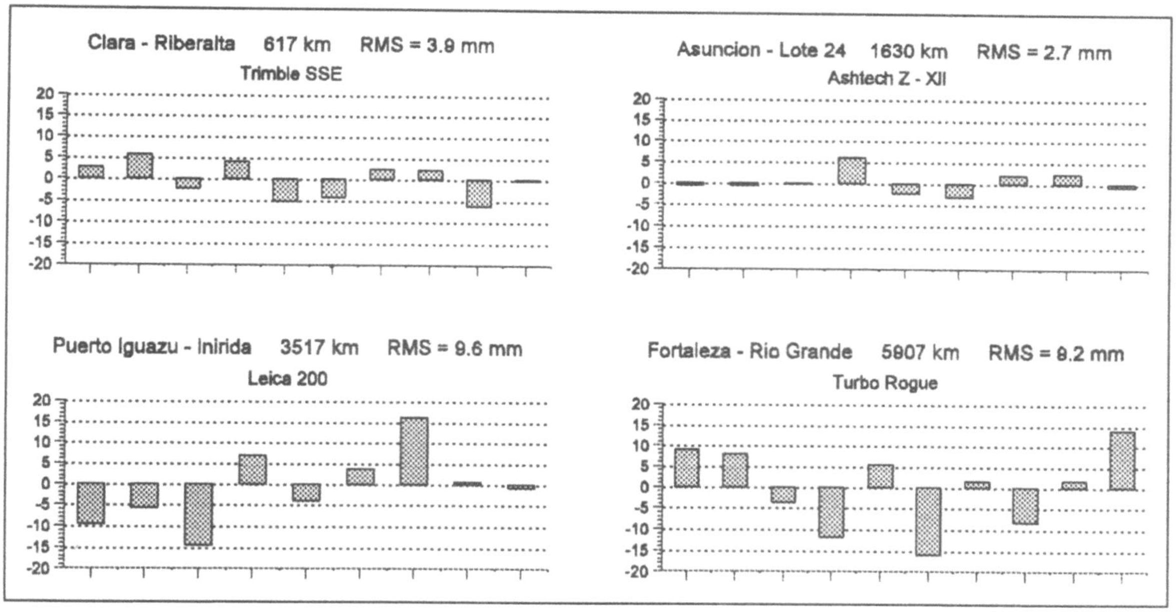

Fig. 3 Differences of daily baseline length estimates from the mean for various baseline lengths and receivers [mm]

REFERENCES

Davis, J.C., T.A. Herring, I.I. Shapiro, A.E.E. Rogers and G. Elgered (1985): Geodesy by Radio Interferometry: Effects of Atmospheric Modelling Errors on Estimates of Baseline Length. Radio Science 20, 1593-1607.

Mervart, L. (1995): Ambiguity Resolution Techniques in Geodetic and Geodynamic Applications of the Global Positioning System. Geodätisch-geophysikalische Arbeiten in der Schweiz, Bd. 53.

Rothacher, M., G. Beutler, W. Gurtner, E. Brockmann and L. Mervart (1993): The Bernese GPS Software Version 3.4. Astronomical Institute, University of Berne.

Rothacher, M. and G. Mader (1996): Combination of Antenna Phase Center Offsets and Variations; Antenna Calibration Set IGS-01. International GPS Service for Geodynamics (IGS).

Saastamoinen, J. (1973): Contributions to the Theory of Atmospheric Refraction, Part II: Refraction Corrections in Satellite Geodesy. Bull. Geod. 107,13-34.

SIRGAS (1997): SIRGAS Final Report, Working Groups I and II. Edited by Instituto Brasileiro de Geografia e Estatística (IBGE), Rio de Janeiro, 1997.

THE INTEGRATION OF BRAZILIAN GEODETIC NETWORK INTO SIRGAS

Costa, Sonia Maria Alves
Pereira, Kátia Duarte
e-mail : soniamaria@ibge.gov.br
Brazilian Institute of Geography and Statistics
Av. Brasil 15671 - Parada de Lucas
Rio de Janeiro - RJ - CEP : 21241-051, Brazil

Beattie, Don
e-mail : beattie@geod.emr.ca
Geomatics Canada
615 Booth Street - Ottawa, Ontario - K1A 0E9, Canada

Abstract

Cooperating as part of South American Geocentric Reference System (SIRGAS), whose purpose is to establish a geocentric reference system for the whole of South America, Brazil will take the next step to continue this change.

The integration of Brazilian Geodetic Network to SIRGAS will be carried out through a simultaneous adjustment of the network using the original data and the software Geodetic Adjustment using Helmert Blocking of Space and Terrestrial data (GHOST). In the adjustment will be adopted the GRS80 elipsoid and the Horizontal Network will be constrained to fit the coordinate values of the eleven SIRGAS stations in the national area.

The installation and operation of Brazilian Network for Continuous Monitoring of GPS (RBMC), concluded in 1997, will provide a continuous link to SIRGAS for future users.

Introduction

Being responsible for geodesy and mapping in Brazil, the Brazilian Institute of Geography and Statistics (IBGE) started the establishment of the Brazilian geodetic network in the 1940's. Since this time, the methods of surveying and processing surffered through great transformations mainly on the respect of accuracy. Focusing on homogeneous results, the geodetic net had various adjustment in differents horizontal datums. Due to computational system limitations at the moment, only partial adjustment of the net were carried out.

Definition and establishing of the South American Datum 1969 (SAD69), targeted the last partial adjustment made. From that time, a continental adjustment software was looking for, and efforts were put on a first global and simultaneous adjustment of the Brazilian geodetic net.

Next step to be performed till the end of 1997, its the integration of the Brazilian net into SIRGAS System. With this aim the 11 SIRGAS stations in the country were connected to existing geodetic stations of the net. With 493 stations, the Brazilian GPS

Network is already adjusted and constrained into SIRGAS stations, but more campaigns have to be included. The combination of GPS and classical geodetic network step, have 5225 stations already simultaneously adjusted through the Helmert Blocking method.

This paper present a historical abstract of all adjustments made and geodetic systems used in Brazil, also the efforts in preparation and processing of both classical and GPS nets, in order to integrate them into SIRGAS system.

History of Horizontal Geodetic Control in Brazil - Establishment, Adjustments and Geodetic Reference Systems

A triangulation project in the south of the country in the 1940's, started the establishing of the high accuracy geodetic network. Since that time, the triangulation network spread over meridians and paralels, allowing to use the polygonal method for further densification.

In the 1970's the availability and accuracy of the TRANSIT satellite geodetic positioning, transported the horizontal net to areas where conventional methods were difficult to use, as Amazonic region.

The densification method by GPS process was adopted when four TRIMBLE GPS, 4000 sst model came in 1991, marking a new tridimensional geodetic network era. Establishing the Active Control GPS Network, named Brazilian Network for Continuous Monitoring of GPS – RBMC, with 9 stations (same as SIRGAS Project's ones occupied) was a very important step for the geodesy in Brazil. At the moment, the geodetic net have 6318 stations being of them, 3498 triangulation stations, 26 trilateration stations, 1158 traverses stations, 1143 Doppler stations and 493 GPS stations.

It is importat to pointed out the fact that coordinates were adjusted using various methods and geodetics systems during this time. Brazilian network had three main adjustments.

First adjustment was made with the condition equations method and adopting the Córrego Alegre Geodetic System as reference.

Second adjustment had two computer phases, first one carried out by DMA/IAGS (Inter American Geodetic Survey), adopting SAD69 as horizontal datum and using the computer system HAVOC (Horizontal Adjustment by Variation of Coordinates). The next phase performed at IBGE, adjusted the remained network using USHER program (Users System for Horizontal Evaluation and Reduction). During the 20 years application of the adjustment in parts, unavoidable net distortions were introduced. It was felt the needed for a computational system, able to perform a simultaneous adjustment of great amount of geodetic data and at the same time using a tridimensional model in order to include classical and GPS observations. At the end of the 1980's GHOST (Geodetic adjustment using Helmert blocking Of Space and Terrestrial data) system for DOS and UNIX platforms was installed at IBGE. The whole classical observations were passed through a validation and magnetic storage process.

It was then in the middle of 1996, when the third geodetic adjustment was finished, but this time, was a global and simultaneous one. Adjustment was made on the SAD69 geodetic system because it is the official one in the country. Adoption of geocentric reference system is another step to be carried out. With this aim, Brazil has an active participation in the SIRGAS Project, through technical cooperation, being one of the project data center and applying the projects' recomendations made, one of them refering the integration of geodetic network into SIRGAS geodetic system.

Terrestrial Data

The classical network observations used in the global adjustment of 1996 were: 16913 horizontal directions, 1534 geodetic distances, 389 astronomic azimuts and 379 astronomic coordinates. Same observations will be used in the adjustment into SIRGAS system. The validation and storage observations steps were development in three different projects.

A first project was the horizontal directions treatment, being the longest due to the amount of data to be analysed.

The second one was the reprocessing of astronomic observation (latitude, longitude and azimut) in FK5 system was made in the project "Pro-Astro".

The third project aimed the reprocessing of geodetic bases.

All the three projects were carried out with computational systems developed at IBGE. The data of 144 traverses were through the validation and reprocessing phase as well. Once the project was finished, it was started an analysis of the combination of data adjusted. The classical network was divided in 13 areas, separately adjusted and during a later step the classical network was adjusted using the Helmert Blocking method.

Spatial Data

During the period 1973-1991, IBGE made use of the satellite's TRANSIT geodetic positioning system, mainly in difficult areas for classical field works, e.g. Amazon region. The more used observational method for stations determinations, was point positioning. For a more accurate treatment of DOPPLER observations, a precise ephemeris postprocessing with program GEODOP V was made. During the adjustment only 179 DOPPLER stations were used, those which are coincident with classical ones.

GPS technology was first used at IBGE, with the purchase of four geodetic receivers in 1991. Till 1994 a total of 187 GPS stations, were processed with TRIMVEC Plus, D version and using broadcast ephemeris. Later on data was processed with Bernese version 4.0 and using now, precise ephemeris. In both cases, the station coordinates and its own variance-covariance matrix were generated in the final processing solution and later on used in the GPS network adjustment. The mathematical correlations were handled correctly for the data processed with Bernese software. In order to prepare the adjustment, the data of each individual campaign were put in GHOST format and a minimun constrain adjustment was performed. The variance factor obtained on adjustment will be used to scale the matrixes. For each campaign this procedure is repeated and in a second step it is included in the final adjustment file.

For SIRGAS integration we have participating a total of 1565 GPS baselines, as can be seen in figure 1.

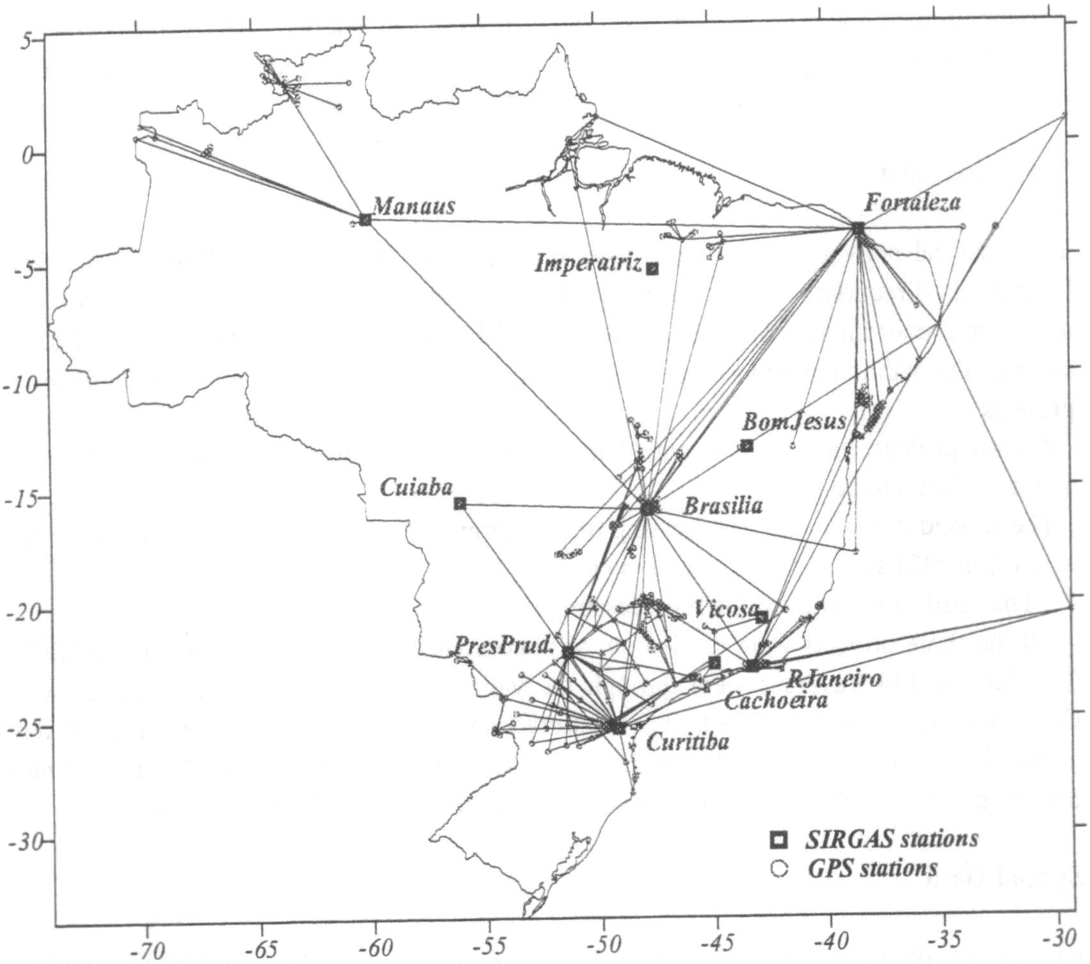

FIGURE 1 - Brazilian GPS Network - 1997

Adjustment Methodology - SIRGAS Integration

For the adjustment of large geodetic nets the numerical procedure of partitioning the total net in several blocks in a very useful tool. The subdivision into blocks can be brought about according to diferent points of view: with respect to regional aspects and special kind of observations. It is not only useful for the partitioning of a geodetic net, moreover plays an important role for *the rigorous fusion of triangulation nets with other geodetic systems*. This is why GHOST software was chosen to carry the brazilian network adjustment out.

GHOST software is composed by a set of programs using a tridimensional modeled system for least squares adjustment. Its main features is to employ the Helmert Blocking method for partitioning a continental size geodetic network as the brazilian one is.

The defined strategy for block partitioning is based on *a minimun number of junction station between blocks*, minimizing in that way the computational effort in the adjustment.

As seen in figure 2, the geodetic network was splited into 8 blocks, each of them with three different levels. The adjustment task was divided into three tasks. After finished all data set validation of classical network, it was performed a Helmert blocking adjustment. The second task was perform the adjustment with GPS data. The third and last task was performed an adjustment combining GPS and classical observations. The eleven SIRGAS station complement the integration with the introduction of the sigmas' weighted coordinates.

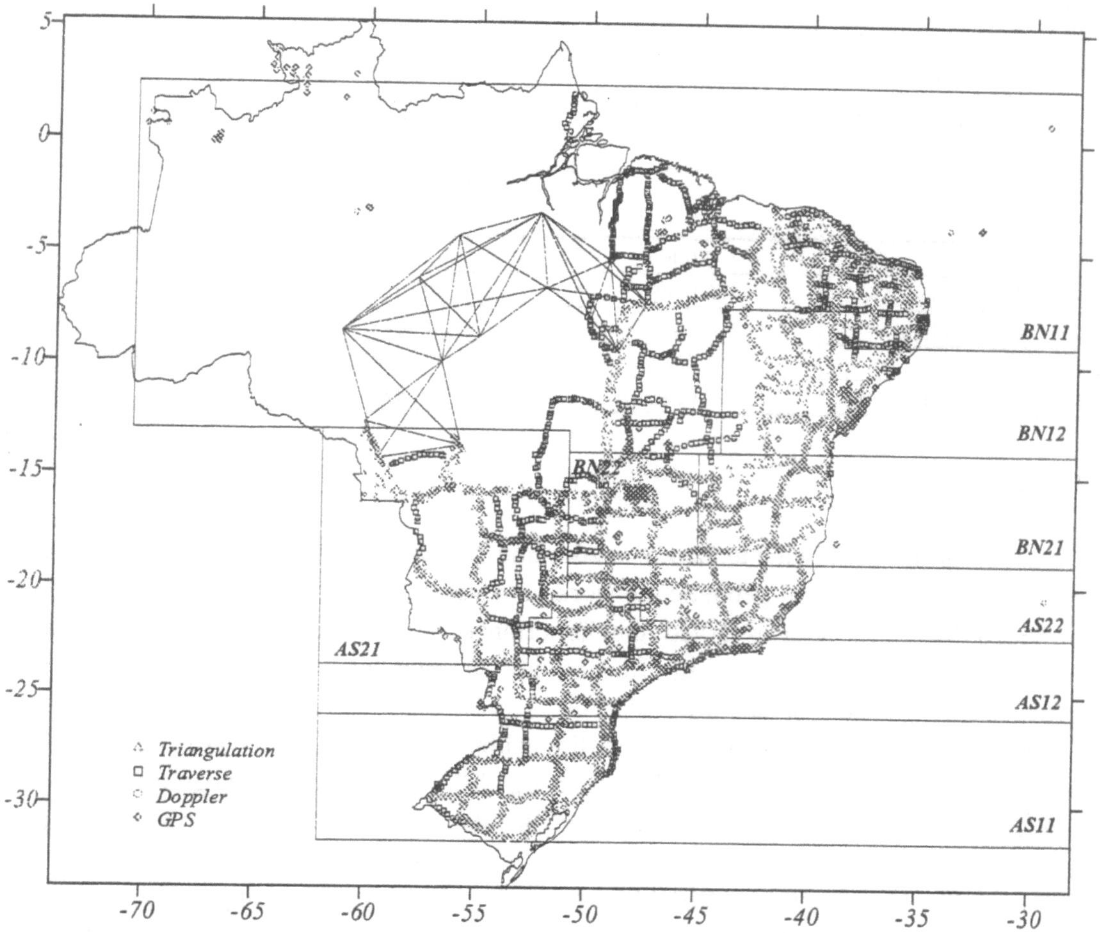

FIGURE 2 - Brazilian Geodetic Network 1997 and Helmert Blocking division

Relation of SAD-69 to SIRGAS

Established on the end of 1960's, the SAD-69 was officially used in Brazil in the 1970's. With the satellite positioning techniques adopted, it was inferred the necessity for solving transformation parameters between various geocentric referece systems, as NWL-10D, NSWC-9Z2 and WGS-84, aiming the agreement with SAD-69.

From this eleven SIRGAS station, only five already had SAD-69 coordinates known, generated from the simultaneous and global adjustment finished in 1996. In a first test, there were determined a set of seven Helmert transformation parameters. During the results analysis was find the rotations as one hundred part of a second.

In the second test with translations only, high Fortaleza station residuals were found as compared with the remained stations. Taking this into account and aiming the most accurate results, Fortaleza station was removed for the computation step, leaving the following ones: Curitiba, Cachoeira Paulista, Brasília, Presidente Prudente, and having this final results.

Number of Parameters : 3
Number of Coordinates : 12
rms of transformation : 0.0716 M
Parameters:
Translation in x : 67.327 +- 0.036 M
Translation in y : -3.899 +- 0.036 M
Translation in z : 38.292 +- 0.036 M

NAME	RESIDUALS IN METERS X	Y	Z
Brasília	0.0064	-0.0468	-0.0977
Pres. Prudente	0.0148	-0.0123	0.0842
Cachoeira Paulista	-0.0299	0.0017	0.1125
Curitiba	0.0145	0.0558	-0.0998
RMS / COMPONENT	0.0213	0.0427	0.1144

Conclusion

Considering the worldwide tendency in adopting geocentric reference systems, bustled by the bigger necessity of geodetic network integration and the technique accessibility provided by GPS, its was noted the importance of the Brazilian Geodetic Network integration into SIRGAS system, judged at the moment one of the most accurate geocentric reference system worldwide.

According to the development project timetable, it is scheduled for next december the activities conclusion, with some still processing GPS campaign final integration. The new SIRGAS system referred coordinates will be available to the comunity at the begining of 1998.

Some studies for adoption of SIRGAS system for mapping are already being made. The first step was the determination of transformation parameters in order to become automatic the SIRGAS system refered digital maps elaboration. It is known that is this process is slow because great part of the maps sets is not in a digital way, but efforts have been made to conclude this step as soon as possible.

Bibliography

- **Beattie, D.S.** - ***Program GHOST User Documentation***, Geodetic Survey of Canada, Ottawa, 1987.
- **Cadess, H. et al** - ***InterAmericam Geodetic Integration*** - PAIGH, 1991.
- **Costa, S.M. & Fortes L.P.S.** - ***Resultados Preliminares do Ajustamento da Rede Planimétrica do Sistema Geodésico Brasileiro***, Rio de Janeiro, 1993.
- **NGS**, ***North American Datum of 1983***, NOAA Professional Paper NOS2, National Geodetic Information Center, NOAA, 1983.
- **Rothacher M.& Mervart L.** - *Bernese GPS Software Version 4.0* - Astronomical Institute University of Berne - Berne - september 1996

ADJUSTMENT OF THE NEW VENEZUELAN NATIONAL GPS NETWORK WITHIN THE SIRGAS REFERENCE FRAME

Hermann Drewes and Herbert Tremel
Deutsches Geodätisches Forschungsinstitut, Abt. I
Marstallplatz 8, D-80539 München/Germany

José Napoleón Hernández
Servicio Autónomo de Geografía y Cartografía Nacional
Edif. Camejo, CSB, Caracas 1010/Venezuela

ABSTRACT

The new Venezuelan national geodetic network (Red Geodésica Venezolana, REGVEN) was observed nearly simultaneously with the GPS campaign for the establishment of a geocentric reference system for South America, Sistema de Referencia Geocéntrico para América del Sur (SIRGAS). This paper describes the REGVEN data, its processing and the connection to the SIRGAS reference frame. The accuracy of the geocentric station coordinates is in the order of ±2 cm.

INTRODUCTION

The GPS campaign for establishing an accurate geocentric geodetic reference system for South America, SIRGAS, has been carried out in May/June 1995. The Deutsches Geodätisches Forschungsinstitut, Abt.I, München (DGFI/I) was acting as a data center and as an analysis center processing the complete data set of this project. Besides the national institutions of South America, several institutions in Europe and North America participated in the SIRGAS observation campaign.

All South American countries expressed their intention to connect their national reference systems to SIRGAS, or to establish new national reference systems within the SIRGAS frame by GPS, respectively. The Venezuelan Servicio Autónomo de Geografía y Cartografía Nacional (SAGECAN) installed a completely new geodetic network (Red Geodésica Venezolana, REGVEN). It consists of 67 stations covering the entire territory of Venezuela, except the sparsely settled southern area (Territorio Federal Amazonas) and was observed quasi simultaneously with the SIRGAS network. REGVEN is connected to SIRGAS by the five Venezuelan SIRGAS stations Maracaibo, Junquito, Agua Linda, La Canoa and Kama. As SIRGAS refers to the ITRF94 at epoch 1995.4, REGVEN is thus defined in the same reference system.

OBSERVATIONS

The REGVEN GPS campaign was carried out during the period May 20 - June 16, 1995. The observation period covers thus 28 days. Four of the five SIRGAS reference stations in Venezuela were attended and supported by DGFI/I, all the other stations, including the SIRGAS reference station Junquito were observed by SAGECAN. The GPS systems Trimble SSE / SST and Leica 200 (only during the SIRGAS campaign at station Kama) have been used for the measurements.

The daily occupation of stations was as follows:

- May 20 - 25, 1995 10 stations,
- May 26 - June 4, 1995 13 stations and
- June 5 - 16, 1995 5 to 7 stations.

Apart from a few exceptions all stations observed at least two or three different days (sessions), operating six hours per day. All the SIRGAS reference stations recorded 23 or 24 hours daily during the SIRGAS observation campaign. Outside the SIRGAS campaign at least three SIRGAS reference stations were occupied simultaneously. Generally the sampling rate was 15 seconds. The observations included data down to an elevation cut off angle of 10°. The location of all sites occupied during the REGVEN campaign is shown in figure 1. The names of stations correspond to the conventions of SAGECAN.

DATA PROCESSING

For the processing of the REGVEN data mainly two program systems were used:

- the Bernese GPS software system version 3.4 (BV3.4) (Rothacher et al., 1993) with some modifications implemented at DGFI/I;
- the DGFI/I program ACCSOL (ACCumulate and SOLve) for the accumulation and solution of normal equations.

The processing started with the RINEX (Receiver INdependent EXchange format) files containing the observation data, which had to be verified and edited in a first step. This included in particular the checking of the compatibility of file headers with the RINEX format and the reduction of the antenna phase centers to the defined reference points.

A generalized scheme of the complete GPS data processing with BV3.4 and ACCSOL is given in the description of the DGFI processing of the SIRGAS '95 GPS network in the official SIRGAS Report (SIRGAS, 1997).

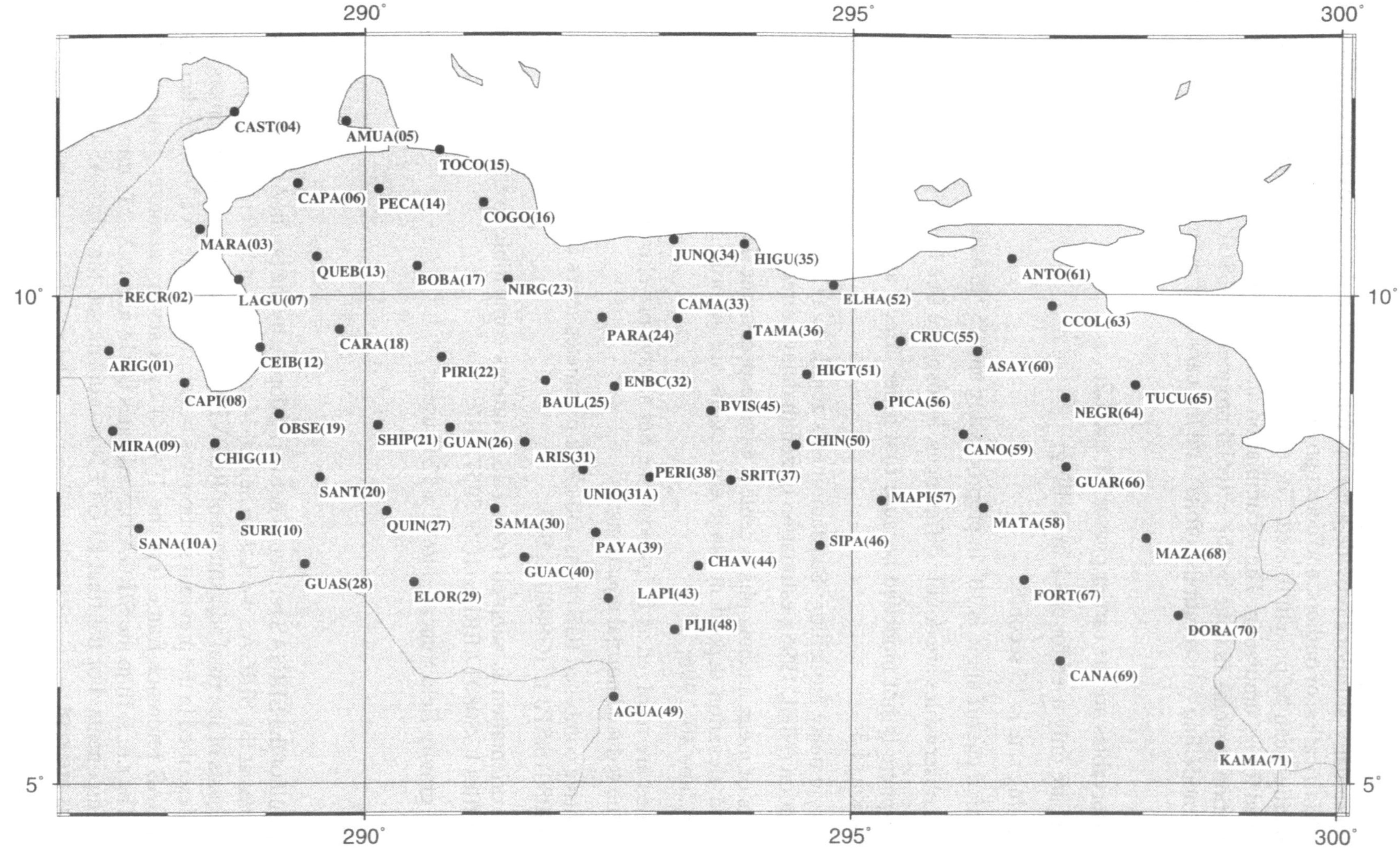

Figure 1: Location of sites of the new Venezuelan reference network (REGVEN)

Similar to the SIRGAS processing daily solutions for the 28 days of REGVEN were performed. These daily solutions provide a means for assessing the day by day consistency, allow the identification of outliers, and detect processing errors; in addition local parameters (e.g. ambiguities) can be pre-eliminated. The reduced and unconstrained normal equations are then saved for accumulation. The accumulation and solution of all the 28 unconstrained normal equation systems is done by the external program ACCSOL. The main characteristics of the modelling (observations, orbits) and the parameter estimation are described in the sequel:

- basic observables are the carrier phases L1 and L2;
- code data are only used for clock modelling;
- the sampling rate is 15 seconds;
- the elevation cut off angle is 10° in preprocessing and 15° in adjustment;
- modelled observables are double differences, ionosphere free linear combination L3;
- the ionosphere is not modelled because first order effects are eliminated by the linear combinations L3;
- tropospheric delay modelling: Saastamoinen's zenith delay prediction and Niell's mapping function (Niell, 1996); estimation of zenith delay corrections every two hours;
- the offsets of mean antenna phase centers with respect to the defined antenna reference points and elevation dependent range corrections are according to the model IGS-01, issued on June 30, 1996;
- ambiguities are resolved as far as possible, but always pre-eliminated in the daily solutions of the reduced normal equations;
- the definitive combined IGS orbits and clock parameter series are used, according to the files IGS08016.SP3 to IGS08056.SP3;
- the earth orientation is adopted from IGS solutions compatible to the satellite orbits, given in files IGS08017.ERP to IGS08057.ERP;
- GEM T3 gravity field truncated to 12 x 12 is used.

RESULTS

The final result of the SIRGAS reference network adopted by the South American countries' representatives at the SIRGAS workshop held at Margarita Island, Venezuela, April 8 - 11, 1997, is the basis of the final adjustment of REGVEN. The unconstrained station coordinate solution as described in the previous section was transformed to the five Venezuelan stations of the SIRGAS reference frame (epoche 1995.4) by a seven parameter similarity transformation. The r.m.s. fit to the SIRGAS coordinates is ±14 mm in X, ±11 mm in Y, and ±15 mm in Z components. For the final REGVEN coordinate set, the SIRGAS coordinates of these stations are kept.

The overall accuracy of the REGVEN adjustment may be estimated from the daily solutions and the 28 days final solution. For their comparision the numerical rank defects are removed by performing similarity transformations. The r.m.s. differences between the daily adjustments and the final solution range from ±2.9 mm to ±23.6 mm in the horizontal components and ±4.2 mm to ±28.8 mm in the vertical component. Estimating the external accuracy one has to consider, that some stations have been occupied only one day. As the stations are situated close to the equator, the observations are strongly influenced by high fluctuations of the ionosphere.

A further possibility to control parts of the REGVEN coordinates indepentently, are the results of the CASA (Central and South America) project (Kaniuth et al, 1997). Eight stations of the CASA network are identical with REGVEN stations. In table 1 the residuals after a similarity transformation between the two networks at epoch 1995.4 are given. They are nearly all less then two centimeters.

As a summary one may estimate the accuracy of the station coordinates of the REGVEN network to be in the level of ±2 cm or better for all components. This is a very good basis to serve as a national reference frame for the future. The maintenance of the network will include repeated observation campaigns allowing to derive station coordinate changes which have to be expected due to the tectonically active region along the Caribbean - South American plate boundary crossing the northern part of Venezuela (see also Kaniuth et al. 1997).

Table 1: Residuals (mm) in North (N), East (E) and Height (H) between identical stations of REGVEN and CASA after similarity transformation and number of sessions in the two independent campaigns

Station	No. of Sessions		Residuals [mm]		
	REGVEN	CASA	N	E	H
AMUA(5)	2	4	16	-11	-1
NIRG(23)	4	4	-9	22	-11
ENBC(32)	3	4	4	16	-13
KAMA(71)	21	13	3	-8	0
MARA(3)	16	13	-2	-7	-4
JUNQ(34)	28	13	-7	-3	17
CANO(59)	27	13	-2	-1	-1
OBSE(19)	2	4	-3	-9	13

REFERENCES

Kaniuth K., H. Drewes, K. Stuber, H. Tremel, N. Hernández, M. Hoyer, E. Wildermann,H.G. Kahle, Ch. Straub (1997): Crustal Deformation along the Caribbean-South American Plate Boundary Derived from the CASA GPS Project. IAG Scientific Assembly, Rio de Janeiro, Sept. 3-9, 1997.

Niell A.E. (1996): Global mapping functions for the atmospheric delay at radio wavelength. J. Geophys. Res., 101, B2, 3227-3246.

Rothacher M., G. Beutler, W. Gurtner, E. Brockmann, L. Mervart (1993): Bernese GPS Software Version 3.4 - Documentation. Astronomical Institute, University Bern, 1993.

SIRGAS (1997): SIRGAS Final Report, working groups I and II. Edited by Instituto Brasileiro de Geografia e Estatística (IBGE), Rio de Janeiro, 1997.

REALIZATION OF A GEOCENTRIC REFERENCE SYSTEM IN ARGENTINA IN CONNECTION WITH SIRGAS

J. Moirano[1] and C. Brunini
Fac. de Cs. Astronómicas y Geofisicas (FCAG), UNLP.
Paseo del bosque s/n, 1900 La Plata, Argentina

H. Drewes and K. Kaniuth
Deutsches Geodätisches Forschungsinstitut Abt. 1 (DGFI/1)
Marstallplatz 8, D-80539 Munich, Germany

ABSTRACT

Between 1993 and 1994 the POSiciones Geodésicas ARgentinas (POSGAR) GPS network was observed, and early in 1995 the POSGAR'94 reference frame was established. It consists of 127 evenly distributed stations providing a relative precision better than 1 ppm in the WGS84 reference system.

Since 1995, the DGFI/1 and the FCAG are working together to obtain a Geocentric Reference Frame for Argentina, consistent with the one realized by the SIstema de Referencia Geocéntrico para América del Sur (SIRGAS) network.

This work, which involved a cooperation in the SIRGAS reference network computation and then the re-processing of Argentina's reference network is scheduled to be finished in early 1998.

INTRODUCTION

In 1993 there was a growing need in Argentina of a national reference frame in accordance with the positioning accuracy of the new space techniques. In this context, a GPS geodetic network with a relative precision of 1 ppm was proposed.

Between 1993 and 1994, *POSGAR* was observed mainly by the IGM (Instituto Geográfico Militar) with the collaboration of the UNAVCO consortium through the Central Andes geodynamic Project (CAP). The network was computed by the FCAG

[1] Fellow of the Consejo Nacional de Investigaciones Científicas y Técnicas

(Facultad de Ciencias Astronómicas y Geofisicas) on contract with the 'Programa de desarrollo económico y saneamiento financiero de las provincias argentinas', a development project managed by the Ministry of the Interior and partially funded by the World Bank.

The observations were processed with a commercial software and the reference frame was realized through 19 CAP stations and one DORIS point all of them transformed to WGS84.

The *POSGAR94* reference frame was delivered in early 1995 and published by IGM some months later [IGM, 1995]. It consists of 127 evenly distributed points forming a grid of approximately 200 km cell size covering the whole country.

Meanwhile, in 1993, the SIRGAS project had begun, its main goals being the establishment of a high precision geocentric reference frame and the integration of all the South American countries to this frame. Within this project, DGFI/I and FCAG have been working in cooperation, first on the measurement and computation of the reference network, and now on the integration of Argentina to the SIRGAS frame. This integration is being done following the SIRGAS WGII recommendations delivered to the countries by the end of 1996 [SIRGAS, 1996][SIRGAS WGII, 1997].

Finally, in 1997, FCAG and IGM signed an agreement to cooperate in the maintenance of the national reference system. This constitutes an important step towards a future adoption of SIRGAS as the official zero level reference frame for Argentina.

The *POSGAR94* frame is briefly described here from the observations to the final adjustment. Indications of the relative and geocentric quality of the network are included. The status and strategy of the integration of POSGAR into SIRGAS are also described. Finally, some results showing the difference between POSGAR94 and its recomputation are also included.

THE OBSERVATIONS

The measurements of POSGAR stem from three different GPS campaigns: POSGAR'93 and CAP, carried out simultaneously, and POSGAR'94, observed in the following year. The POSGAR'93 observations were made between 10th February to 12th April 1993, whereas the CAP observations were done between 11th February and 11th March 1993. The POSGAR'93 sessions last for 6 hours and the CAP ones for 22 hours.

The receivers used were TRIMBLE 4000 SST and TOPCON both with double frequency and squaring L2 recovery. The number of receivers varies from 5 for the pure POSGAR'93 sessions up to 9 for the ones when POSGAR and CAP were measured simultaneously.

The POSGAR'94 campaign was carried out between 8th March and 5th May 1994. The sessions lasted for 6 hours and 3 TOPCON receivers were used.

COMPUTATIONS AND INTRODUCTION OF THE REFERENCE SYSTEM

All the possible vectors between the simultaneously measuring stations on each session were computed by means of the program GPPS V5.1 from Ashtech.

The adjustment was carried out by means of the program *oalpnet*, developed in FCAG [Usandivaras and Brunini, 1992]. It treats the vector components as observations and allows for the introduction of a priori coordinates and weights for some stations.

The reference system was introduced by means of the coordinates of 20 stations:
19 CAP stations located on the north-west and mid-west of the country.
The EARG (Estación Astronómica de Río Grande) station, on Tierra del Fuego, whose coordinates came from the DORIS beacon which operates on the site.

The CAP coordinates were obtained through a personal communication from Robert Smalley, from the Centre for Earthquake Research and Information at Memphis State University. In both cases the original coordinates were transformed to WGS84 and then introduced into the final adjustment with an a priori error of 3 cm [Usandivaras and Brunini; 1995].

THE POSGAR'94 FRAME

Figure 1 shows the number of ties for each point as seen by the adjustment software, the source being the 660 vectors included in the final adjustment in 1994. However, it should be kept in mind that this number of ties does not reflect reality in the sense that the inter-baseline correlations were not taken into account.

A better idea of the redundancy within the network would be given by the number of times each station was occupied. This information is shown on figure 2.

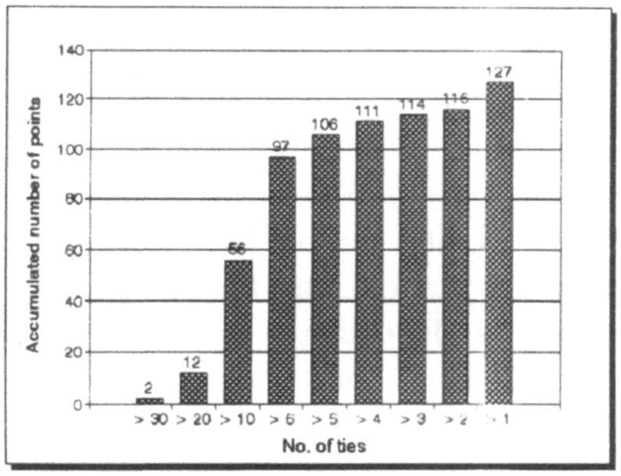

Figure 1: Number of ties per point

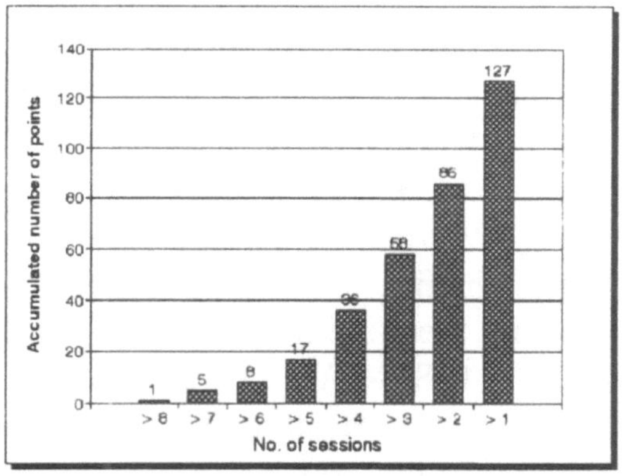

Figure 2: Number of sessions per point

In 80% of all cases the errors of the final coordinates are below 30 cm at a confidence level of 95%. This is shown in figure 3.

From the errors in the final coordinates, the relative errors were computed for the 660 baselines that were included in the adjustment. The result is shown in figure 4, where it can be seen that for a confidence level of 64% more than 75% of the baselines have errors of less than 1 ppm and more than 90% lie below 2 ppm.

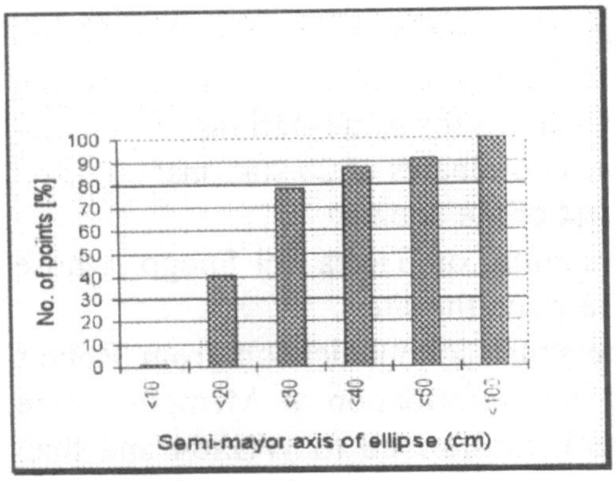

Figure 3: Errors in absolute coordinates (95% significance ellipses).

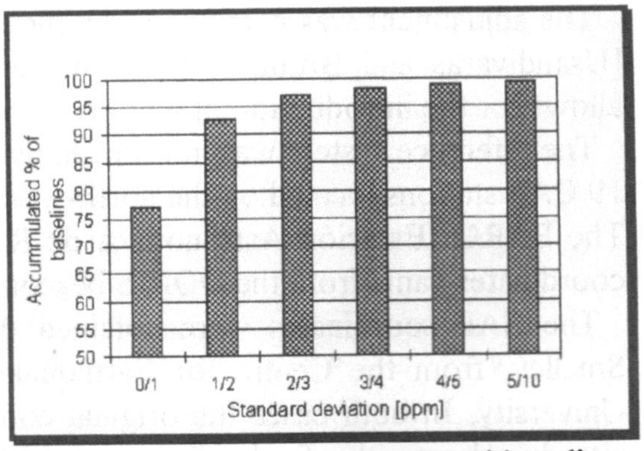

Figure 4: Std. Errors in computed baselines.

As an external check, POSGAR'94 and SAGA (South American Geodynamic Activities) networks where compared in four points in northern Patagonia. The comparison is summarized in table 1.

Another comparison, for one point in Buenos Aires province was made between POSGAR'94 and SAGA [Del Cogliano et al., 1996]. This is shown in table 2.

Table 1

SAGA-POSGAR94	*Mean (m)*	*RMS (m)*
Latitude	-0320	0.024
Longitude	-0.125	0.017
Height	0.422	0.089

Table 2

SAGA-POSGAR94	*Difference (m)*
Latitude	-0.065 m
Longitude	-0.264 m
Height	1.016 m

POSGAR'94 and the SIRGAS network. In April 1997, during the SIRGAS meeting in Margarita, the final solution of the SIRGAS network was presented. This final solution has a geocentric precision of better than 1 cm and realizes the ITRF94 reference system for the epoch 1995.4 [SIRGAS, 1997].

There are 6 identical points in the SIRGAS and POSGAR'94 frames. They were compared directly (no transformations involved) to have an overall external assessment of geocentric accuracy and relative precision of the POSGAR'94 frame. The results are summarized in tables 3 and 4, which show the relative precision of the network to be better than 1 ppm and the geocentric accuracy to be better than 1 m.

THE NEW COMPUTATION AND INTEGRATION OF POSGAR INTO SIRGAS

During the first half of 1996 some processing tests showed that a re-computation of the POSGAR network with the Bernese software would significantly improve the quality of the national reference frame. This improvement would allow a better definition of the geocentric reference frame when connecting POSGAR to SIRGAS.

Since August 1996, the FCAG is working with the DGFI/I in the re-computation of the POSGAR network as a first step before the connection to SIRGAS is made. At this time, adding four CAP stations from the Chile campaign plus the Santiago IGS (also SIRGAS) station were judged useful.

Table 3 *Geocentric coordinates*

POINT	*LAT [cm]*	*LON [cm]*	*H [cm]*
LOTE 10B	48	53	-14
LOTE 24	49	14	-17
EL MAITEN	40	26	34
MORRO	50	24	41
V. ROBLES	52	23	4
RIO GRANDE	83	33	-128

Table 4 *Baselines*

BASELINE LOTE10B to	*LENGTH [KM]*	*DIF. [PPM]*
LOTE 24	900	0.04
EL MAITEN	498	-0.01
MORRO	1438	0.04
V. ROBLES	2040	-0.01
RIO GRANDE	862	0.31

For the data processing, the Bernese software V3.5 is used [Rothacher and Mervart; 1996]. During the quite lengthy phase pre-processing, a careful data editing was necessary to eliminate noisy data that would otherwise be misinterpreted as series of cycle slips. This implies a first processing of each vector alone to make finding data problems easier. CODE orbits and rotation parameters were used for all the POSGAR campaigns since there are no IGS precise orbits available for the 1993 campaigns. The elevation mask in processing was set to 15 degrees. The tropospheric delay is computed with the Saastamoinen prediction model using standard atmospheric data plus a zenith delay correction obtained from the GPS data with a 1/cos(z) mapping function. These parameters are set for each station to be valid for at most 5 hours.

After cleaning all vectors, each session was processed, solving for coordinates, tropospheric parameters, and ambiguities. In these runs, ambiguities were eliminated after inversion of the normal equations which were also saved. No a priori weights for the coordinates were applied, leaving then the definition of the reference system to the orbits. Finally, the normal equations from all sessions were added and the complete network was adjusted. Firstly, only a free adjustment was computed in order to perform quality testing. In the near future, the reference frame will be introduced through the SIRGAS stations.

The epoch for the fiducial coordinates will be set to the middle of the POSGAR campaigns, which is 1993.8. The best choice for the velocity model to be used for the transformation will be analyzed. The final POSGAR coordinates will be transformed to epoch 1995.4 so as them to be in the SIRGAS system.

The link between POSGAR and SIRGAS. The SIRGAS points in Argentina not included in POSGAR have been integrated in three cases: LPGS (La Plata, IGS), CRIC (Cricyt) and IGUA (Iguazu). They were connected during 1995 and 1996 with at least one POSGAR station by at least two sessions of more than 6 hours. The computations will be finished soon this year. The coordinates will be obtained in ITRF94, epoch of the observations, and then transformed as stated above to be used as fiducials in the final adjustment.

POSGAR'94 and POSGAR'98. Here we show the comparisons between the SIRGAS results and the POSGAR coordinates for both POSGAR'94 and the latest new computation. Table 5 shows a summary of the residuals from similarity transformations computed between the frames for 5 out of the 6 points common to the three frames.

Table 5 **POSGAR'94-SIRGAS** **POSGAR'98-SIRGAS**

RMS	*7 parameter*	*3-D shift.*	*7 parameter*	*3-D shift.*	*Ratio 7 P.*	*Ratio 3 P.*
North (m)	0.030	0.026	0.008	0.010	3.8	2.6
East (m)	0.042	0.086	0.017	0.014	2.5	6.1
Up (m)	0.176	0.289	0.007	0.027	25.1	10.70
General	0.130	0.175	0.014	0.018	9.3	9.72

It is clear from the ratios between the transformations that the new computation is about one order of magnitude more consistent with SIRGAS than POSGAR'94. It is also clear that most of the improvement is achieved in the heights, the greatest weakness of POSGAR'94 as can be readily seen from tables 1, 2 and 3.

CONCLUSION

POSGAR'94 has a relative precision of 1 ppm and an accuracy of roughly 1 m. The new computation being carried out by the FCAG in cooperation with the DGFI/I would improve the quality of the frame by a factor of ten and at the same time integrate Argentina to the continent through the SIRGAS frame. This work will be finished early in 1998.

REFERENCES

SIRGAS, 1997. Final Report of the SIRGAS project, Working Groups I and II. Presented at the IAG Scientific Assembly, Rio de Janeiro, Sept 3-9, 1997.

SIRGAS WG II, 1997. Recommendations for the integration of national geodetic networks into SIRGAS; Personal communication.

SIRGAS, 1996. Boletín informativo n° 4.

Rothacher M. and Mervart L., 1996. Bernese GPS Software Version 4.0; Astronomical Institute, University of Bern.

Del Cogliano D., Perdomo R., Di Croche N. and Napal E., 1996. Control de referencias con GPS en Bahía Blanca, FCAG; La Plata.

Usandivaras J., Brunini C., Canosa D., Mondinalli C., Gende M., Moirano J. and Alvarez J. 1995. Cálculo de la red geodésica nacional argentina, estrategia y resultados; Proceedings of the XVII Congreso Brasileño de Cartografía, Salvador, Bahía.

IGM, Instituto Geográfico Militar, 1995. La Red POSGAR; IGM, Buenos Aires.

Usandivaras J. and Brunini C. 1992. Programa general de compensación por mínimos cuadrados; 17th meeting of the AAGG; Buenos Aires.

SOME CONSIDERATIONS RELATED TO THE NEW REALIZATION OF SAD-69 IN BRAZIL

Leonardo C. Oliveira[1] ; João F. G. Monico[2] ; Marcelo C. Santos[3] ; Denizar Blitzkow[4]

[1] IME - Departamento de Engenharia Cartográfica - Praça General Tibúrcio, 80 Rio de Janeiro - RJ - Brazil - 22290-270 - (*leonardo@aquarius.ime.eb.br*)

[2] UNESP - Departamento de Cartografia - Rua Roberto Simonsen, 305 Presidente Prudente - SP - Brazil - 19060-900 - (*galera@stetnet.com.br*)

[3] UFPR - Departamento de Geociências - Caixa Postal 19011 Curitiba - PR - Brazil - 81531-970 - (*mcsantos@geoc.ufpr.br*)

[4] USP - Departamento de Transportes - Av. Prof. Almeida Peado, s/n São Paulo - SP - Brazil - 05508-900 - (*dblitzko@usp.br*)

ABSTRACT

IBGE - the institution responsible for the Brazilian Geodetic System (BGS) worked for more than ten years in order to readjust the observables belonging to this system. This work was finished at the end of 1996. The main reason for this paper is that there is the need for complementary analysis of the results. A few aspects related to the new realization of the SAD-69 in Brazil will be shown, taking into account its accuracy and aspects related to the conversion between the new and the old realizations. For the last case, a studied was carried out by using either the seven or three parameters transformation model. A new proposal for the transformation of coordinates is presented. Further investigations should still be carried out. The analysis of the accuracy was based on 34 GPS stations, which are coincident with the conventional network, being their coordinates considered as true values. A total of 34 adjustments were carried out, in such way that the GPS baseline vectors of each one of the stations were removed in the process. The estimated coordinates were compared with the true ones. The results shown an improvement of the new realization. For a more conclusive analysis, further GPS stations would be required.

Introduction

The establishment of the South American Datum 69 (SAD-69) had as its major objective to provide the whole continent with a uniform geodetic system, capable of controlling both surveying and cartographic activities (Fisher, 1973). In Brazil SAD-69 was adopted late 1970's. Since then all geodetic positioning has been refered to it. Its first adjustment was carried out by Defense Mapping Agency (DMA), by sequentially adjusting triangulation blocks, being this the first realization of SAD-69 in Brazil. Hereafter this realization is simply referred to as SAD-69.

The readjustment of SAD-69 took part of a ten years long projetc called REPLAN, which was concluded in 1996. This task was carried out using the software GHOST - Geodetic adjustment using Helmert blocking Of Space and Terrestrial data, involving 4,939 geodetic points. The estimated variance factor was equal to 1.493, refering to the adjustment of 16,913 horizontal directions, 389 astronomical azimuths, 378 astronomical stations, 257 geodetic baselines (triangulation), 1,277 geodetic baselines (traverses), 179 doppler coordinates (point positioning), and 1,198 GPS baselines. In order to better deal with network deformations, 11 auxiliary parameters were included, namely 7 scale parameters for the classical network, 1 orientation parameter for the astronomical azimuths and 3 translation parametres for the doppler stations (Costa, 1996). This realization is hereafter simply referred to as SAD-69/96.

The origin point of the network - Chuá, did not have its coordinates, nor its vertical components and nor its associated geoidal undulation altered. Also unchanged remained the origin azimuth Chuá-Uberaba. In this way, the difference between SAD-69 and SAD-69/96 coordinates represent the improvement as a consequence of (a) the insertion of GPS control points into the geodetic network; (b) the new methodology of computation; (c) the simultaneous adjustment of all observations. This paper deals with the new realization of the South American Datum in Brazil.

Assessment of SAD-69/96 accuracy

An experiment was carried out in order to assess the accuracy of SAD-69/96. For this purpose the IBGE Brazilian GPS network was taken as a reference. This network is composed of 187 GPS stations, of which 49 coincident with the classical network. A number of 15 GPS stations were not used due to connecting problems with the classical network.

This experiment involved the computation of geodetic distance and difference in orthometric height between station positions as given by the adjustment of the GPS network and its corresponding ones derived from a number of 34 "global adjustments". In each one of these global adjustments the GPS observations of the station under scrutiny were removed, meaning that this sation was been considered as a classical station only.

The largest planimetric difference is equal to 2.889 m; the smallest equal to 0.015 m; with an average value equal to 0.319 ± 0.550 m. The largest altimetric difference is equal to -4.377 m; the smallest equal to -0.132 m; with an average absolute value equal to 1.239 ± 1.075 m. As far as Chuá is concerned, the planimetric and altimetric differences are equal to 0.396 m and 0.619 m, respectively.

By inspecting standard deviations with respect to the avarage values, it can be noted that there are values with significant magnitude within the sample, indicating a non-homogeneity of the data. These results seem to indicated a non-parallelism between SAD-69/96 and WGS-84, even though dealing with a small sample size. Probable causes for that may be:

- the non-uniform distribution of the GPS stations used for controlling the classical network;
- the use of different processing strategies in the computation of GPS stations as far as modelling is concerned. For example, the use of triple difference for processing long baselines.

Transformation between SAD-69 and SAD-69/96

Two geometric transformation have been tested, one using only three translation parameters - like the officially adopted in Brazil by IBGE for the transformation between SAD-69 and WGS-84, and the other one the seven-parameter similarity transformation, indicated by most of the literature elsewhere. For the three-parameters model (translations ΔX, ΔY, ΔZ along the X-, Y-, Z-axis), the following results were obtained:

ΔX = - 0,248 m
ΔY = 1,202 m
ΔZ = - 1,908 m.

Taking the significance level α equal 100% for the Chi-square test, the "confidence interval" for the estimated variance factor has to be equal to 1.016. Taking as hypothesis in the adjustment that all coordinates are of same precision, the standard deviation value for the acceptance of the test on the estimated variance factor for the estipulated significance level is equal to 2.769 m. Table 1 shows the values (in metres) of the largest and smallest residual (V), and its rms value for the X, Y and Z coordinates:

	Vmax	Vmin	rms
X	10.680	- 7.689	2.062
Y	9.138	-13.009	1.830
Z	17.314	-30.385	3.949

Table 1 - Residual and rms values for the three-parameters model.

For the seven-parameters model (translations ΔX, ΔY, ΔZ along the X-, Y-, Z-axis, scale κ, and rotations Rx, Ry,Rz along the X-, Y-, Z-axis) follows the results:

ΔX = -12,841 m
ΔY = 14,366 m
ΔZ = 4,990 m
κ = 0.00000308
Rx = -0,1991″
Ry = 0,1353″
Rz = 0,0670″.

Applying the same procedure as before,the standard deviation is equal to 1.979 m. Likewise the three-parameter case, table 2 can be build:

	Vmax	Vmin	rms
X	8.200	- 9.730	1.281
Y	8.631	-12.083	1.474
Z	19.643	-25.630	2.833

Table 2 - Residual and rms values for the seven-parameters model.

Obviously, the seven-parameter model allows a better mapping of the coordinate transformation from SAD-69/96 to SAD-69, such as the results seem to indicate. However, it is evident that the use of this model is not enough to handle the situation treated in this paper because it cannot absorb all deformations existing in the network.

A new proposal for the transformation between SAD-69 and SAD-69/96

The new proposal consists in dividing the network in regions, applying geometric transformation to the set of neighbouring stations, as opposed to the several solutions found elsewhere in the literature (Junkins, 1990; Nakiboglu et el., 1994; Brunini et al., 1996; Sillard & Boucher, 1996; Vanicek & Steeves, 1996). The transformation should allow the evaluation of all elements which define the relative positioning of the coordinate systems (translations, rotations, non-orthogonality and scales). This transformation is the Afinne Transformation in which every axis has a specific treatment.

The metodology applied for the regionalization of the network was the Delaunay Triangulation, since it maximizes the inner angles of the tetrahedra formed (in the case of a 3D space). In this case, the best geometry the region can be used. For every station of the network there is a number of tetrahedra associated with it. The region of interest for this particular station is defined by both its neighbours and by itself. There is a huge number of material concerning Delaunay Triangulation available electronically (Dubois, 1997).

The Affinne Transformation is used only after defining the regions. In the first step, to generate preliminary coordinates; in the second step, a new transformation is then applied in order to map with sufficient precision the coordinates from SAD-69/96 to SAD-69, or vice-versa. In the second step, instead of transforming coordinates what is really being done is transforming corrections. For that, it is necessary to identify the tetrahedron which contains the point of interest. The solution given for this is to work with homogeneous coordinates, since in this case the coordinates of the point belonging to the tetrahedron have values always positives and never larger than one.

The methodology was tested on the set of stations for the transformation from SAD-69/96 to SAD-69. The larger difference found in absolue value is equal to $4*10^{-9}$. This behaviour was already expected since the determination of the corrections imply a deterministic solution, because the degrees of freedom is equal to zero. Preliminary testes with non-coincident stations of the network have shown good results, either with the direct and the inverse transformation.

Due to the coordinate difference between SAD-69 e SAD-69/96, the building of the Delaunay Triangulation yielded different (triangulation) regions for both systems. For example, the station number 10022 belongs to 28 tetrahedra in the SAD-69/96 and to 12 tetrahedra in the SAD-69. In this way, the parameters derived from the Affinne Transformation related to the mapping of coordinates and corrections may be different for certain regions of the network.

It is important to stress that this methodology provides good results probably due to the large number of stations involved, all belonging to the same geodetic system - in the definition level, for two different realizations in distinct time. Another important point to mention is that the methodology can be directly applied for points within the tetraedra. For points outside them, an extrapolation should not be used. Other approach still to be worked on could be applied

Conclusions

The use of geometric transformations alone cannot absorb all the deformations existing in the geodetic network. This fact can be degrade for those applications which require to map coordinates from one system into another. The methodology under consideration is based on the alternative that the transformation can be done with respect to a certain region of the network associated with the possibility that all degrees of freedom in space are quantified, namely, 3 translations, 3 scale factors, 3 rotations and 3 parameters due to the relative orientation of the axis (non-orthogonality). This can yield more expressive results as far as modelling network deformations is concerned.

It can be shown the improvement brought by inserting GPS baselines into the classical networks. For the case presented in this paper, it is obviuos that more GPS stations would be desirable. However, the densification with GPS stations should follow a carefull study.

Finally, the transformation between geodetic systems should be regarded as a very important topic due to its implications into other sciences, such as Cartography, specially in the Brazilian case in which all mapping activities remain attached to the South American Datum.

References

Brunini, C.; Olondriz, J.; Rodriguez, R.C. Determinacion de Parametros de Transformacion entre los Sistemas de Campo Inchauspe 1969 y WGS-84 en Argentina. Presented to IV Congresso Internacional de Ciencias de la Tierra. Santiago. August, 5-9, 1996.

Costa, S.M.A. Ajustamento da Rede Planimétrica do Sistema Geodesico Brasileiro. Rio de Janeiro, Fundacao Instituto Brasileiro de Geografia e Estatistica. Technical Report. July, 1996. Not published.

Dubois, G. The AI-GEOSTATS WEB SITE. URL: http://java.ei.jre.it/rem/gregoire. August, 1997.

Fischer, I. The Basic Framework of the South American Datum of 1969. Presented to XII Pan American Cosultation on Cartography. Panama, 1973

Junkins, D. The National Transformation for Converting Between NAD27 and NAD83 in Canada. D. C. Barnes (Ed.). In Moving to NAD'83, The Canadian Institute of Surveying and mapping. Ottawa, 1990.

Nakiboglu, S.M.; Eren, K. & Shedayed, A.M. Analysis of distortions in the National Geodetic Network of Saudi Arabia. Bulletin Geodesique (1994) 68:220-229.

Sillard, P. & Boucher, C. Improvement of the transformation between ITRF and Doppler-Realized WGS-84. Journal of Geodesy (1996) 70:768-780.

Vanicek, P. & Steeves, R.R. Transformation of coordinates between two horizontal geodetic datums. Journal of Geodesy (1996) 70:740-745.

THE IGS REGIONAL NETWORK ASSOCIATE ANALYSIS CENTER FOR SOUTH AMERICA AT DGFI/I

Wolfgang Seemüller and Hermann Drewes
Deutsches Geodätisches Forschungsinstitut, Abt.I
Marstallplatz 8, D-80539 München/Germany

ABSTRACT

In response to IGS's call for participation in the pilot project for the densification of the ITRF, DGFI/I is acting on behalf of the SIRGAS project as a RNAAC for South America. The analysis results of the first year of activity are presented here in terms of the internal stability of the station coordinates and the discrepancies with respect to global solutions. The r.m.s. deviations for both internal and external comparisons are in the centimeter level.

INTRODUCTION

Early in 1996 the International GPS Service for Geodynamics (IGS) called for participation in a pilot project to densify the IERS Terrestrial Reference Frame (ITRF) by regional networks and a distributed data processing by Regional Network Associate Analysis Centers (RNAAC). DGFI/I is acting on behalf of the SIRGAS project (Sistema de Referencia Geocéntrico para América del Sur) as a RNAAC for South America. Beginning in mid of 1996 all available data of permanently observing GPS stations in the mainland of South America and surrounding regions are processed routinely and forewarded as SINEX files to the IGS Global Data Centers.

RNAAC-SIR NETWORK

At the beginning of the pilot project the SIR network consisted of 12 stations; due to the sparse coverage of South America with permanent GPS stations, all these stations were also processed by at least one of the global analysis centers. Recently the data of 22 stations are processed (Figure 1). Stations exclusively included in the RNAAC SIR network are the new Brazilean permanent stations Bom Jesus Lapa, Cuiaba, Curitiba, Imperatriz, Manaus, Presidente Prudente and Viçosa.

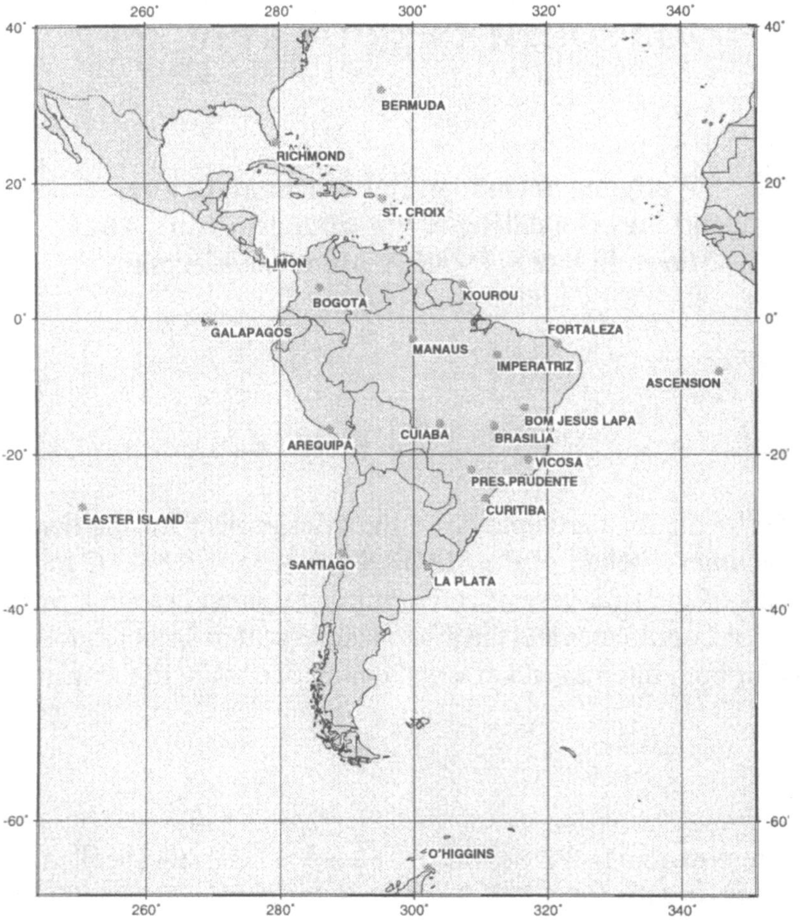

Figure 1: Regional Network processed by RNAAC-SIR

DATA PROCESSING

The RNAAC-SIR coordinate solutions are generated weekly using the automated Bernese software version 4.0, the so-called Bernese Processing Engine (BPE, Rothacher and Mervart, 1996). Up to the end of 1996 a HP workstation was used, since early 1997 the software is running on an IBM AIX workstation. Characterizing features of the performed solutions are:

- Final IGS orbits and Earth orientation parameters applied;
- Cut off elevation angle 15°, sampling rate 2 minutes for single day adjustments;
- Residual tropospheric zenith delays every four hours, a priori sigmas applied with respect to the prediction model: first parameter ± 5 m absolute,
  following parameters ± 10 cm relative;
- Ambiguities partly resolved according to statistical tests,
  remaining ones estimated as real numbers;
- Station coordinates estimated in IGS orbit frame, applying a priori sigmas of ± 1 m.

The weekly solutions are forwarded as R-SINEX (Software INdependent EXchange Format) files to the IGS Global Data Center. They are then combined by the IGS Global Associate Analysis Centers (GNAAC) with the global solutions (G-SINEX) to the polyhedron solution (P-SINEX).

RESULTS

Figures 2, 3 and 4 demonstrate the stability of the processing results. Figure 2 shows the internal stability of the weekly RNAAC SIR solutions for one regional and one global station. The coordinate variations are in the centimeter level. Some deviations - similar in all stations - are obviously due to reference frame effects.

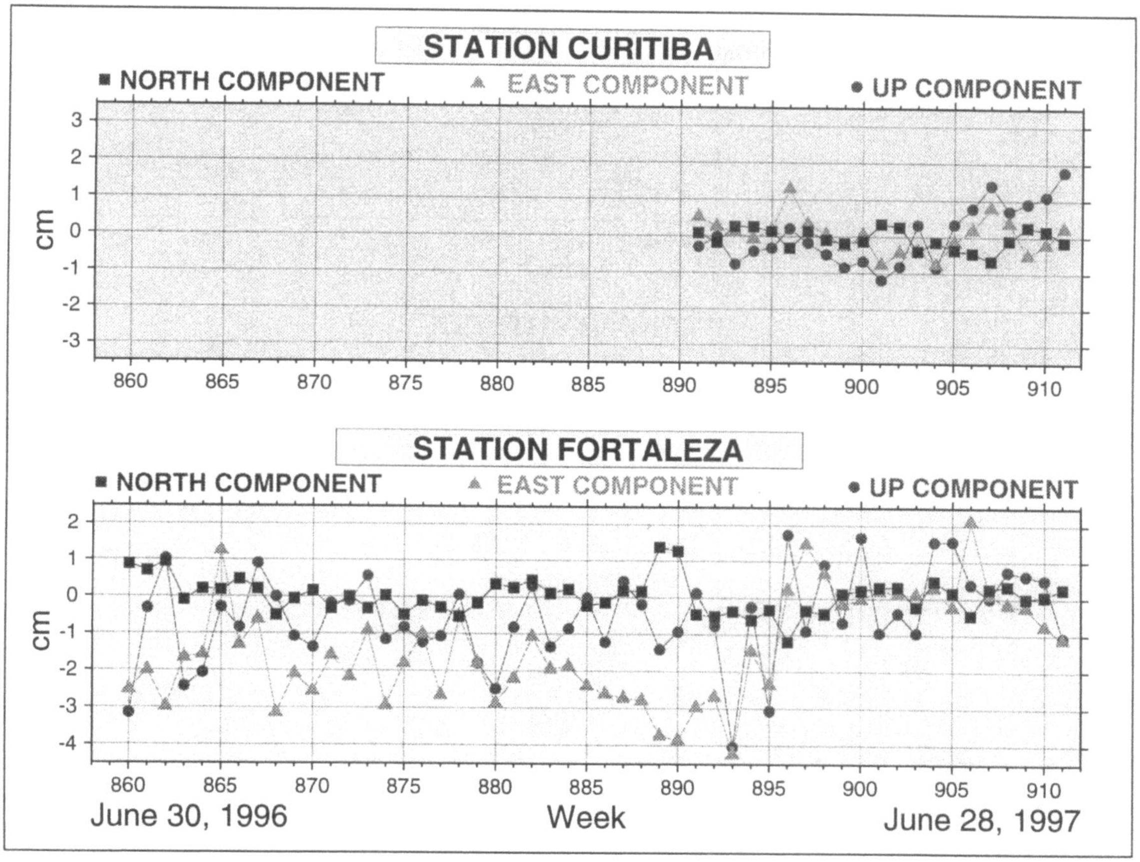

Figure 2: Internal stability of RNAAC results: Coordinate variations for a regional (Curitiba) and a global (Fortaleza) station

Figure 3 shows the deviation of the weekly RNAAC solution with respect to the combined global polyhedron solution of the GNAACs at MIT (USA) and Newcastle University (GB), respectively. It is obvious that the discrepancy of the regional station, which is only processed by RNAAC-SIR, is smaller than the discrepancy of the global station, which is also processed by the GNAACs independently.

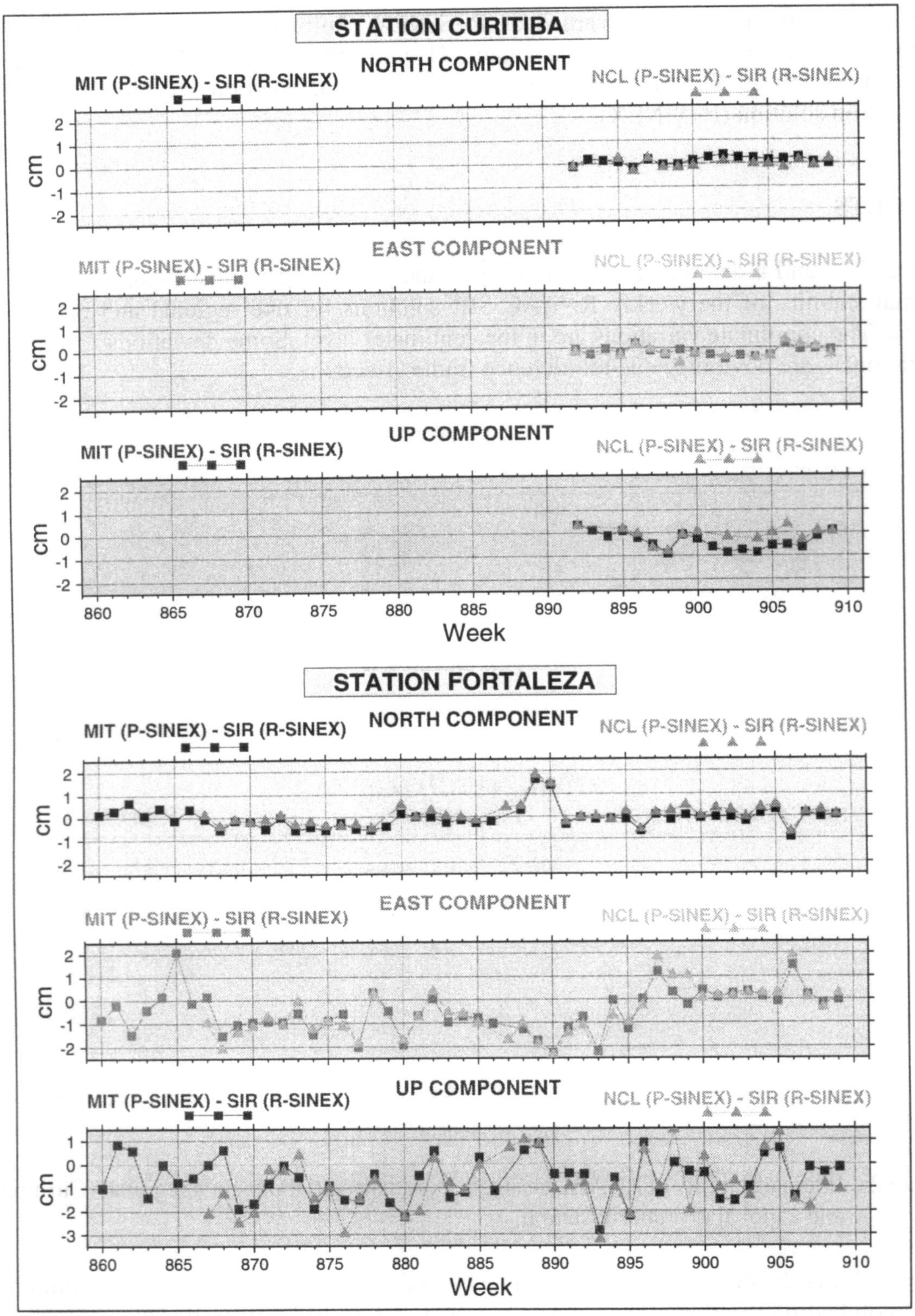

Figure 3: Differences of RNAAC-SIR coordinates with respect to GNAAC global polyhedron solutions after seven parameter (Helmert) transformation

Figure 4 gives the variations of the weekly MIT solutions for two selected SIR stations. This is to demonstrate the external stability of station coordinates.

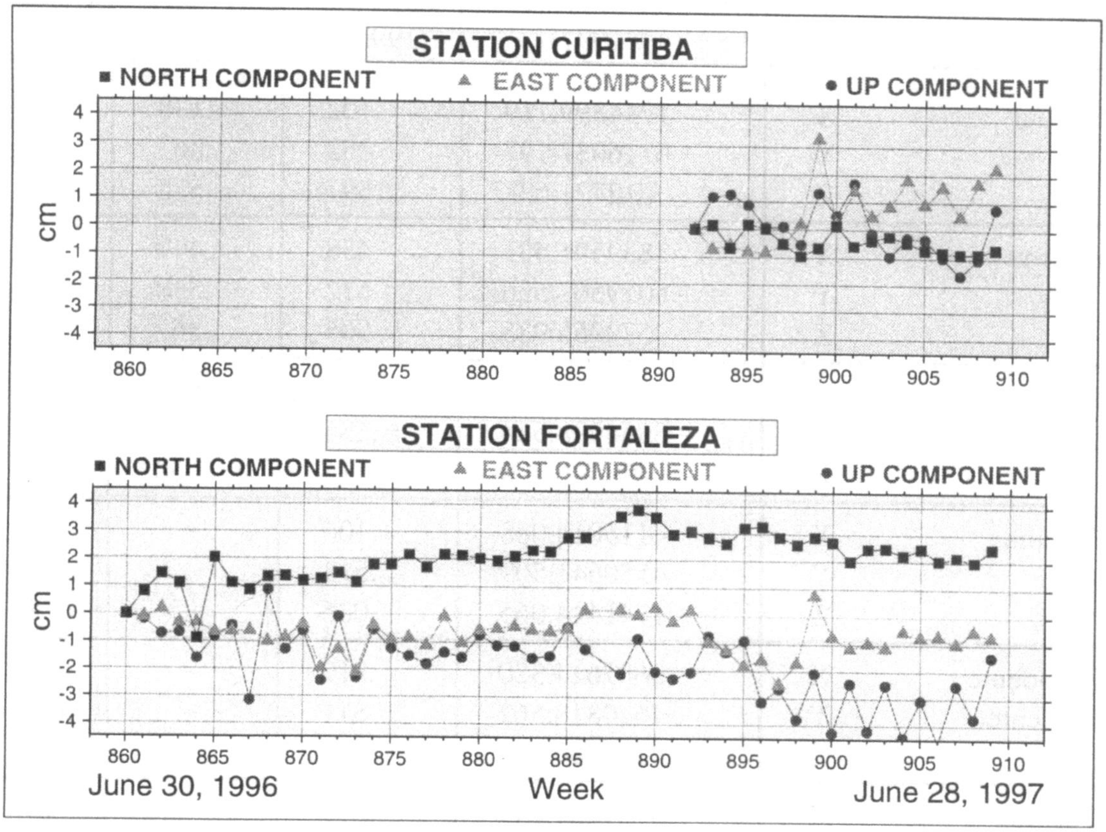

Figure 4: Consistency of GNAAC MIT weekly polyhedron solutions

Table 1 shows a comparison of coordinates of selected stations on the South American mainland from independent solutions. Considering the station velocities (V) we find an agreement in the centimeter level.

CONCLUSION

The first analysis of RNAAC SIR processing demonstrates satisfying results. The South American network has grown rapidly in a short time interval. The number of stations will further increase in the near future and may serve as a permanent control net in South America. Thanks are due to the contributing countries, in particular to the Instituto Brasileiro de Geografía e Estatistica (IBGE/DEGED) for its efforts in providing the data of the Brazilean network with highest reliability.

Table 1: Comparison of independent coordinate solutions

Station Name		IGS P-SINEX (a/v 1997)	SIRGAS (1995.4)	ITRF94 (1993.0)	V (IGS) (mm/a)
Bogotá	X	1744399.119	.102	.138	0
	Y	- 6116037.797	.852	.863	10
	Z	512731.622	.597	.575	25
Kourou	X	3839591.431	.438	.447	- 1
	Y	- 5059567.536	.532	.538	- 6
	Z	579956.925	.906	.867	15
Fortaleza	X	4985386.629	.661	.643	1
	Y	-3954998.564	.585	.560	-10
	Z	- 428426.490	.515	.550	14
Brasilia	X	4115014.088	.106	-	- 1
	Y	- 4550641.509	.513	-	-14
	Z	- 1741444.065	.086	-	12
Presidente Prudente	X	3687624.320	.310	-	(0)
	Y	- 4620818.562	.571	-	(-6)
	Z	- 2386880.388	.407	-	(10)
Curitiba	X	3763751.647	.639	-	(0)
	Y	- 4365113.753	.768	-	(-6)
	Z	- 2724404.735	.755	-	(10)
Arequipa	X	1942826.731	.727	.693	14
	Y	- 5804070.240	.249	.243	- 10
	Z	- 1796893.971	.994	4.035	16
Santiago	X	1769693.350	.314	.258	22
	Y	- 5044574.135	.146	.114	-17
	Z	- 3468321.050	.077	.112	11
La Plata	X	2780102.976	.979	-	4
	Y	- 4437418.854	.845	-	-10
	Z	- 3629404.586	.604	-	14

REFERENCE

Rothacher, M. and Mervart, L. (eds): Bernese GPS software version 4.0. Astronomical Institute, University of Berne, 1996.

THE REDEFINITION OF THE GEODETIC REFERENCE SYSTEM OF URUGUAY INTO SIRGAS FRAME

Subiza, Walter H.
Servicio Geográfico Militar
8 de Octubre 3255, Montevideo 11600, Uruguay

Pérez Rodino, R., Barbato, F.
Instituto de Agrimensura, Facultad de Ingeniería
Universidad de la República, Julio H. Reissig 565
Montevideo 11600, Uruguay

Alves Costa, Sonia M.
Instituto Brasileiro de Geografía e Estatística
Av. Brasil 15671, Parada Lucas, 21241-051
Rio de Janeiro, RJ, Brasil

Abstract

In 1908, a systematic geodetic surveys started with cadastral and mapping objetives. Classical operations were carried out along main paralel and meridian chains till 1960's when the Fundamental Geodetic Network was finished.

A common adjustment using the variation coordinates method was performed in 1963, by the US Army Map Service. The whole 1st . Order data available was evaluated giving a mean direction error of 0″.71 and 0″.89 misclosure triangle. The data included 15 bases, 12 astronomical stations and 248 geodetic stations. Hayford Ellipsoid was used and the Yacare Datum was established.

After that, normal geodetic densification of the 1st. Order was performed. The introduction of the EDM in the 1960's, allowed for electronic polygons, getting field operations faster than classical method.

During 1992 the GPS stroke on in our country and quickly remained as the principal tool to get positioning. The 1/50.000 scale cartography was rapidly finished with a couple of submeter receivers.

Two years later, the SGM bought three Ashtech Z-XII receivers, and the Facultad de Ingeniería (Universidad de la República) got Wild-Leica ones. Later on others institutions went into GPS world. The fact allowed Uruguay to participate in SIRGAS Project, and densify the network.

With all this information, a new redefinition of the National Geodetic Reference System was felt needed, in order to adopt a worldwide geodetic and cartography system. GHOST (Geodetic adjustment using Helmert Blocking of Space and Terrestrial data) software from Geomatics of Canada was used to carry the task out in the frame given by SIRGAS Project. Major steps taken and some results are presented here.

Historical development of the Uruguayan Geodetic Network

The geodesy as a cartographic and positioning tool was born in *1908* with the first systematic surveys in the center of the country. Bamberg and Huetz theodolites and directions methods were used.

In 1914 a triangulation network plan was designed, including four meridian chains and five paralel ones. Clarke ellipsoid was adopted as reference surface.

The Laguna Merim International Chain with Brasil was measured in 1912-1918. In 1917-1919 a survey of the capital of the country was made.The International Río Uruguay Chain with Argentina was observed between 1918 and 1937. In 1946 the first triangulation adjustment was calculated, using Least Squares Method. The complete boundary chain with Brasil first, and the argentinian counterpart later were calculated using 1930 International Ellipsoid as surface reference. In the 1940's a complete Río Negro Valley survey, in the middle of the country, was carried out, with approximately 400 km long. Densification of the geodetic net continues, using better equipments for directions and distances (Wild T4 and T3 theodolites, invar bases and electronics distance meters), until *1961 when the fundamental network planned in 1914 was finished.*

First major adjustment was performed in USA in 1965, using coordinates variation method, involving 248 geodetic station and using the 1930 International Reference Ellipsoid. In 1969 the main geodetic network was included in the South American Datum 69, adding at this time 4 new astronomics stations and 7 geodetic bases in order to improve the net. In 1990 approximately 420 stations have been survey and calculated for the first order. The satellite technology for positioning began in 1993, and Sirgas Project put the net at regional and global level.

SIRGAS Project

Three main geodetic stations (LaPlace stations) were measured with Ashtech ZXII, during a ten days-long campaign in 1995 processing (95/147, 148 and 149 Jdays). At the same time five more stations were observed during three days long with Wild-Leica receivers.

In 1997 a two days campaign was performed (97/116 and 117 Jdays), adding three more stations. Montevideo and Bella Vista stations were used as a link with Sirgas campaign. (Figure 1)

Satellite data processing

Software used for processing GPS data was Bernese version 4.0, developed by Berna University.

GPS data to process comprised two campaigns, observed with both double frecuency/code, Ashtech Z-XII and Leica receivers. 1995 GPS campaign was simultaneous with Sirgas one, using same observation parameters and having eigth stations (three Sirgas), it was processed with 15 ° elevation and 30 seconds recording time. 1997 campaign, included five stations measured in two days long, processed at 15 ° elevation and 60 seconds recording time.

As baselines distances were ranging from 100-250 km, the pre-processing and processing was developed on modeled observable L3 (ionosphere free linear combination), the files that contain local ionosphere parameters was not necessary be generated. As a manual suggestion was used Saastamoinen troposfere model without meteorological values observed. The receiver phase center correction was applied according file IGS_PHAS.01 .

The strategy to create single diferences files was shortest in both cases.(Figure 2) It means that only number of stations - 1 independent single differences were formed, choosing the shortest paths. The cycle slips automatic detection and correction was made in triple diference, The ambiguities resolution was developed in double diference, solving for 90% of ambiguities in a linear combination L5 (wide-lane). The solved ambiguities were introduced as known values in the final solution.

The final combination of 95 and 97 results were made using coordinates and variance-covariance matrix of coordinates from session/day processings, showing *4 mm maximun difference on the baselines and sigmas less than 3 mm on final cartesian coordinates* Montevideo was used as fixed station. These informations were converted in Ghost input format to integrate in the same adjustment extraterrestrial and terrestrial data. A seven parameters Helmert transformation between Bernese and Ghost GPS software results was made. *Residual less than 3 mm and a RMS of 3 mm was obtained, showing a good agreement between campaigns/software.* An scaled adjustment of both 95 & 97 campaigns in Ghost result on a 1.017 variance factor.

Geodetic Network Adjustment. Results and comments

The final adjustment was made scaling GPS campaign according the preprocessing variance factor obtained (100.7 for 1995 and 108.8 for 1997), resulting on a 1.655 variance factor. *Sirgas stations were included as weighted observed positions* with Sirgas's sigmas.

From this first approximation, a computer program was adapted in order to get geoidal altitudes for each station and improve the Z component. Regional geoidal model

URUGUAYAN GEODETIC NETWORK WITH GEOIDAL MODEL WGS84

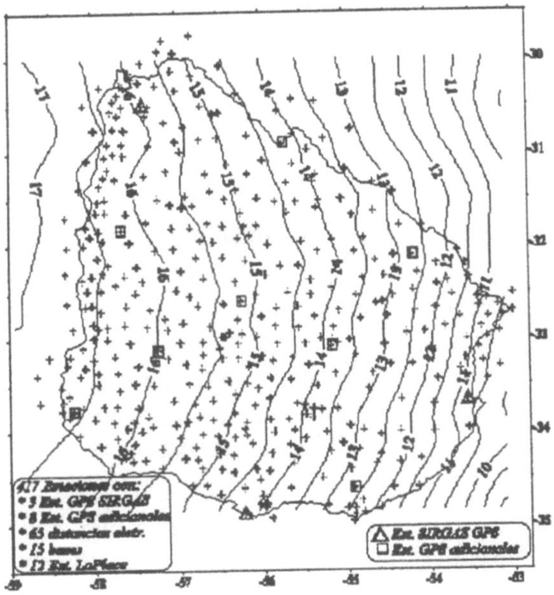

Figure 1

GPS LINES MOUNTED FOR PREPROCESSING 95 & 97 Campaigns

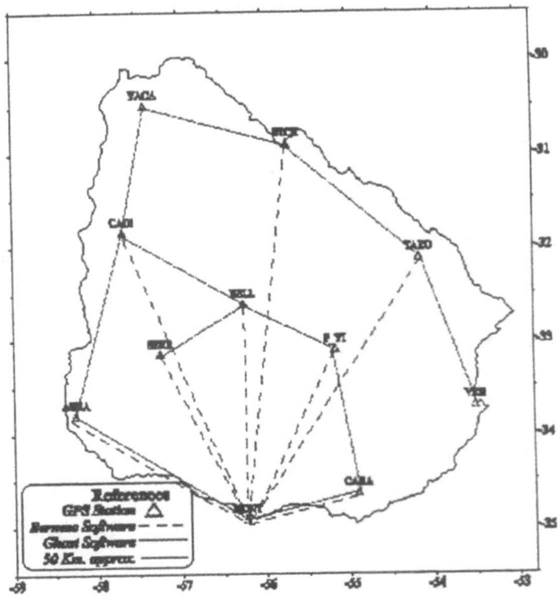

Figure 2

HORIZONTAL VECTORS OF DISTORTIONS YACARE DATUM - GRS 80

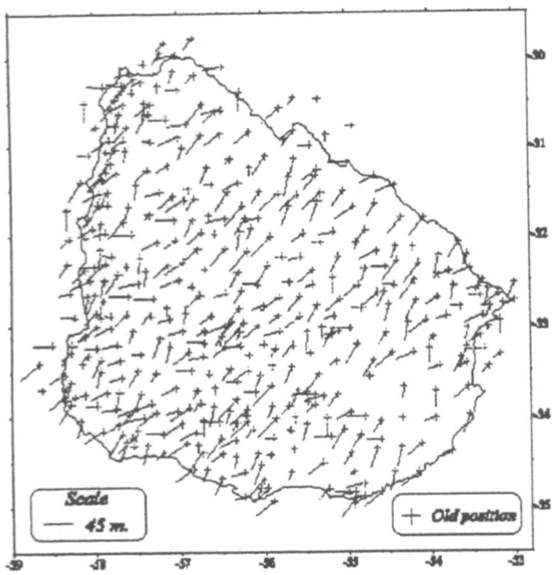

Figure 3

CONFIDENCE ELIPSE ERRORS GPS NETWORK STATIONS

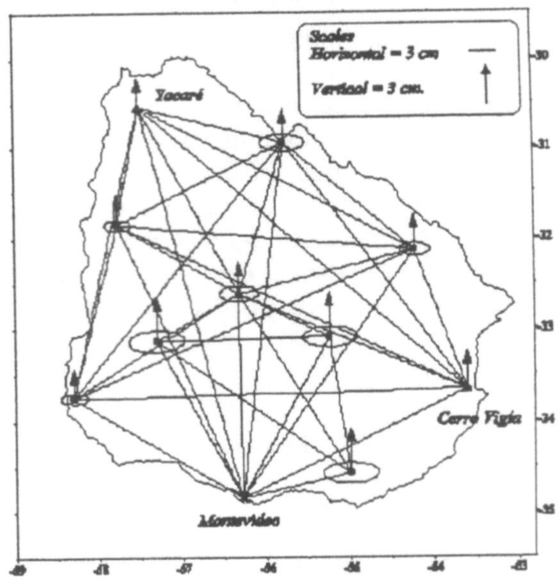

Figure 4

developed for Uruguay in 1993, with a thousand of gravimetric stations and GEMT2 model as reference , was used.

Geoidal altitudes were added on the adjustment file later.

Adjustment data is at follows:

* 417 geodetic stations with geoidal altitudes (including 11 GPS stations)
* 2337 horizontal directions
* 66 distances
* 11 astronomic azimuths
* 4 parameters to determine (3 electronic distances and 1 azimuth)
* 1341 normal equations generated.
* 1162 system degrees of freedom
* 0 fixed station
* 3 SIRGAS weighted position equations
* 65 position difference equations

After adjustment a 1.65 variance factor was reached and the maximum horizontal residual was ± 4".

A total of 181 (43.4 %) geodetic stations were classified as First Order, 232 (55.6 %) as Second, 4 as Third and Four Order (1 %).

The weighted contributions of different kind of observations were: directions 92.31 %, distances 3.12 %, azimuths 1.07 %, position difference equations 3.49 %, weighted positions 0.01 %.

Raw azimuths without any reductions, had a maximum residual of ± 1"., mostly of them below of their a priori sigmas.

Distances were classified in 5 groups according the ppm showed (30 up to 3 ppm, 21 3-10 ppm and 15 more than 10-35 ppm). The old invar bases were in first group, meanwhile some electronic ones had to be eliminated due to high residuals, maybe by lack of correct equipment calibration.

Horizontal distortions vectors of the network (Figure 3) due to change of reference system, showed a general direction of 260° and a mean scalar value of 2 m in latitude and -37 m in longitude, with *a maximum of –48 m* in the last one. It is clear the readjustment in longitude caused by GPS observations.

Latitude-longitude standard deviation have values from 0.07-0.44 m, and confidence elipse errors shown a value ranging from 0.2 to 1.2 m for classical observations and some tens of mm for GPS stations (Graphic 4).

Conclusions

A major readjustment of the geodetic net was succesfully performed in the Sirgas Frame last July/97. Combination of tridimensional and clasical geodetic observations were possible through the Ghost software use.

A new set of coordinates in ITRF/WGS84, is available to local or international institutions/researches, and mainly for positioning or cartographic use. Clasical stations showed *confidence errors between 0.2-1.2 m and GPS ones, some cm.*

Local cartographic system of reference will change to standard UTM 21 & 22 zones, with WGS84 ellipsoid and coordinates

Further steps will be the development of a more accurate geoidal model introducing some one thousand gravimetric stations and GPS data observed over the levelling net.

Acknowledgements

Special thanks to Don Beattie from Geomatics Canada for fully and patient support. Also to Sirgas Project and Geomatics of Canada/IBGE, Brazil & Bern University, Switzerland for kindly allow to use the adjust and GPS processing software.

References

Geodetic Survey Data Dictionary. 1987, Ghost documentation. Geomatics of Canada.

Bernese Manual Reference 4.0. 1997 Bern University, Switzerland.

GPS AMBIGUITY RESOLUTION FOR NAVIGATION, RAPID STATIC SURVEYING, AND REGIONAL NETWORKS

Paul J. de Jonge
Delft University of Technology, Faculty of Geodetic Engineering
Thijsseweg 11, 2629 JA Delft, The Netherlands
E-mail: p.j.dejonge@geo.tudelft.nl

ABSTRACT

In this contribution an overview is given of the capabilities of methods for ambiguity resolution in use nowadays. This is done in the context of SSG 1.157, "Ambiguity Resolution and Validation", although the overview is not restricted to the algorithms used within this group.

Methods from the fields of navigation, rapid static surveying, and regional networks are compared, and similarities and differences are indicated. It is shown that the subdivision of the algorithms by their applications is a rather arbitrary one, and that some of the ambiguity resolution algorithms can be applied in all three fields.

The capabilities of state of the art algorithms for ambiguity resolution are indicated in terms of maximum distances over which the ambiguities can be resolved and validated. In particular the role of the observational time span and the set of observables used in the estimation (single versus dual frequency, phase-only versus phase and pseudo range) is addressed. As the ionospheric delay is an important reason for being unable to resolve the ambiguities, the consequences of the next solar activity maximum at the end of the century are briefly discussed.

INTRODUCTION

Traditionally, algorithms for the estimation of the integer GPS double difference (DD) ambiguities have been developed for two different fields. On the one hand methods have been devised for applications where a multiple of stations are occupied for several hours until several days, and maximum inter-station distance can be of the order of thousands of kilometers, (Dong and Bock 1989), (Blewitt 1989), (Mervart 1995). On the other hand methods have been developed for rapid-static and navigation applications, where usually only two stations are involved, the maximum distance is some tens of kilometers, and time of occupation is of the order of seconds to minutes, or the receiver is moving: AFM (Counselman and Gourevitch 1981), LSAST (Hatch 1990), FARA (Frei 1991), optimized search using Cholesky (Euler and Landau 1992), LAMBDA (Teunissen 1993), GASP (Corbett 1994), null-space method (Martin-Neira et. al. 1995).

At first sight it seems that ambiguity estimation for large regional networks is treated as an entirely different problem, although conceptually there is no real difference with the short baselines as far as the estimation of the integer ambiguities is concerned. Of course, for longer inter-station distances more adequate mathematical models for e.g. tropospheric and ionospheric delays, as well for the orbital parameters have to be employed.

If one looks more carefully, one may observe similarities between some of the algorithms applied to the regional networks, and some of those applied to the short baselines. The methods described in Dong and Bock (1989), Blewitt (1989) as well as the 'sigma' and QIF (Quasi-Ionosphere Free) method in the Bernese software (Rothacher et. al. 1996) all use a sequential conditioning (fixing) of ambiguities. The conditioning makes that subsequent ambiguities become more precise, and usually pushes the values of the ambiguities *conditioned* on the previous ambiguities towards integer values. The schemes differ in the criteria that are used for selecting the next ambiguity to be fixed, and in the way parameters are modeled and constrained in the preceding float solution.

This principle of conditioning of ambiguities one also finds with (Talbot 1991) and in the sequential conditional least-squares adjustment which is part of the LAMBDA method. The latter differs from the methods above in the sense that if the resolution or adjustment is concluded with a complete or partial vector of integer ambiguities $\breve{a}$, it is guaranteed that that this vector minimizes the integer least-squares criterion $(\hat{a}-\breve{a})^t\, Q_{\hat{a}}^{-1}(\hat{a}-\breve{a})$, with $\hat{a}$ the vector of float ambiguities, and $Q_{\hat{a}}$ its variance-covariance matrix.

Due to the hybrid character of regional networks with inter-station distances from some kilometers to several thousands of kilometers, and with some receiver-satellite combinations observed less well than others, the chance to find a valid integer solution for all ambiguities together is decreased, and thus one has to resort to a proper subset for which such a solution exists.

Another difference between the regional network and the short baseline algorithms, is that in the former no validation step is involved, other than comparing several solutions in time. The baseline algorithms that are used in rapid static and navigation applications have some sort of validation step which is based on the data itself, and thus can be computed and evaluated as soon as the data are collected and processed.

Decorrelation.

In Teunissen (1993) it was shown that the DD ambiguities are strongly correlated, especially when the observational time span is short, due to the small change in the receiver-satellite geometry. Since this correlation makes the estimation of the integer ambiguities via a sequential conditioning of the ambiguities far from efficient, a method to construct a decorrelating transformation for the ambiguities was proposed in ibid. Alternative methods can be found in Li and Gao (1997), and in Han and Rizos (1995), where it is also shown that decorrelating the ambiguities often has a favorable influence on the efficiency of other resolution methods too.

The ambiguities of the one day solutions for the large regional networks are less correlated, but still can be improved upon. This most likely will produce more ambiguities that will pass the criteria that allow them to be fixed, thereby further improving the eventual fixed solution. Possibly it could also reduce the time span for which a fixed solution can be computed. For monitoring coseismic and postseismic deformation with magnitudes up to several decimeters, and several mm/day, respectively, shortening these time spans would be very welcome.

MEASURING PERFORMANCE OF AMBIGUITY RESOLUTION SCHEMES

Not all ambiguity resolution algorithms have been described in sufficient detail, such that they can be implemented and tested, and for only a few algorithms the source code is available. A complicating factor is that often 'improved' and 'combined' versions of existing algorithms are proposed.

The algorithms that are used inside the large scientific GPS softwares like Bernese, GAMIT and GIPSY, for resolution in regional networks are well described, and source code is in principle available. Other methods that are described in detail, and for which source code is available are the FARA (implemented in Bernese s/w) and the LAMBDA method (de Jonge and Tiberius, 1996). The reason for this lack of willingness of making algorithms available in the public domain, usually has its roots in the protection of the commercial interests of the proprietor of the software. Needless to say that this makes a full comparison of methods not very feasible. Despite these difficulties, Han and Rizos (1995), and Hein and Werner (1995) show a comparison of some methods.

In the absence of having an accurate description of the various methods one can still make bilateral comparisons of methods on the basis of an exchange of e.g. the float vector of ambiguities and its variance-covariance matrix. Exchange of e.g. RINEX data itself has to be discouraged since often one will end up comparing ways to compute the float solution, instead of comparing the integer ambiguity estimation process itself.

Ground truth.
When evaluating the performance of ambiguity estimation one has to have some sort of reliable (accurate) 'ground truth' available. This ground truth usually refers to the coordinates of the points surveyed, or functions thereof. For stationary points one can look at the repeatability of discrepancies of the coordinates with respect to a well surveyed point.

For moving points, the discrepancies between the coordinates computed with respect to a well surveyed point, and those that have another well surveyed point as reference, can be used. Another approach is to look at the discrepancies between the computed length between two moving points, and the a priori known length. For example the case of two or more antennas on a moving vehicle.

As far as a threshold for the discrepancies is concerned, one should take care to stay at the conservative side. If the ground truth is established with a sufficient accuracy level, the ambiguities are constrained to the correct integers, and (in case of longer distances) the troposphere and ionosphere are modeled correctly, the discrepancies should be at the few centimeter level, and not at the sub-cycle length level as sometimes is proposed. Fixing one of a set of ambiguities at a value wrong by one cycle does not necessarily cause a shift in coordinates of the level of one cycle.

One can also inspect the continuity of the ambiguities over a not too short time span; the time span should not be too short since the ambiguities often show a remarkable persistence in sticking to the wrong integer values. So preferably this approach should be used in combination with one of the strategies above. Some ambiguity resolution algorithms use the continuity of the ambiguities as a measure for validation; i.e. if the ambiguities stay constant for a predefined period of time after convergence, they are believed to be the correct ones. A disadvantage of this validation approach is that there is no clear link with probability theory, and that the minimum time for a reliable fix can never be less than the time needed for convergence of the ambiguities.

Things get more complicated once the distance between the reference point and the point of which the coordinates are to be established, becomes larger. To get a sufficient good accuracy for the ground truth of the new point it will have to be occupied for a longer period, but in setting up an experiment one is usually inclined to put most effort and time in the actual experiment while neglecting the establishment of a solid ground truth. This is often also the case for data sets that are not especially collected for the purpose of testing a resolution scheme.

For the regional networks, one usually looks at the repeatability of the coordinates over the days, and at the a posteriori variance of the observables after fixing the ambiguities.

How well can we resolve the ambiguities?

The answer to this question depends on to whom you ask it. The answers from people in practice that are engaged in making ‘ambiguity resolution engines’ which will give with a minimum of user interference a near 100 percent correct answer, will be far more pessimistic than the answer from a research person. For instantaneous ambiguity resolution, i.e. using dual frequency phase and pseudo range data of one epoch, answers vary between a maximum baseline distance of 5 km using at least 7 satellites and a maximum baseline distance of 10 to 15 kilometer with at least 6 satellites and a validated solution for 98-100 percent of the time.

For kinematic solutions successful resolution is reported for single frequency phase and pseudo range over distances of 20 to 30 kilometer for time spans of some minutes using at least 6 satellites, while for distances less than 20 kilometers using dual frequency phase and pseudo range data, time spans can be as short as 1 minute.

Results that were reported for longer baselines (but that only could be validated using external information) were e.g. correct resolution over a baseline of 88 km using dual frequency phase and pseudorange on L1, for a time span of 1 minute for 20 out of 120 cases, and correct resolution over a baseline of 70 km using 15 minutes of dual frequency phase data. These are rather individual cases, but they indicate where the frontiers of ambiguity resolution lie.

As far as the large regional networks are concerned, correct ambiguity resolution over baselines up to 2000 kilometer using a time span of one day using dual frequency phase observables is routinely performed for 85-95 percent of the baselines. In the case where the ambiguities are resolved in a network approach, a similar percentage of the ambiguities can be resolved.

... and when is the ambiguity resolution becoming difficult?
Signal multipath decreases the success rate of ambiguity resolution, and can render resolution impossible even over short (several tens of meters) baselines. Just before and after data gaps, the success rate is also decreased.

In areas with a large ionospheric activity, as e.g. near the geomagnetic poles, or near the geomagnetic equator, the chances of successful ambiguity resolution are also greatly reduced. As we are heading again towards a maximum sunspot number which is predicted to occur in late 1999 or early 2000 (Joselyn et. al. 1997) resolving the ambiguities will probably become more difficult. During the period from 1999 through 2005, severe geomagnetic storms are likely to occur. A comparison with the last sunspot maximum is difficult, because since then the capabilities of receivers and algorithms have seen large improvements, and at present a whole new generation of GPS satellites (Block IIR) is being deployed.

ACTIVITIES OF THE SPECIAL STUDY GROUP AND OUTLOOK

A start has been made in exchanging data in the form of float vectors and their variance-covariance matrices. As far as the exchange of source code is concerned, until now only the LAMBDA method has been made general available, and some bilateral exchange of prototype software has been established. A more general exchange of source code will be further encouraged in the future.

The theory of ambiguity resolution and validation is still not finished; for validation purposes more research into e.g. the discrete probability distribution of the integer ambiguities is needed. Research is ongoing into ambiguity resolution for GNSS2 with possibly a third frequency and for GLONASS possibly in combination with GPS.

Papers produced by the members of SSG 1.157 over the last two years, can be found at the web page of the SSG: http://www.geo.tudelft.nl/mgp/people/paul/ssg_1157.html. Also at these pages a still growing bibliography of more than 200 papers concerned with ambiguity resolution and validation can be found: http://www.geo.tudelft.nl/mgp/people/paul/amb.html

The contributions of Simon Corbett, Mattia Crespi, Hans-Jürgen Euler, Shaowei Han, Hans-Jörg Kutterer, Herbert Landau, Zuofa Li, Manuel Martin-Neira, João Galera Monico, Daniela Morujão, Benjamin Remondi, Stefan Schaer, Peter Teunissen, Christian Tiberius, Ming Yang, and Wolfgang Werner are acknowledged.

REFERENCES

Blewitt, G. (1989): Carrier phase ambiguity resolution for the Global Positioning System applied to baselines up to 2000 km, *Journal of Geophysical Research*, Vol. 94, No. B8, pp. 10187-10203.

Counselman, C.C., and S.A. Gourevitch (1981): Miniature interferometer terminals for earth surveying: ambiguity and multipath with Global Positioning System, *IEEE Trans. on Geoscience and Remote Sensing*, Vol. 19, No. 4, pp. 244-252.

Corbett, S.J. (1994): *GPS single epoch ambiguity resolution for airborne positioning and orientation*, PhD thesis, Dept. of Surveying, University of Newcastle upon Tyne, UK.

Dong, D.-N., and Y. Bock (1989): Global Positioning System network analysis with phase ambiguity resolution applied to crustal deformation studies in California, *Journal of Geophysical Research*, Vol. 94, No. B4, pp. 3949-3966.

Euler, H.-J., and H. Landau (1992): Fast GPS ambiguity resolution on-the-fly for real-time applications, *Proceedings of Sixth International Geodetic Symposium on Satellite Positioning*, Columbus, OH, March 17-20, pp. 650-659.

Frei, E. (1991): *Rapid differential positioning with the Global Positioning System*, PhD thesis, University of Berne, Switzerland, Geodetic and Geophysical Studies in Switzerland, Vol. 44.

Han, S., and C. Rizos (1995): A new method of constructing multi-satellite ambiguity combinations for improved ambiguity resolution, *Proceedings of ION GPS-95*, Palm Springs, 12-15 September, pp. 1145-1153.

Hatch, R. (1990): Instantaneous ambiguity resolution, *Proceedings KIS90*, Banff, Canada, September 10-13, Springer Verlag, pp. 299-308

Hein, G.W., and W. Werner (1995): Comparison of different on-the-fly ambiguity resolution techniques, *Proceedings ION GPS-95*, Palm Springs, 12-15 September, pp. 1137-1144.

Jonge, P.J. de, and C.C.J.M. Tiberius (1996): The LAMBDA method for integer estimation: implementation aspects, *Delft Geodetic Computer Centre LGR series*, No. 12.

Joselyn, J.A., J.B. Anderson, H. Coffey, K. Harvey, D. Hathaway, G. Heckman, E. Hildner, W. Mende, K. Schatten, R. Thompson, A.W.P. Thomsom, and O.R. White (1997): Panel achieves consensus prediction of solar cycle 23, *Eos Trans. AGU*, Vol. 78, No. 20, pp. 205, 211-212.

Li, Z., and Y. Gao (1997): Construction of high dimensional ambiguity transformation for the LAMBDA method, *Proceedings KIS97*, Banff, Canada, 3-6 June, pp. 409-416.

Martin-Neira, M., M. Toledo, and A. Pelaez (1995): The null space method for GPS ambiguity resolution. *Proceedings of DSNS'95*, Bergen, April 24-28, Paper No. 31, 8 pp.

Mervart, L. (1995): *Ambiguity resolution techniques in geodetic and geodynamic applications of the Global Positioning System*, PhD thesis, Astronomical Institute of the University of Berne, Switzerland.

Rothacher, M., G. Beutler, E. Brockmann, S. Fankhauser, W. Gurtner, J. Johnson, L. Mervart, S. Schaer, T. Springer, and R. Weber (1996): *The Bernese GPS software Version 4.0*, edited by M. Rothacher and L. Mervart, Astronomical Institute, University of Berne, Switzerland.

Talbot, N.C. (1991): Sequential phase ambiguity resolution for real-time static differential GPS positioning, *Manuscripta Geodaetica*, Vol. 16, pp. 274-282.

Teunissen, P.J.G. (1993): Least-squares estimation of the integer GPS ambiguities, Invited lecture, Section IV Theory and Methodology, IAG General Meeting, Beijing, August, also in *Delft Geodetic Computing Centre LGR series,* No. 6, 16 pp.

GPS ANTENNA AND SITE EFFECTS

Jan M. Johansson

Onsala Space Observatory, Chalmers University of Technology, S-43992 Onsala, Sweden

Abstract

The improvement in precision obtained from GPS observations over recent years has revealed problems related to the local conditions at the GPS sites. In order to further improve high precision GPS positioning, orbit determination, and estimation of atmospheric parameters, investigations of site dependent effects are required. Concerns have been raised regarding the antennas and the monuments used and the long- and short-term mechanical and electromagnetic stability of the sites. Here, we review the problems associated with site-specific errors and present recommendations on how to eliminate or minimize these effects.

Introduction

During the last few years, an increasing number of permanent GPS sites have been established. The demonstrated repeatability of horizontal position estimates for regional networks is currently of the order of 2 mm and typically a factor of 3–5 greater for the vertical component. There are many advantages to continuously operating GPS networks. Stable pillars with fixed antennas eliminate errors associated with variations in the measurement of the local vector from the reference marker to the phase reference point of the antenna. For fixed pillars in a continuously operating network, the reference marker is usually a fixed, well-defined point on the antenna. In addition, denser position estimates (spatially and temporally) decrease the statistical uncertainty of the results. Continuously operating networks may also serve as a global or regional reference frame for different types of regional and local surveys. Another essential advantage is the increased ability to study and eliminate unmodeled systematic effects on daily estimates of site positions, both short- and long-term effects.

To be able to constrain the common mode of motion, sometimes in the submillimeter range, in a regional or local network a strong reference network is needed. The origin of the reference frame must be maintained with a high degree of robustness. In addition, orbits must be compatible with the reference frame. For this purpose data from the IGS network and other permanent network need to be regularly examined in detail. Site-specific errors at permanent stations may introduce errors in the determination of satellite orbit parameters and in the estimate of site positions. In the following we adress the problem of site-specific errors and present some recommendations on how to handle these errors.

International Association of Geodesy Symposia, Vol. 118
Brunner (ed.), Advances in Positioning and Reference Frames

Site Specific Errors

We have chosen to divide the site specific error sources into three subgroups. The first group consists of problems associated with the receiver, antenna, radome, and the signal. These are effects that will not, in general, change on a day-to-day basis. However, they might introduce biases in the solution. As long as nothing changes the effect stays the same. If something changes, such as the satellite constellation or the elevation cut-off angle, the results will be affected. The influence is especially obvious looking at the estimates of the vertical components and precipitable water vapour where bias terms can be introduced (e.g. [*Elósegui et al.*, 1995]; [*Niell et al.*, 1994]; [*Niell*, 1996]). Such a bias could seriously affect the interpretation of the GPS data.

The second group represents areas of site effects that will vary, but will only periodically affect the measurements. Precipitation, multipath, atmospheric pressure loading, and atmospheric gradients are probably the most important of these, but others may be discovered.

Finally, the third group consists of errors that might affect the long-term stability of a site such as the location of the site, ground, and the monument. Most of the material related to this specific group of site errors are rather new. These errors may seriously affect the reference frame and the geodynamical projects.

GPS Antennas

It has been found that antenna-to-antenna phase differences can introduce range biases at the several centimeter level, which may limit the precision of the measurements [*Rocken*, 1992]. Differential phase errors due to GPS antennas will not only affect the precision in GPS networks with different types of antennas, but also in networks using identical antennas if the network covers a large spatial area (baseline lengths $\gtrsim$1000 km) [*Schupler and Clark*, 1991]; [*Schupler et al.*, 1994]. Differential phase errors in regional networks (baseline lengths $\lesssim$1000 km) using identical antennas are dependent on the electromagnetic environment around each individual antenna.

The problem of antenna mixing was addressed at the IGS Analysis Center Workshop in Silver Spring, 1996. Two sets of phase calibration corrections (PCC) tables have been put together based on material presented by *Mader and MacKay* [1996], *Rothacher and Schär* [1996], and *Meertens et al.* [1996a] to be used by the IGS Analysis Centers and others in the GPS community: (1) a set of "mean" phase center offsets and (2) a set of elevation-dependent PCC and offsets relative to the Dorne Margolin T antenna.

Since the PCC values are all relative to the Dorne Margolin T antenna some effects of antenna mixing still remain. Even with the same type of antenna the variation in the apparent phase center as a function of elevation angle will influence the results on longer baselines. Therefore the task of getting absolute calibration of the antennas through, e.g., chamber measurements or simulation software may be essential for some applications even though these calibration values most likely will change when the antenna is deployed in the field.

Effects like these can of course be reduced by utilizing antennas less sensitive to scattering from external structures. One way to achieve this is to reduce the side- and back-lobe levels of the amplitude patterns by means of well designed ground-planes. For this purpose new antenna designs have been proposed (see e.g., [*Alber*, 1996]; [*Ware et al.*, 1997]; [*Jaldehag*, 1995]; and [*Clark et al.*, 1996]). Futhermore, several groups are currently developing methods to perform absolute field calibration of antennas (see e.g., [*Wübbena et al.*, 1996]) and insitu calibration of antenna/pillar systems.

Antenna-Pillar System and the Signal

Here we concentrate on the site-dependent error associated with the electromagnetic coupling between the antenna and its nearby environment (e.g., [*Tranquilla*, 1986]; [*Tranquilla and Colpitts*, 1988]). The total electromagnetic field of an antenna which radiates a signal in the presence of conducting structures may be expressed as a superposition of the transmitted field and the fields scattered (i.e., reflected and diffracted) by the structures. By reciprocity, the same is true for a receiving antenna. The significance of the scattered field depends on the degree of electromagnetic coupling between the antenna and the scatterer, that is, the distance to the scatterer and the size and reflectivity of the scatterer. Signal scattering affects both the amplitude and phase of the received GPS signal, presumably independently at each site in a network. This independence creates differential phase errors.

Scattering from structures in the vicinity of the antenna effectively changes the antenna phase pattern, and, thus, affects the precision of the carrier phase measurements of the GPS signal. In studies by *Elósegui et al.* [1995] and *Jaldehag et al.* [1996a] it was shown that estimates of the vertical component of baselines formed between sites using identical antennas were dependent on the minimum elevation angle of the data processed. Both studies found that the elevation-angle- dependent systematic effect was associated with non-identical pillar arrangements, causing differential phase errors due to scattering from structures associated with the mounting of the antenna to the pillar, and with the pillar itself. Even the most perfectly calibrated antenna the antenna phase pattern will change when attached to a pillar.

Jaldehag et al. [1996a] demonstrate that estimates of the vertical component of many baselines strongly depend on the minimum elevation angle (elevation cutoff angle) of the data analyzed. A significant part was found to be due to differential phase errors caused by scattering from structures associated with the mounting of the antenna to the pillar and with the pillar itself. As the precision and accuracy of GPS measurements improve in general, antenna phase pattern variations due to different pillars and antenna mounts could be the major error source in just a few years, if not now. Modeling of the scattering effect, or rather the complete phase response of the antenna system, including the pillar, is an important issue for future improvements of the GPS technique.

Radomes - Protective Covers

At several permanent GPS sites located in areas with periodically severe environmental conditions (snow, rain) radomes have been employed. Until recently, most radomes in use have had a conical shape.

All materials have some effect on a electromagnetic wave. Radomes appear to delay and refract the GPS-signal in a similar way as snow [*Jaldehag et al.*, 1996b]. Several groups have recently been investigating effects due to the excess signal path delay through the radome. Different radomes have been tested in anechoic chambers [*Clark et al.*, 1996]; [*Meertens et al.*, 1996b] as well as in field tests [*Meertens et al.*, 1996b]; [*Jaldehag et al.*, 1996c]. All tests show that a conical cover may cause cm-level vertical errors when the tropospheric delay parameter is estimated. The recently employed hemispheric radomes seems to show much less elevation dependence. The influence on the tropospheric wet delay estimates and subsequently, the vertical component will only be on the 1-2 mm level. We can conclude that all radomes effect the GPS signal at some level and in form of an excess signal path delay which will map into other parameters in the GPS software. The effect of the protective covers can most likely be misinterpreted as a tropospheric effect in a similar way as snow. The effect is more or less constant and may be calibrated or modeled.

Precipitation

Signal propagation delay during snow storms has been investigated by, e.g., *Tranquilla and Al-Rizzo* [1993] and *Tranquilla and Al-Rizzo* [1994] who demonstrated that due to the localized nature of many snow storms differential effects may cause systematic variations at the centimeter level in estimates of the vertical coordinate of site position. Systematic variations introduced by snow storms may, however, if short-lived (minutes to hours), be reduced to a high degree by data averaging. A potentially more serious effect of heavy snow precipitation is the accumulation of snow on the top of the GPS antenna and on its surroundings, such as on the top of the GPS pillar or, when present, on the radome covering the antenna. This accumulation may last for days, weeks, or months. *Webb et al.* [1995] reported variations on the order of 0.4 m in estimates of the vertical coordinate of site position. The variations were correlated with the accumulation of snow over the antenna. Variations at the several centimeter level in estimates of the vertical coordinate of site position strongly correlated with changes in the accumulation of snow on top of GPS antennas have also been observed by others [*Jaldehag et al.*, 1996b]; [*BIFROST project members*, 1996]; [*Meertens et al.*, 1996a]. The results indicate that the variations in the vertical coordinate of site position can be fully explained by reasonable accumulations of snow which retard the GPS signals and enhance signal scattering effects.

Horizontal Atmospheric Gradients and Air Pressure Loading Effects

In the data processing the atmosphere is normally considered to be spherically stratified. We assume that one equivalent zenith wet delay value determines the wet delay in any direction, given a certain elevation angle. More advanced models, using more parameters to describe the atmosphere, have been proposed as alternatives to this very simplified model (e.g., [*Davis et al.*, 1993]; [*MacMillan*, 1995]). Several groups are now implementing possibilities to estimate horizontal gradients in the software [*Bar-Sever and Kroger*, 1996]; [*Chen and Herring*, 1996].

The lack of pressure data available during the GPS analysis can be the reason for different errors. During the entire GPS processing we have to model many external and internal effects on the crust of the earth. One effect currently not modeled is the pressure loading. The vertical position of the GPS receiver changes due to different atmospheric pressure loading the Earth [*vanDam and Herring*, 1994]. Extreme values could affect the vertical component of the GPS estimates on the cm level. These effects are of course related more to the general presssure field in the region rather than to a specific site. To properly model this effect a grid of pressure data has to be available. Unfortunately, it is very difficult to isolate these effects from other elevation-angle-dependent effects (multipath, scattering, snow/ice, etc.). Small variations in the vertical component are also caused by these other errors. We are thus not in the position of being able to correct for horizontal atmospheric gradients and loading errors optimally. At this point, theoretical studies are needed to quantify these effects, and to understand how we can best deal with these problems.

Local Stability and Monumentation

As GPS measurements have become more precise and are more frequently acquired, the issue of monumentation and site stability has become more important. The long-term contribution to the maintainance and densification of the global reference frame could be seriously affected by instable sites. The IGS network consists of a large variety of monuments established on top of everything from solid bedrock to buildings. The long-term stability of the reference frame and products associated with it, such as the orbits, are at issue here. Much attention is currently focused towards motions of geodetic monuments. These motions have been found by some researchers to be random-walk-like (e.g. [*Johnson and Agnew*, 1995]; [*Johnson et al.*, 1996]) while others find no evidence for random walk behavior (e.g. [*Mao et al.*, 1996]; [*Davis et al.*, 1996]). An ideal GPS monument would move in response only to the tectonic motion of the Earth. However, location, ground, and the environment at ground surface can have dramtic impact on the long-term stability of a site. The implication of this type of power-law noise is serious if the data are used to estimate low-frequency characterisitics of a time series such as the slope (deformation rate). *Mao et al.* [1996] and *Davis et al.* [1996] found no tendency of a random-walk like behavior possibly because the records where not long enough to see a random walk component above the noise in the low portion of the signal. It is also quite possible that monument motion may depend critically on the monument design and the site locations.

Nevertheless these investigations will continue and are most effectively addressed using continuous GPS measurements gathering data in a large variety of local conditions and GPS satellite configurations. There are design techniques which can be employed to mitigate this unwanted influence, most of which involve anchoring the monument to several points at depth and isolating the monument from surface material. Detailed spectral analyses and examination of the long time series available for some of the global sites. Monument and local stability problems could also manifest themselves with a periodic behavior, and be correlated with atmospheric conditions and precipitation.

Conclusion

Site-specific errors cannot be separated out when data from the global IGS sites are being used to determine orbits and reference frame. To be able to constrain the common mode of motion, sometimes in the submillimeter range, in a regional or local network a strong reference network is needed. The origin of the reference frame must be maintained with a high degree of robustness. In addition, orbits must be compatible with the reference frame. For this purpose the IGS sites need to be better examined. We especially found that the problems associated with the antenna-pillar system and the signal distortions have to be addressed. The effect of the antenna and signal related errors are constant from day-to-day but are biasing products like the orbit determination, station time series, and precipitable water vapor time series. Any changes either at a station or in the GPS-data analysis strategy might change this bias and thereby affect the daily products and the reference frame. The other important issue that needs attention is the long-term stability of the sites and the monuments used in the IGS network. This is especially important bearing in mind that local and regional continuously operating GPS networks are now used to detect motion at the level of 1 mm/yr or less.

Acknowledgements

Many thanks to all members of IAG SSG 1.158. Special thanks to Jim Zumberge, Jim Davis and Chuck Meertens for valuable contributions to the manuscript. This research was in part supported by the EC Environment and Climate Research Programme, the Swedish Natural Research Council, and the Swedish National Space Board.

References

Alber, C., Millimeter Precision GPS Surveying and GPS Sensing of Slant Path Water Vapor, Ph.D.Thesis, Univ. Colorado, Dec. 1996.

BIFROST project members, GPS Measurements to Constrain Geodynamic Processes in Fennoscandia, *EOS, Transactions, American Geophysical Union, 35*, 1996.

Bar-Sever, Y., and P. M. Kroger,Strategies for GPS-Based Estimates of Troposphere Delay, *Proc. of the 9th International Technical Meeting of the Satellite Division of the Institute of Navigation*, vol. 1, 615–623, 1996.

Chen, G., and Herring, T.A., Effects of Atmospheric Azimuth Asymmetry in the Analysis of Space Geodetic Data, *Eos Trans. AGU, 77*(46), Fall Meet. Suppl., F453, 1996.

Clark, T.A., B. Schupler, C. Kodak, and R. Allshouse, GPS Antennas: Towards Improving Our Understanding of Factors Affecting Geodetic Performance, *Eos Trans. AGU, 77*, Fall Met. Suppl. 1996

Davis, J. L., G. Elgered, A. E. Niell, and C. E. Kuehn, Ground-based measurement of gradients in the "wet" radio refractivity of air, *Radio Sci.,28*, 1003–1018, 1993.

Davis, J.L., R.A. Bennet, J.M. Johansson, H.G. Scherneck, and J.X Mitrovica, Spectral Analysis of Site Position Variations from Project BIFROST, *Eos Trans. AGU, 77*(46), Fall Meet. Suppl. F453, 1996.

Elósegui, P., J.L. Davis, R.T.K. Jaldehag, J.M. Johansson, and A.E. Niell, Effects of Signal Multipath on GPS Estimates of the Atmospheric Propagation Delay, *Eos Trans. AGU, 75*(44), p. 173, Fall Meeting Suppl., 1994.

Elósegui, P., J. L. Davis, R. T. K. Jaldehag, J. M. Johansson, A. E. Niell, and I. I. Shapiro, Geodesy Using the Global Positioning System: The Effects of Signal Scattering on Estimates of Site Position, *J. Geophys. Res.*, *100*, 9921–9934, 1995.

Jaldehag, R. T. K., Space Geodesy Techniques: An Experimental and Theoretical Study of Antenna Related Error Sources, Ph. D. Thesis, Tech. Rept. No. 276, School of Electrical and Coputer Engineering, Chalmers University of Technology, Gothenburg, Sweden, 1995.

Jaldehag, R. T. K., J. M. Johansson, P. Elósegui, J. L. Davis, A. E. Niell, B. O. Rönnäng, and I. I. Shapiro, Geodesy Using the Swedish Permanent GPS Network: Effects of Signal Scattering on Estimates of Relative Site Positions, *J. Geophys. Res.*, *101*, 17814–17860, 1996a.

Jaldehag, R. T. K., J.M. Johansson, J.L. Davis, and P. Elósegui, Geodesy using the Swedish Permanent GPS Network: Effects of snow accumulation on estimates of site positions, *Geophys. Res. Lett.*, *23*, 1601–1604, 1996b.

Jaldehag, R.T.K., J.M. Johansson, and J.L. Davis, Environmental Effects on the Swedish Permanent Network, *Eos Trans. AGU, 77*(46), Fall Meet. Suppl., F453, 1996c.

Johnson H.O.and D.C.Agnew, Monument Motion and Measurements Crustal Velocities, *Geophys. Res. Lett., 22*, 3533-3536, 1995.

Johnson H.O., F.K. Wyatt, D.C.Agnew, and J. Langbein, Evidence for Power-Law Behavior in Geodetic Time Series: Should we care, and if so, Why?, *Eos Trans. AGU, 77*(46), Fall Meet. Suppl. F453, 1996.

Mao A., C.G.S. Harison, and T. Dixon, Monument Stability of the Permanent GPS stations and Tide Gauges, *Eos Trans. AGU, 77*(46), Fall Meet. Suppl., F453, 1996.

MacMillan, D. S., Atmospheric gradients from very long baseline interferometry observations, *Geophys. Res. Letters, 22*(9), 1041–1044, 1995.

Mader, G.L., and MacKay, J.R., Calibration of GPS Antennas, *Proc. IGS Analysis Center Workshop*, Silver Sping, MD, 1996.

Meertens C., Rocken, C., Braun, J., Exner, M., Stephens, B., and Ware, R., Field and Anechoic Chamber Tests of GPS Antennas, *Proc. IGS Analysis Center Workshop*, Silver Sping, MD, 1996a.

Meertens C., Braun, J., Alber C., Johnson J., Rocken C., Van Hove, T., Stephens, B., and Ware, R., Antenna and Antenna Mounting Effects in GPS Surveying and Atmospheric Sensing, *EOS Trans. AGU, 77*(46), Fall Meet. Suppl., F453, 1996b.

Niell, A. E., P. Elósegui, J.L. Davis, I.I. Shapiro, R.T.K. Jaldehag, and J.M. Johansson, Reduction of Signal Multipath Effects on GPS Estimates of Site Position (abstract), *Eos Trans. AGU, 75*(44), p. 171 Fall Meeting Suppl., 1994.

Niell A.E., Reducing elvation-dependent errors for ground-based measurements, *EOS Trans. AGU, 77*(46), Fall Meet. Suppl., F453, 1996.

Rocken, C., GPS antenna mixing problems, *UNAVCO Memo*, UNAVCO, Boulder, Colorado, 1992.

Rothacher, M., and Schär, S., Antenna Phase Center Offsets and Variations Estimated from GPS Data, *Proc. IGS Analysis Center Workshop*, Silver Sping, MD, 1996.

Schupler, B. R. and T. A. Clark, How different antennas affect the GPS observable, *GPS World*, 32–36, Nov./Dec., 1991.

Schupler, B. R., R. L. Allshouse, and T. A. Clark, Signal characteristics of GPS user antennas, *Navigation, 41*, No. 3, 277–295, 1994.

Tranquilla, J.M., Multipath and imaging problems in GPS receiver antennas, paper presented at Fourth International Symposium of Satellite Positioning, DMA, Austin, Tex., 1986.

Tranquilla, J.M., and B.G. Colpitts, GPS antenna design characteristics for high precision applications, paper presented at ASCE Specialty Conference GPS-88: Engineering Applications of GPS Satellite Surveying Technology, Am. Soc. Civ. Eng., Nashville, Tenn., May 11–14, 1988.

Tranquilla, J.M. and H.M. Al-Rizzo, Theoretical and experimental evaluation of precise relative positioning during periods of snowfall precipitation using the global positioning system, *Manuscripta Geodaetica, 18*, 362–379, 1993.

Tranquilla, J.M. and H.M. Al-Rizzo, Range errors in Global Positioning System during ice cloud and snowfall periods, *IEEE Trans. Antennas and Propagat., 42*, 157–165, 1994.

vanDam, T. M., and T. A. Herring, Detection of Atmospheric Pressure loading using Very Long Baseline Interferometry Measurements, *J. Geophys. Res., 99*, 4505–4517, 1994.

Ware, R., C. Alber, C. Rocken, and F. Solheim, Sensing integrated water vapor along GPS ray paths, *Geophys. Res. Lett 24*, pp 417–420, 1997.

Webb, F.H., M. Bursik, T. Dixon, F. Farina, G. Marshall, and R.S. Stein, Inflation of Long Valley Caldera from one year of continuous GPS observations, *Geophy. Res. Lett., 22*, 195–198, 1995.

Wübbena G., Menge F., Schmitz, M., Seeber, G., and Völksen, C., A New Approach for Field Calibrations of Absolute Antenna Phase Center Variations, *Proc. ION GPS-96*, 1996

PRECISE GPS POSITIONING IMPROVEMENTS BY REDUCING ANTENNA AND SITE DEPENDENT EFFECTS

Günter Seeber, Falko Menge, Christof Völksen
Institut für Erdmessung, Universität Hannover, D-30167 Hannover, Germany

Gerhard Wübbena, Martin Schmitz
Geo++, D-30827 Garbsen, Germany

ABSTRACT

In order to estimate the characteristics of GPS antennas this paper describes a new approach for the estimation of absolute phase center variations (PCV) in a field calibration. The main objective of this approach is the elimination of multipath from the GPS observable and the elimination of any influence of the reference antenna. This is achieved by forming the so called mean sidereal day time difference between observations of successive days. Since the satellite geometry of GPS satellites repeats every mean sidereal day multipath does as well. Therefore, by differentiating observations of two different days multipath can be eliminated. To gain information about the PCV special rotations have to be applied to the antenna of interest. The paper shows the results for three different antenna types.

A second application of the mean sidereal day time difference shows the estimation of small position changes between consecutive days. The mean sidereal day time difference is formed in the same manner as for antenna calibration, however, without any rotation. Therefore multipath and PCV are eliminated from the observable. We achieved with observation periods of 10 minutes a horizontal position resolution of only 1.2 mm which is better than results from standard data processing.

INTRODUCTION

Precise positioning with GPS has now reached a level of accuracy where the remaining error sources are identified as effects caused by phase center variations (PCV) of the antenna and site effects like multipath. Phase center corrections are generally important for GPS observations with mixed antenna design to take into account the different phase patterns of each antenna type. In addition it is not easy to separate between tropospheric errors and phase center biases. Multipath can have a significant influence on precise positioning depending on the site. Therefore the impact of multipath on the position estimation should be considered carefully.

So far the PCV are estimated by two different methods. In an absolute sense it is possible to determine the PCV in an anechoic chamber [1] and in a relative sense by a field procedure [2] with a known antenna as a reference, whose PCV are estimated based on the first method. The main disadvantage of the relative field calibration is the influence of the used sites. One can never assume that the field calibration is free of multipath. Therefore it is very likely that the determined PCV highly depend on the multipath effects and are correlated with the selected site. With a new method [3] we have shown a field procedure for the calibration of antennas in an absolute sense. This new method is based on the elimination of multipath by using observation data of different days.

ELIMINATION OF MULTIPATH

A site which is totally unaffected by multipath does not exist. Thus, field procedures for an antenna calibration as well as short time GPS observations are disturbed by this effect. The results are incorrectly estimated PCV values which are site dependent and a decreased resolution for the point positioning with short observation times.

A short summary of the basic concept for the method to eliminate multipath is given in this paragraph. It is based on the repeatability of the satellite geometry and therefore of the multipath signals after one mean sidereal day [4, 5]. New investigations concerning the exact period for the repeatability of the satellite constellation during a four days test in 1997 revealed slightly different values for each satellite in a range between 240 s and 256 s instead of the generally assumed value of 3 min 56 s (236 s). The period for the constellation repeatability was estimated by three methods. First of all, double differences of two successive days were cross-correlated in the time domain. The maxima show the time lags and clearly indicate the significant periodical appearance of multipath after a mean sidereal day (see [3]).

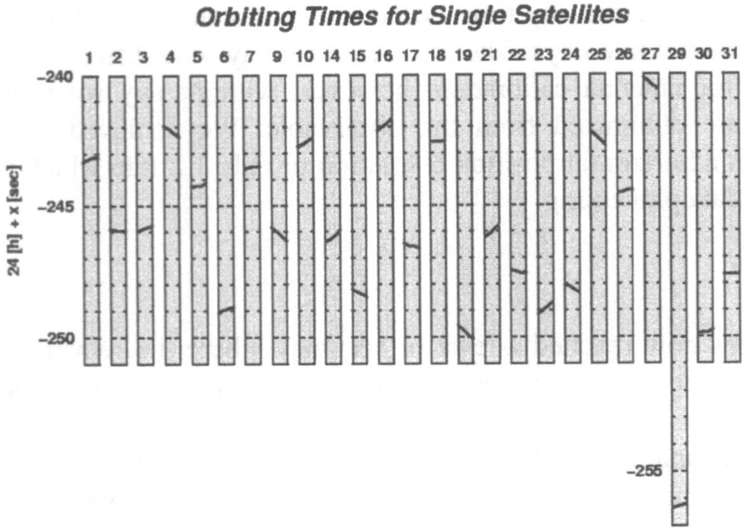

Figure 1: The orbiting times for different satellites.

Furthermore the periods are calculated from the correlation of elevation/ azimuth time series and from individual ephemeris. The results of these three methods are in good agreement. Figure1 shows the orbiting times for different satellites computed from the ephemeris of four days. The periodical appearance is used to greatly reduce the effect of multipath by subtracting the un-differenced GPS observable of two successive days, taking into account the evaluated time lag for the difference between a mean solar and a mean sidereal day, which should be calculated from the data sets. The following is a simplified linearized notation of the phase observation equations l_Φ in meters containing the design matrix subvector **a**; the receiver coordinate corrections **x**; the receiver and satellite clock error dt and dT, respectively, scaled to meters by the speed of light c_0; the ambiguity N scaled to meters by the wavelength λ; the error terms d for ionosphere (ION), troposphere (TROP), multipath (MP), and phase center variations (PCV); and the noise of the phase ε_Φ:

$$l_{\Phi i}^{\ j} = \mathbf{a}_i^j \cdot \mathbf{x}_i + c_0 \cdot (dt_i - dT^j) - \lambda \cdot N_i^j - d_{ION\,i}^{\ \ j} + d_{TROP\,i}^{\ \ j} + d_{MP\,i}^{\ \ j} + d_{PCV\,i}^{\ \ j} + \varepsilon_\Phi . \qquad (1)$$

The subscript i and superscript j stand for receivers and satellites, respectively. Building a mean sidereal day time difference δ^{SID} eliminates the multipath, phase center variation and the complete geometric information. The following observation equation does not contain any information about geometry, since the design elements **a** are almost identical on two successive days:

$$\delta^{SID} l_{\Phi i}^{\ j} = c_0 \cdot (\delta^{SID} dt_i - \delta^{SID} dT^j) - \lambda \cdot \delta^{SID} N_i^j - \delta^{SID} d_{ION\,i}^{\ \ j} + \delta^{SID} d_{TROP\,i}^{\ \ j} + \delta^{SID} \varepsilon_\Phi . \qquad (2)$$

The remaining terms comprise the mean sidereal time differences of every component, which are small for a short baseline (i.e. atmospheric errors) and/or are correctly modeled in the GPS processing package GEONAP [6] (i.e. clock errors). The noise level of the observable changed due to error propagation to $\delta^{SID}\varepsilon_\Phi$. In our paper [3] we demonstrated the elimination of multipath applying the mean sidereal day time difference δ^{SID}. Together with the dramatically reduced multipath the resolution of the phase measurements improved by a factor of 1.5.

MODELLING AND ESTIMATION OF ABSOLUTE PHASE CENTER VARIATIONS (PCV)

Although the mean sidereal day time difference eliminates site dependent errors, this approach can be used for the determination of absolute phase center variations. Observations are carried out on two days. One antenna is identically orientated on both days, whereas the antenna to be calibrated performs vertical and horizontal rotations on the second day only. Now, in extension to equation (2), the linearized observation equation (3) for the time difference δ^{SID} of the rotated antenna contains PCV of two different orientations:

$$\begin{aligned} \delta^{SID} l_{\Phi i}^{\ j} = {} & c_0 \cdot (\delta^{SID} dt_i - \delta^{SID} dT^j) - \lambda \cdot \delta^{SID} N_i^j - \delta^{SID} d_{ION\,i}^{\ \ j} + \delta^{SID} d_{TROP\,i}^{\ \ j} \\ & + d_{PCV}^{(\alpha_0, z_0)\,j}{}_i - d_{PCV}^{(\alpha_0+\Delta\alpha, z_0+\Delta z)\,j}{}_i + \delta^{SID} \varepsilon_\Phi . \end{aligned} \qquad (3)$$

The observable for the estimation of the absolute PCV is the difference in the PCV of two antenna orientations to an identical satellite:

$$\Delta d_{PCV}(\alpha_0, z_0, \Delta\alpha, \Delta z) = d_{PCV}(\alpha_0, z_0) - d_{PCV}(\alpha_0 + \Delta\alpha, z_0 + \Delta z) . \qquad (4)$$

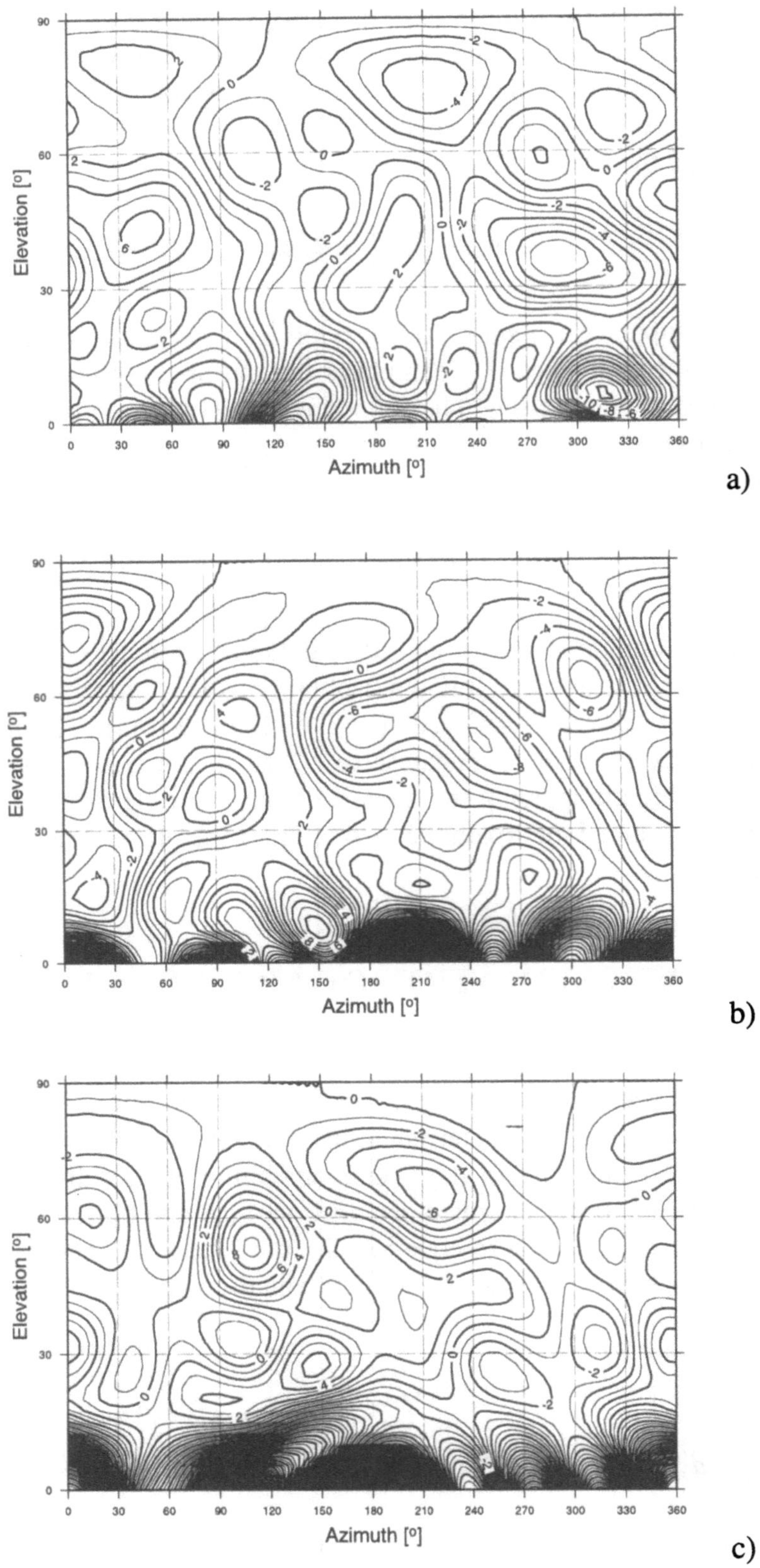

Fig. 2: L1-PCV [mm], a) Ashtech 700228 Rev. B, b) Trimble 4000 ST L1/2 Geod, c) Trimble Dorne Margolin Choke Ring.

In equation (4) α_0 und z_0 are the orientation parameter of the first day, while $\Delta\alpha$ and Δz represent the applied orientation changes on the second day.

As noted before, relative observables are used for the generation of the absolute phase pattern. Thus, only the topology of the pattern can be described. The absolute size is not known, but will be treated as a constant clock error or a hardware delay in the GPS position solution. The term "absolute antenna calibration", however, is still valid for the approach, because the phase center variations are determined independently from the phase pattern of a reference antenna. For the estimation of the elevation (and azimuth) dependent PCV serves a spherical harmonic function [3]:

$$d_{PCV}(\alpha, z) = \sum_{n=0}^{n_{\max}} \sum_{m=0}^{n} (A_{nm} \cos m\alpha + B_{nm} \sin m\alpha) P_{nm}(\cos z). \tag{5}$$

P_{nm} are normalized associated Legendre functions. Azimuth α and zenith angle z refer to the position of a particular satellite in the antenna coordinate system. To enable horizontal rotations and vertical tilts of the GPS antenna, a special mount must be used. Through the tilts a reception of undisturbed satellite signals from higher elevations contribute to a reliable PCV determination at low elevations in the antenna coordinate system, even at elevation zero.

Using this approach three different antenna types have been calibrated so far (Ashtech 700228 Rev. B Notches, Trimble 4000 ST L1/L2 Geod, Trimble Dorne Margolin Choke Ring). The L1- and L2-PCV are determined by a spherical harmonic function of degree 10 and order 5. Only the L1-PCV of these antenna types are shown in figure 2. The calibration for the Ashtech and the Trimble Choke Ring antenna was verified in a second independent experiment showing good agreement. Clearly, elevation and azimuth dependent variations are visible. All antennas have very large variations at low elevations in common, especially for the Trimble Choke Ring reaching up to several cm. The performance of the antennas seems to be rather poor at very low elevations.

DETECTION OF SMALL DEFORMATIONS

Another application of the proposed observable δ^{SID} is feasible for the determination of small position changes. A baseline is observed on two different days. The linearisation, identical on both days for each station, is performed in the same manner as shown in equation (1). Using the reduced observable, i.e. the mean sidereal day time difference, one should yield the baseline components with a higher precision due to the improved phase measurement resolution. In an experiment for the verification of this approach a short baseline was observed on two different days. One of the antennas was mounted to a special mechanical sledge (figure 3) which allows shifts in two horizontal directions normal to each other. These can be performed with high precision (better than 0.1 mm). The observations on the first day were carried out without changing the position of the sledge. Small position changes at a rate of 2 mm were applied during the second day. The sledge was shifted every 10 minutes, first to the north and then to the east. This procedure was repeated several times. The dashed lines in figure 4 and 5 show the performed shifts.

Fig. 3: Sledge

Figure 4 shows the displacements computed from the observations of the second day only. The positions were estimated with a simultaneous adjustment of L1 and L2, called LX. It is clearly visible that the agreement with the known shifts is rather poor. In figure 5 one can see the estimated positions after applying the mean sidereal day time difference with the GEONAP software module GNSDIF. The agreement between the computed displacements and the actual displacements is very good. Therefore the shifts of the sledge can easily be monitored by the mean siderial day time difference. The accuracy of the horizontal position estimation is in the order of 1.2 mm for an observation windows of 10 minutes only. Hence, with this method it is possible to monitor rather small deformations at the 1mm-level, even with very short observation times. Although observations on two days with an identical setup (mount, antenna etc.) are required, a number of applications are conceivable (e.g. dam monitoring).

CONCLUSIONS

Multipath is still the most limiting factor for precise positioning. With the proposed approach, the mean siderial day time difference, it is possible to eliminate multipath from the GPS observables. This derived observable can be used to observe small position changes from one day to another or estimate absolute PCV of an antenna.

The main advantages of the absolute antenna calibration are that it is a field procedure, free of multipath, and independent from a reference antenna's phase pattern. Through the rotations and tilts of the antenna areas without any observations are avoided, i.e. northern hole or low elevations, and the antenna is completely covered. Still, further examinations are requisite, e.g. errors introduced by the mount (errors in the rotation), influence of the

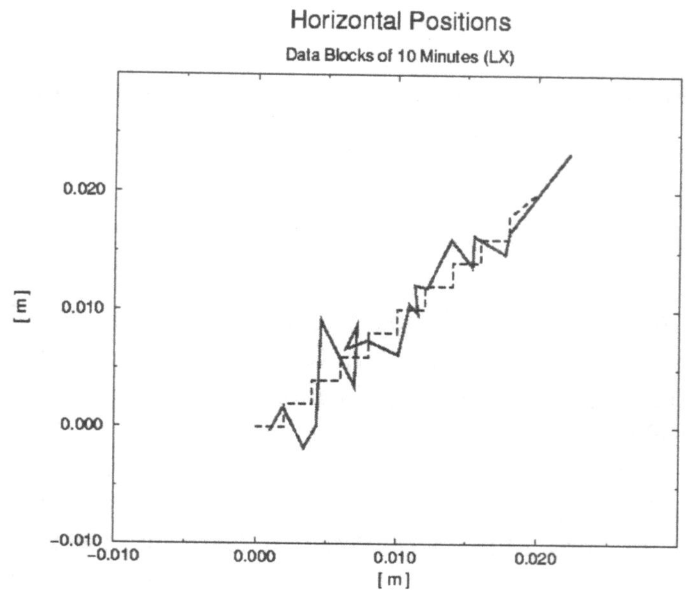

Fig. 4: True positions (dashed line) versus estimated positions (solid line) without δ^{SID}.

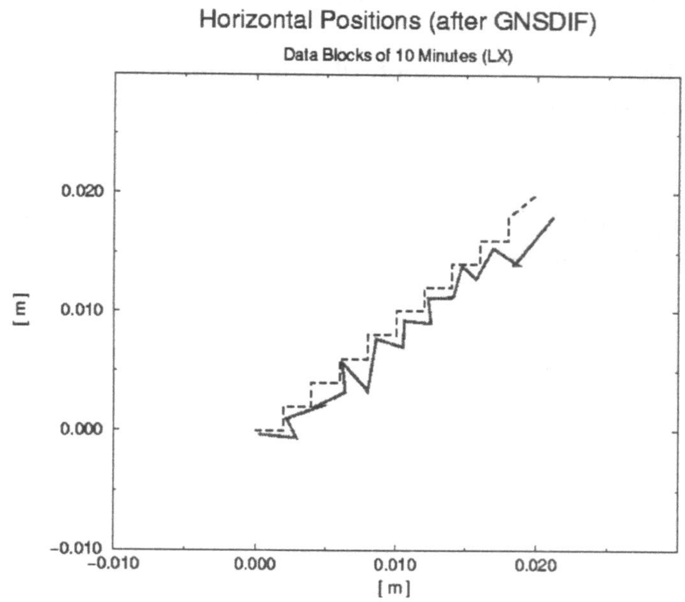

Fig. 5: True positions (dashed line) versus estimated positions (solid line) using δ^{SID}.

mount on the multipath environment, and comparison with other sources by applying the PCV in operational GPS evaluations. A future goal is the automatic calibration with a very precise robot to avoid instrumental erros of the mount and save observation time.

The first experiments with the mechanical sledge for the determination of small deformations are already promising. The resolution of the positioning is clearly improved. Further experiments have to be carried out in order to improve this method for different applications, e.g. deformation monitoring in permanent GPS arrays.

Acknowledgment: Parts of this work are funded by the German Bundesministerium für Bildung, Wissenschaft, Forschung und Technologie (BMBF, No. 03PL022B).

REFERENCES

1. Schupler, B.R., T.A. Clark, R.L. Allshouse: *Characterizations of GPS User Antennas: Reanalysis and New Results,* In: Beutler, G. et al. (Eds.). GPS Trends in Precise Terrestrial, Airborne, and Spaceborne Applications. IAG Symposium, No. 113, Boulder, Colorado, USA, 1995.
2. Rothacher, M., S. Schaer, L. Mervart, G. Beutler: *Determination of Antenna Phase Center Variations Using GPS Data,* Workshop Proceedings, IGS Workshop Special Topics And New Directions, May 15-18, Potsdam, Germany, 1995.
3. Wübbena, G., F. Menge, M. Schmitz, G. Seeber, C. Völksen: *A New Approach for Field Calibration of Absolute Antenna Phase Center Variations*, Proceedings of ION GPS-96, 9th International Technical Meeting, September 10-17, Kansas City, Missouri, USA, 1996.
4. Bock, Y.: *Continuous Monitoring of Crustal Deformation*, GPS World, June 1991, pp. 40-47, 1991.
5. Genrich, J.F.,Y. Bock: *Rapid Resolution of Crustal Motion at Short Ranges With the Global Positioning System*, Journal of Geophysical Research, Vo. 97, No. B3, pp. 3261-3269, 1992.
6. Wübbena, G.: *The GPS Adjustment Software Package -GEONAP- Concepts and Models*, Proceedings of the Fifth International Symposium on Satellite Positioning, Las Cruces, New Mexico, pp. 452-461, 1989.

IGS ORBIT, CLOCK AND EOP COMBINED PRODUCTS: AN UPDATE

Jan Kouba and Yves Mireault
Geodetic Survey Division
Natural Resources Canada
Ottawa, Ont., Canada

Abstract

Since January 1994 when the International GPS Service for Geodynamics (IGS) became an IAG service, several classes of IGS products have been generated on daily and weekly basis. IGS orbits, clocks and EOPs, based on weighted averages of the seven IGS Analysis Center (AC) daily solutions, are among these products. Throughout 1996 and 1997, a number of improvements, both in quality and product delivery times were realized resulting in enhanced quality of IGS combined products. The Final and Rapid orbit, clock and EOP delays were reduced significantly on June 30, 1996. Delays were reduced from one month to eleven days for the Final products and from eleven days to 24 hours for the Rapid products. New improved modeling and conventions, such as subdaily EOP modeling, ITRF94 and EOP rate solutions, were introduced on June 30, 1996 by all seven IGS ACs. These enhancements, along with the improved geometry of the IGS tracking network, resulted in steady orbit, clock and EOP solution improvements. The IGS combined orbit, clock and EOP products are currently approaching the 5 cm, 0.5 ns and 0.1 mas precision level respectively. Since March 1997, a new IGS orbit prediction product, also based on a weighted average, has been introduced. The new IGS orbit prediction is available in real-time and is significantly better than the broadcast GPS orbits. Typically, the IGS orbit prediction, when compared to the IGS Rapid orbits, is at or below the 1 metre precision level, while broadcast orbits compare at the 3-5m RMS level. Also, in March 1997, a new IGS LOD/UT1 combination based on a weighted average of AC LOD solutions was officially introduced. Encouraging results were obtained and are presented hereafter.

Introduction

Since 1994, the International GPS Service for Geodynamics (IGS) has been producing combined products of GPS orbits/clocks and Earth Orientation Parameters (EOP) (Kouba et al., 1995, Kouba and Mireault, 1996; 1997). On June 30, 1996 (GPS Week 860), a number of significant changes and improvements have been implemented. The former IGS Rapid orbit/clock/EOP combination (IGR), based on EOP(IGS) and produced within 11 days after the last observation, became the IGS Final combination (IGS) replacing the former IGS Final combination (Bulletin B (ITRF93)) produced about two months after the last observation. New IGR orbit/clock/EOP products, produced within 24 hours from the last observation, were introduced on GPS Week 860 (ITRF94) replacing the former IGS Preliminary combinations (ITRF93) run on a trial basis only. The new IGR combination is generated daily as opposed to a weekly cycle for the IGS Final products. In addition, two new IGS combined products have been introduced: the IGS 2-day orbit prediction (IGP) and the IGS LOD/UT1 combination. All the changes that have occurred since January 1994 (GPS Week 734) are summarized in Table 1, including the product names, orbit/EOP reference frame, orientation and availability after last observation.

Table 1: History of the IGS Rapid and Final Products

GPS Weeks	Short Name	IGS Name	Reference/Orientation Frame	Availability	Status	Cycle
730-781	IGS	IGS Final	ITRF92/IERS Bull.B-ITRF92	2 months	Official	Weekly
	IGR	IGS Rapid	ITRF92/IERS Bull.A	14 days	Official	Weekly
782-802	IGS	IGS Final	ITRF93/IERS Bull.B-ITRF93	2 months	Official	Weekly
	IGR	IGS Rapid	ITRF93/IERS Bull.A	14 days	Official	Weekly
803-859	IGS	IGS Final	ITRF93/IERS Bull.B-ITRF93	2 months	Official	Weekly
	IGR	IGS Rapid	ITRF93/ITRF93	11 days	Official	Weekly
834-859	IGP	IGS Prelim.	ITRF93/ITRF93	38 hours	Pilot	Daily
860-...	IGS	IGS Final	ITRF94/ITRF94	11 days	Official	Weekly
860-901	IGR	IGS Rapid	ITRF94/ITRF94	24:00 UT	Official	Daily
902-...	IGR	IGS Rapid	ITRF94/ITRF94	***22:00 UT***	Official	Daily
895-...	IGP	IGS Pred.	ITRF94/IERS Bull. A	~23:30 UT	Official	Daily

Orbit and Clock Evaluations

The long arc orbit evaluation was implemented for the IGS Final orbits to detect problems that could affect the daily weighted average combinations and to assess the consistency of individual AC solutions, including IGS combined orbits, over a one week period. Ephemerides are analyzed for individual AC independently from the combination process. The Long arc (La) orbit evaluation is described in more detail in the IGS 1994 Annual Report (Kouba et al., 1995). LaRMS are summarized in Table 2.

Starting with GPS Week 834 (Dec. 31, 1995), the IGS Final/Rapid combined orbits/clocks as well as all AC solutions which contain both the orbit and clock data, are further evaluated by an independent single point positioning program (navigation mode) developed at NRCan. This is done to verify clock solution precision and orbit/clock consistency. Pseudorange data from three stations are used daily and their corresponding position RMS (with respect to ITRF93 prior to GPS Week 860 and to ITRF94 since GPS Week 860) are summarized in the weekly/daily IGS Final/Rapid combination summary reports. The three stations are Brussels in Belgium (BRUS), Usuda in Japan (USUD) and Williams Lake in Canada (WILL). On average, horizontal and vertical point positioning RMS are below the 50 cm and the 100 cm level respectively.

1996-97 IGS Rapid and Final Combination Results

AC orbits, clocks and EOPs are routinely combined into IGS official products on a daily (Rapid products) and weekly cycles (Final products). Orbit, clock and EOP combinations are described in more detail in the IGS 1994, 1995 and 1996 Analysis Coordinator Reports (Kouba et al.,1995; Kouba and Mireault, 1996; 1997) and will not be described here. Since GPS Week 803 (May 28, 1995), Polar Motion (PM) x and y coordinates and since GPS Week 857, PM x and y rates have also been combined as weighted averages from available AC PM values using orbit weights. Orbit, clock and PM overall RMS with respect to the Final combination results for the period covering June 30, 1996 to August 9, 1997 (GPS weeks 860 to 917) are summarized in Table 2.

Three types of orbit RMS are included in Table 2: the weighted combination RMS (WRMS), the combination RMS, and the Long arc evaluation RMS (LaRMS). ACs used in the Final clock combination are COD, EMR, ESA, GFZ, and JPL. NGS provides broadcast clock corrections and SIO provides no clock corrections at all. ACs used in the Rapid clock combination are EMR, ESA, GFZ and JPL. Starting on GPS Week 902/day 3 (April 23, 1997), USNO's Rapid products (USN) have been used in the IGS Rapid combinations. NGS and SIO are excluded for the reasons mentioned above and COD, as NGS, is excluded because it provides broadcast clock corrections in its Rapid submissions. However, the clock information these ACs provide is still compared with the combination results. The IGR orbit, clock and EOP results are included in the Final combination for comparisons only.

Bad satellite orbit and/or clock solutions are excluded from the combination if they bias the IGS combined solution but are included in the RMS computations. All exclusions are reported in the IGS weekly/daily summary reports. For the IGS Final combination, the best clock RMS have now reached the 0.5 ns level for more than one AC and the best orbit position RMS have been approaching the 5 cm level (see Table 2). For the Rapid combination (IGR), the best clock RMS have also reached the 0.5 ns level and the best orbit position RMS vary between 5 and 10 cm which is quite remarkable considering that these are available within a day after the last observation.

Table 2: Final IGS Orbit/Clock/PM Combination RMS

GPS Weeks: 860-917; WRMS: Weighted Rms; LaRMS: Long arc RMS
cm: centimetres; ns: nanoseconds; mas: milliarc-seconds

AC/Products	Orbit			Clock	PM RMS		PM rates RMS	
	RMS (cm)	WRMS (cm)	LaRMS (cm)	RMS (ns)	x (mas)	y (mas)	x (mas)	y (mas)
IGS	--	--	9	---	---	---	---	---
COD	7	5	9	0.9	.14	.32	.21	.21
EMR	10	10	13	0.5	.20	.33	.73	.64
ESA	10	7	11	1.7	.19	.27	.51	.48
GFZ	7	7	11	0.5	.12	.13	.23	.24
JPL	8	6	11	0.5	.13	.18	.41	.31
NGS	15	15	17	---	.47	.52	.91	1.08
SIO	9	8	11	---	.74	.70	.44	.52
IGR	8	6	11	0.7	.12	.22	.37	.33

New IGS Products

Following recommendations of the 1996 AC Workshop (Neilan et al., 1996), the IGS 2-day predicted orbit (IGP) and the IGS LOD/UT1 combinations were introduced on March 2/97 (GPS Week 895). It took about a year to implement and the quality of both products is still under evaluation.

IGS 2-Day Orbit Prediction Combination (IGP)

Work on the generation of the IGS Predicted orbit combinations (IGP) started in the Summer of 1996. ACs were encouraged to take part, to develop and test their long arc orbit prediction scheme. After several months of testing, it became apparent that a combination of AC predictions would also be more reliable and in most cases more precise than the best individual AC orbit predictions. Since March 2, 1997, daily AC orbit predictions, available by 23:00UT, are combined and made available to the IGS DCs and IGS CB no latter than 23:30UT, so that the IGP orbits are available for real time applications for the following day. Extrapolated broadcast clocks are also included in IGP orbit files. Within 22 hours after the end of the day, the IGP and the corresponding broadcast (BRD) orbits are compared with the IGR orbits. The statistics and transformation parameters are also included in the daily IGR report files to provide timely quality evaluation of IGP and BRD orbits and the transformations to ITRF94. Figure 1 shows the daily WRMS, RMS and median RMS (with respect to IGR after a 7-parameter Helmert transformation) for the broadcast orbits (BRD) and the predicted orbit combinations (IGP) since GPS Week 895. With the exception of occasional high RMS for certain satellites usually affecting both types of orbits, the IGP precision is much better than the broadcast (~ 50 cm median RMS for IGP compared to ~ 250-300 cm for BRD). All but one AC are contributing to IGP.

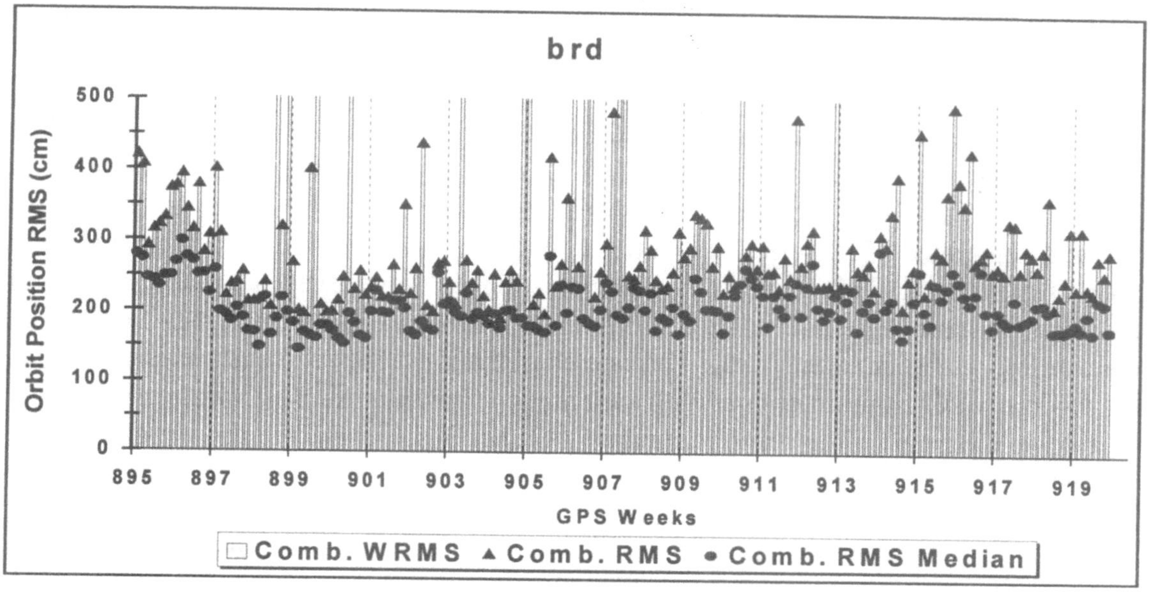

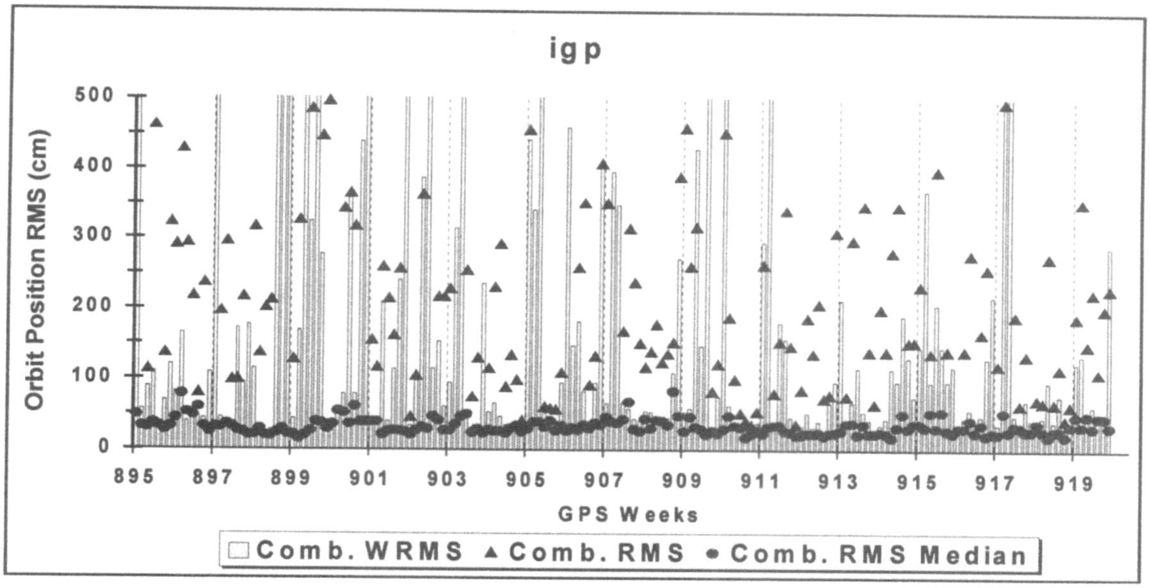

Figure 1. Daily Mean Orbit Prediction Position RMS for BRD and IGP (GPS weeks 895-919)

IGS LOD/UT1 Combinations

The need for IGS LOD/UT1 combinations was discussed at the 1996 AC Workshop (Ray, 1996b; Kouba, 1996). By the end of 1996, an approach proposed by Ray (1996a) has been implemented with some modifications and improvements. It is based on a weighted average of AC LOD solutions, which are calibrated with respect to 21 days of Bulletin A non predicted UT1 values. This calibrated LOD series is then integrated into IGS UT1. Tests from November 1996 to March 1997 indicated that the 2-6 day Bulletin

A UT1 predictions (used in IGR) with sigmas of about 600 μs could be significantly reduced to about 170 μs using the LOD/UT1 combinations.

The IGS combined LOD and integrated UT1 series were officially adopted for the IGS Rapid (EOP(IGS)96P02) and Final (EOP(IGS)95P02) combinations on GPS Week 895 (MJD 50509) and GPS Week 894 (MJD 50502) respectively. Since GPS Week 908/day 3 (MJD 50603) for IGR and GPS Week 907 (MJD 50593) for IGS, the IGS combined LOD and integrated UT1 series are initiated with the five day old non predicted UT1 values of Bulletin A. This reduces the effects of higher noise levels so characteristic for the latest Bulletin A non predicted UT1 values usually based on small number of observations. As in the IGS PM and PM rate combinations, the AC orbit weights are also used for the LOD combinations. Well determined Bulletin A UT1 values are usually available at the time of the IGS Final combinations. Significant UT1 improvements are expected for IGR LOD/UT1. As of the GPS Week 895/894 both EOP(IGS) series (95P02 and 96P02) include consistent PM, LOD and UT1 combinations.

Figure 2 shows comparisons of two UT1 series used for IGR combinations since March 2, 1997 with respect to Bulletin A's August 21, 1997 UT1 update. The one labeled "old" represents Bulletin A UT1 values available at the time of the IGR combination including Bulletin A predictions. The second one, labeled "new", is the IGR UT1 series obtained by integrating IGR combined LODs. The "new" LOD/UT1 combination strategy shows noticeable improvement.

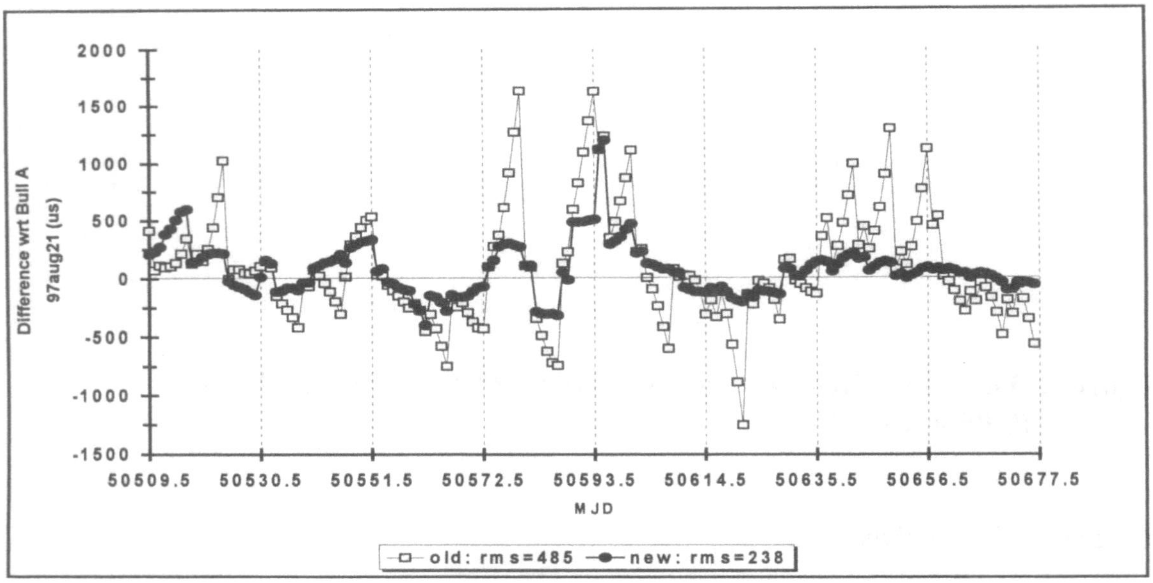

Figure 2. Comparison Between Bulletin A Predictions ("Old Strategy"), IGR Integrated UT1 ("New strategy") and Bulletin A's 97Aug21 Update.

Acknowledgments

The authors gratefully acknowledge and thank all IGS AC colleagues for their continuous assistance and cooperation without which the IGS product combinations and enhancements would not be possible.

References

Kouba, J., 1996, IGS combination of GPS Earth orientation Parameters (EOP), IGS AC Workshop Proceedings, Silver Spring, Md., March 19-21, 1996. pp. 33-43.

Kouba, J, Y. Mireault and F. Lahaye, 1995, 1994 IGS Orbit/Clock Combination and Evaluation, Appendix of the Analysis Coordinator Report, International GPS Service for Geodynamics (IGS) 1994 Annual Report, pp. 70-94.

Kouba, J. and Y. Mireault, 1996, Analysis Coordinator Report, International GPS Service for Geodynamics (IGS) 1995 Annual Report, pp. 45-76.

Kouba, J. and Y. Mireault, 1997, Analysis Coordinator Report, International GPS Service for Geodynamics (IGS) 1996 Annual Report, pp. 55-100.

Neilan, R.E., P.A. Van Scoy and J.F. Zumberge (ed.), 1996, Proceeding of the IGS AC Workshop Proceedings, Silver Spring, Md., March 19-21, 1996.

Ray, J., 1996a, Measurements of Length-of-Day using the Global Positioning System, Journal of Geophysical Research (JGR), Vol. 102, No. B9, 20141-20149.

Ray, J., 1996b, GPS Measurements of Length-of-Day: Comparisons with VLBI and Consequences for UT1, IGS AC Workshop Proceedings, Silver Spring, Md., March 19-21, 1996, pp. 43-60.

THE USE OF GPS FOR MONITORING OF THE IONOSPHERIC DISTURBANCES

L.W.Baran
Institute of Geodesy, Olsztyn University
of Agriculture and Technology
Oczapowski St.1, 10-957 Olsztyn, Poland

I.I. Shagimuratov
West Department of the IZMIRAN
Pobeda St.41, 236017 Kaliningrad, Russia

ABSTRACT

Since January 1995 GPS observations performed by Polish IGS network stations have been used for monitoring the ionosphere. During January-May 1995 we revealed a number of violations of the ionospheric structure from regularity. The disturbances resulted from magnetic storms with sudden onsets. The storms caused an increase of ionospheric delay by a factor of 1.5-3.0. During disturbances gradients and TID's effects are also increased. We found that such severe ionospheric conditions have an effect on the determination of GPS equipment biases. The biases determined from code data for disturbed days are different from those for quiet days. The differences can reach up to 0.3-0.5 m in differential delay. It can be expected that the effects will be even more appreciable during the next solar maximum.

INTRODUCTION

GPS observations performed at stations of the IGS network may be used for monitoring of the ionosphere on both time and space scales. GPS observations have provided the possibility of regular studies of ionospheric total electron content (TEC) variations on a diurnal, seasonal and solar activity cycle basis. These investigations provide the basis for developing, calibrating and improving ionospheric models for different applications.

Precise GPS positioning requires the resolution of the double differenced phase ambiguities. In such a case, the geometry-free linear combination is analyzed, which contains the ionospheric information. Ionospheric disturbances cause severe conditions for

International Association of Geodesy Symposia, Vol. 118
Brunner (ed.), Advances in Positioning and Reference Frames

GPS measurements and complicate the ambiguity resolution process. The large-scale horizontal gradients in the ionosphere produce errors in precise relative GPS positioning (Wanninger, 1995).

In this paper the storm-time effects in the ionosphere at mid latitudes during solar minimum activity are presented. The analysis of the results of determination of GPS equipment biases during disturbances are also discussed. For the study, GPS measurements recorded at Polish stations of the IGS network: Lamkowko (53.9N, 20.7E) and Borowiec (52.1N, 17.1E), have been used.

ALGORITHM USED FOR DETERMINATION OF TEC FROM GPS OBSERVABLES.

To determine the ionospheric TEC, a geometry-free linear combination is used. It contains the ionospheric delay and the ambiguities for phase measurements and equipment biases for code measurements.

The relationship between ionospheric delay and the TEC, and the difference between dual-frequency code (P) and phase (Φ) measurements may be written as (Baran et al., 1997):

$$\Delta P\ [m] = M \cdot TEC/\cos z + A_P + \varepsilon_P, \qquad (1a)$$

$$\Delta \Phi\ [m] = M \cdot TEC/\cos z + A_\Phi + \varepsilon_\Phi. \qquad (1b)$$

Here TEC is the vertical electron content, M is a scale factor, ε_P, ε_Φ are noise terms, A_p and A_Φ are equipment biases (A_Φ contains the phase ambiguity), z is the zenith angle of the ray at the sub-ionospheric point. In the algorithm, the following ionospheric model of the diurnal variation of TEC is used (Georgiadiou, 1994):

$$TEC = \sum_{n=0}^{6} a_n \cos(ns) + b_n \sin(ns) + a_z\, \Delta\varphi + a_s\, \Delta\varphi\, s, \qquad s = \pi(LT-14)/12. \qquad (2)$$

Here LT is the local solar time and $\Delta\varphi$ is the latitudinal difference between the coordinates of the receiver and the sub-ionospheric point. The TEC values are estimated simultaneously with equipment biases over a 24 hour period using a least squares procedure. We have employed such an algorithm where the code and phase measurements are treated as independent of one another. The TEC obtained from both observables should coincide. The difference is less than 0.1 - 0.2 of the TEC unit. The following estimates are presented in the form of diurnal variations of TEC.

STORM-TIME VARIATIONS OF TEC

For the period of January - May 1995, some moderate but sudden ionospheric disturbances occurred. For all storms, the geomagnetic conditions were nearly similar. In

Fig.1 storm-time variations in the ionosphere for the period 28 - 31 January, 1995 are presented.

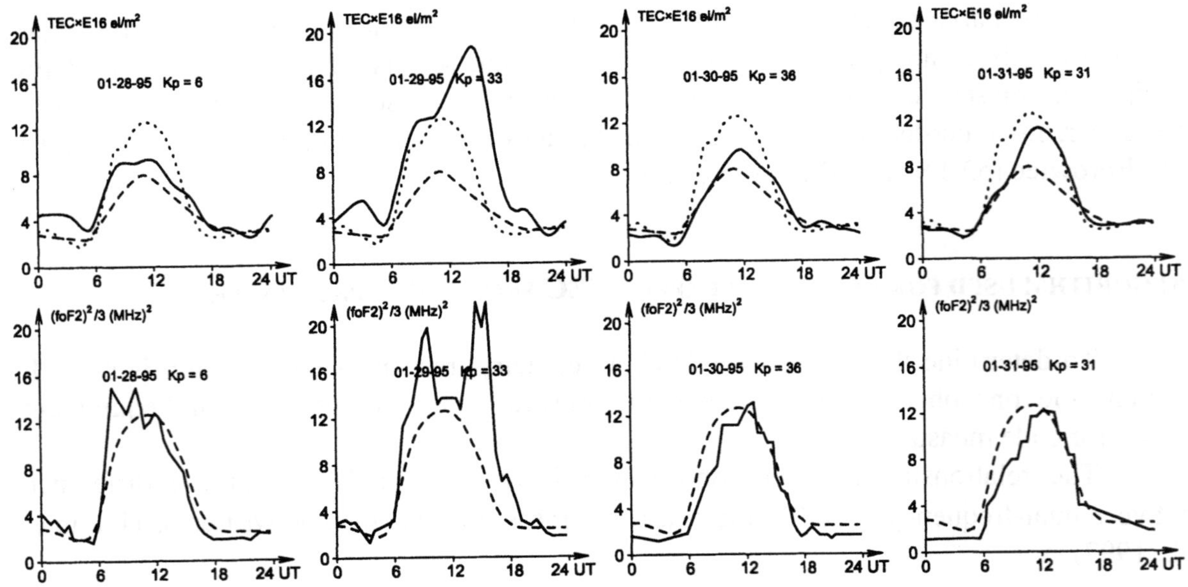

Fig.1: Storm-time effects in the ionosphere. Top panel: variations of TEC from GPS measurements (solid), TEC of the IRI model (dotted), the monthly mean of GPS TEC (dashed). Bottom panel: variations of $(foF2)^2$ (solid), monthly mean of $(foF2)^2$ at Warsaw (dashed).

On the top panel, the day by day ionosphere TEC variations are shown. For comparison, the average daily GPS TEC and IRI model derived TEC are shown. Model TEC was estimated using the International Reference Ionosphere (IRI) model with correction by the monthly mean frequency foF2 from ionosonde measurements near Warsaw. For winter the IRI model typically overestimates values of TEC compared with GPS estimates.

The day that preceded the storm (28 January) was very quiet, with a summary index of magnetic disturbances (Kp) was equal to 6. In the first day of the storm, on 29 January (Kp=33) TEC had increased by a factor of 1.5 - 2.6 relative to the previous, undisturbed day. In the diurnal variation two humps appear with a pronounced maximum after noon. During the following days, the negative phase of the storm began. The ionosphere TEC dropped to a level lower than the undisturbed one. The negative phase of the storm is less marked in the diurnal variation.

The ionospheric TEC values show a good correlation with the electron density of the F2 ionospheric layer. The maximum density (NmF2) may be obtained from ionosonde measurements: $NmF2 = k \times (foF2)^2$, where k is a constant. On the bottom panel the variation of $(foF2)^2$ at Warsaw is plotted. For comparison, the monthly mean is also given. The increasing value $(foF2)^2$, the same as for TEC, occurred on 29 January. Here also are two pronounced humps. The abnormal variation of foF2 may be seen simultaneously at Kaliningrad (54.7 N, 20.4 E) and Warsaw (52.1 N, 21.1 E) (Fig.2).

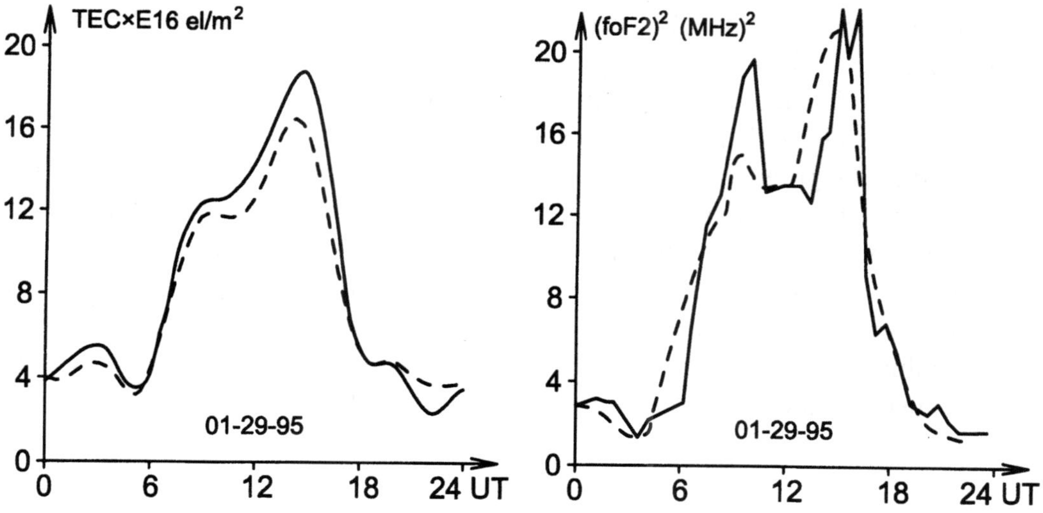

Fig. 2: The latitude behaviour of TEC and $(foF2)^2$ for the disturbed day of 29 January 1995. The TEC at Lamkowko (dashed) and Borowiec (solid). The foF2 at Kaliningrad (dashed) and Warsaw (solid).

For the more southerly Warsaw station, the first hump has a pronounced character. The same as for foF2, the ionosphere TEC value is higher at Borowiec, located to the south of Lamkowko. Some time delay may be seen between Warsaw and Kalinigrad events (Fig. 2). The delay amounts to about 30 min. On assuming that the disturbance moves from north to south, we may estimate the speed of the disturbance. It is about 700 m/sec.

The differences of TEC on quiet and disturbed days can be revealed by studying the behaviour of the electron content for individual satellite passes. In Fig. 3 the vertical TEC variation for satellite passes of PRN 5 and 2, on January 28 and 29, 1995 respectively, are given.

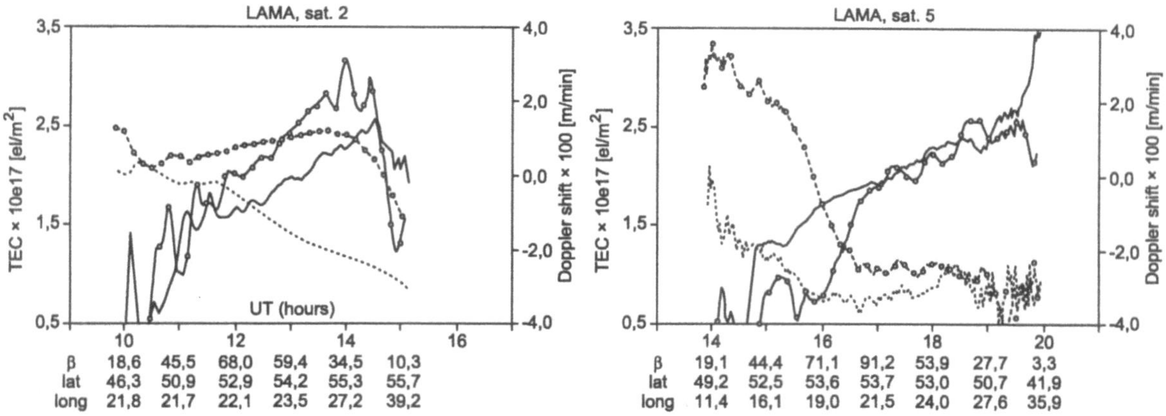

Fig. 3: Variation of vertical TEC and ionospheric Doppler shifts during satellite passes of PRN # 5 and PRN # 2 on quiet and disturbed days. The TEC of 28 January 1995 (·········), TEC of 29 January 1995 (———). Doppler shift of 28 January 1995 (··o··o··o··), and 29 January 1995(—o—o—o—).

For PRN 5 the absolute TEC is determined with code measurements corrected for equipment biases. For PRN 2 we used phase measurements corrected by the initial phase ambiguities. The code data includes multipath effects. In Fig.3 the ionospheric Doppler shifts obtained from phase data are also presented.

Storm-time Doppler shifts indicate increasing TID's effects. Ionospheric Doppler shifts characterize the gradients of TEC along the satellite trajectory. The large-scale gradients on a disturbed day are distinguished from quiet ones. In disturbed periods, the Doppler shift may even be of opposite sign compared to that of the quiet time (see Fig. 3). The horizontal large-scale gradients affect the determination of instrumental biases and ambiguity resolution. The pronounced difference in Doppler shifts for two days (Fig.3) seems to be caused by the variations of the biases during disturbances.

THE NEGATIVE PHASE OF STORM

On the second day of the storm, the ionosphere again reaches almost its ordinary state. In the diurnal variation the negative phase is poorly distinguished.

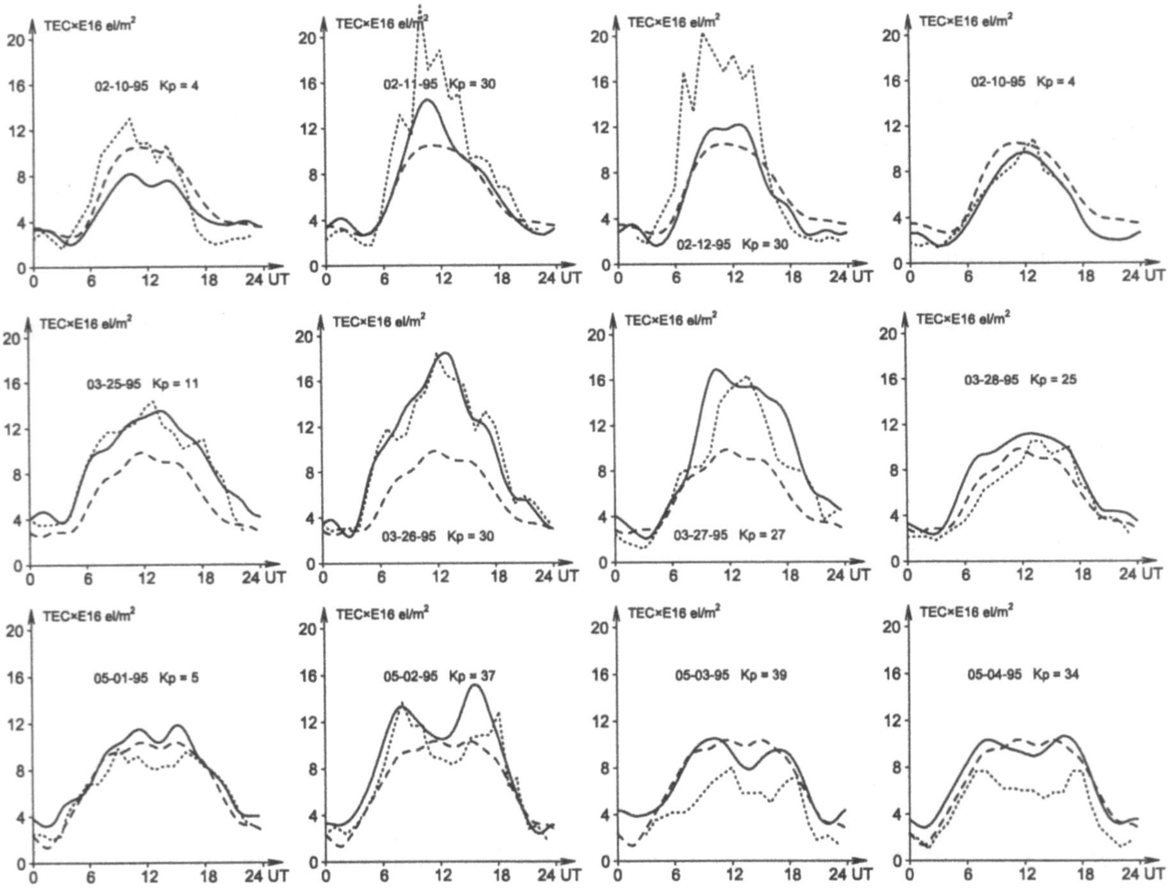

Fig.4: Storm-time variations in the ionosphere on February, March and May 1995. The GPS TEC at Lamkowko (solid), the $(foF2)^2$ at Warsaw (dotted) and the monthly mean of TEC (dashed).

For the analysis of behaviour of the TEC, we used the TEC variations on individual satellites. In that case, the code data were corrected for equipment biases obtained during quiet days. The procedure enabled the determination of the changes in TEC more accurately. The analyses show that for most passes the total electron content on the second day is lower than that for the quiet day. The decrease was about 20 - 30 % in comparison to the undisturbed level. For other satellite passes at the same time the changes are not marked relative to the quiet day. So the response of ionosphere to storm recovery stage is ambiguous.

The response of the ionospheric electron content during storms at other periods may be seen in Fig. 4. The general behaviour of TEC for storms with a sudden onset is still there. Always there is a pronounced increase in the TEC on the first days and recovery to previous level following one-two days.

VARIATIONS OF EQUIPMENT BIAS VALUES

Differential biases (Ap) have been computed day after day simultaneously with the ionospheric TEC from full 24 hours of measurements. Results of the determination of satellite/receiver biases for some satellites over the period of 28-31 January 1995 are presented in Table 1. The data refer to the Lamkowko receiver.

Table 1: Satellite/receiver biases

	Quiet days, Ap (m)					Disturbed days, Ap (m)		
PRN	25.01	26.01	28.01	Mean	RMS	29.01	30.01	31.01
12	0.47	0.48	0.37	0.44	0.04	0.44	0.33	0.53
14	-0.59	-0.49	-0.61	-0.56	0.05	-0.75	-0.79	-0.62
17	-0.39	-0.44	-0.49	-0.44	0.03	-0.59	-0.48	-0.44
18	-1.05	-1.03	-1.05	-1.04	0.01	-1.16	-1.27	-1.13
25	-1.54	-1.55	-1.50	-1.53	0.02	-1.71	-1.70	-1.58
29	-1.49	-1.46	-1.45	-1.46	0.01	-1.57	-1.69	-1.55
31	-1.43	-1,36	-1.32	-1.37	0.04	-1.83	-1.61	-1.39
ΣKp	11	6	6			33	36	31
$\overline{Ap}$/L	-0.84	-0.82	-0.87			-1.01	-0.91	-0.79
$\overline{Ap}$/B	-0.25	-0.28	-0.37			-0.50	-0.37	-0.39

The biases are expressed in units of meters. In Table 1 there are listed the biases averaged over quiet days (Mean) and RMS respectively. In the second part of Table 1, the summary index of magnetic disturbances (ΣKp) and the biases averaged for all satellites observed over 24 hours ($\overline{Ap}$) are given. It can be seen that for the highly ionosphere disturbed day of 29 January, 1995, the $\overline{Ap}$ are different from those for the quiet days. The difference of the mean biases amounts to about 0.2 m (2TEC). For individual satellites the difference may reach 0.3 - 0.5m. In the last line on the same period the mean biases for Borowiec ($\overline{Ap}$/B) are shown. The behaviour of $\overline{Ap}$ day by day for Borowiec is similar to that for Lamkowko

($\bar{A}p/L$). The equipment biases, as for the ionospheric TEC, are a maximum on the first day of the storm. In Table 2, the results for variation of biases are summarised for all disturbances over the January-May 1995 period.

Table 2: Comparison of biases for quiet and disturbed days

	Quiet days		Disturbed days		
January	-0.82	-0.87	-1.01	-0.91	-0.79
February	-0.65	-0.65	-0.60	-0.87	-0.85
March	-0.86	-0.82	-1.11	-1.34	-1.28
April	-1.17	-1.12	-1.34	-1.23	-1.11
May	-1.08	-1.11	-1.11	-1.34	-1.10

Here the biases are presented for two days before the disturbances and three days after the onset of the storm. The behaviour of biases on all storm-time days is similar.

The large - scale horizontal gradients substantially influence the determination of the equipment biases as well as the phase ambiguities. In Fig. 3 in the records of Doppler shifts it is seen how the gradients may be markedly distinguished between quiet and disturbed conditions. In addition to this, during storm-time the TID's effects are increased. In turn, the errors in determining biases are increased. We note that the TEC obtained independently from code and phase data are similar.

CONCLUSIONS

During ionospheric disturbances at solar minimum, at mid latitudes, the ionospheric electron content may be increased by a factor of 1.5 - 3.0 relative to quiet conditions. During storm-time, the abnormal large-medium-scale horizontal gradients in the ionosphere appear and the TID's effects are also increased. Storm ionospheric conditions cause the equipment biases determined from code measurements on disturbed days to be different from those obtained for quiet periods. Variations in the ionosphere may give rise to errors in the ambiguity resolution when using the geometry-free combination. The ionospheric effects will increase in the next ten years. Thus studies of storm-time effects that cause errors in precise GPS positioning must be continued.

REFERENCES

Baran L.W., I.I.Shagimuratov, N.J.Tepenitzina, (1997). The use of GPS for ionospheric studies. *Artificial Satellites. Journal of Planetary Geodesy.* V. 32, N 1, pp 49-60.

Georgiadiou J., (1994). Modelling the Ionosphere for an Active Control Networks of GPS Stations. In *LGP Series, Publications of the Delft Geodetic Computing Centre*, No 7.

Wanninger L. (1995). Monitoring ionospheric disturbances using the IGS network. In *Proceedings of the 1995 IGS Workshop*, May 1995, Potsdam, Germany.

AN INTEGRATED GPS MONITORING SYSTEM FOR SITE INVESTIGATION OF NUCLEAR WASTE DISPOSAL

Ruizhi Chen and Juhani Kakkuri
The Finnish Geodetic Institute,
Geodeetinrinne 2 ,
02430, MASALA,
FINLAND

Abstract

An integrated GPS monitoring system has been in operation in the about 10-km^2 study area at Olkiluoto, Finland since October, 1994. The system includes a permanent GPS station and a local GPS monitoring network. The permanent station is used to investigate regional crustal deformations in Finnish territory, while the local monitoring network is used to monitor relative deformations within the study area. The analysis of 2.8 years of continuous GPS tracking data shows that the horizontal components of the regional crustal deformations are about 2 mm/yr, while the vertical component is about 8 mm/yr. The local GPS monitoring network has been measured six times at half-yearly intervals. The horizontal components of the local movements obtained from the repeated measurements are less than 1.0 mm/yr. The GPS result indicates that the investigation area at Olkiluoto is rather stable.

1. Introduction

According to a decision by the Finnish government, the Finnish Posiva Oy is responsible for preparations for the permanent disposal of spent nuclear fuel in Finland. A special programme including studies of geology and geophysics has been carried out since 1983. According to the generic studies, three areas, namely Olkiluoto on the west coast, Kivetty in Central Finland and Romuvaara in northeast Finland, were selected in 1992 for detailed investigation. The detailed studies were begun in 1993, and will be continued until the year 2000. One of the areas will then be selected as the location for permanent disposal of the spent fuel. Complementary investigations will be continued in the area selected until the year 2010. In co-operation with other geophysical studies, the Finnish Geodetic Institute is now taking responsibility for operating a GPS monitoring system to monitor the

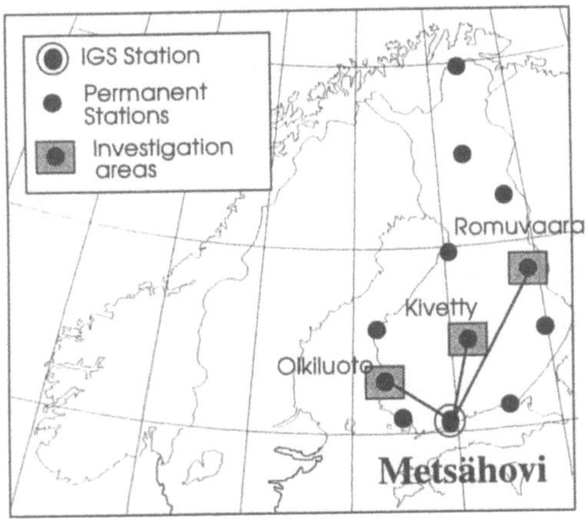

Fig. 1. The Finnish permanent GPS array and the investigation areas for nuclear waste disposal in Finland.

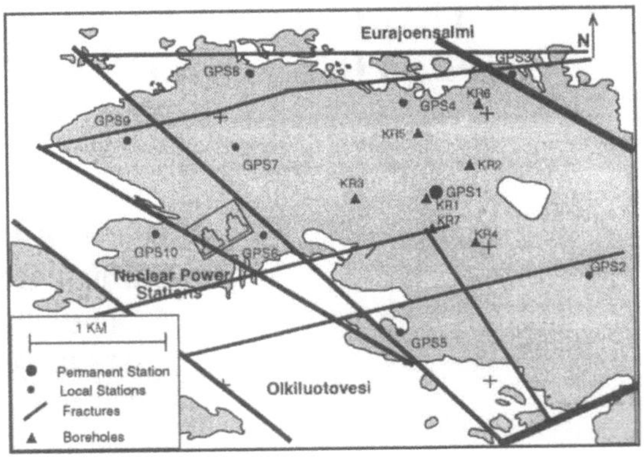

Fig. 2. The local GPS monitoring network at Olkiluoto. Locations of the fractures are obtained from *Anttila et al.* 1992.

crustal deformations at the study areas (*Chen and Kakkuri*, 1993; 1994; 1996; 1997). The system includes:

- three permanent GPS stations which operate continuously, and
- three high-precision local GPS monitoring networks at the study areas.

The permanent stations belong to the Finnish permanent GPS array which consists of 12 stations as shown in Fig. 1. The network is connected to the IGS network through Metsähovi. Each station is installed with an Ashtech Z-XII receiver and a choke-ring antenna.

The first local GPS monitoring network was established at the Olkiluoto study area in October, 1994 as shown in Fig. 2, while the other two networks for the Kivetty and Romuvaara study areas were established in September, 1995. Each network includes a permanent station such that the local GPS monitoring networks are in fact connected to the IGS network.

The reinforced pillars need about half a year to settle completely, therefore the first measurement was performed in May, 1995 for Olkiluoto, while those for the other two investigation areas were performed in May, 1996. For this reason, we have observations for only about one year from the networks at Kivetty and Romuvaara. This is too short a time to make any deformation analysis for these two areas, therefore we will show only the results from the Olkiluoto investigation area. The results are based on 2.8 years of continuous GPS observations for the permanent station, and six repeated measurements of the local monitoring network.

2. Regional deformations

Regional deformations are relative deformations between the permanent GPS stations and the IGS station at Metsähovi. GPS data collected from the permanent stations are processed with the Bernese GPS software by a standard procedure. A set of batch programs has been developed for the automatic processing of daily observations. Only a single DOS command, as simple as "*C:\AUTOSESS\ATP*", is required to process all the data stored on a CD-ROM. The system can identify the days of GPS data on the CD-ROM and create corresponding preprocessing batch files. The daily normal equation is stored for the combination adjustments which are performed for every five days. The GPS data are processed with the IGS precise orbits. The IGS station at Metsähovi is used as the fixed station in the daily solutions. Fig. 3 shows the velocity components obtained from 2.5 years of continuous GPS observations.

The horizontal components indicate that Olkiluoto is moving by 2 mm/yr in a southeasterly direction relative to the IGS station at Metsähovi. The compression direction agrees very well with global tectonic models (*DeMets et al.* 1990, 1994). The borehole

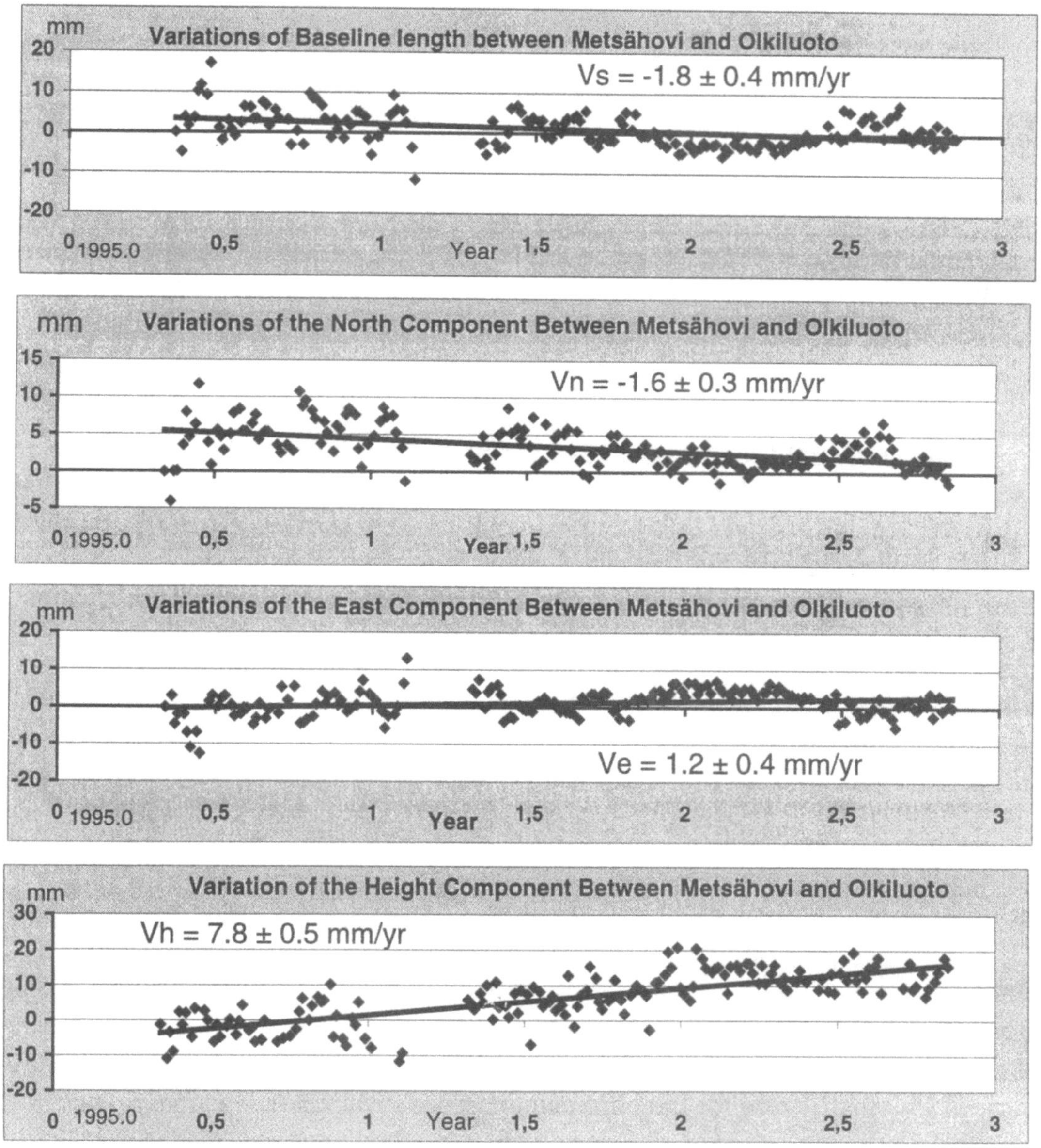

Fig. 3. Velocity components of the GPS permanent station at Olkiluoto. The result is obtained from five-day solutions. The zero-epoch in the figure is 1995.0.

stress measurements performed in the investigation area indicate that the maximum stress orientation is in the east-west direction (*Anttila, et al.* 1992). The discrepancy can be explained in such a manner that the GPS result represents regional stress orientation for the zone between Metsähovi and Olkiluoto, while the borehole measurements represent the local stress orientation at Olkiluoto.

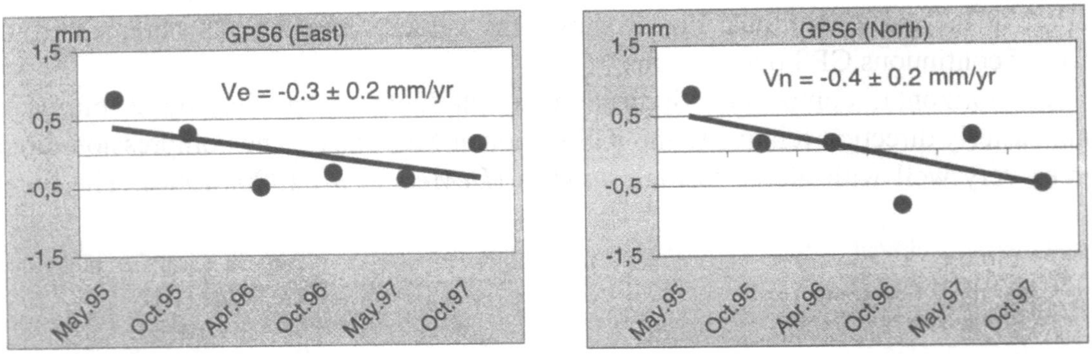

Fig. 4. Result obtained from five measurements at the local GPS network at Olkiluoto. The measurements are repeated at a half-yearly interval beginning in May, 1995. The velocities are relative to the permanent station GPS1, which is considered as the fixed station, as shown in Fig. 2. Similar pictures can be found for other stations.

For the vertical component, the uplift rate obtained from GPS is 7.8 mm/yr, which is larger than the uplift value of 4.1 mm/yr obtained from precise levelling (*Kääriänen*, 1966). By taking the uplift of the geoid into account, the uplift rate of the ellipsoidal height obtained from precise levelling becomes 4.2 mm/yr as the difference of the uplift of the geoid between Metsähovi and Olkiluoto is about 0.1 mm/yr (*Kakkuri*, 1997; *Kakkuri and Chen*, 1997). Although the GPS result is rather scattered, it indicates the correct trend, therefore a more reliable result can be obtained with GPS over a longer observation time span.

3. Local deformations

The permanent stations in the investigation areas are used as reference stations for the corresponding local monitoring network. Therefore, local movements are referred to the corresponding permanent station.

The local GPS monitoring network at Olkiluoto has been observed for six times since 1995 at half-yearly intervals. The stations were occupied for about 10 hours with Ashtech Z-XII receivers and choke-ring antennae. The GPS data are analyzed carefully with a processing procedure which has been studied comprehensively in our earlier study (*Chen and Kakkuri*, 1993). In an optimal observation environment, we are able to achieve an accuracy of less than 1 mm for the horizontal components (*Chen and Kakkuri,* 1993). Fig. 4 shows the result obtained from six repeated measurements for station GPS 6.

The rotation errors of the network should not be significant because

- the IGS precise orbits are used in processing the GPS data, and

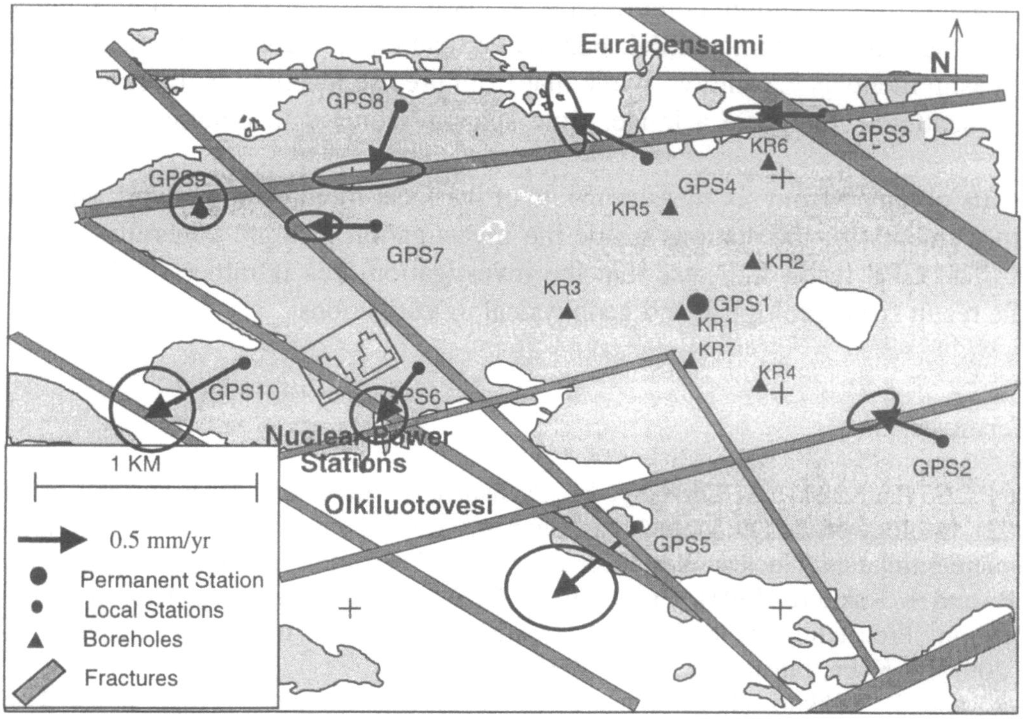

Fig. 5. Local movements in the surroundings of the nuclear power stations at Olkiluoto. The error ellipses of the displacement vectors are at confidence level of 95%.

- the fixed station is a permanent station, therefore the accuracy of the fixed coordinates is high.

It is therefore reasonable to depict the relative deformations with displacement vectors though it is theoretically not strict. The displacement vectors give a more direct view of the deformation pattern than the strain parameters.

Fig. 5 shows the relative velocity vectors at the stations of the local monitoring network. According to geological and geophysical investigations (*Anttila et al.*, 1992), the investigation area is one of the most stable areas in Finland. It is therefore not surprising that the relative velocities obtained inside the about 10-km^2 investigation area are less than 1.0 mm/yr.

We can conclude conservatively that the relative movements among the GPS stations in the investigation area are less than 1.0 mm/yr. As the movements are so small, a longer period is required to detect such small movements.

4. Conclusions

An integrated GPS monitoring system has been in operation since October, 1994 for site investigation of nuclear waste disposal in Finland. Based on the analysis of 2.8 years of continuous GPS observations, regional deformations for the investigation area were obtained. It has been shown that the permanent station in the Olkiluoto investigation area is moving in a southeasterly direction by 2 mm/yr relative to the IGS station at Metsähovi.

This indicates that maximum compression occurs in the southeast direction. It agrees very well with the global tectonic model. For the vertical component, the result obtained from GPS measurements is 7.8 mm/yr, while that from precise levelling is 4.2 mm/yr. The GPS result is scattered, therefore a longer time span is required to obtained a more reliable result.

Results obtained from six measurements of the local monitoring network show that the local movements for the stations inside the investigation area are generally less than 0.5 mm/yr. The GPS result indicates that the investigation area is rather stable. This agrees with the result from geological and geophysical investigations.

5. References

Anttila P., S. Paulamäki, A. Lindberg, M. Paananen, T. Koistinen, K. Front and P. Pitkänen (1992). Geology of the Olkiluoto Areas, Summary Report. Report YJT-92-28. Voimayhtiöiden Ydinjätetoimikunta (Nuclear Waste Commission of Finnish Power Companies), Helsinki.

Chen R. and J. Kakkuri (1993). Capability of GPS Technique for Local Crustal Deformation Detection. Proceedings of the Eighth International Symposium on Recent Crustal Movements. Pp. 209-213. Kobe, Japan, Dec. 6 -11, 1993.

Chen R. and J. Kakkuri (1994). Feasibility Study and Technical Proposal for Long-term Observations of Bedrock Stability with GPS. Report YJT-94-02. Nuclear Waste Commission of Finnish Power Companies. Helsinki.

Chen R. and Kakkuri (1996). GPS Operations at Olkiluoto, Kivetty and Romuvaara for 1995. Project report, PATU -96-07e, POSIVA OY, 1996, Helsinki.

Chen R. and Kakkuri (1997). GPS Operations at Olkiluoto, Kivetty and Romuvaara for 1996. Project report, PATU -96-65e, POSIVA OY, 1996, Helsinki.

DeMets C., R.G. Gordon, D.F. Argus, S. Stein, 1990. Current plate motions. *Geophys. J. Int., 101,* 425-478.

DeMets C., R.G. Gordon, D.F. Argus, S. Stein, 1994. Effect of recent revisions to the geomagnetic reversal time scale on estimates of current plate motions. *Geophys. Res. Lett., 21,* 2191-2194.

Kakkuri J, 1997. Postglacial deformation of the Fennoscandian crust. *Geophysica* 33, pp. 99-109.

Kakkuri J. and R . Chen, 1997. Postglacial deformation of the Fennoscandian crust. Latest results from the geodetic measurements. Proceedings of the IAG Regional Symposium. Deformations and Crustal Movement Investigations using Geodetic Techniques. Pp. 8-15. 31 Aug. - 05 Sept. 1996, Szekesfehervar, Hungury. 1997.

Kääriäinen E. 1966. The second levelling of Finland in 1935-1955. Publication of the Finnish Geodetic Institute. Nr. 61. Helsinki.

DETERMINATION OF STOCHASTIC MODELS OF GPS BASELINES IN GPS NETWORK ADJUSTMENT

Xiaoli Ding
Department of Land Surveying and Geo-Informatics
Hong Kong Polytechnic University, Kowloon, Hong Kong

Michael Stewart
School of Surveying and Land Information, Curtin University of Technology
GPO Box U1987, Perth, WA 6001, Australia

Jason Chao
Department of Land Surveying and Geo-Informatics
Hong Kong Polytechnic University, Kowloon, Hong Kong

Abstract

It is well known that the variance covariance matrices of GPS baselines (vectors) as obtained from the processing of GPS baseline observations are in general too optimistic. Therefore, it is a common practice, in the adjustment of GPS networks, to scale up the variance covariance matrices by for example multiplying the matrices with a constant factor. Though the method can always manage to get an adjustment to pass the necessary statistical tests, the method is not rigorous and often arbitrary.

This paper uses the newly completed Western Australian state wide GPS network STATEFIX as an example to look at the effects of the different ways of determining the scale factors for the variance covariance matrices of GPS baselines. The method of variance component estimation is used to assist the analysis. It is found that the adjustment results can vary significantly when different methods are used to determine the scale factors. For most GPS networks, it is recommended that a scale factor be estimated for each independent GPS observation session using the technique of variance component estimation.

1. Introduction

The variance covariance matrices of GPS baselines (vectors) can be determined either using certain functional models formulated according to the error properties of GPS surveys (e.g., Ananga et al., 1994) or directly from GPS baseline processing.

In least squares adjustment of GPS networks, when using the variance covariance matrices of baselines obtained from baseline processing, it is a common practice to scale up the matrices using certain scale factors as they are usually too optimistic. This paper looks at, using the data from the newly completed Western Australian state wide GPS network STATEFIX as an example, the effects of different ways of determining the scale factors when using the method of variance component estimation.

2. The STATEFIX Network and the Initial Adjustment

The STATEFIX network is a GPS network that densifies the ANN (Australian Fiducial Network) in the State of Western Australia. The STATEFIX network consists of over 200 observed GPS baselines and 80 stations, with 17 of these stations being the ANN stations (Figure 1a). The baseline lengths range from as short as 45km to distances of longer than 450km. The mean baseline length is about 200km.

2.1 Baseline Processing

The STATEFIX network was divided into 9 interconnected individual cells (sub-networks) (Figure 1b). Each cell contains a number of GPS baselines observed in different GPS observation sessions.

Cells 1, 2 and 3 were processed by dividing the observation time spans into two independent sessions and computing two solutions for each baseline. Consequently, two baseline vectors were generated for each baseline in the three cells. This was done mainly for the purpose of quality control. The results thus obtained for the baselines are very close to those obtained by processing all the data together. Besides, some baselines in the network were observed twice as they constituted the cell boundaries.

All data in a session were processed together. Therefore, each session produced one or more baselines depending on the number of receivers used for that session. 232 baseline vectors in total and their variance covariance matrices were produced from the processing of 141 independent sessions (see Table 1). The data processing strategies and further details about the STATEFIX network can be found in Stewart et al. (1997a; 1997b).

2.2 Cell Adjustment

All the baseline vectors within a cell were combined in a least squares adjustment. A minimum constraint datum was used in the adjustment by fixing one of the points in the cell. The adjustment results for each cell were checked for any problems in baseline processing. The variance covariance (VCV) matrices of all the baseline vectors in a cell was scaled up, in the cell adjustment, using a scale factor to make the estimated variance factor to be close to 1. The scale factors used in the adjustment are listed in Table 1. The GEOLAB (version 2.4) software was used for all the adjustment computations.

Table 1: Scale factor for the variance covariance matrices
Determined from cell adjustment

Cell	Sessions	Baselines	Scale Factor for VCV Matrix	Cell	Sessions	Baselines	Scale Factor for VCV Matrix
1	24	68	45.6	6	8	18	168.7
2	18	50	68.3	7	13	26	150.1
3	34	66	46.6	8	10	20	83.9
4	10	26	110.7	9	10	28	192.5
5	14	30	83.7				

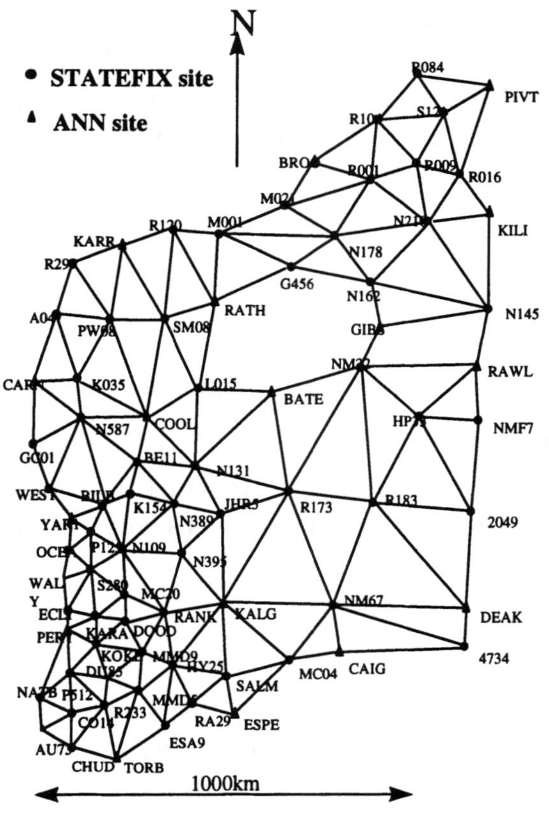

Figure 1a The STATEFIX Network

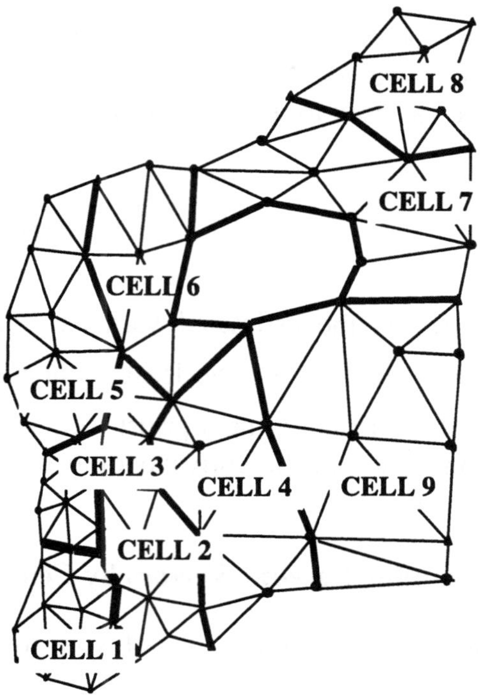

Figure 1b Diagram showing division of STATEFIX Network into 9 cells

2.3 Full Network Adjustment

The baseline vectors and their variance covariance matrices from all the 9 cells were combined in the final full network adjustment. The coordinates of all the 17 ANN points, except the height of point DEAK due to the problems with its height value, were used as controls in the adjustment. The published uncertainty values of 3 cm in horizontal directions and 5 cm in height (at 95% confidence level) were used for all the ANN coordinates.

The estimated variance factor from the combined adjustment passed the standard Chi-square test (at 95% confidence level), though all the estimated scale factors listed in Table 1 were scaled further up by 10% to make the variance factor almost exactly equal 1.

3. Variance Component Estimation

Some further tests were carried out to assess the appropriateness of the stochastic models used in the above full adjustment and to examine the effects of using different scale factors for the variance covariance matrices. The methods for variance component estimation (e.g., Grafarend, et al., 1980; Welsch, 1981; Sahin et al., 1992) can be used for determining the scale factors for the variance covariance matrices. Since the adjustment computation was carried out using the GEOLAB software and the variance component estimation was done based on the output from the software, the following approximate method was used for estimating the variance components,

$$s_{0g}^2 = \frac{v_g^T P_g v_g}{tr(Q_{v_g} P_g)} = \frac{v_g^T P_g v_g}{f_g} \tag{1}$$

where s_{0g}^2 is the estimated variance component for the gth group of baselines; and v_g, P_g, Q_{v_g}, and f_g are the least squares residuals of the baseline vectors from the last epoch of adjustment, the a priori weight matrix, the cofactor matrix of the residuals and the redundancy in the corresponding group. A few iterations were usually required for computation.

3.1 Grouping Baseline Vectors by Cells

All the baseline vectors in a cell were first considered to belong to the same observation group. The ANN coordinates were also considered as a group of observations in the adjustment and variance component estimation. There were therefore totally ten observations groups formed, one for each of the 9 cells in the network and one for the ANN coordinates. A variance component was estimated for each of the ten groups. The estimated values are given in Table 2.

It can be seen from the results that the estimated variance components (scale factors) for the the GPS baselines are fairly close to those obtained from the individual cell adjustments. In fact the values would be even closer if the scale factors determined from the cell adjustments have not been scaled up by 10%. This indicates that the scale factors used in the full network adjustment discussed in Section 2.3 should not have been increased. Instead, the uncertainty values of the ANN coordinates should be scaled up.

When the new scale factors were used in the adjustment, the estimated variance factor from the adjustment still remained to be equal to 1. The estimated coordinates and their variance covariance matrix differ to a certain extent from the results obtained in the initial adjustment. The changes in the coordinates are summarised in Table 3. The sizes of all the absolute error ellipses for the horizontal point positions and of the error bars for the height positions increased noticeably due to the scale factor of 1.6 for the variance covariance matrix of the ANN coordinates.

Table 2: The estimated variance components when observations are grouped by cells

Observation Group	Original Scale Factor for VCV Matrix	Estimated Variance Component	New Scale Factor for VCV Matrix
Cell 1	50.6	0.86	43.5
Cell 2	75.9	0.84	63.8
Cell 3	51.8	1.00	51.8
Cell 4	123.0	0.77	94.7
Cell 5	93.2	1.60	149.1
Cell 6	187.4	1.10	206.1
Cell 7	166.8	1.00	166.8
Cell 8	93.2	0.86	80.2
Cell 9	213.9	1.00	213.9
ANN Coordinates	1.0	1.60	1.6

3.2 Grouping Baseline Vectors by Observation Sessions

The baseline vectors were next grouped by the observation sessions for the purpose of variance component estimation. A variance component was estimated for each observation session on the basis of the results from Section 2.3. The 141 variance components for the GPS baselines vary from 0.01 to 9.34 with a mean value of 1.039 and a standard deviation of ±1.328.

The estimated coordinates from the adjustment using the new variance factors also changed fairly significantly from those obtained in the initial adjustment as seen in Table 3.

Table 3: Changes in coordinates when new variance components were used (unit = m)

	Grouping baselines by cells				Grouping baselines by observation sessions			
	Spatial	N/S	E/W	U/D	Spatial	N/S	E/W	U/D
Min	0.00034	0.00003	0.00001	0.00000	0.00080	0.00000	0.00000	0.00025
Max	0.00500	0.00105	0.00067	0.00488	0.02049	0.00312	0.00437	0.02032
RMS	0.00013	0.00005	0.00003	0.00011	0.00092	0.00012	0.00016	0.00090

Grouping GPS baselines by observation sessions allows in general a better determination of the stochastic models of the GPS baselines than grouping the baselines into larger sub-networks such as the network cells in the STATEFIX. However small observation groups may also mean lower reliability in the estimated variance components as the estimation would be easily affected by the errors in the observations. This may be responsible for some of the very small and very large scale factors obtained. Besides, as a general condition for reliable variance component estimation, the network should have sufficient and well distributed redundancy.

The histogram diagram of the standardised residuals of all the GPS baselines and the ANN coordinates, generated from the adjustment using the new variance components, looks to follow the normal distribution more closely than the residuals obtained from the initial adjustment discussed in Section 2.3.

4. Conclusions

The variance covariance matrices of GPS baselines as obtained from GPS baseline processing are usually too optimistic. Therefore the variance covariance matrices need to be scaled up using certain scale factors. This paper has looked at, using the STATEFIX GPS network as an example, the method of determining the scale factors using the technique of variance component estimation. The effects of different ways of grouping the observed GPS baseline vectors in the estimation of the variance components have also been demonstrated using the example.

It has been seen that the method of dividing a large network into smaller sub-networks can lead to reasonably good estimation of the scale factors. The method of grouping observations according to the observation sessions in the estimation of the variance components is preferred though care should be taken as the estimation is more easily affected by observation errors.

Acknowledgements
The work is partly supported by a research grant from the Hong Kong Polytechnic University (Grant No. 351/614).

References

Ananga, N., Coleman, R. and Rizos, C. (1994) Variance-covariance estimation of GPS networks. Bull. Geod. 68:77-87.

Grafarend, E., Kleusberg, A. and Schaffrin, B. (1980) An introduction to the variance covariance components estimation of Helmert type, zfv 105:129-137.

Sahin, M., Cross, P.A. and Sellers, P.C. (1992) Variance component estimation applied to satellite laser ranging. Bull. Geod. 66:284-295.

Stewart, M.P., Ding, X., Tsakiri, M. and Featherstone, W.E.. (1997a) 1996 STATEFIX project, final report. Report Submitted to West Australian Department of Land Administration. School of Surveying and Land Information. Curtin University of Technology, 47pp.

Stewart, M.P., Houghton, H. and Ding, X. (1997b) The STATEFIX Western Australian GPS network. Proc. of IAG97.

Welsch, W. (1981) Estimation of variances and covariances of geodetic observations. Aust. J. Geod. Photo. Surv., No. 34. Pp.1-14.

THE ASSISI LANDSLIDE GPS NETWORK

D. Dominici (*), F.Radicioni (**), S. Selli (**), A. Stoppini (**)

***DISTART - University of Bologna**
Viale Risorgimento 2, 40136 Bologna - Italy
**** Istituto di Ingegneria Ambientale - University of Perugia**
Via Duranti 1, 06125 Perugia - Italy

Abstract. *The surface movements of a landslide involving an urban area of the Assisi town, in central Italy, have been measured since 1995 using the GPS technique. This paper presents some design aspects of the GPS network, and a statistical analysis of three annual survey campaigns.*

Introduction

The control networks are set up to monitor natural and anthropic phenomena such as landslides, subsidence and geotectonic movements. The GPS monitoring networks have certain advantages compared with the traditional geodetic networks: visibility between points not required, rapid and weather-independent surveying, etc.

A GPS network for the monitoring of the Assisi landslide, in central Italy, was established in 1995. Since then, three measurement campaigns have been carried out of this network. The monitoring zone is an urban area of about 2.5 km^2. The movements of the landslide are relatively slow (a few centimetres per year); therefore great accuracy is necessary for the three-dimensional positioning of the control points. All measurements have been made by GPS double-frequency receivers, in static mode, using a forced-centring system for all points (*Dominici et al.*, 1996).

This paper presents an analysis of the results obtained so far. The quality of each solution is discussed and a comparison is made between different computation modes (single baseline vs. multibaseline, etc.). Some design aspects are also considered and a comparison is made between simulated and real results.

Finally, a two-epoch statistical analysis is performed, in order to: a) verify the stability of the "fiducial" markers; b) highlight the likely ground surface movements and evaluate their significance.

International Association of Geodesy Symposia, Vol. 118
Brunner (ed.), Advances in Positioning and Reference Frames

GPS control network at Assisi

A recently builded area of the Assisi town, in central Italy, is located on a slope where a landslide has caused, in the last 20-25 years, consistent damage to several buildings. The area has been monitored since the early '80s using geotechnical instruments (inclinometers, piezometers, etc.), and by means of a traditional topographic survey of the surface movements for some control points. About the end of the '80s, the topographic survey was interrupted for economical reasons. Meanwhile, some stabilisation measures had been applied, so that the landslide motion has slowed down.

In 1995, a GPS network was monumented by the University of Perugia with the financial support of CNR (Italian National Research Council), aiming to: a) newly monitoring the possible ground surface movements in the landslide area; b) testing the effective accuracy in the three-dimensional positioning achievable with a local GPS network.

The network (fig.1) consists of six reference or "fiducial" points located in geologically stable sites, and fourteen control points situated in the landslide area. The monumentation of the vertices was done using stainless steel centring devices on concrete foundations. The network was set up in spring 1995, and three measurement campaigns were carried out in June 1995, July 1996 and May 1997. Most of the observations were made using Trimble 4000 SSE receivers, with geodetic L1/L2 antennas. Only in the 1997 campaign, an Ashtech Z-Surveyor receiver was used in some points having radio noise troubles.

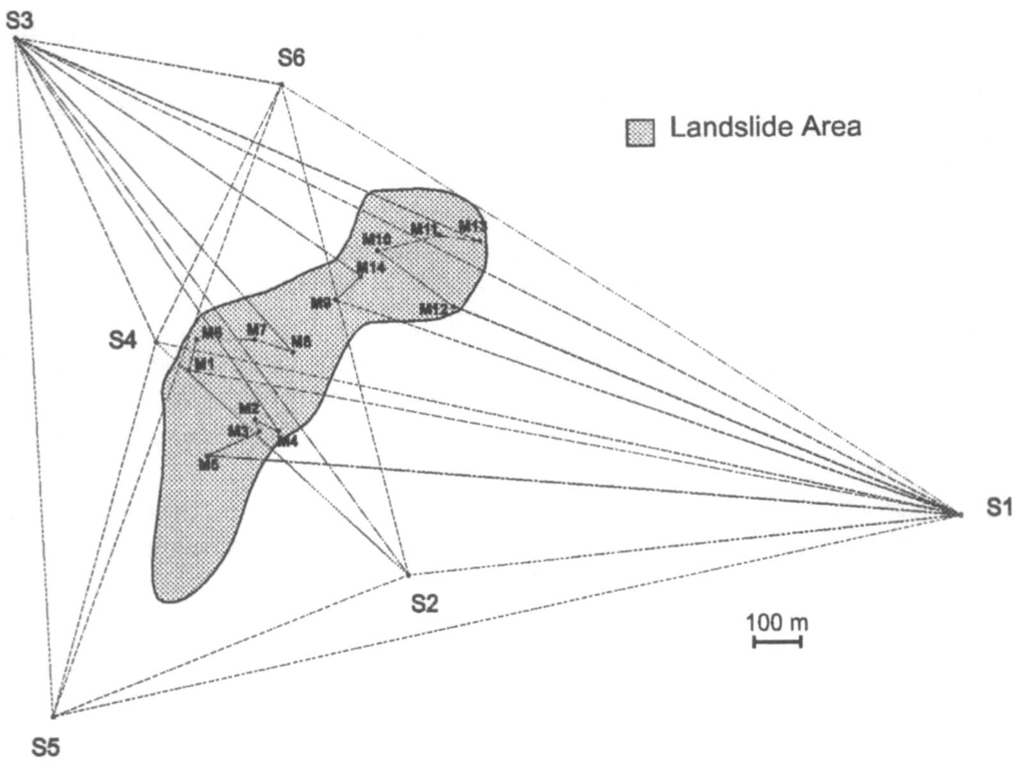

Fig. 1: The Assisi GPS network.

The survey was organised in sessions, in order to obtain all the independent baselines for the fiducial points network, and adequate connections between those and the control points. The solution for each annual campaign was calculated using the *Geotracer Ver. 2.25* processing software, adopting the "single-baseline" procedure and L1+L2 frequency (achieving fixed ambiguities solutions), and then performing a minimum constraint adjustment for the whole network (with the S2 reference station kept fixed).

To control the quality of the results, further solutions were obtained for each campaign with the *Bernese Ver. 4.0* software, both in single and multi-baseline mode. Table 1 shows a comparison between the adjusted coordinates of the 1996 survey obtained with two different methods and softwares.

Points	Vx (mm)	Vy (mm)	Vz (mm)	Vn (mm)	Ve (mm)	Vh (mm)
S1	0.0	-2.2	-1.7	-0.9	-2.2	-1.5
S3	-1.5	-2.1	-2.7	-0.7	-1.8	-3.3
S4	0.5	2.9	-0.3	-1.0	2.7	0.7
S5	-1.1	1.4	2.2	2.1	1.6	0.9
S6	2.0	0.1	2.5	0.5	-0.3	3.2

Table 1: Fiducial network; campaign 1996.
Residuals of a 6-parameter transformations between:
a) Geotracer single baseline solution; b) Bernese multibaseline solution

Another solution for each campaign was computed by means of a robust method in the adjustment phase, adopting a suitable influence function. The differences on the adjusted coordinates with respect to the previous solutions resulted to be less than 2 millimeters.

Design aspects

To investigate *a priori* the quality of the network in the design phase, a simulation was done, using a simplified approach (*Dominici et al.*, 1994) which takes into account only the network configuration on the ground, referring to "standard" observation conditions.

Considering a baseline expressed in the local system by NEH components, a simple *a priori* stochastic model is adopted, assuming that:

- the correlations between the components N, E, H are negligible;
- the variances of the planimetric components N and E are equal and given by:

$$\sigma_p^2 = \sigma_n^2 = \sigma_e^2 = a^2 + (bL)^2 \qquad (1)$$

- the variance of the height component H is higher than those of the planimetric ones, and given by:

$$\sigma_h^2 = c^2 + (dL)^2 \qquad (2)$$

In the above expressions, L is the baseline length, while a , b , c , d are coefficients whose values take into account the observation conditions, and are assigned on the basis of previous experience. The error ellipsoids obtained with this simulation model have the major axis in the H direction and the N-E section degenerates into a circle.

In Table 2, the results of the simulation (assuming a= 2 mm; b= 1 ppm; c= 3 mm; d= 1.5 ppm) are summarised and compared with those of the real solution. The comparison shows the simulation values being slightly "optimistic", especially for the height component.

	Planimetric components		Height component	
	(s)	(r)	(s)	(r)
Fiducial points	3.3 ÷ 3.5	2.0 ÷ 3.7	4.0 ÷ 4.2	3.9 ÷ 4.9
Control points	5.5 ÷ 6.0	3.7 ÷ 7.1	6.6 ÷ 7.2	8.2 ÷ 13.3

Table 2: Results of the simulation compared with those of the real solution.
2D (planimetric) and 1D (height comp.) confidence regions at 95% level; values in mm
(s) = Simulation; (r) = using real observations (campaign 1995)

Reference frame

For a local control network, like the one presented here, it does not seem necessary to operate with a very accurate definition of the geocentric reference frame, since the results are substantially relative positions in a small area. The problem of the reference system stability, indeed, is more involved with the GPS networks of great extension.

Nevertheless, due to the high accuracy requirements, it is always recommended to perform the computations in a "correct" reference frame, which practically involves:

- the connection of the network with one or more stations with known ITRF co-ordinates (such as the IGS permanent stations);
- the use of precise ephemerides (such as the IGS ones).

Thus, the Assisi network data were processed using precise IGS ephemerides and taking "good" ITRF coordinates from the Perugia University roof marker. This marker is scheduled to host a GPS permanent station, and was connected to IGS permanent stations previously.

For the analysis of the possible movements of the control points, a "local" reference system - at a few millimetre level accuracy - is then defined by the coordinates of the fiducial points, which must be tested if they can be assumed as fixed by means of the congruency test between the results of each campaign (see next paragraph).

Stability of the fiducial network

The solutions obtained for the fiducial points from each measurement campaign (1995-1996-1997) have been first compared using a 6-parameter Helmert transformation (1996 on 1995 and 1997 on 1995). As expected, the residuals were a few millimeters in magnitude only.

A more detailed analysis was then performed by means of a congruency test between two epochs (*Caspary*, 1987). The aim of the congruency test is to check whether the reference points have remained stable.

The null hypothesis H_0 is represented by the equation:

$$H_0: \ E[\mathbf{x}_i]=E[\mathbf{x}_j]=\mathbf{x} \ \Rightarrow \mathbf{H}\cdot\tilde{\mathbf{x}}=\mathbf{0}; \ \mathbf{H}=[\mathbf{I} \quad -\mathbf{I}] \ , \ \tilde{\mathbf{x}}=\begin{bmatrix}\mathbf{x}_i\\ \mathbf{x}_j\end{bmatrix} \qquad (3)$$

From a practical point of view, a suitable test statistic is expressed by:

$$T=\frac{q_\Delta / f_\Delta}{q/f} \qquad (4)$$

$$q_\Delta=(\hat{\mathbf{x}}_i-\hat{\mathbf{x}}_j)^T\cdot(\mathbf{Q}_{\hat{\mathbf{x}}_i\hat{\mathbf{x}}_i}+\mathbf{Q}_{\hat{\mathbf{x}}_j\hat{\mathbf{x}}_j})^{-1}\cdot(\hat{\mathbf{x}}_i-\hat{\mathbf{x}}_j) \ , \ q=\mathbf{v}_i{}^T\cdot\mathbf{Q}_{\hat{\mathbf{x}}_i\hat{\mathbf{x}}_i}^{-1}\cdot\mathbf{v}_i+\mathbf{v}_j{}^T\cdot\mathbf{Q}_{\hat{\mathbf{x}}_j\hat{\mathbf{x}}_j}^{-1}\cdot\mathbf{v}_j \qquad (5)$$

Where f_Δ represents the independent rows of the H_0 system, f represents the degree of freedom of the two-epoch system and T is distributed as a central Fisher distribution if H_0 is true.

The results of the test between the 1995 and 1997 epochs are summarised in Table 3. For the congruency test between the 1995 and 1996 solutions, the reader is referred to a previous paper (*Dominici et al.*, 1996).

Point	Adj. Coord.	WGS 84 coordinates 1997 sol.	Displacements 1997-1995 (mm)	Test value $t_{30}(95\%)=1.697$ $t_{30}(99\%)=2.457$	Test results
S1	ϕ	43°03'40.95537"	-1.7	0.715	not
	λ	12°38'45.89612"	-1.3	0.769	not
	h	803.415 m	-8.7	1.884	?
S3	ϕ	43°04'24.04899"	1.8	0.322	not
	λ	12°36'51.97882"	4.1	0.713	not
	h	548.001 m	-0.4	0.067	not
S4	ϕ	43°03'56.75781"	2.8	0.51	not
	λ	12°37'08.08233"	4.4	0.807	not
	h	431.219 m	0.5	0.095	not
S5	ϕ	43°03'24.18942"	-6.8	0.807	not
	λ	12°36'55.36042"	8.6	1.010	not
	h	281.804 m	8.8	1.045	not
S6	ϕ	43°04'19.48823"	8.5	1.384	not
	λ	12°37'23.43907"	4.3	0.698	not
	h	486.317 m	-11.1	1.800	?

Table 3: Summary of the results of the congruency test between 1995 and 1997 solutions (N.B.: Point S2 was fixed in the minimal constraint adjustment)

Testing movements of the control points

Finally, for those points that are expected to move (the control points), it is necessary to compute vector displacements and related confidence regions, and to verify the significance of the movements.

This analysis was performed comparing the 1996 and 1997 solutions with the initial 1995 positions. The movements found in the 2-year period (1÷2 cm values) result to be "at the border" of the significance. However, a general trend could be seen: planimetric movements towards south-west – directed downhill - and lower height components for most of points, indicating a downhill motion .

Obviously, the analyses will be performed also for the next campaigns scheduled from 1998 to 2000. Some movements, indeed, are reasonably expected during the five-year period 1995-2000, although some stabilisation work has already been carried out and will be continued during the next few years.

Conclusions

The results shown have substantially confirmed the possibility of using the GPS technique for monitoring surface movements at the cm accuracy level, and the correctness of the statistical testing procedures adopted.

Future research should explore the application of robust estimation methods to the deformation analysis.

References

Ayan T., Hekimoglu S., Özlüdemir M.T., "Analysis of landslide deformation measurements by robust estimation methods", 1st Turkish Intern. Symposium on Deformation (1994).

Caspary W.F., "Concepts and deformation analysis", Monograph 11 School of Surveying, The University of New South Wales, Kensington, Australia (1987).

Chrzanowski A., Chen Y.Q., Secord J.M., "Geometrical analysis of deformation surveys" Deformation Measurements Workshop, MIT (1986).

Chrzanowski A., "FIG Commission 6 activities in deformation monitoring and analysis", 1st Turkish International Symposium on Deformation (1994).

Dominici D., Radicioni F., Stoppini A., Unguendoli M., "Testing on the redundancy effect in GPS networks: some examples", 1st Turkish International Symposium on Deformation (1994).

Dominici D., Radicioni F., Selli S., "Statistical analysis of GPS monitoring network", Proc. IAG Regional Symposium, Székesfehérvàr, Hungary (1996).

Gruendig L., "Adjustment and design methodology", Deformation Measurements Workshop, MIT (1986).

Kleusberg A.,Teunissen P.J.G.(Eds.), "GPS for Geodesy", Lecture Notes in Earth Sciences, 60, Springer-Verlag, Berlin, Heidelberg (1996).

Schaffrin B., "Aspects of network design", Optimization and design of geodetic networks, Springer-Verlag (1985).

USE OF A REGIONAL IONOSPHERIC MODEL IN GPS GEODETIC APPLICATIONS

M. Gende[1], C. Brunini[1] and A. Kleusberg[2]

[1] Facultad de Ciencias Astronomicas y Geofisicas, Universidad Nacional de La Plata, Argentina

[2] Department of Geodesy and Geomatic Engineering, University of New Brunswick, Canada

Abstract

A regional ionospheric model was developed using data from dual-frequency GPS receivers aiming to correct single frequency observations. The data analyzed correspond to an intermediate southern latitude band in a period of stable ionospheric conditions. The efficiency of the ionospheric model was tested by comparing L3, L1 plus the ionospheric model and L1 solutions. This was done using different numbers of GPS stations, and different session lengths were used. These comparisons were performed by similarity transformations, paying particular attention to the scale factor. Advantages and limitations of the model are discussed.

Introduction

This study analyzes the ability of a regional ionospheric model to correct the residual ionospheric error. This error produces a contraction in the baseline length by up to 0.06 PPM per TECU of ionospheric electron content (Santerre, 1991).

Data obtained from the survey of a 1st order geodetic network covering most of the Chubut Province, Argentina, was used. This province is located in an intermediate latitude band (42° to 46° south) which is usually a region with stable ionospheric conditions. The period

---

[1]Fellow of the CONICET Consejo Nacional de Investigaciones Científicas y Técnicas. Argentina

analyzed (27-28 April 1995) corresponds to a period of low solar activity. The network was measured with six dual-frequency GPS receivers. The distances between stations ranged from 200 to 400 Km. The observations were collected during a period of 15 hours. These observations were used to adjust a regional ionospheric model that was incorporated within the Bernese GPS, V 3.5, software (Rothacher et al., 1993). Two sets of network coordinates were calculated with this software using L1 observations only, with and without ionospheric corrections.

The Geometry-Free Observable

In order to preserve the ionospheric delay information, the geometry-free observable P4 (difference between P code pseudorange in L1 and L2) was computed. Since P4 eliminates all frequency-independent biases (selective availability, clock errors, ephemerides errors and tropospheric delay) this linear combination was used to adjust the ionospheric models (Beutler, 1989):

$$P4 = P1 - P2 = \kappa\, TEC + c\ (\Delta\tau_R + \Delta\tau^S) + \nu \qquad (1)$$

where TEC (total electron content) is the total amount of electrons contained in a cross section of 1 m^2 of a cylinder, having as its axis the signal path between the satellite and the receiver. When the TEC is expressed in TECU (TEC Unit; 1 TECU= 10^{16} electrons/ m^2), the constant k takes the value -0.105 m/ TECU; c is the speed of light, $\Delta\tau_R$ and $\Delta\tau^S$ represent the bias due to differential hardware delay in the receiver ($\Delta\tau_R$) and in the satellites ($\Delta\tau^S$) respectively; ν is the combination of the noise of measurement and of the multipath in P1 and P2.

The Ionospheric Model

In this model it has been assumed that the temporary variations of the vertical electron content, VEC (Vertical Electron Content) are slow when they are described in a coordinate system in which the sun remains approximately fixed (Brunini and Kleusberg, 1995), being a geocentric system whose Z-axis points to the north pole and the system rotates about the Z-axis, maintaining the X-axis on the meridian that contains the sun.

In order to relate the TEC to the VEC, the mapping function M=TEC/VEC has been introduced. To model this function a simplification was made: the ionosphere was represented as a thin spherical layer located at a height of 400 kilometers (the height which approximately corresponds to the F1 layer).

Figure 1 represents the basic geometry of the model: the incoming signal from a satellite S goes through the thin layer at the point P and arrives at the station E at a zenith distance z. The projection of the point P on to the terrestrial surface is the subionospheric point Q. The spherical coordinates of the subionospheric point in the sun-fixed system are the latitude ϕ and the hour angle h.

We adopted a geometric mapping function equal to the ratio between the oblique and vertical distance, to the H height of the thin ionospheric layer:

$$M(z') = \frac{TEC}{VEC} \cong \operatorname{cosec}(z') \tag{2}$$

where z' is the zenith distance of the satellite to the height of the ionospheric layer. This approach ignores the contribution of the horizontal gradients in the distribution of free electrons.

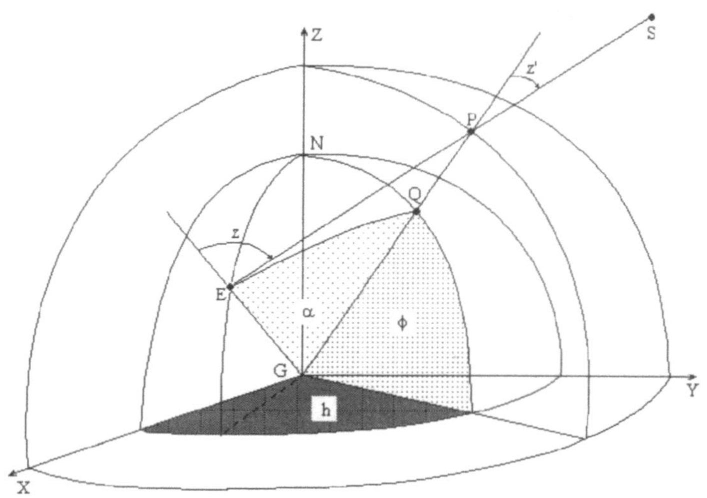

S: satellite, P: layer point, E: station, X: points to the sun

Figure 1: Basic Geometry for a single layer model in a sun-fixed system

Parametrization and Adjustment of VEC

A polynomial equation was used to describe the spatial variations of the VEC in the sun-fixed system:

$$\mathrm{VEC}(h,\phi) = \sum_{l=0}^{L}\sum_{m=0}^{M} a_{lm}(h-h_0)^l(\phi-\phi_0)^m \tag{3}$$

where h and ϕ are the coordinates of the subionospheric point in the sun-fixed system. Using equation (2) and (3), equation (1) becomes:

$$P4 = \kappa\,\mathrm{cosec}(z')\sum_{l=0}^{L}\sum_{m=0}^{M} a_{lm}(h-h_0)^l(\phi-\phi_0)^m + c\,(\Delta\tau_R + \Delta\tau^S) + \nu \tag{4}$$

Equation (4) is the basic observation equation of the problem. The unknown a_{lm} and $\Delta\tau_R + \Delta\tau^S$ are determined by a least squares adjustment of the P4 observations. Figure 2 shows an ionospheric delay map in meters obtained between 22.5 hours (local time) on 27 April 1995 and 8.5 hours on 28 April 1995.

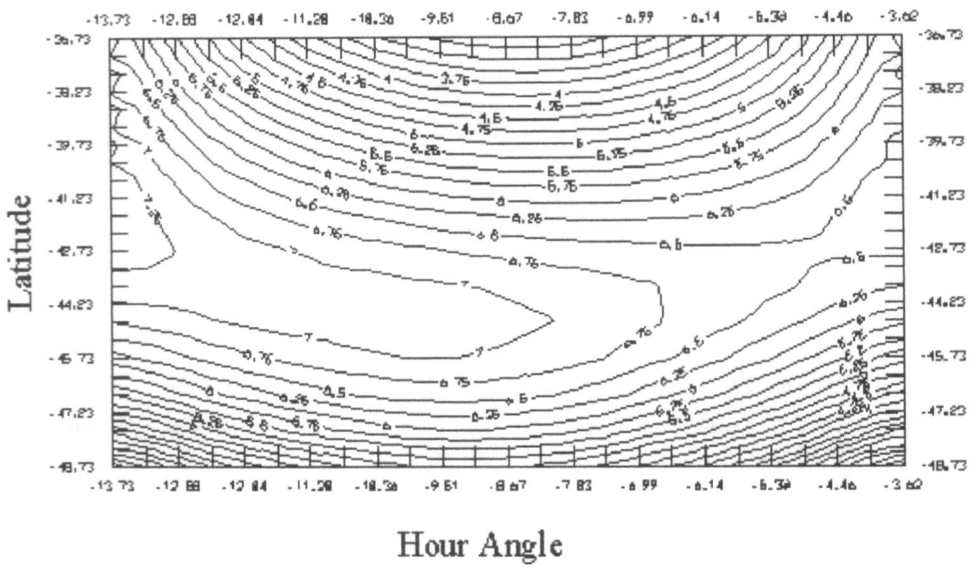

Figure 2. Regional ionospheric map

Computing the Model

We computed the ionospheric model using different numbers of GPS stations finding that it does not change significantly with the number of stations used.

Due to the North-South dimension of the network a development of 2nd degree in latitude was chosen; the longest sessions had a development of 2 in hour angle, and the shorter ones of 1 in hour angle. Other developments were tested by increasing the number of terms in the series, but the results got worse due to the Runge effect.

The session lengths were shortened, from 16 to 3 hours. The limiting factor for reducing the session length was the deterioration of the results as a consequence of the worsening of the parameters (ambiguities and coordinates) estimated from the shorter observation sessions. Three hours was the optimum session length for our baseline solutions.

Evaluation of Model

The following procedure has been used in order to evaluate the results obtained after applying the ionospheric corrections. The observations of all the stations of the network were processed using the ionospheric-free combination (L3). This linear combination will reduce the first order ionospheric effect, leaving only about 2% of it (Kleusberg et al., 1996). Since L3 gives the most precise set of coordinates (considering the dimensions of our network) we will designate these as the "true coordinate set".

L1 frequency observations only were processed, in which case the ionospheric delay was not corrected at all, and from this solution a different set of coordinates was obtained. Finally we processed the L1 frequency observations applying our ionospheric model. Now the effect of the ionosphere was partially absorbed although it was not entirely removed. A third set of coordinates was obtained. We evaluated the efficiency of our ionospheric model by comparing different similarity transformations. The first one was between the "true coordinates" and those obtained using the L1 data only; the second one was between the "true coordinates" and the coordinates calculated with the L1 data corrected by the ionospheric model. In this way two sets of 7 parameters were obtained and it was possible to compare the influences that the ionospheric model produced on the network stations. This procedure was repeated for the four sessions of different length.

Session length hrs	Scale without ionospheric model	RMS without ionospheric model	Scale with ionospheric model	RMS with ionospheric model.
16	0.147 ± 0.073	0.0407	0.016 ± 0.073	0.0407
8	0.257 ± 0.119	0.0663	0.084 ± 0.116	0.0646
4	0.241 ± 0.210	0.1174	0.031 ± 0.214	0.1195
3	0.287 ± 0.181	0.1012	0.016 ± 0.190	0.1064

Table 1: Results for 7 parameter transformation. Residuals in Local System (North, East, Up). Taylor series development: 2 latitude terms, 2 hour angle terms and 2 mixed terms.

Conclusion

The results in Table 1 show the utility of the model in minimizing the scale factor of the 7 parameter transformation.

Once L1 was processed without using the ionospheric model, a scale factor similar to the theoretical value suggested by Santerre (1991) was obtained, while the factor diminished by at least one order of magnitude when the model was applied. This fact demonstrates the success of the model that we have used, consistent with the previously mentioned fact that the ionosphere biases result in a baseline contraction. Appreciable differences in the other parameters do not arise from the present results.

Residuals (RMS) do not change with the application of the model, which indicates that only low frequency information could be estimated. Residuals also grow as the session length decreases.

Bibliography

Beutler G., Gurtner W., Rothacher M., Wild U. and Frei E. (1989), Relative static positioning with the Global Positioning System: basic technical considerations. In: The IAG General Meeting, Edinburgh

Brunini C. and Kleusberg A. (1995), Mapas globales de retardo ionosférico vertical a partir de observaciones GPS. In: Actas de V Congreso Internacional Ciencias de la Tierra, Santiago, Chile.

Georgiadou Y. and Kleusberg A. (1988), On the effect of ionospheric delay on geodetic relative GPS positioning. Manuscripta Geodaetica, Vol. 13. 1-8.

Kleusberg A., Teunissen P. (Eds.) (1996), GPS for Geodesy. Springer. 175 -218.

Rothacher M., Beutler G., Gurtner W., Schildknecht T. and Wild U. (1993), In: Bernese GPS Software Version 3.4. Documentation, Astronomical Institute, University of Berne.

Santerre R. (1991), Impact of GPS satellite sky distribution. In: Manuscripta Geodaetica, Vol. 16, 28-53.

INSTANTANEOUS AMBIGUITY RESOLUTION FOR MEDIUM-RANGE GPS KINEMATIC POSITIONING USING MULTIPLE REFERENCE STATIONS

Shaowei Han & Chris Rizos
School of Geomatic Engineering
The University of New South Wales
Sydney NSW 2052, Australia

ABSTRACT

A linear combination functional model, formed from the single-differenced functional equation for baselines from a mobile receiver to three or more reference receivers, is proposed, in which the orbit bias and ionospheric delay can be eliminated, and, in addition, the tropospheric delay, multipath and observation noise can be significantly reduced. As a result, the ambiguity resolution technique that can be employed for medium-range GPS kinematic positioning is similar to that used for the short-range case. Kinematic tests have been carried out in Sydney, Australia, with separations from the nearest reference stations greater than 30km. The carrier phase ambiguities can be resolved for every epoch and the success rate (correct identification of the integer ambiguities) was 100%. This technique is well suited to real-time precise GPS kinematic positioning.

INTRODUCTION

On-the-fly ambiguity resolution for short-range kinematic positioning assumes that the orbit bias and differential ionospheric delay can be ignored, and the integer ambiguities can be resolved easily (Han, 1997a). For medium-range GPS kinematic positioning, the residual biases after double-differencing the data have been investigated over the last few years. It was shown that using more than three reference stations with known coordinates, the orbit bias can be eliminated for medium-range applications (less than 100 km to the nearest reference station) through the use of a linear combinations of single-differenced observations (Han & Rizos, 1996; Wu, 1994). The ionospheric delay relative to one reference station can also be interpolated if the relative ionospheric delay for three or more reference stations are known (Han & Rizos, 1996; Wanninger, 1995; Webster & Kleusberg, 1992). In this paper a linear combination method has been used to account for orbit bias and ionospheric delay. The tropospheric delay, multipath, and observation noise will also be mitigated using this method. After the distance-dependent biases are eliminated or mitigated, an ambiguity resolution method incorporating a three-step quality control procedure is used, which has been succesfully used for short-range GPS kinematic positioning (Han, 1997a).

LINEAR COMBINATION MODEL

Figure 1 shows a three reference station network with one roving receiver. It is preferable that one of the reference stations is connected to the IGS network in order to obtain precise positions in the global ITRF frame. Data processing for the reference stations is necessary in order to determine the double-differenced integer ambiguities between them.

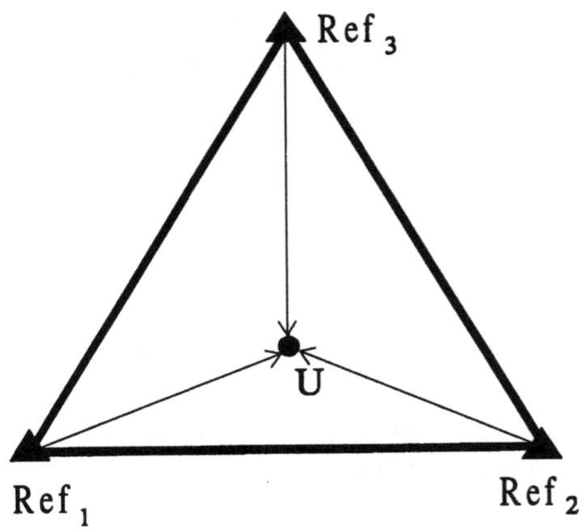

Figure 1. Configuration of the reference stations and the roving receiver

The single-differenced carrier phase observation can be written (Han, 1997b):

$$\Delta\phi_i = \Delta\rho_i + \Delta d\rho_i - c\cdot\Delta dT_i + \lambda\cdot\Delta N_i - \Delta d_{ion,i} + \Delta d_{trop,i} + \Delta d^{\phi}_{mp,i} + \varepsilon_{\Delta\phi_i} \tag{1}$$

where $\Delta(\cdot)_i = (\cdot)_u - (\cdot)_i$; i indicates the reference station i, and u the user station; ϕ: the carrier phase observation in unit of metres; ρ: $=\left\|X^s - X\right\|$, X^s is the satellite position vector, X is the station position vector; $d\rho$: the effect of ephemeris errors, including S/A effects; dT: the receiver clock error with respect to GPS time; d_{ion}: the ionospheric delay; d_{trop}: the tropospheric delay after model correction; d^{ϕ}_{mp}: the multipath on the carrier phase; ε_{ϕ}: the carrier phase observation noise; λ: the wavelength of the carrier phase; and, N: the integer ambiguity.

A set of parameters α_i can be determined, based on the conditions given in Han & Rizos (1996):

$$\sum_{i=1}^{3}\alpha_i = 1 \qquad \text{and} \qquad \sum_{i=1}^{3}\alpha_i\left(\vec{X}_u - \vec{X}_i\right) = 0 \tag{2}$$

where $\vec{X}_u$ and $\vec{X}_i$ (i=1,2,3) are the position in the Gauss plane coordinate system. If more than 3 reference stations are used, $\sum_{i=1}^{3} \alpha_i^2 = \min$ should be introduced in roder to reduce the observation noise (see equation (8)).

The linear combination of the single-differenced observations can be formed as:

$$\sum_{i=1}^{3} \alpha_i \cdot \Delta\phi_i = \sum_{i=1}^{3} \alpha_i \cdot \Delta\rho_i + \sum_{i=1}^{3} \alpha_i \cdot \Delta d\rho_i - c \cdot \sum_{i=1}^{3} \alpha_i \cdot \Delta dT_i + \lambda \cdot \sum_{i=1}^{3} \alpha_i \cdot \Delta N_i - \\ -\sum_{i=1}^{3} \alpha_i \cdot \Delta d_{ion,i} + \sum_{i=1}^{3} \alpha_i \cdot \Delta d_{trop,i} + \sum_{i=1}^{3} \alpha_i \cdot \Delta d_{mp,i}^{\phi} + \varepsilon_{\sum_{i=1}^{3} \alpha_i \cdot \Delta\phi_i} \tag{3}$$

The orbit bias term has been proved to be (Han, 1997b):

$$\sum_{i=1}^{3} \alpha_i \cdot \Delta d\rho_i \approx 0 \tag{4}$$

The epoch-by-epoch and satellite-by-satellite ionospheric model can be applied to estimate the single-differenced ionospheric delay relative to reference receiver 3. It has been proved that the interpolation of the single-differenced TEC between receivers, or double-differenced TEC based on an ionosphere layer can be simplified to an the interpolation dependent on the receiver position in the Gauss plane coordinate system (Han, 1997b). The ionospheric delay term can be deduced as (ibid, 1997b):

$$\sum_{i=1}^{3} \alpha_i \cdot \Delta d_{ion,i} = d_{ion,u} - d_{ion,3} - \begin{bmatrix} \alpha_1 \\ \alpha_2 \end{bmatrix}^T \begin{bmatrix} d_{ion,1} - d_{ion,3} \\ d_{ion,2} - d_{ion,3} \end{bmatrix} = 0 \tag{5}$$

When the distance between the reference stations increases, the residual error will become greater due to the ionospheric delay interpolation.

If it can be assumed that the tropospheric delay can be interpolated from the residual tropospheric delay at the reference stations, the residual tropospheric delay can be represented as:

$$\sum_{i=1}^{3} \alpha_i \cdot \Delta d_{trop,i} = d_{trop,u} - d_{trop,3} - \begin{bmatrix} x_u & y_u \end{bmatrix} \begin{bmatrix} x_1 & y_1 \\ x_2 & y_2 \end{bmatrix}^{-1} \begin{bmatrix} d_{trop,1} - d_{trop,3} \\ d_{trop,2} - d_{trop,3} \end{bmatrix} \tag{6}$$

where (x_u, y_u), (x_1, y_1) and (x_2, y_2) are North and East coordinates relative to reference station 3 in Gauss plan coordinate system. How closely it should be to zero depends on the spatial correction of tropospheric delay. The residual tropospheric delay is mostly contributed to by the wet component of the troposphere, which shows strong variation with height, time and location. But it can only be expected the term $\sum_{i=1}^{3}\alpha_i \cdot \Delta d_{trop,i}$ to be mitigated to some extent dependent on the distance between stations.

The multipath term can be rewritten as:

$$\sum_{i=1}^{3}\alpha_i \cdot \Delta d^{\phi}_{mp,i} = d^{\phi}_{mp,u} - \sum_{i=1}^{3}\alpha_i \cdot d^{\phi}_{mp,i} \tag{7}$$

The last term $\sum_{i=1}^{3}\alpha_i \cdot d^{\phi}_{mp,i}$ on the right hand side of equation (7) is the weighted mean value of the multipath values at the three reference stations for this satellite. Due to the random nature of multipath at the different stations, the weighted mean value will be significantly reduced if all α_i (i=1,2,3) are positive and less than 1, although the weight α_i is not derived from its standard deviation. On the other hand, the multipath at the roving station will become a high frequency bias, and mostly will be close to random noise (Zhang & Schwarz, 1996). Therefore, the multipath term has been significantly reduced and will be ignored in the functional model.

The standard deviation of the one-way carrier phase observation can be approximated as a function of the elevation angle. Because all stations are located within a region of about 100km radius, the elevation of a satellite is approximately the same. The standard deviation of the one-way carrier phase observation can also be approximated by σ^j and then the standard deviation of the linear combination of single-differenced observations $\varepsilon_{\sum_{i=1}^{3}\alpha_i \cdot \Delta\phi_i}$ can be expressed as:

$$\sigma_{\sum_{i=1}^{3}\alpha_i \cdot \Delta\phi_i} = \sqrt{1+\alpha_1^2+\alpha_2^2+\alpha_3^2} \cdot \sigma^j \tag{8}$$

Compared to the standard deviation of the single-differenced carrier phase observation $\sqrt{2} \cdot \sigma^j$, the standard deviation will become smaller if the roving station is located within the triangle formed by the reference stations (Han, 1997b).

The single-differenced carrier phase observation functional model can be simplified to:

$$\sum_{i=1}^{3}\alpha_i \cdot \Delta\phi_i = \sum_{i=1}^{3}\alpha_i \cdot \Delta\rho_i - c \cdot \sum_{i=1}^{3}\alpha_i \cdot \Delta dT_i + \lambda \cdot \sum_{i=1}^{3}\alpha_i \cdot \Delta N_i + \varepsilon_{\sum_{i=1}^{3}\alpha_i \cdot \Delta\phi_i} \tag{9}$$

DOUBLE-DIFFERENCED FUNCTIONAL MODEL

Equation (14) can be written as:

$$\Delta\phi_{u,3} - \left[\alpha_1 \cdot \Delta\phi_{1,3} + \alpha_2 \cdot \Delta\phi_{2,3}\right] = \Delta\rho_{u,3} - \left[\alpha_1 \cdot \Delta\rho_{1,3} + \alpha_2 \cdot \Delta\rho_{2,3}\right] - c \cdot \sum_{i=1}^{3} \alpha_i \cdot \Delta dT_i + + \lambda \cdot \Delta N_{u,3} - \left[\alpha_1 \cdot \Delta N_{1,3} + \alpha_2 \cdot \Delta N_{2,3}\right] + \varepsilon_{\sum_{i=1}^{3} \alpha_i \cdot \Delta\phi_i} \quad (10)$$

Define the residual vectors:

$$V_{1,3} = \nabla\Delta\phi_{1,3} - \nabla\Delta N_{1,3} - \nabla\Delta\rho_{1,3} \quad (11)$$

$$V_{2,3} = \nabla\Delta\phi_{2,3} - \nabla\Delta N_{2,3} - \nabla\Delta\rho_{2,3} \quad (12)$$

The double-differenced observation model can then be written as:

$$\nabla\Delta\phi_{u,3} - \left[\alpha_1 \cdot V_{1,3} + \alpha_2 \cdot V_{2,3}\right] = \nabla\Delta\rho_{u,3} + \lambda \cdot \nabla\Delta N_{u,3} + \varepsilon_{\sum_{i=1}^{3} \alpha_i \cdot \nabla\Delta\phi_i} \quad (13)$$

Using the known coordinates and known integer ambiguities, the correction vectors $V_{1,3}$, $V_{2,3}$. For real-time applications, the correction vectors, together with the carrier phase and pseudo-range data at reference station 3, can be sent to the roving receiver in real-time.

EXPERIMENTS

The experiment was carried out on 14 December, 1996, using four Ashtech Z12 GPS receivers. Three reference receivers were set up with separation of 40.0 km, 65.6 km and 70.7 km, and a roving receiver was mounted on a car from a site 31.44 km, 34.11 km and 46.5 km distant from the reference receivers. The roving receiver started to move a distance of about 7 km and back to the start point, during a period of about 32 minutes. The data rate was 1 Hz and a total of 1903 epochs were collected.

The coordinates have been precisely determined at the three reference stations, and the correction terms $\left[\alpha_1 \cdot V_{1,3} + \alpha_2 \cdot V_{2,3}\right]$ in equation (13) for the L1 and L2 carrier phase observations were derived and, as an example, plotted in Figure 2 for satellite pair PRNs 9 & 24. Using the proposed functional model and an ambiguity resolution method with three-step improvements suggested in Han (1997a), integer ambiguities have been resolved using single-epoch data with 100% success rate.

CONCLUDING REMARKS

A linear combination functional model is proposed, formed from the single-differenced functional equation for baselines, from the user roving receiver to three or more reference stations, in which the orbit biases and ionospheric delay terms can be eliminated and, in addition, the tropospheric delay, multipath and the observation noise can be reduced.

Using this model, the integrated method with three-step quality control procedure can resolve integer ambiguity instantaneously. The medium-range experiment described here indicates a 100% success rate. This experiment has demonstrated the feasibility of the proposed technique for medium-range kinematic positioning. Greater separation between the roving receiver and reference stations will be tested in the near future. This algorithm has been designed for real-time applications.

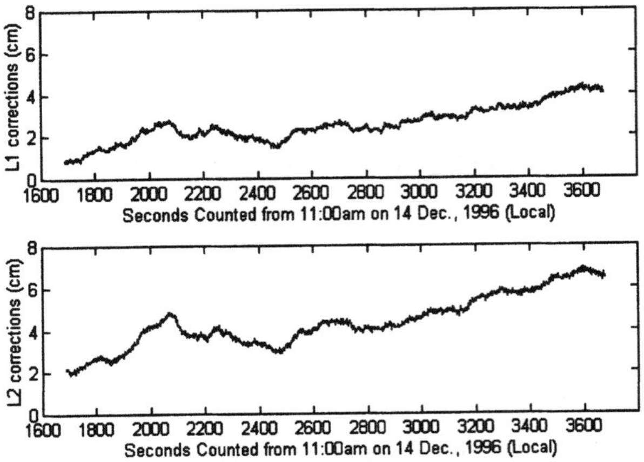

Figure 2. Example of the corrections for double-differenced carrier phase observations for satellite pair PRNs 24 & 9

REFERENCES

Han, S. (1997a). Quality Control Issues Relating to Ambiguity Resolution for Real-Time GPS Kinematic Positioning. Journal of Geodesy, 71(6):351-361.

Han, S. (1997b). Carrier Phase-Based Long-Range GPS Kinematic Positioning. PhD Dissertation, UNISURV S-49, School of Geomatic Engineering, The University of New South Wales, Sydney, Australia, 185pp.

Han, S. and C. Rizos (1996). GPS Network Design and Error Mitigation for Real-Time Continuous Array Monitoring Systems. Proc. 9th Int. Tech. Meeting of The Sat. Div. of The U.S. Inst. of Navigation, Kansas City, Missouri, 17-20 Sept., 1827-1836.

Wanninger, L. (1995). Improved Ambiguity Resolution by Regional Differential Modelling of the Ionosphere. Proc. 8th Int. Tech. Meeting of The Sat. Div. of The U.S. Inst. of Navigation, Palm Springs, California, 12-15 Sept., 55-62.

Webster, I. and A. Kleusberg (1992). Regional Modelling of the Ionosphere for Single Frequency Users of the Global Positioning System. Proc. 6th Int. Geodetic Symp. on Satellite Positioning, Columbus, Ohio, 17-20 March, 230-239.

Wu, J. T. (1994). Weighted Differential GPS Method for Reducing Ephemeris Error. Manuscripta Geodaetica, 20, 1-7.

Zhang, Q. J. and Schwarz, K. P. (1996). Estimating Double Difference GPS Multipath under Kinematic Conditions. Proc. IEEE Position Location & Navigation Symp. PLANS'96, Atlanta, Georgia, 22-26 April, 285-291.

EXPLOITING THE SIRGAS COLOCATIONS FOR DETERMINING ELEVATION DEPENDENT PHASE CENTER VARIATIONS OF GEODETIC GPS ANTENNAS

Klaus Kaniuth
Deutsches Geodätisches Forschungsinstitut, Abt. I (DGFI/I)
Marstallplatz 8, D-80539 München, Germany

ABSTRACT

During the SIRGAS 95 GPS campaign several sites were continuously occupied during ten days by two different receiver systems. From the data acquired at all those colocation sites providing sufficiently accurate ground truth, either from terrestrial geodetic observations or from SIRGAS independent GPS measurements, the elevation dependent phase center differences between the involved antennas have been determined. The modelling approach and the achieved results are presented.

INTRODUCTION

Besides multipath and tropospheric effects the antenna phase center variations are probably the most important error contribution when aiming at few millimeters accuracies in GPS applications. Moreover, parameters accounting for phase center biases and for tropospheric delays are highly correlated, and both errors will affect the height estimates. For determining phase center variations two different approaches have been applied so far:

- Absolute calibrations of antenna phase patterns by anechoic chamber measurements (Schupler et al. 1995);
- Determination of relative phase center variations from field calibration experiments (e.g. Breuer et al. 1995, Rothacher et al. 1995).

Within the International GPS Service for Geodynamics (IGS) frame a compilation combining all available calibration results has recently been realized and suggested for application (Rothacher and Mader 1996).

Field calibration experiments in local networks extending over some tens of meters only generally require that

- the eccentricity vectors in the network are very accurately established or can be determined by properly interchanging the antennas during the experiment,
- any systematic effects such as orbit errors, satellite clock biases and tropospheric errors cancel over such short distances.

The elevation dependent phase center variations presented in this analysis do not result from a dedicated calibration experiment, instead they are derived from colocation data acquired during a GPS field campaign.

DATA SET

In May / June 1995 a GPS campaign for establishing a geocentric reference system for the South American continent, SIstema de Referencia Geocéntrico para America del Sur (SIRGAS), has been carried out. Consisting of some sixty stations occupied during ten days, different precise GPS receiver systems were involved. As at the planning stage of that project no commonly accepted antenna phase center models were available, different receivers were operated in colocation at a number of sites allowing to process receiver specific subnetworks and to combine these at the normal equation level by including independently established local ties. Table 1 displays all colocations providing at least four days of observations, the majority of sites were occupied during the whole campaign.

Table 1 Colocations of different receiver systems during the SIRGAS 95 GPS campaign

Site (Country)	Involved Receiver / Antenna Systems
Bogotá (Colombia)	- TurboRogue / Dorne Margolin T Leica 200 / Internal
Curitiba (Brazil)	- Trimble SSE / 4000ST L1/L2 Geo Ashtech Z-XII3 /Geodetic L1/L2 700 228
El Maiten (Argentina)	- Trimble SSE / TR Geod L1/L2 GP Leica 200 / Internal
Maracaibo (Venezuela)	- Trimble SSE / TR Geod L1/L2 GP Leica 200 / Internal
Montevideo (Uruguay)	- Leica 200 / Internal Ashtech Z-XII3 / Geodetic L1/L2 700 718
Puerto Iguazu (Argentina)	- Ashtech Z-XII3 / Geodetic L1/L2 700 228 Leica 200 / Internal
Rio Grande (Argentina)	- TurboRogue / Dorne Margolin T Leica 200 / Internal

The selection of these colocation sites could not be adequately based on theoretical considerations but was mainly depending on equipment resources, infrastructure and logistic aspects. This is the reason for the overrepresentation of the Leica 200 Internal compared to all the other systems.

The ground truth required for estimating the phase center differences is based on GPS measurements performed independently of the SIRGAS campaign with identical receiver/-antenna systems. At some of the sites also levelled height differences between the GPS markers are available. Unfortunately, compared to the large amount of colocation data the local ties cannot be considered equally accurate.

ANALYSIS

This analysis is restricted to elevation dependent phase center variations because due to the objective of the performed colocations the antennas were not rotated and at some sites the vertical ground truth was more accurate than the horizontal. For estimating phase center variations the following approach has been implemented in the Bernese GPS software version 3.4 (Rothacher et al. 1993) and in DGFI's program ACCSOL which accumulates and solves normal equations and allows to introduce a variety of fiducial informations and constraints:

- Considering that the main part of offsets with respect to the applied a priori phase center locations (table 2) is expected in the vertical and that some software systems are not capable of applying elevation dependent phase center corrections, we firstly estimate a mean vertical phase center offset between the involved antennas; this estimate is the least squares fit to all included data down to the selected cutoff angle and maps to any elevation angle E with sine E.

- Relative to these mean vertical phase center offsets elevation dependent range corrections are estimated in a second step; either one polynomial of arbitrary degree is fitted to the entire elevation range, or piecewise linear functions for elevation bins constrained to continuity at the bin boundaries are estimated. The results to be presented are based on the latter approach.

- All fiducial information such as the local baseline vectors or height differences are introduced in program ACCSOL as individually weighted observation equations in both adjustment steps; in particular, the elevation dependent corrections relative to the mean vertical phase center are solved applying the minimum condition

  $$\|y - Ax\|^2 + \lambda \|x_p\|^2 = \text{minimum},$$

  where y is the observation vector, A the design matrix, x_s the vector of station coordinates, x_p the vector of phase center parameters, $x^T = (x_s, x_p)$, and λ a relative weighting factor which was set to one in this analysis.

Table 2 Applied a priori phase center offsets in L1 and L2 [mm] with respect to antenna reference point (ARP); BPA = Bottom of preamplifier, TOP = Top of pole (350 mm above height hook)

Antenna		ARP	L1	L2
TurboRogue	Dorne Margolin T	BPA	110.0	128.0
Leica	Internal	TOP	91.0	91.0
Trimble	4000ST L1/L2 Geo	BPA	70.0	68.0
	TR Geod L1/L2 GP	BPA	62.5	62.5
Ashtech	Geodetic L1/L2 700 228	BPA	64.0	64.0
	Geodetic L1/L2 700 718	BPA	72.0	72.0

RESULTS

The discussed approach for estimating mean vertical antenna phase center differences as well as elevation dependent phase center variation differences has been applied to the SIRGAS colocation data. From all the performed computations the following results are presented:

- The mean vertical phase center differences between the colocated antennas with respect to the a priori effects (table 3);
- An example of the consistency of the elevation dependent range correction estimates (figure 2);
- The resulting elevation dependent range corrections in L1 and L2 for each antenna combination (figures 3 and 4).

It should be noted that the piecewise linear functions are represented by cubic splines in figures 2, 3 and 4.

Table 3 Resulting mean vertical phase center offset differences [mm] between colocated antennas relative to the applied a priori offsets

Antenna I	- Antenna II	L1	L2
Leica Internal	- TurboRogue Dorne Margolin T	21.7	25.9
Leica Internal	- Ashtech Geodetic L1/L2 700 228	11.6	11.4
Leica Internal	- Ashtech Geodetic L1/L2 700 718	7.1	18.4
Leica Internal	- Trimble TR Geod L1/L2 GP	13.6	16.1
Trimble 4000ST L1/L2 Geo	- Ashtech Geodetic L1/L2 700 228	-19.3	-12.5

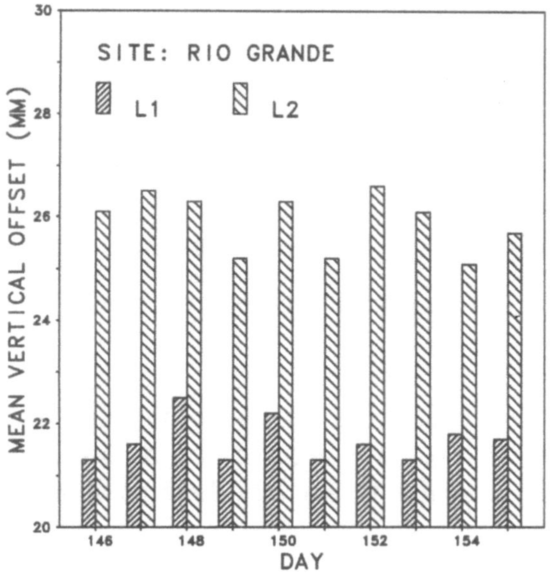

Fig. 1 Consistency of daily estimates of the mean vertical phase center offset difference Leica Internal – TurboRogue Dorne Margolin T for both frequencies L1 and L2

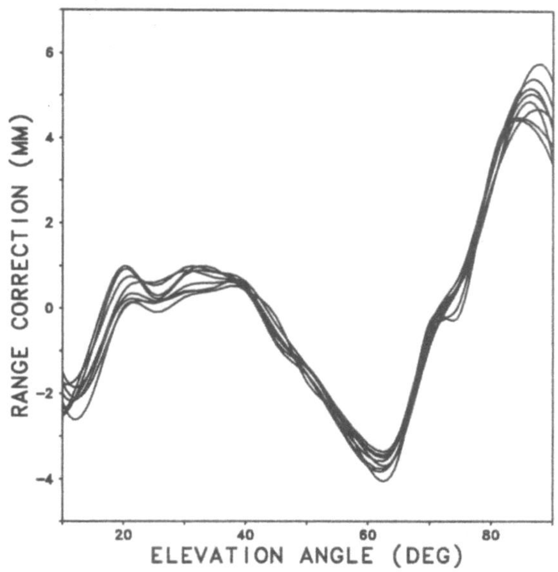

Fig. 2 Consistency of daily estimates of the elevation dependent range correction differences Leica Internal - Trimble SSE TR Geod L1/L2 GP for L1

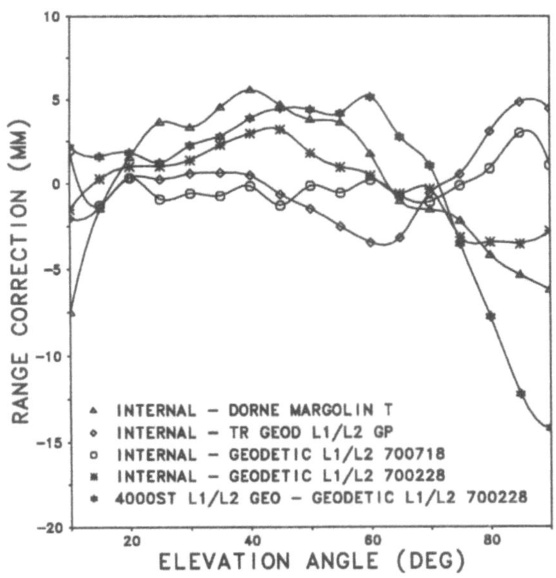

Fig. 3 Resulting elevation dependent range correction differences for all analyzed antenna pairs (L1)

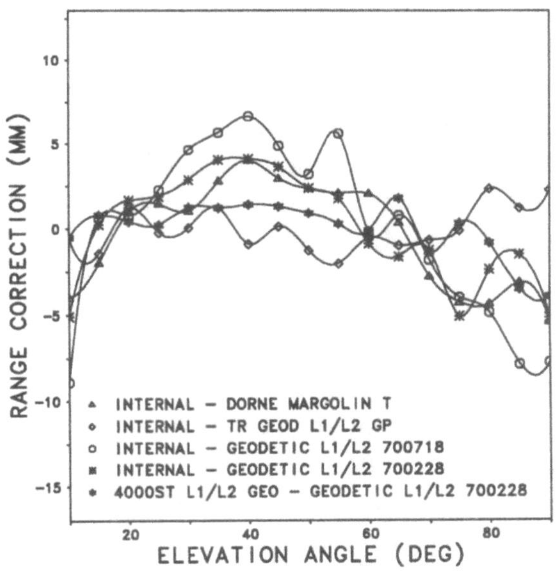

Fig. 4 Resulting elevation dependent range correction differences for all analyzed antenna pairs (L2)

CONCLUSION

Data acquired during the SIRGAS 95 GPS campaign by colocated receivers have been exploited for determining the differences between the elevation dependent phase center variations of the involved antennas. The modelling approach includes

- the fitting of a mean vertical phase center offset to the whole elevation range, and
- a subsequent estimation of elevation dependent range corrections applying a minimum condition.

The consistency of daily estimates of both the mean phase center as well as the elevation dependent correction terms is on the one millimeter level. The accuracy of the resulting variations depends on the reliability of the available ground truth, and may therefore in some cases be slightly worse.

ACKNOWLEDGEMENT

The support given by several institutions providing local tie measurements at the SIRGAS colocation sites is gratefully acknowledged.

REFERENCES

Breuer, B., J. Campbell, B. Görres, J. Hawig and R. Wohlleben (1995): Kalibrierung von GPS-Antennen für hochgenaue geodätische Anwendungen. Journal of Satellite Based Positioning, Navigation and Communication (SPN), 2/1995, 49-59.

Rothacher, M., G. Beutler, W. Gurtner, E. Brockmann, and L. Mervart (1993): The Bernese GPS Software Version 3.4. Astronomical Institute, University of Berne.

Rothacher, M., S. Schaer, L. Mervart and G. Beutler (1995): Determination of Antenna Phase Center Variations Using GPS Data. In Gendt, G. and G. Dick (Eds): Proceedings of the IGS Workshop on Special Topics and New Directions, GeoForschungsZentrum Potsdam, May 15-18, 1995, 205-220.

Rothacher, M. and G. Mader (1996): Combination of Antenna Phase Center Offsets and Variations; Antenna Calibration Set IGS-01. International GPS Service for Geodynamics (IGS).

Schupler, B.R., T.A. Clark and R.C. Allshouse (1995): Characterization of GPS User Antennas: Reanalysis and New Results. In Beutler, G. et al. (Eds): GPS Trends in Precise Terrestrial, Airborne and Spaceborne Applications, IAG Symp. No 115, July 3-4, 1995, Boulder, Co., 328-332.

REAL-TIME FAILURE DETECTION AND REPAIR IN IONOSPHERIC DELAY ESTIMATION USING GPS BY ROBUST AND CONVENTIONAL KALMAN FILTER STATE ESTIMATES

Lao-Sheng Lin & Chris Rizos
School of Geomatic Engineering, The University of New South Wales
Sydney NSW 2052, Australia

Stelios Mertikas
Department of Mineral Resources Engineering
Technical University of Crete
GR-73 100 Chania, Crete, Greece

ABSTRACT

The ionospheric delay is one of the main error sources for precise GPS positioning and navigation. Dual-frequency GPS receivers can be used to eliminate the ionospheric delay (to the first order) through a linear combination of L1 and L2 observations (Hofmann-Wellenhof et al., 1994). Taking advantage of this physical dispersive principle, one or more dual-frequency GPS receivers can be used to determine a model of the ionospheric delays over a region of interest and, if implemented in real-time, can support single-frequency GPS positioning and navigation applications.

There are, however, several factors which may affect the accuracy of real-time ionospheric delay estimation. One of the dominant factors, the failure in the carrier phase observations, such as the occurrence of cycle slips (for example, Engler et al., 1995), is the topic of this paper. In order to improve the real-time ionospheric delay estimation performance a novel approach to addressing this problem is presented here.

The concept and methodology of the proposed algorithm is introduced. Preliminary test results are presented to indicate the performance of this approach for real-time ionospheric delay estimation.

REAL-TIME FAILURE DETECTION AND REPAIR

The Kalman filter, referred to as the *conventional Kalman filter* here, is a recursive estimation procedure that solves the state vector estimation problem for a linear dynamic system with the linear observation model (Gelb, 1974). For a linear discrete-time stochastic system with Gaussian white noises the conventional Kalman filter produces the minimum variance estimate of the states. However, when the noise processes are non-Gaussian the filter performance is very sensitive and the state estimates will be easily biased, yielding poor accuracy of its estimates. In order to obtain a robust estimate of states a *robust Kalman filter* of the Masreliez-Martin type (Mertikas & Rizos, 1998) is adopted in this paper.

Further details concerning robust Kalman filters can be found in, for example, Wang et al. (1993), Mertikas & Rizos (1998), and other Kalman filtering textbooks

Concept of Real-Time Failure Detection and Repair

The differences between the conventional (non-robust) and the robust Kalman filter estimates of the states are considered as being "failure-indicating signals" (Ibid, 1998). These signals are assumed to follow a normal distribution with mean value μ and variance σ^2, provided the assumptions underpinning the Kalman filter are satisfied. Measurement departures from a conforming class (process in-control) will be indicated as shifts in the mean value μ and the standard deviation σ of the process. Further, when a gross error is present in the measurements this can be detected by readily implementing an outlier screening procedure in a sequential form. The following steps describe the method of detecting failure of Total Electron Content (TEC) estimates using GPS measurements in real-time (Lin, 1997).

The failure-indicating signals X are defined as:

$$X_i = TEC_i - TEC_i^*, \qquad i = 1,2,\cdots \tag{1}$$

where TEC_i and TEC_i^* are the TEC states estimated by the conventional and robust Kalman filter at time i respectively.

The sample mean value $\overline{X}_r$ and the sample variance S_r^2 of the first r observations $\{X_1, X_2, \cdots, X_r\}$ can be obtained sequentially using the following expressions:

$$\begin{aligned} \overline{X}_r &= \frac{1}{r}[(r-1)\overline{X}_{r-1} + X_r] \\ S_r^2 &= \frac{r-2}{r-1}S_{r-1}^2 + \frac{1}{r}(X_r - \overline{X}_{r-1})^2 \end{aligned} \tag{2}$$

In order to downgrade the influence of outliers on the estimation of the mean and standard deviation of the process, the Winsorization process is used to *robustify* the estimation procedure. In this case the raw measurements X_r have been replaced by pseudo-observations X_r^* such that:

$$X_r^* = \begin{cases} X_r & \text{if } \dfrac{|X_r - \overline{X}_r|}{S_r} < c \\ \overline{X}_r + cS_r & \text{if } \dfrac{X_r - \overline{X}_r}{S_r} > c \\ \overline{X}_r - cS_r & \text{if } \dfrac{X_r - \overline{X}_r}{S_r} < -c \end{cases} \tag{3}$$

These truncated data values X^* are then used to produce estimates of the mean $\overline{X}_r^*$ and the standard deviation S_r^* of the process. The constant c regulates the amount of robustness in the Winsorization process (in this paper c=3, a value derived empirically). To test the compatibility of the new random variable X_r^* with its predecessors $\{X_1^*, X_2^*, \cdots, X_{r-1}^*\}$, the standardised variate t, which is a good approximate to the Student's t-distribution, is defined as:

$$t_r = \sqrt{\frac{(r-1)}{r}}\left[\frac{(X^{\bullet}_r - \overline{X}^{\bullet}_{r-1})}{S^{\bullet}_{r-1}}\right] \tag{4}$$

The new variables are then transformed to a more convenient standard normal variable $u(t_r)$ using the following approximation:

$$u(t_r) = \text{sign}(X^{\bullet}_r - \overline{X}^{\bullet}_{r-1})\left[\frac{8(r-2)+1}{8(r-2)+3}\right]\sqrt{(r-2)\log\left(1+\frac{t_r^2}{(r-2)}\right)} \sim N(0,1) \tag{5}$$

To screen the outliers in this sequential data set, a discordance test using the well-known three-sigma standard deviation identification rule is used:

$$\text{Declare } X_r \text{ as failure if } |u(t_r)| > 3 \tag{6}$$

A disagreement between the conventional state estimates and its robust ones indicates that a failure has been detected. The following example illustrates the operation of the failure detection and repair algorithm.

Detection and Identification Step

Two Kalman filters, one conventional and one robust, are used to estimate the TEC for each tracked GPS satellite (in this example, only PRN1 satellite is considered). If the $|u(t_r)|$ value of PRN1 is greater than 3, then it is assumed that a carrier phase failure has occurred.

Repair Step

Let TS be the phase-derived TEC, D the linear combination of the integer ambiguity parameters for L1 and L2 carrier phase observations (λ_1, λ_2 being the wavelengths), ΔTS the failure correction to TS, $\overline{TS}$ the corrected TS, and CN the cycle slip number (see Lin, 1997, for more details). ΔTS and CN are set to zero at the start of tracking of PRN1. The relationship between TS and $\overline{TS}$ is:

$$\overline{TS} = TS + \Delta TS \tag{7}$$

Let Inno(TS) be the innovation value of TS, which can be computed from the conventional Kalman filter as:

$$\text{Inno}(TS) = \overline{TS} - [\,TEC(-) + D(-)\,] \tag{8}$$

where TEC(-) and D(-) are estimates of TEC and D before measurement update.

Let dN_1 and dN_2 be the carrier phase cycle slips on the L1 and L2 signals respectively (in units of cycles). The combined effects of dN_1 and dN_2 on TS are:

$$dTS = 9.5196 \times (\lambda_1 dN_1 - \lambda_2 dN_2) \tag{9}$$

Considering the fact that the minimum possible value of dTS is ±0.24 TECU (TEC units), when dN_1=-5 and dN_2=-4, or dN_1=5 and dN_2=4, the critical value of 0.25 TECU was selected as the criteria for invoking the repair action. For example, if the failure occurred at epoch k, then the innovation Inno(TS) is computed by equation (8) and the failure correction ΔTS is determined according to the steps listed in Table 1. The predicted states of epoch k are adopted as estimated states. From epoch k+1 on, the TS is corrected by equation (7).

Table 1. Main steps of the failure detection and repair algorithm.

```
CN=0, and ΔTS=0
while PRN1 is tracked:
      read data
        compute TR, TS
        T̄S̄= TS+ΔTS
        estimate TEC, D, etc.
            if |u(t_r)|> 3
                if |Inno(TS)| ≥ 0.25TECU
                    if CN=0
                       ΔTS=0
                       ΔTS=ΔTS-Inno(TS)
                       CN=CN+1
                    endif
                  ΔTS=ΔTS-Inno(TS)
                  CN=CN+1
                  X(+)=X(-)
                endif
            endif
endwhile
```

TEST RESULTS AND DISCUSSION

Data Description

The UNSW data set was originally collected from 28 April, 1995 (day 118) to 1 May, 1995 (day 121) on the roof of the Geography & Surveying Building, The University of New South Wales, using Leica System 200 GPS receivers. Four astronomical pillars, EC02, EC05, EC14 and EC19, were selected as station sites. Each GPS receiver collected 2 hours of data on four consecutive days. The data rate was 1 second and the elevation cut-off angle was set to 20 degrees.

From earlier tests using this data it was found that there were measurement failures (that is, cycle slips) in the data collected from pillar EC05. Therefore, the data sets from this site were chosen to test the above algorithm.

Cycle Slip Detection and Repair

Post-processing mode

In order to verify the performance of the proposed procedure, a carrier phase data post-processing algorithm involving widelane and $\Phi_{-7,9}$ phase combinations (Han, 1995) was used to detect and repair the cycle slips. The data, after application of the cycle slip repair procedure, was processed using the carrier phase levelling process (see, for example, Lin et al., 1996). The carrier phase levelled TEC derived from the cycle slip free data (line B in Figure 1) was used to verify the performance of the procedure.

Real-time mode

A real-time TEC estimation software package, known as REALTEC, was also developed. REALTEC was used to process the raw data from day 121, 1995. The test results are shown in Figure 1. The real-time TEC estimates without applying the multipath template (Lin, 1997) are represented by line A. The post-processed TEC estimates with cycle slip free data are represented by line B. Line C represents the real-time TEC estimates after applying the multipath template generated from the current day's data. Line D indicates the post-processed TEC estimates from the raw data. The range of elevation angle of the satellite (PRN21) is also shown. (Note that line C has been shifted by +0.5 TECU from line B for clarity, otherwise there appears to be no difference in the cycle slip estimates whether made in post-processed mode or in real-time.) From these test results it is concluded that: (1) REALTEC is capable of detecting and repairing the carrier phase cycle slips in real-time, and (2) the achieved TEC accuracy from REALTEC is identical to that obtained from post-processing *if the multipath template is used.*

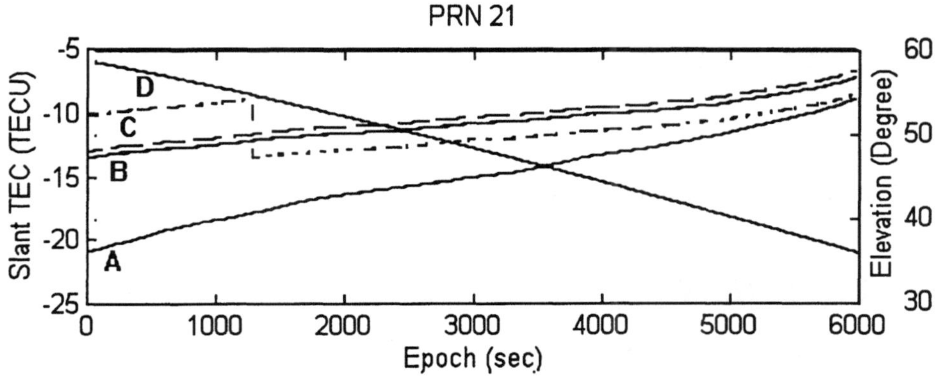

Figure 1. Post-processed versus real-time TEC estimates for PRN21, for day 121, 1995.

Another example is taken from data from day 119, 1995. The test results are shown in Figure 2. The post-processing result indicates that there are three cycle slips on PRN21, as indicated by line C. The post-processed TEC estimates with cycle slip free data are are shown as line A. The real-time TEC estimates using REALTEC with raw data are indicated by line B. (As before, line B has been shifted from line A by +0.5 TECU for the purposes of clarity.) The current day's multipath template is applied. Again, it can be concluded that REALTEC is able to generate good quality TEC estimates (when compared to post-processed TEC estimates) in real-time, even though there are multiple cycle slips in the raw data.

CONCLUDING REMARKS

In order to improve the accuracy of real-time ionospheric delay estimation an integrated approach to the processing of phase and pseudo-range data was developed. On the basis of several tests, it appears feasible that carrier phase derived TEC "failure", caused by cycle slips, can be detected and repaired in real-time. Further tests to refine the algorithm are necessary.

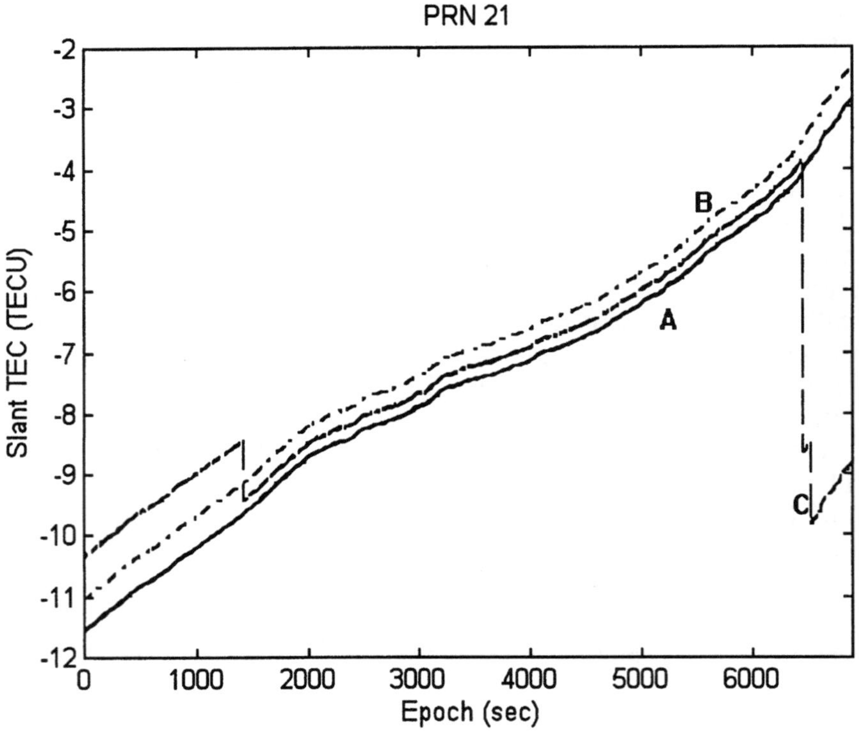

Figure 2. Post-processed versus real-time TEC estimates for PRN21, for day 119, 1995.

ACKNOWLEDGEMENTS

The post-processing cycle slip detection and repair software was kindly provided by Dr. Han (UNSW). The first author would also like to acknowledge the financial support of the Government of Taiwan, Republic of China.

REFERENCES

Engler, E., Sardon, E., Jakowski, N., Jungstand, A. & D. Klahn, 1995. Real-time monitoring of the ionosphere, Proc. 1995 IGS Workshop, Postdam, Germany, May 15-18, 67-76.

Gelb, A., 1974. *Applied Optimal Estimation.*, The M.I.T. Press, Cambridge, Massachusetts, 374 pp.

Han, S., 1995. Ambiguity recovery for GPS long range kinematic positioning, *8th Int. Tech. Meeting of the Sat. Div. of the U.S. Inst. of Navigation*, Palm Springs, California, Sept. 12-15, 349-360.

Hofmann-Wellenhof, B., Lichtenegger, H. & J. Collins, 1994. *Global Positioning System: Theory and Practice*, 3rd edition, Springer-Verlag Wien, New York, 355 pp.

Lin, L.S., 1997. A novel approach to improving the accuracy of real-time ionospheric delay estimation using GPS, *10th Int. Tech. Meeting of the Sat. Div. of the U.S. Inst. of Navigation*, Kansas City, Missouri, Sept. 16-19, 169-178.

Lin, L.S., Rizos, C. & Y.J. Wang, 1996. Real-time estimation of ionospheric delays using GPS, Proc. 1996 Int. Conf. on GPS, Taipei, Taiwan, June 12-13, 117-127.

Mertikas, S.P. & C. Rizos, 1998. Real-time failure detection in the carrier phase measurements of GPS by robust and conventional Kalman state estimates, *Marine Geodesy*, Vol.21,41-65.

Wang, Y.J. & K.K. Kubik, 1993. Robust Kalman filter and its geodetic applications, *Manuscripta Geodaetica*, 18, 349 - 354.

GPS LEVELLING RESULTS FROM TWO TEST AREAS IN FINLAND

Matti Ollikainen
Finnish Geodetic Institute
Geodeetinrinne 2, FIN-02430 Masala, Finland
E-mail: Matti.Ollikainen@fgi.fi

Abstract

GPS levelling was used in two test areas in Finland in order to investigate the capability of GPS measurements for orthometric height determinations. Both test areas are surrounded by the spirit levelling loops of the Precise Levelling network. Besides, the test areas are covered by a dense network of lower order levelling lines that are connected to the Precise Levelling network. The GPS networks measured in both areas consisted of 51 and 45 GPS sites that were located at spirit levelling benchmarks. The average side length was 15 km.

Four different geoid models, two global models (OSU91A, EGM96) and two local models (NKG89, FIN95), were used to convert the GPS ellipsoidal heights to orthometric heights, after which the results were compared to the spirit levelled heights. In all cases the height differences showed a clear tilt which was removed by fitting first and second order surfaces to the differences. When the global geoid models were used for the conversion, the RMS of the height differences varied from ±40 mm to ±80 mm. The RMS of the differences was reduced to appr. ±15 mm when the local geoid models were used. When the errors caused by the spirit levelling in the computed height differences were removed, the estimation of the average accuracy of the GPS levelling on the test areas gave the result ±12 mm.

Research area

Two areas limited by 2 levelling loops of the Precise Levelling Network were chosen as research areas. The size of the first area, observed with GPS in 1994, is about 9000 km^2, and its location is between longitudes 24.8°E and 27.0°E and latitudes 60.0°N and 61.0°N. The second area, where GPS observations were performed in 1995, is located on the southwestern coast of Finland, between longitudes 21.8°E and 24.0°E, and latitudes 60.4°Nand 61.5°N; the size of the area is about 10000 km^2.

The height differences in both areas are small; the heights of the GPS sites vary from 0 to 120 m in Area I and from 7 to 122 m in Area II.

We tried to choose the research areas such that the shape of the geoid would be different in the two areas. The shape of the geoid is reasonably flat throughout the country; the steepest slopes are located in northern Finland, but in these areas the levelling network is sparsest. The areas chosen for research are surrounded by Precise Levelling loops, while a dense network of lower-order levelling lines occurs inside the loops. In Area I *N* varies 15 - 19 m running east to west, but in Area II 18.8 - 19.6 m.

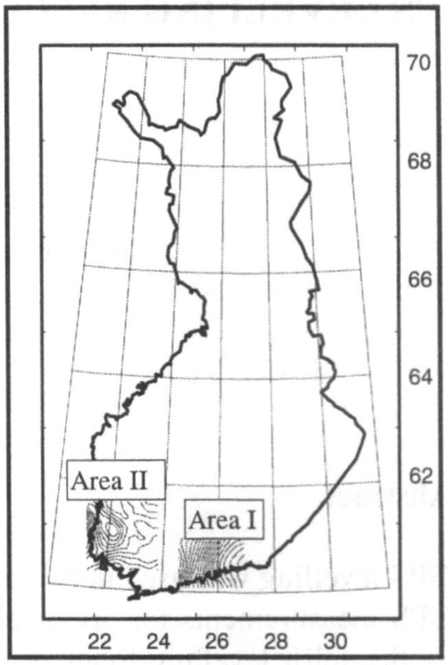

Fig. 1 Location of the research areas in Finland.

Spirit Levelling lines

The Precise Levelling loops limiting the areas discussed have been levelled three times. In Area I, which is surrounded by the levelling Loop III, the epochs were 1899.7, 1937.7, 1979.3 for the First, Second and Third Precise Levelling, respectively. In the second research area, Area II, surrounded by the levelling Loop VI, the respective levelling epochs were: 1902.6, 1946.6 and 1981.4. According to the three precise levellings the postglacial land uplift can be computed and the height of the levelling benchmarks can be reduced to any epoch, e.g. to the epoch of the GPS observations discussed here. The accuracy of precise levelling was evaluated, according to misclosures of the levelling loops in southern Finland, to be ±0.73 mm/$\sqrt{\text{km}}$ in the First, ±0.42 mm/$\sqrt{\text{km}}$ in the Second and ±0.88mm/$\sqrt{\text{km}}$ in the Third Precise Levelling, respectively (KAKKURI and VERMEER 1985). In addition to the Precise Levelling Network, several lower-order levelling lines occur inside the Precise Levelling loops. These lines have been levelled by the National Land Survey (NLS). The levelling accuracy performed by NLS is according to (OKSANEN 1991) ±0.5-±0.8 mm/$\sqrt{\text{km}}$ in Class Ib and ±1.5 mm/$\sqrt{\text{km}}$ in Class II levellings. The levelling lines are shown in Fig. 2.

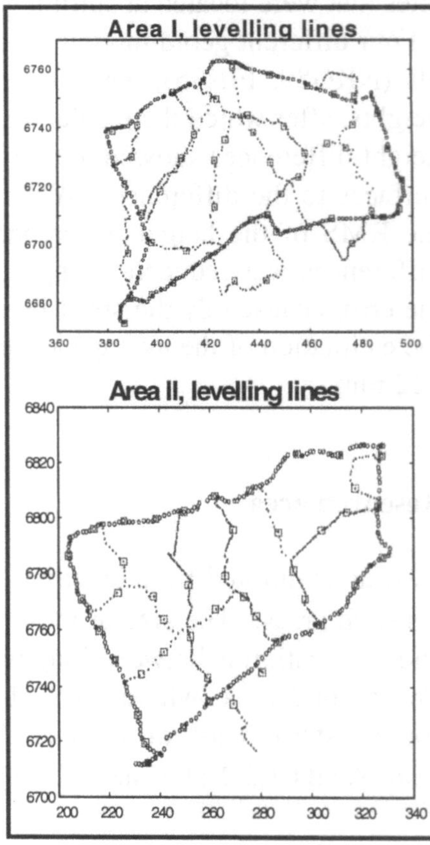

Fig. 2 Levelling lines in Area I and II.
o Precise Levelling line
+ Class Ib and II levelling lines
□ GPS site

Geoidal heights

A total of four different geoid models were used to convert the GPS ellipsoidal heights to orthometric heights e.g. OSU91A (RAPP *et al.* 1991), EGM96, NKG89 (FORSBERG 1990) and FIN95 (VERMEER 1995). As an example of the geoidal height in the research areas, the geoidal heights according to the FIN95 geoid are shown in Fig. 3.

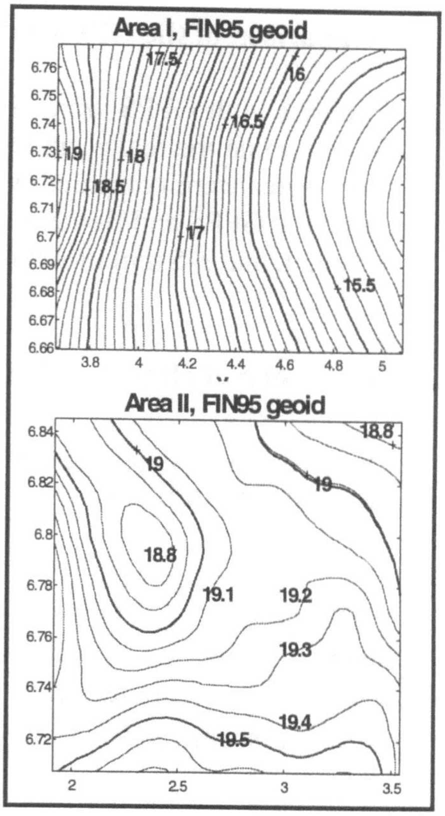

Fig. 3 Geoidal heights of the FIN95 geoid in Area I and Are II Contour interval: 0.1 m.

GPS observations

The GPS sites observed in Areas I and II are shown in Fig. 2. The number of sites observed in Area I was 51, and the area of the network was approx. 9000 km^2 or one GPS site per 158 km^2; the average distance between stations is thus 13 km. The size of Area II was approx. 10 000 km^2, and the number of sites was 45, or one GPS site per 222 km^2; the average distance between sites was 15 km.

The GPS observations were made using 6 Ashtech Z-12™ type receivers. In Area I the observations were performed during a period of 6 days. The lengths of the observation sessions were two hours only. In Area II, the distances between observation sites were slightly greater than in the first area, which is why the observation sessions were prolonged to three hours. In Area II the observations were performed during a period of 8 days.

Solution of the GPS observations

The GPS observations were processed using two software packages: GPPS Ver 5.0 software (ASHTECH 1993) and Bernese Ver 3.5 software (ROTHACHER *et al.* 1993).

The final solution with the GPPS software was obtained using Precise Ephemeris obtained from IGS/Center for Orbit Determination in Europe (1994-95), the Lc option (ionospheric-free solution), and widelane solution. All possible baselines were solved in each session, and finally all baselines with fixed ambiguities were adjusted by the FILLNET program (ASHTECH 1992). The strategy used in GPS solutions with the Bernese software is presented in Table 1.

Table 1. The strategy used in GPS-solutions with the Bernese software.

1. L3, ionospheric-free solution of the baselines.
2. Evaluation of the ionospheric models.
3. Wide-lane solution of L5 ambiguities.
4. L1&L2 solution of L1 and L2 ambiguities
5. Final L3 solution introducing the ambiguities from the previous step.
6. Troposphere zenith corrections were not estimated.

Comparison of the GPS solutions

Solutions with GPPS were undertaken to obtain preliminary coordinates for the stations. To determine the real accuracy obtainable with a commercial software package, the GPPS coordinates were compared with the Bernese solutions. The largest height differences between the GPPS and Bernese solutions are -20 mm. The RMS of the differences is ± 4 mm in Area I, and ± 5 mm in Area II.

Conversion of ellipsoidal height differences to orthometric height differences

The orthometric height differences between the benchmarks were computed according to the following formula:

$$\Delta H_{GPS} = \Delta h_{GPS} - \Delta N \quad (1)$$

where Δh_{GPS} is the ellipsoidal height difference observed by GPS and ΔN the geoidal height difference. To determine the orthometric heights obtained with GPS levelling (H_{GPS}) in the same reference frame as the levelled orthometric heights (H_{Lev}) the height differences were added to the orthometric height of the initial benchmark, H_o:

$$H_{GPS} = H_o + \Delta H_{GPS} \quad (2)$$

The height differences obtained with GPS were reduced for the land uplift from the epochs of the observations to the epoch of the spirit levelling (1960.0) using the following formula:

$$dH_U = (1960 - T) \cdot \lambda \quad (3)$$

where T is the epoch of GPS observations and λ the land uplift (mgpu/year). λ values were obtained from the adjustment of the consecutive Precise Levellings of the loops surrouding the research areas.

Comparison of GPS-levelled heights with spirit-levelled heights

The GPS levelling results obtained with four different geoids were compared with the spirit levelled heights. The height differences computed showed in all cases a clear tilt, which is why the systematic part of the differences was removed by fitting first- and second-order surfaces to the differences. The following conclusions may be drawn according to the RMS of the residuals of the plane fit:

The RMS values of the residuals clearly describe the errors in the geoidal heights used. In both areas the best result was achieved by using the FIN95 geoidal heights. When the FIN95 geoid was used, the RMS of the residuals was ± 17 mm in Area I, and ± 14 mm in Area II. The result obtained with the NKG89 geoid was very good, (< ± 20 mm on both areas), but the global models resulted in RMS values that varied from ± 40 mm to ± 80 mm. The RMS values of the residuals are shown in Figs. 4-1 and 4-2.

Discussion of the accuracy

Height differences (*ΔH*) between GPS levelling and spirit levelling results contain errors originating from three main sources: GPS determinations (σ_h), spirit levelling (σ_H) and geoidal heights (σ_N). A minor error (σ_U) is caused by the land uplift reduction which is included in the height differences. We estimated the different error components according to the standard deviations of different adjustments. σ_h is slightly dependent on the distance from the initial point of the network. With linear regression we obtained the following formula:

$$\sigma_h[\mathrm{mm}] = \pm\,(8 + 0.02 \cdot d\,[\mathrm{km}]\,) \qquad (4)$$

For estimation of σ_H we performed an common adjustment of the Precise Levelling and lower order levelling data in both research areas. These adjustments resulted in the common value ± $0.8\,\mathrm{mm}/\sqrt{\mathrm{km}}$ for both research areas. The adjustment of the Precise Levelling loops gave for σ_U the value ± $0.02\ \mathrm{mm}/\sqrt{\mathrm{km}}/\mathrm{y}$.

σ_N was estimatd according to the rule of thumb derived by VERMEER (1995) for the FIN95 geoid:

$$\sigma_N = 0.4 \cdot d \cdot \sigma_{\Delta g}\sqrt{\ln\,(D/d)} \qquad (5)$$

where $\sigma_{\Delta g}$ is the accuracy of prediction of gravity anomalies [2 mGal], *d* is gravity point spacing [5 km] and *D* is the spacing of the GPS sites used in the fitting of the FIN95 geoid [~100 km].

The error components may be summarized as follows:

$$\sigma_{\Delta H}{}^2 = \sigma_h^2 + \sigma_H^2 + \sigma_N^2 + \sigma_U^2 \qquad (6)$$

The errors caused by spirit levelling do not, however, affect the actual GPS levelling and may not be considered in the error budget, i.e. estimation of the GPS levelling errors ($\sigma^2_{H(GPS)}$) may be done according to the formula:

$$\sigma^2_{H(GPS)} = \sigma_h^2 + \sigma_N^2 \qquad (7)$$

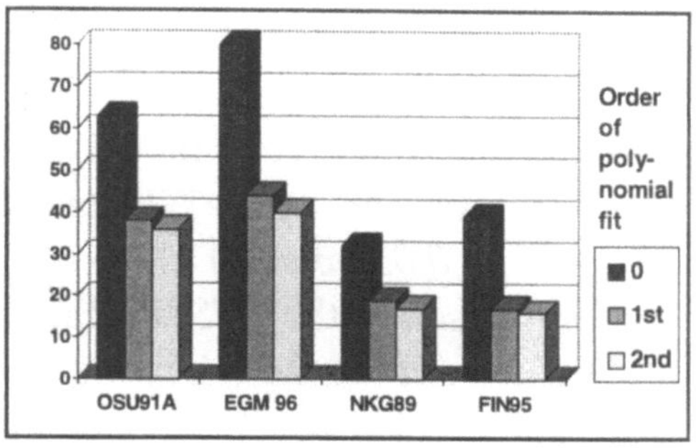

Fig 4-1 RMS of the height differences in mm between GPS and spirit levelling in Area I.

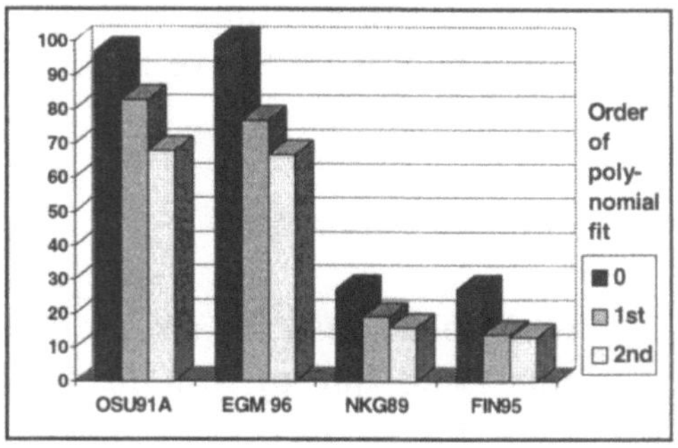

Fig 4-2 RMS of the height differences in mm between GPS and spirit levelling in Area II .

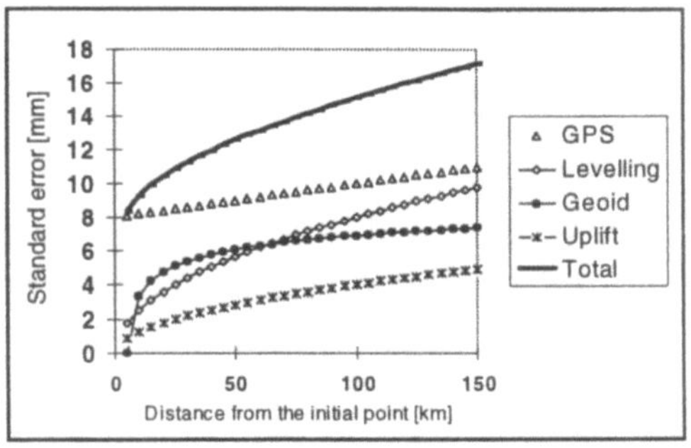

Fig. 5-1 Error budget of height differences H_{GPS}-H_{Lev}

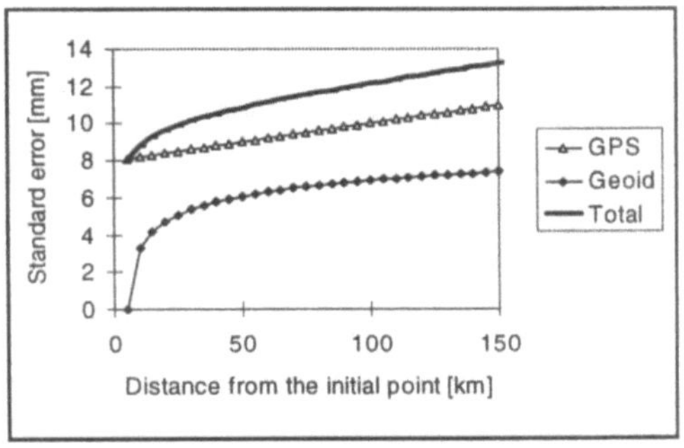

Fig. 5-2 Error budget of GPS levelling.

Conclusions

According to the error model the following rule of thumb was derived:

- *The accuracy of the GPS levelling over moderate distances in Finland is approx. ± (9 mm + 0.03 mm/km).*

Comparison of the GPS levelling accuracy with the spirit levelling accuracy gives some clear answers to the following questions:

- The Precise Levelling cannot be replaced by GPS levelling.
- The lower order levelling (accuracy ~ ± 1.5 mm/$\sqrt{\text{km}}$) could be replaced by GPS levelling on lines that are longer than 50 km.
- If the spirit levelling accuracy is lower than ± 2 mm/$\sqrt{\text{km}}$, better results are probably obtained by GPS levelling.

Acknowledgements

I wish to express my sincere thanks to Mssrs. SEPPO OKSANEN and MATTI MUSTO, National Land Survey, for the spirit levelling data observed by NLS and to Professor Dr. TEUVO PARM, Helsinki University of Technology for loaning two of the GPS receivers used in the fieldwork.

References

ASHTECH, Inc. (1992): *FILLNET, GPS-adjustment program,* version 3.1. Sunnyvale, CA, USA.

ASHTECH, Inc. (1993): *ASHTECH XII GPPS™, The GPS Post-Processing System Manual.* Software version 5.0.00, Document Number 600196, Rev. A. Sunnyvale, CA, USA.

FORSBERG, R. (1990): A new high resolution geoid of the Nordic Area. Proc. IAG Symposium 106: *"Determination of the Geoid, Present and Future"*. (Eds. R. Rapp, F. Sansò). Springer Verlag. New York.

KAKKURI, J. and M. VERMEER (1985): The study of land uplift using the Third Precise Levelling of Finland. Rep. of the FGI, No. 85:1. Helsinki.

OKSANEN, S. (1991): Geodetic Operations of the National Board of Survey. In *Geodetic Operations in Finland 1987-1991*. (Ed. J. Kakkuri). FGI. Helsinki.

RAPP, H.R., Y.M. WANG and N.K. PAVLIS (1991): The Ohio State 1991 Geopotential and Sea Surface Topography Harmonic Coefficient Model. Report No. 410, Department of Geodetic Science and Surveying, The Ohio State University, Columbus, Ohio.

ROTHACHER, M., G. BEUTLER, W. GURTNER, E. BROCKMANN and L. MERVART. (1993): Bernese GPS software, version 3.4, documentation May 1993. Bern.

VERMEER, M. (1995): Two New Geoids Determined at the FGI. Rep. of the FGI, No. 95:5. Helsinki.

QUALITY CONTROL ALGORITHMS FOR PERMANENT GPS RECEIVER APPLICATIONS

Chris Rizos
Lao-Sheng Lin
Shaowei Han
School of Geomatic Engineering, The University of New South Wales
Sydney NSW 2052, Australia

Lienhart Troyer
Department of Engineering Geodesy, Technical University Graz
8010 Graz, Austria

Stelios Mertikas
Department of Mineral Resources Engineering, Technical University of Crete
GR-73 100 Chania, Crete, Greece

ABSTRACT

There has been a burgeoning of activity involving the establishment and operation of permanent GPS stations. The variety of applications is broad, from geodetic reference sites in support of IGS initiatives, to national base stations for real-time differential GPS, "integrity monitoring" and datum maintenance, and even local GPS base stations providing data to surveyors and other users. The level of sophistication of such permanent stations also varies, from systems capable of automatic data collection, transmission and sometimes processing, to relatively simple configurations. All applications, however, share a concern for GPS data "quality". The detection of cycle slips, data "spikes", multipath disturbances, and bad data sequences of GPS measurements is a major problem for many applications, even more so in the case of real-time applications.

Some quality control (QC) strategies may be implemented either on a single receiver basis or using GPS network data. This paper discusses the issue of QC for permanent GPS receivers, and comments on aspects of QC which require investigation.

QC AND PERMANENT GPS RECEIVERS

Concerns about GPS "quality" are shared by all users, from those engaged in the most precise geodetic applications through to the casual navigator. The quality of GPS positioning is dependent on a number of factors. Experience with precise geodetic applications of GPS has shown that sophisticated mathematical modelling, careful field procedures and top-of-the-line GPS hardware are all necessary prerequisites, however great care still has to be applied to ensure that data quality is uniformly high. "Quality control" (QC) must be applied to both receivers, the one located at the "reference" site and the second GPS receiver, the one the user is generally interested in. Increasingly the former may be a *permanent* GPS receiver operated by an agency or organisation on behalf of a wide variety of users, while the latter may be in motion and hence experiencing very different conditions. Although QC for such a "roving" receiver is a significant challenge, the quality of the data from the permanent reference receiver must also be assured. This paper deals exclusively with this issue and hence is concerned with single (reference) receiver QC.

Quality Management Issues

To date the procedures of data screening, computations, result evaluation and quality assurance have generally been an off-line (as well as iterative) process. However, with the development of precise "on-the-fly" GPS positioning, and other real-time techniques, it is no longer possible to analyse (and re-analyse) GPS data in post-mission mode until GPS result quality is assured. The challenge therefore is to develop quality control (QC) and quality assurance (QA) procedures that can also be applied in real-time (or "near-real-time") GPS applications during or shortly after the measurement process.

Quality management, in the context of GPS applications, is concerned with assuring an agreed level of accuracy and reliability for GPS results. The focus is therefore on procedures for defining, measuring and verifying quality. The language of "quality management" is replete with terms such as QC and QA. Although the two terms are often considered synonymous, a good working definition is that: (a) QA refers to the set of practices and procedures which are intended to maximise the chances that the product or service will satisfy the client's requirements, at a reasonable cost, while (b) QC refers to the procedures used to verify the level of quality achieved and, if it is unacceptable, to detect the source of the problem and, if possible, remedy it. The QC/QA practices and procedures include:

- *Procedures for hardware procurement* -- if the GPS instrumentation is inappropriate or somehow faulty, how can the data be expected to be of consistently high quality?
- *Procedures for software verification* -- does the software generate "correct" results?
- *Algorithm definition and testing* -- the data processing algorithm plays a significant role in ensuring high quality results, as it may be, e.g., more or less "robust" in relation to varying quality input measurements.
- *Procedures for site selection* -- local environmental conditions may cause multipath and signal interruption problems.
- *Procedures for instrumental installation* -- e.g. quality of groundmark, ruggedness of cabling, efficiency of power sub-systems, etc.
- *Check procedures* -- how is the overall system quality being monitored (are there backup systems, check surveys, quality indicators, etc.?), is it being carried out on a continuous or periodic basis?

IAG Initiatives

The International Association of Geodesy (IAG) established the Special Study Group (SSG) 1.154[1] on "Quality Issues in Real-Time GPS Positioning" in 1995 to identify practical procedures, as well as mathematical techniques, that can be applied to assure the quality of positioning results obtained from real-time GPS applications. There are several characteristics that distinguish these applications from others:

- the communication of data from GPS receivers to a computing site where may occur transmission interruption, either recovering by duplicated encoding procedure or re-transmitting,

---

[1] The first author is Chairman of this SSG, while the third and fifth authors are members. Web site URL for this SSG is http://www.gmat.unsw.edu.au/ssg_RTQC/.

- data latency due to the transmission, where it is processed with no, or minimum, delay,
- make use of carrier phase data, in addition to pseudo-range data,
- rely on data processing on an epoch-by-epoch basis, or at the very least small "batches" of GPS data,
- do not permit extensive data "pre-processing" or the review of data and results in iterative procedures, and
- may involve kinematic or static positioning.

Poor quality GPS measurements makes the task of assuring high quality GPS results very difficult, and perhaps impossible. The data from a permanent reference receiver must therefore be assured before it is used in any computations.

QC/QA OF GPS MEASUREMENTS

QC/QA procedures must contend with those factors and influences that affect measurements. Restricting our interest to permanent GPS receivers we have the following problems to contend with:

- Measurement "failure" requiring isolated measurements to be discarded.
- Measurement "failure" that causes cycle slips in carrier phase observations, which can be repaired.
- Variable data quality arising from multipath disturbances, which must be mitigated at the single receiver level if at all possible .
- Antenna phase centre variation, which is an antenna-dependent error, which should be corrected for if use is to be made of different kind of GPS receivers,
- Variable data quality due to unusual atmospheric effects such as ionospheric scintillation.

Many procedures and processing strategies have been developed to account for such effects, both mathematically and empirically, however we will only mention several that can be applied directly to the raw data from a single receiver capable of making dual-frequency carrier phase (ϕ_1, ϕ_2 in radians) and pseudo-range (R_1, R_2 in metres) measurements, as is typical of modern GPS reference station receivers.

Failure in GPS Measurements

By "failure" is meant changes in the geometric (or range) information contained within the measurements that are: (a) of unknown magnitude, and (b) occur at unknown instants of time. A failure in GPS measurements includes cycle slips in phase data or outliers in pseudo-range date. Such failures have usually been detected during processing, or pre-processing, of double-differenced (i.e. baseline) data. A repair procedure may then be applied, or the data simply flagged "bad" during residual testing, and the amended dataset is ready for processing. The procedure is applied iteratively until the process converges.

A procedure for scanning the raw observation data based on using a Kalman filter to model the behaviour of phase and phase-rate measurements (and their changes) in discrete time is described in, for example, Cannon et al. (1992), Mertikas & Rizos (1998). This procedure can be applied to each of the data types (ϕ_1, ϕ_2, R_1, R_2) separately. The fact that the GPS receiver is stationary is an important constraint, making it a much easier task than if the receiver were moving. Following the detection step, a subsequent procedure may be applied to repair cycle slips in the phase data. Outliers in the pseudo-range data cannot be repaired and must be eliminated from the dataset. Such an algorithm's application to the problem of determining Total Electron Content (TEC) from single receiver measurements which may be contaminated with cycle slips is reported in, for example, Jungstand et al. (1995), Lin (1997a), and others.

Dual-frequency data offers especially rich opportunities to construct combinations of observables (phase-only, pseudo-range-only, as well as phase and pseudo-range) (e.g., Hofmann-Wellenhof et al., 1994; Rizos, 1997) which may be screened using a number of procedures based on Kalman filters of various types, trend-following polynomials, digital filters, and so on. The following sequences are typically used:

- One-way residual sequences, including receiver clock bias, satellite clock bias, SA effect, atmospheric biases, multipath and observation noise.
- The difference sequences ($\lambda_1.\phi_1 - \lambda_2.\phi_2$) between L1 and L2 carrier phase data in units of metres, which includes ionospheric delay and ionospheric scintillation, and cycle slips.
- $N_{i,j} = i.\phi_1 + j.\phi_2 - \dfrac{9240(i+j)+289i}{2329.\lambda_1}.R_1 + \dfrac{9240(i+j)+289j}{2329.\lambda_2}.R_2$

  especially the wide-lane ambiguity sequences (i=1, j=-1), where λ_1 and λ_2 are the wavelength of the L1 and L2 carrier phase, i and j are integers. These sequences are affected by ionospheric scintillation, cycle slips and pseudo-range biases.
- Multipath on pseudo-range data can also be estimated, though the estimates are affected by cycle slips, ionospheric scintillation and pseudo-range biases.
- Single-differencing (between satellites) may be applied to the above sequences, so as to eliminate the receiver clock bias.

Multipath at Permanent Receiver Sites

Multipath is a signal disturbance arising from the fact that the signal entering the GPS antenna, in addition to containing the direct satellite-receiver component, also includes reflections from buildings, water surfaces and the ground. Multipath and diffraction effects cannot be easily accounted for during data processing. Fortunately the multipath error on carrier phase observations is significantly less than that experienced on pseudo-range data (of the order of several centimetres, compared with metre level disturbance on pseudo-ranges). Furthermore, its effect tends to average out for static baseline determinations with observation sessions of the order of an hour or more. Nevertheless, for the highest precision static and kinematic applications the effect of multipath disturbance must be addressed.

The multipath component in the L1 and L2 pseudo-range data can be estimated (e.g., Hofmann-Wellenhof et al., 1994; Rizos, 1997). If the L1 and L2 carrier phase is measured, the multipath in the pseudo-range data can be estimated. Obviously this approach cannot be used with single-frequency data.

The multipath error in carrier phase cannot be estimated from the raw measurements on a single receiver basis, but may be estimable on a baseline basis from the double-differenced residuals after baseline processing.

In the case of a permanent receiver, the geometry of the satellites, relative to the receiver and surrounding reflective objects, is almost exactly the same after one sidereal day. Hence the multipath disturbance tends to exhibit a daily signature, both in the raw measurements and in the baseline residuals. So-called "multipath-templates" can be constructed to correct pseudo-range measurements, from an analysis of the past one or more day's data. This technique is applied to precise pseudo-range data at permanent GPS sites used, for example, to estimate atmospheric parameters such as TEC (Lin, 1997b).

In the case of carrier phase, multipath *elimination* could be achieved by using a filter tuned to the appropriate frequency/amplitude characteristics of the disturbance. Because of the daily multipath signature at permanent GPS receivers the filter is easier to "tune" than under other circumstances. A Finite Impulse Response (FIR) lowpass filter has been proposed by Han & Rizos (1997).

However, apart from mathematical procedures there are several strategies for overcoming the problem of multipath in the observations at permanent GPS receivers:

- Careful selection of site in order to minimise the multipath environment.
- Use multipath resistant antennas.
- Use special receivers that contain some form of "multipath elimination technology".
- If it is possible, multipath should be "corrected" at each receiver.

Ionospheric Scintillation

The ionospheric delay on GPS measurements are subject to spatial and temporal variations. The spatial variations are usually low in frequency and generally correspond to the various ionospheric latitude zones: tropical, mid-latitude and auroral. The temporal variations can have high frequencies (of the order of minutes to hours), medium frequency (diurnal and seasonal effects), and low frequency (the 11 year solar cycle). TEC is a maximum at low latitudes (the tropical zone) and at the poles (the auroral zone), and is a minimum at mid-latitudes. At night the ionospheric delay is approximately five to ten times less than for day time observations. The diurnal cycle for TEC is such that the maximum occurs two hours after solar noon, and is a minimum before dawn. Ionospheric disturbances, which can occur suddenly and be very severe, affect the amplitude and phase of GPS signals (Wanninger, 1993; Knight & Finn, 1996). One of the phenomena responsible for these are "travelling ionospheric disturbances", another is due to irregularities in the ionosphere causing "scintillations" (especially in the tropical and auroral zones). Under such conditions the ionosphere is so perturbed that single frequency operations may become impossible because the GPS receiver loses lock on the satellite signals. Where tracking is possible, the likelihood of cycle slips and interrupted tracking is increased, both of which, for example, make ambiguity resolution a more difficult and unreliable task.

Knight & Finn (1996) describes an algorithm for determining the so-called S_4 "scintillation index". Empirical filtering techniques will need to be developed to cope with such effects in real-time, particularly when the next solar cycle maximum occurs at the turn of the century.

Other Biases Dependent on Receivers and Antennas

There is an offset bias which is dependent on the antenna used due to phase centre offset and variation. The electrical phase centre of the antenna is the point to which the radio signal measurement is referred and generally is not coincident with the physical antenna centre. The offset depends on the elevation, azimuth, and intensity of the satellite signal, and is different for L1 and L2. The antenna phase centre offsets for different kinds of GPS receivers relative to the Turbo Rogue receiver with the Dorne Margolin T antenna have been determined by the International GPS Service for Geodynamics (IGS report available from http://igscb.jpl.nasa.gov/, 1996). The offset components in the north and east direction can reach 5mm. The antenna component in height can reach 3cm, and changes significantly with elevation angle.

If the reference GPS receivers are generating data for different kinds of "mobile" GPS receivers, the antenna offsets should be considered carefully when seeking to achieve millimetre accuracy. Therefore, GPS antenna offset determination and correction will become necessary for precise GPS positioning.

The bias which is dependent on the type of GPS receiver is due to the manner in which different receivers internally process the carrier phase signals. A zero baseline experiment has been performed by Braun & Rocken (1995) using the same antenna (Trimble 4000ST L1/L2 Geod antenna) and two different types of GPS receivers from the same manufacturer (Trimble 4000SST and Trimble 4000SSE). The results indicate that vertical height error can be up to five centimetres, even when using one hour observation spans. Similar results were reported by Brunner & Tregoning (1994). More research is necessary if millimetre accuracy positioning using mixed GPS receivers/antennas is to become routine.

CONCLUDING REMARKS

Ideally what is required is an algorithm, or "cocktail" of algorithms, capable of automatic, real-time operation, on a single site basis. They must be able to delete obviously "bad data", repair failures such as carrier phase cycle slips, and mitigate multipath in pseudo-range and phase data.

However, the quality of the data either before or after QC procedures are applied cannot be characterised in a simplistic "black and white" fashion. Even after the application of such an algorithm (or algorithms) the data quality will likely be on a "grey scale", ranging from "bad" (but not so "bad" that it is considered an outlier) to "good" inferring a phase (or pseudo-range) noise (typically measured as a measurement variance) that is not constant. Perhaps what is required is a measure of data quality, such as an index between 1 and 5, that ensures that the measurement stochastic model is realistic.

There is still considerable R&D necessary to address this issue, even for the relatively benign QC/QA circumstances of permanent, stationary GPS reference receivers.

REFERENCES

BRAUN, J. & C. ROCKEN, 1995. Vertical height errors when mixing Trimble 4000SST and Trimble 4000SSE observations, UNAVCO report at http://www.unavco.ucar.edu.

BRUNNER, F.K. & P. TREGONING, 1994. Investigation og height repeatability from GPS measurements, *Aust. J. Geod.Photo.Surv.*, 60, 33-48.

CANNON, M.E., SCHWARZ, K.P., WEI, M. & D. DELIKARAOGLOU, 1992. A consistency test of airborne GPS using multiple monitor station, *Bull. Geod.*, 66, 2-11.

HAN, S. & C. RIZOS, 1997. Multipath effects on GPS in mine environments, *Xth Int. Congress of the Int. Soc. for Mine Surveying*, Fremantle, Australia, 2-6 November, 447-457.

KNIGHT, M. & A. FINN, 1996. The impact of ionospheric scintillations on GPS performance, *9th Int. Tech. Meeting of the Sat. Div. of the U.S. Inst. of Navigation*, Kansas City, Missouri, Sept. 17-20, 555-564.

HOFMANN-WELLENHOF, B., LICHTENEGGER, H. & COLLINS, J., 1994. *GPS Theory and Practice*, Springer-Verlag, Vienna New York, 3rd ed., 355pp.

JUNGSTAND, A., ENGLER, E., SARDON, E. & KLAHN, D., 1995. Error separation concept in experimental TEC monitoring network, *U.S. Inst. of Navigation National Tech. Meeing*, Washington, D.C., 323-335.

LIN, L.S., 1997a. A novel approach to improving the accuracy of real-time ionospheric delay estimation using GPS, *10th Int. Tech. Meeting of the Sat. Div. of the U.S. Inst. of Navigation*, Kansas City, Missouri, Sept. 16-19, 169-178.

LIN, L.S., 1997b. Real-time estimation of ionospheric delay using GPS measurements, PhD dissertation, School of Geomatic Engineering, The University of New South Wales, Sydney, Australia.

MERTIKAS, S.P. & C. RIZOS, 1998. Real-time failure detection in the carrier phase measurements of GPS by robust and conventional Kalman filtering, *Journal of Marine Geodesy*, Vol..21, 41-65.

RIZOS, C., 1997. Principles and practice of GPS surveying, Monograph 17, School of Geomatic Engineering, The University of New South Wales, ISBN 0-85839--071-X, approx. 560pp.

WANNINGER, L., 1993. Effects of the equatorial ionosphere on GPS, *GPS World*, 4(7), 48-54.

FAST AMBIGUITY RESOLUTION IN NETWORK MODE

P.J.G. Teunissen, P.J. de Jonge, D. Odijk, C.C.J.M. Tiberius
Delft University of Technology, Faculty of Geodetic Engineering
Thijsseweg 11, 2629 JA Delft, The Netherlands

Abstract

This contribution presents and discusses practical results of network-based instantaneous ambiguity resolution. It is shown that high-dimensional integer solutions can be computed efficiently with current ambiguity resolution methods. For that purpose the least-squares ambiguity decorrelation method has been integrated in the network analysis. We also introduce a new ambiguity dilution of precision measure to express the intrinsic precision of the ambiguities and show why it is preferred over the more traditional measures.

Measurements and model

The experiment on which our data analysis is based, uses a small network of four points. The experiment took place on December 22nd, 1996, about 80 km North-East from Delft. Four dual frequency receivers were used (Trimble 4000 SSI Geodetic Surveyor). Two of the receivers were placed stationary (points 15 and 28), about 12.7 km apart. The antennas of the third and fourth receiver were mounted on the roof of a van (points 38 and 39).

The data were collected at a one second sampling rate, during several sessions; each session took about 50 minutes. In the following we will show the results of only one session. These results are representative for all sessions. To get an impression of the dynamics involved, figure 1 shows a typical example of the velocity curve.

With the exception of some interruptions, the same 7 satellites (PRN's 02, 05, 07, 09, 21, 23 and 26) were tracked by three of the four receivers. The fourth receiver (point 39) failed to track satellite PRN 26. All satellites were tracked at an elevation angle larger than 10 degrees. The corresponding skyplot is shown in Figure 2.

The linearized system of DD observation equations of epoch i can be cast in the matrix-vector form as $y(i) = Aa + B(i)b(i) + e(i)$, $i = 1, \ldots, k$, where the vector $y(i)$ consists of the observed minus computed dual frequency phase and code data of epoch i, a is the time-invariant vector of unknown integer DD ambiguities, $b(i)$ is the unknown increment vector of the remaining parameters (such as network coordinates and possibly ionospheric delays) at epoch i, $e(i)$ is the noise vector and A and $B(i)$ are the appropriate design matrices.

Our analysis is based on using dual frequency phase data and single frequency C/A code data. For the undifferenced observables, the a priori standard deviations were set at 30 cm for the code data and at 3mm for the phase data. Time correlation was assumed absent. Due to the relatively short baselines and small height differences between the points, the tropospheric and ionospheric delays were initially assumed absent as unknown parameters. A priori corrections for the tropospheric delays were applied on the basis of the Saastamoinen model.

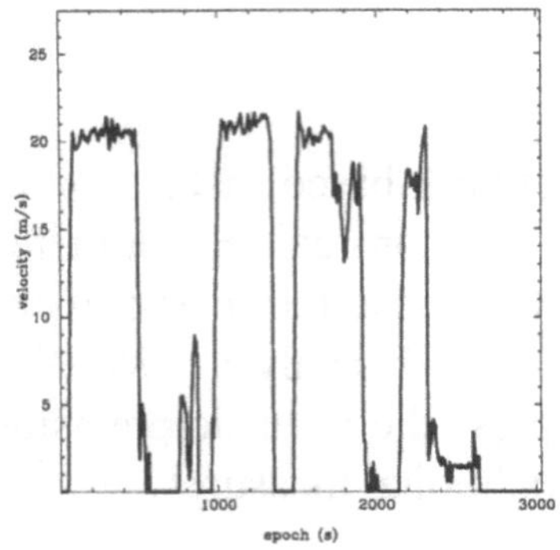

Figure 1: Velocity curve of van

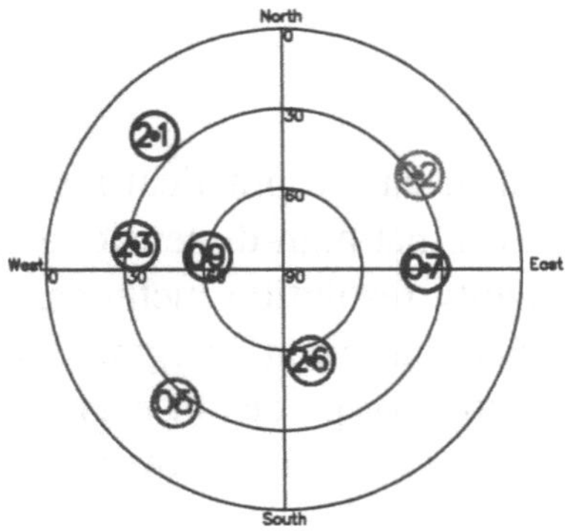

Figure 2: Skyplot 12-22-96 $\varphi = 52°26'$N $\lambda = 5°14'$E 14:30 UTC

The data of the four receivers were directly adjusted in a network-mode, instead of using a baseline-by-baseline approach. An example of a small-scale network with rapid static GPS for crustal deformation analysis can be found in *Genrich et al.* (1997). The vector $b(i)$ consists of the coordinate increments of all receivers, but relative to the reference point 15. The vector a consists then of the DD ambiguities belonging to all observed DD carrier phases. A computationally efficient method enables fast and strict integer least-squares estimation of the ambiguities. In the following, when we speak of ambiguity resolution, we refer to solutions which were obtained by solving all ambiguities of the complete network in one step. For our small network of four points this implies that the total number of ambiguities equals 34 (=2*[7+7+6]-2*3). This number is based on using dual-frequency data with three receivers tracking 7 satellites and the fourth receiver tracking one satellite less.

Ambiguity resolution

The process of ambiguity resolution can conveniently be separated into two parts: (1) the ambiguity estimation problem; (2) the ambiguity validation problem. The first part addresses the problem of finding the least-squares estimates of the integer ambiguities. The second part is concerned with the question whether one is willing to accept this integer least-squares solution. Conceptually one can break the first part into three different steps. In the first step one simply disregards the integer constraints. Hence the problem has now become a standard least-squares solution. As a result one obtains the (real-valued) estimates of a and $b(i)$ (the so-called float solution), together with their variance matrices. In the second step, the 'float' ambiguity vector and its variance matrix are used to solve for the integer least-squares ambiguities. These integer ambiguities are then finally used in the third step to obtain the 'fixed' solution of $b(i)$.

The difficult part in the above procedure is the computation of the integer least-squares ambiguities. The LAMBDA method provides a very efficient way of solving this problem, see e.g. *Teunissen* (1993), *de Jonge and Tiberius* (1996) or *Kleusberg and Teunissen* (1996). In short it consists of the following steps. First a local search region, the ambiguity search space, is defined: $(\hat{a}-a)^T Q_{\hat{a}}^{-1}(\hat{a}-a) \leq \chi^2, a \in Z^n$, with $\hat{a}$ the (real-valued) DD least-squares ambiguity vector and where χ^2 is an appropriate chosen constant. Before the actual search is performed, the DD ambiguities and their search space are first transformed by means of a decorrelating transformation. As a result one obtains transformed ambiguities that are largely decorrelated, more precise than the original ambiguities and that have a search space which is closer to a hyper-sphere. These properties make it possible to perform the search in an efficient manner. For setting the value of χ^2 one can make also a good use of the high precision of the transformed ambiguities. The value χ^2 should namely be not too small but also not too large. A too large value would imply a search space with an abundance of unnecessary grid points, while a too small value could result in an empty search space. Once the size and shape of the search space are set, the actual search can commence. It is based on a sequential conditional least-squares adjustment which results in (sharp) bounds on the individual (transformed) ambiguities. This search will produce the integer least-squares ambiguities and if needed for validation purposes, the second best solution as well.

In our present application we were interested in the performance of *instantaneous* ambiguity resolution using the *network* mode of operation. Hence for the time span of about 50 minutes this amounts to about 3000 single-epoch solutions in which each solution was based on a search within a 34 dimensional search space. Due to the efficiency of the current integer least-squares procedures, this does not pose any difficulties from a computational point of view. The computational cycle times for such single epoch solutions are typically below the 0.2 sec level. This includes all steps after formation of the normal equations. Hence it includes: (a) solving the normal equations; (b) construction and application of the decorrelating ambiguity transformation; (c) the actual search; and (d) the computation of the 'fixed' network parameters. Timing was done on a PC486-66Mhz.

Ambiguity dilution of precision.

For the validation of the ambiguities, we would like to get an impression whether successful ambiguity resolution is to be expected. In many applications of GPS it is already common practice to make use of various *dilution of precision (DOP)* measures. These measures are based on applying the *trace* operator to relevant parts of the variance-covariance matrices. For the ambiguities however, we believe that the trace cannot be considered adequate. There are two main reasons for this: (1) the trace only operates on the diagonal entries of the variance matrix; (2) the trace lacks some important properties of invariance. The fact that the trace operates only on the diagonal entries of the variance matrix would not be too harmful if the correlations were small to moderate and if they only played a secondary role in the process of ambiguity estimation and validation. This is not the case however. The DD ambiguities are usually highly correlated and these correlations have an important impact in shaping the ambiguity search space (*Teunissen*, 1997a).

Also the lack of invariance of the trace, when applying members from the class of admissible ambiguity transformations, makes the trace unsuitable. The trace of the variance matrix remains namely only invariant when orthogonal transformation are applied. But the admissible ambiguity transformations are not orthogonal. They only share the volume preserving property with the orthogonal transformations.

Based on these considerations we make use of the determinant of the ambiguity variance matrix. The ambiguity dilution of precision (*ADOP*) is therefore defined as $ADOP = \sqrt[2n]{\det Q_{\hat{a}}}$ (cycle), where n is the order of the ambiguity variance matrix $Q_{\hat{a}}$ (in our case n=34). The *ADOP* is invariant for the class of admissible ambiguity transformations. Thus the same value is obtained, irrespective of which satellite is chosen as reference in the DD definition. The same value is also obtained when one uses the variance matrix of the transformed ambiguities instead of the variance matrix of the original DD ambiguities. Thus the *ADOP* better measures the *intrinsic* precision characteristics of the ambiguities. It can be shown to approximate the (geometric) average of the precision of the least-squares ambiguities.

The computation of the *ADOP* can be incorporated efficiently into the ambiguity resolution process. The various options available together with the class of GPS models for which closed form expressions exist, are given and discussed in (*Teunissen*, 1997a,b).

Figure 3 shows the *ADOP*s for the single epoch solutions over the complete time span of about 50 minutes. With the exception of the five 'interruptions', the value stays constant at the level of about 0.08 cycle. The five 'interruptions' are due to partial loss of lock of the receivers on the van or due to short occasional loss of the L2 signal (obstacles). The small values of the *ADOP* indicate that successful single epoch ambiguity resolution should be feasible with the present set up.

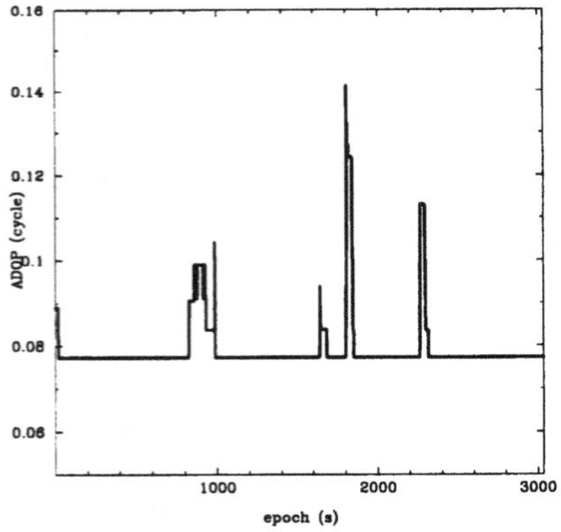

Figure 3: The single epoch based ADOPs for the complete observation time span

Validation.

Various tests were executed in order to get an indication of how well the single-epoch solutions fit the model. These tests consisted of checks on the internal and external consistency of the results. The checks on the external consistency were possible for the stationary baseline 15-28 (using given coordinates) and to a somewhat lesser extent for the periods that the van was stationary (using stationarity). For the internal consistency, we relied on the one hand on the available redundancy of the single epoch solutions and on the other hand on how these single epoch solutions compared to the overall solution based on the data of the complete time span.

The overall picture that emerged from these comparisons is as follows: (1) the float solutions, based on the model without the ionospheric delays included, performed according to expectation; (2) again without the ionospheric delays included, almost all of the computed integer ambiguities could be validated successfully; the success rate was 99.6% (of over 3000 single-epoch solutions); (3) the fixed solutions performed rather poorly when the ionospheric delays were excluded, but improved considerably when the ionospheric delays were included.

As a typical example of these results, the histograms of the single-epoch-based overall model tests statistics are shown in Figure 4. The overall model test statistic is defined as the ratio of the a posteriori variance factor and its a priori counterpart. It has a central *F*-distribution $F(r, \infty, 0)$ under the null hypothesis, where r denotes the redundancy (note: for the fixed solutions one has to make the additional assumption that the integer ambiguities are non-stochastic). Two features can be seen from these histograms. First note the 'outlying' character of the second histogram. As pointed out above, its rather discontinuous behavior is thought to be caused by the erroneous assumption that the ionospheric delays were sufficiently small. This behavior is absent in the first and third histogram. This indicates an interesting feature of integer least-squares estimation. As far as the integer estimation process is concerned, the data are permitted to be corrupted with some bias. That is, despite some biases in the float solution, the integer least-squares process will still pull the float solution to the correct integer value. But once the integer ambiguities are assumed fixed and are treated as such, the bias, still present in the data, becomes predominantly visible in the fixed solution. The fact that these relative small biases are not noticable in the float solution is due to the fact that, in the single epoch case, the phase data fail to contribute. Hence, these solutions are dominated by the relatively poor precision of the code data.

The second feature concerns the expectation of the distribution. Under the null hypothesis it should to be equal to one. The two remaining histograms however show a mean which is much smaller than one. It is our belief, which is also based on past experience, that this is caused by the still too pessimistic assumptions about the stochastic model.

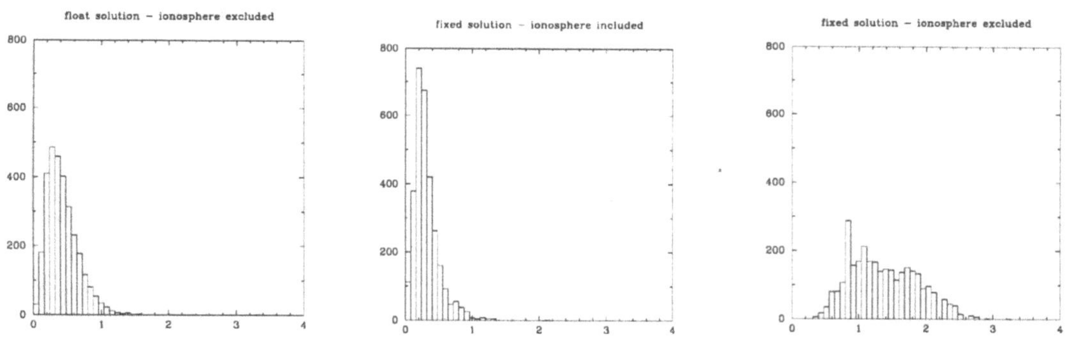

Figure 4: Histograms of the single epoch based overall model tests statistics

To conclude a typical example of the repeatability in the position domain is shown in Figure 5. It shows the histograms of the about 3000 single epoch based coordinate solutions of baseline 15-28. The top row refers to the float solution assuming the ionospheric delays to be absent, while the bottom row refers to the fixed solution assuming the ionospheric delays present. The float solution performs quite well with only a slight bias in the East component. Note the well known GPS phenomenon that the height component is determined worst, while - for mid latitude regions (φ=52°) - the East component is determined better than the North component. The same can be seen for the fixed solution. This solution shows however less homogeneity than the float solution despite the fact that the ionospheric delays were solved for. This indicates that there are still some small, unidentified biases left in the data. Nevertheless note the tremendous improvement in dispersion when compared with the float solution. The improvement, which is solely due to the integer ambiguity resolution, is about a factor of one hundred.

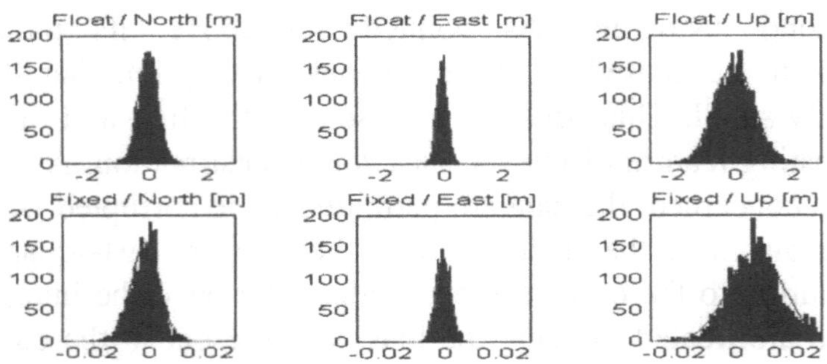

Figure 5: North, East, Up histograms of baseline 15-28: (top row) float solution without solving for the ionospheric delays; (bottom row) fixed solution with the ionospheric delays solved for.

Acknowledgements.

The contribution of the third author was done under contract with the Triangulation Department of the Dutch Cadastre, and the fourth with the Survey Department of the Rijkswaterstaat. Also the computational assistance of Peter Joosten is acknowledged.

References

Genrich, J.F., Y. Bock and R.G. Mason (1997): Crustal deformation across the Imperial Fault: results from kinematic GPS surveys and trilateration of a densely spaced, small-aperture network. *Journal of Geophysical Research* (102): 4985-5004.

Jonge de, P.J., C.C.J.M. Tiberius (1996): The LAMBDA method for integer ambiguity estimation: implementation aspects. *LGR Series*, No. 12, Delft Geodetic Computing Centre. Also available through the internet: http://www.geo.tudelft.nl/mgp/

Kleusberg, A, P.J.G. Teunissen (Eds.)(1996): *GPS for Geodesy*, Lecture Notes in Earth Sciences, Vol. 60. Springer Verlag.

Teunissen, P.J.G. (1993): Least-squares estimation of the integer GPS ambiguities. Invited Lecture, Section IV Theory and Methodology, IAG General Meeting, Beijing, China. Also in: *LGR Series* No. 6, Delft Geodetic Computing Centre.

Teunissen, P.J.G. (1997a): A Canonical Theory for Short GPS Baselines; Part IV: Precision versus Reliability. *Journal of Geodesy* (71): 513-525.

Teunissen, P.J.G. (1997b): Closed Form Expressions for the Volume of the GPS Ambiguity Search Spaces. *Artificial Satellites*, Vol. 32, No.1, pp. 5-20.

HIGH PRECISION GPS KINEMATIC POSITIONING: PROGRESS AND OUTLOOK

Chris Rizos & Shaowei Han

School of Geomatic Engineering, The University of New South Wales
Sydney NSW 2052, Australia

ABSTRACT

High precision GPS kinematic positioning in the post-processed or in the real-time mode is now increasingly used for many surveying and navigation applications on land, at sea and in the air. The distance from the mobile receiver to the nearest reference receiver may range from a few kilometres to hundreds of kilometres. As the receiver separation increases, the problems of accounting for distance-dependent biases grow and, as a consequence, reliable ambiguity resolution becomes an even greater challenge. In this paper, the challenges, progress and outlook for high precision GPS kinematic positioning for the short-range, medium-range and long-range cases, in both the post-processing and real-time modes, will be presented.

INTRODUCTION

The standard mode of precise *differential* positioning is for one reference receiver to be located at a station whose coordinates are known, while the second receiver's coordinates are determined relative to this reference receiver. In addition, carrier phase measurements must be used to assure high positioning accuracy. However, the use of carrier phase data comes at a cost in terms of overall system complexity because the measurements are *ambiguous*, requiring that *ambiguity resolution* (AR) algorithms be incorporated as an integral part of the data processing software.

Such high accuracy techniques are the result of progressive R&D innovations, which have been subsequently implemented by the GPS manufacturers in their top-of-the-line "GPS surveying" products. Over the last half decade or so several significant developments have resulted in this high accuracy performance also being available in "real-time" -- that is, in the field, immediately following the making of measurements, and after the data from the reference receiver has been transmitted to the (second) field receiver for processing. Real-time precise positioning is even possible when the GPS receiver is in motion. These systems are commonly referred to as RTK systems ("real-time-kinematic"), and make feasible the use of GPS-RTK for many *time-critical* applications such as machine control, GPS-guided earthworks/excavations, automated haul truck operations, and other autonomous robotic navigation applications.

CHALLENGES IN PRECISE GPS KINEMATIC POSITIONING

If GPS signals were tracked and loss-of-lock never occurred, the integer ambiguities resolved at the beginning of a survey could be kept for the whole GPS kinematic positioning span. However, the GPS satellite signals are occasionally shaded (e.g. due to buildings in "urban canyon" environments), or momentarily blocked (e.g. when the receiver passes under a bridge or through a tunnel), and in most cases the integer ambiguity values are "lost" and must be *redetermined*. This process can take from a few tens of seconds up to several minutes with present commercial GPS systems for short-range applications. *During this "re-initialisation" period the GPS carrier-range data cannot be obtained, and hence there is "dead" time until sufficient data has been collected to resolve the ambiguities*. If interruptions to the GPS signals occur repeatedly, ambiguity re-initialisation is, at the very least, an irritation, and at worse a significant weakness of commercial GPS-RTK positioning systems. In addition, the longer the period of tracking required to ensure reliable ambiguity resolution on the fly (OTF-AR), the greater the risk that cycle slips will occur during this crucial (re-)initialisation period. These shortcomings are also present in any system based on data post-processing as well.

The goal of all GPS manufacturers is therefore to develop the *ideal* real-time precise GPS positioning system, able to deliver positioning results, on demand, in as easy and transparent a manner as is presently the case using pseudo-range-based differential GPS (DGPS) techniques, which typically deliver positioning accuracies of 1-10 metres. For example, the DGPS technique is robust, implemented in real-time via the transmission of correction data, and there is negligible delay in obtaining results. However, there are significant challenges for the developers of a similarly reliable "plug-and-play" positioning system that is capable of sub-decimetre accuracy:

- Residual biases or errors after double-differencing can only be neglected for AR purposes when the distance between two receivers is less than 15-20km. For medium-range or long-range precise GPS kinematic positioning, the distance-dependent biases, such as orbit bias, ionospheric delay and tropospheric delay, will become significant problems.
- Determining how long the observation span should be for reliable AR is a challenge for real-time GPS kinematic positioning. The longer the observation span is required, the longer the "dead" time during which precise positioning is not possible. This can happen at the ambiguity initialisation step if the GPS survey is just starting, or at the ambiguity re-initialisation step if the GPS signals are blocked, such as cycle slips or data interruptions.
- AR techniques normally require five or more visible satellites and expensive dual-frequency GPS instrumentation in which the geometric constraints and combination of dual-frequency observations make AR easier.
- Data latency is a challenge for many *time-critical* applications. The data latency is normally caused by the data transmission and the data processing, both of which cannot be avoided. Even if the data latency is only of the order of a few tenths of seconds, it may restrict many applications. If the receivers at the reference and mobile stations were outputting raw GPS observation data at 38400 baud rate, the data link rate was 9600 baud, and both differential carrier phase and kinematic records were output at the 9600 baud rate, the kinematic position output latency could be 0.9-1.4 seconds (Lapucha et al., 1995).
- Quality control of the GPS kinematic positioning results is a critical issue and is necessary during all processes: data collection, data processing and data transmission. Quality control procedures are not only applied for carrier phase-based GPS kinematic positioning, but also for pseudo-range-based DGPS positioning. However, the quality control or validation criteria for AR, for precise GPS kinematic positioning, is a significant challenge.

PROGRESS IN PRECISE GPS KINEMATIC POSITIONING

Over the last few years several important developments have occurred that appear to have overcome some of these constraints:

(a) Under certain conditions decimetre level positioning accuracy has been possible even when the baseline lengths have been up to hundreds of kilometres in length.

(b) Reliable OTF-AR in the shortest period of time possible, following just one measurement epoch, has been demonstrated.
(c) Given very short periods of time-to-AR the notion of cycle slips, or having to re-initialise the ambiguities, has no meaning because so-called "instantaneous" OTF (IOTF) is the normal mode of kinematic positioning for *all epochs*.
(d) Improved multipath mitigation *within* the GPS receivers themselves.
(e) For certain applications single-frequency GPS instrumentation can be used.
(f) The release of several commercial integrated GPS-GLONASS receivers.

The two most significant algorithm improvements therefore have been in: (a) overcoming the baseline length constraint, and (b) shortening the "time-to-AR" to just one epoch of data. However, advances in receiver hardware have had to be made in concert.

Short-range (<15km) IOTF-AR has been reported by Han (1997) and others. Developments in fast ambiguity resolution algorithms and validation criteria procedures, together with improvements in stochastic modelling and the application of careful quality control procedures, have generally been responsible for this increased level of performance.

Carrier phase-based medium-range GPS kinematic positioning has been reported for baselines several tens of kilometres in length (Wanninger, 1995; Wübbena, et al., 1996). IOTF-AR has also been reported for medium-range GPS kinematic positioning (Han & Rizos, 1997). Such *medium-range* performance requires the use of multiple reference stations in order to mitigate the orbit bias, as well as the ionospheric and tropospheric biases. These are exciting developments that will require testing and implementation in operational positioning systems.

In the case of long-range kinematic positioning several innovative concepts have been reported. Colombo & Rizos (1996) report results of decimetre accuracy navigation over baselines up to a thousand kilometres in length. Although it is not yet possible to resolve ambiguities OTF for baselines of several hundreds of kilometres in length, ambiguity re-initialisation or *ambiguity recovery* is achievable (Han, 1995). In other words, if loss-of-lock occurs, the AR algorithm can recover the ambiguities as long as any data "gap" is less than a minute or so. Initial AR must be carried out using traditional techniques, including static initialisation. A new long-range precise positioning technique that does not require AR has also been suggested by Han & Rizos (1996b). The technique is best described as *"GPS traversing"*, in which the *relative* positions of successive GPS stations are determined to high accuracy, not the positions in relation to a distant reference receiver.

The development of integrated GPS-GLONASS receivers offers special challenges, not the least being that the signals to the different GLONASS satellites are of different frequency making the standard GPS data processing strategies based on double-differencing inappropriate. However, the extra satellites that can be tracked should make precise positioning a more robust procedure. During the last few years, several research groups has been trying to develop optimal GLONASS data processing techniques. One technique is to determine GLONASS carrier phase ambiguity "float" solution and to then try to "fix" the GPS carrier phase ambiguities (Rossbach & Hein, 1996; Landau & Vollath, 1996). Another method is determine the GLONASS carrier phase ambiguities by either correcting the receiver clock bias, or estimating the small component of the ambiguities (e.g. Raby & Daly, 1993; Leica et al., 1995).

Data latency problems can be resolved in either of the following two ways: (a) synchronise reference data and mobile receiver data (which gives the maximum precision but a substantial delay), or (b) use the latest reference data and extrapolate them to the time of the mobile receiver data (which will cause some additional error). The former is better for the carrier phase AR process, as all errors have to be minimised for maximum reliability and performance. However, the kinematic position will suffer due to a time delay of up to 1-2 seconds (which may be crucial for some real-time applications). The latter solution will introduce additional errors due to observation extrapolation. Experimental results show that the linear extrapolation model will introduce an additional double-differenced error of about 2cm for a 1 second delay and about 8cm for a 2 second delay. A quadratic extrapolation model will introduce an additional double-differenced error of about 4cm for a 2 second delay (Landau et al., 1995).

OUTLOOK FOR IMPROVEMENTS IN PRECISE KINEMATIC GPS

The dramatic improvements in precise GPS kinematic positioning are the result of improvements and innovations in several areas. Some of these are highlighted below.

Improvements in AR Techniques

Several ambiguity search procedures for OTF-AR have been suggested during the last five years, including the FARA, FASF, Cholesky, Hatch, and U-D decomposition methods (Han & Rizos, 1996d). The U-D decomposition method was found to be the fastest. However, the most optimal procedure uses the LAMBDA transformation (Teunissen, 1994; Han & Rizos, 1995b) in combination with the U-D decomposition search procedure. Although these are all search techniques in the estimated ambiguity domain, when combined with search procedures in the measurement and coordinate domain, single-epoch IOTF-AR is possible. Suggestions on how to improve the Ambiguity Function Method (AFM) of AR (generally referred to as searching in the coordinate domain) were made by Han & Rizos (1996a). Although new search algorithms will continue to be developed, the most significant improvements to AR must come from increasing the *reliability* of AR as well as the speed of the AR algorithm, or minimising the time-to-AR. This requires careful attention to quality issues such as statistical testing and other AR validation procedures.

Improvements in Quality Assurance and Validation Techniques

At the heart of reliable IOTF-AR are several innovations in quality assurance, stochastic modelling and result validation. For example, an online stochastic model determination method and an adaptive procedure are described in Han (1997). A factor to scale the over-optimised variance-covariance matrix, which will make the ambiguity search region more reliable, has been suggested by Han & Rizos (1995a). The validation procedure for AR in the ambiguity search domain was proposed in Han (1997), and the validation procedure for the AFM was described in Han & Mok (1997). The application of the quality assurance and validation procedures should ensure correct results, and at the same time the correct results should not be rejected. Further research should be carried out on the refinement of the stochastic model and an adaptive procedure which can automatically adjust with changing environment.

Bias Mitigation and the Use of Multiple Reference Stations

Medium-range kinematic positioning based on OTF-AR requires that baseline length dependent biases be mitigated. The most important of these are satellite orbit, ionospheric and tropospheric biases. Multiple reference stations surrounding the area of survey serve to generate empirical correction terms for the moving GPS receiver (Wanninger, 1995; Wübbena, et al., 1996). A *linear combination model* was proposed (Han & Rizos, 1997) for instantaneous ambiguity resolution, which can eliminate orbit bias, ionospheric delay, as well as mitigate tropospheric delay, multipath and measurement noise across the area of survey. Raquet & Lachapelle (1997) also report the use of reference receiver networks for long-range kinematic positioning. Further research needs to be undertaken in order to determine how closely spaced the network of reference stations should be in order to derive bias corrections with sufficient accuracy to resolve integer ambiguities, and also to perform carrier phase-based GPS kinematic positioning using WADGPS or WAAS differential corrections.

Instrumentation Issues

GPS equipment has undergone rapid improvement, and full wavelength L2 carrier phase and precise pseudo-range data can now be obtained from the new generation of GPS receivers, such as Ashtech Z12, Leica SR 399, Trimble 4000Ssi and NovAtel Millennium. The C/A pseudo-range accuracy can be derived at the 10cm level (Fenton et al., 1991). Multipath can also be reduced through the use of new antenna and improved receiver tracking loop design, such as Multipath Elimination Technology (MET) and the Multipath Elimination Delay Lock Loop (Townsend & Fenton, 1994). With respect to the GPS system itself, an additional civilian frequency could be transmitted by the Block IIF satellites, and Selective Availability (SA) may be turned off. The additional civilian signal will significantly improve the reliability of AR. Without SA, double-differenced carrier phase extrapolation will be possible to a higher accuracy, and the extrapolation period may be much longer than the current 1-2 seconds. GPS and GLONASS receivers such as the Ashtech GG24 will increase satellite availability, improve integrity and accuracy. Further research is needed to improve the GPS receiver technology in order to reduce multipath, to access the GLONASS L2 signals, and to develop optimal algorithms for processing GPS/GLONASS data.

On the other hand, improvement can also be made in operations. The most dramatic improvements in AR have been reported when using the latest generation of dual-frequency GPS receivers, capable of both precise pseudo-range and carrier phase measurements on both L1 and L2. These instruments permit IOTF-AR for short and medium length baselines, and ambiguity recovery in the case of long baselines. However, single-frequency instrumentation has a role to play in the "GPS traversing" technique (Han & Rizos, 1996b) and for GPS-based attitude determination (Han et al., 1997). Further research may develop different operational modes that avoid the need for AR.

Real-Time Implementation

Real-time GPS kinematic positioning products (RTK) has been developed by several GPS manufacturers, with a few decimetres accuracy during integer ambiguity initialisation. The use of IOTF-AR when employed within such RTK systems can improve performance, as is evident from Figure 1. The general issues relating to real-time GPS kinematic positioning and the requirements for AR have been discussed in Han & Rizos (1996c), and Rizos et al. (1997). Further research needs to be undertaken in the general area of quality control, and to investigate the operational requirements of potential applications such as machine control and guidance, robotics, precision farming, etc.

CONCLUDING REMARKS

The future of precise GPS kinematic positioning is dependent on a number of factors, including developments in receiver hardware, changes in official GPS policy, and also the augmentation of GPS with pseudolites or inertial navigation systems, WAAS system, and the combination of GPS with GLONASS. All these will significantly improve the reliability, integrity, and accuracy of the position results. Algorithm development will play an important role and significant performance improvements can be expected in the near future.

REFERENCES

Colombo, O.L. & C. Rizos, 1996. Testing high accuracy long range carrier phase DGPS in Australasia. *Proc. IAG Symp. 115, "GPS Trends in Precise Terrestrial, Airborne, and Spaceborne Applications"*, pub. Springer, 226-230.

Fenton, P.C., W.H. Falkenberg, T.J. Ford, K.K. Ng & A.J. Van Dierendock, 1991. NovAtel's GPS receiver - the high performance OEM sensor of the future. *Proc. ION GPS-91*,

Albuquerque, New Mexico, 11-13 Sept., 49-58.
Han, S., 1995. Ambiguity recovery for GPS long range kinematic positioning. *Proc. ION GPS-95*, Palm Springs, California, 12-15 Sept., 349-360.
Han, S., 1997. Quality control issues relating to ambiguity resolution for real-time GPS kinematic positioning. *J. of Geodesy*, 71(6), 351-361.
Han, S. & E. Mok, 1997. Validation criteria and accuracy estimation of the ambiguity function method. *Geomatics Research Australasia,* 67, 67-82.
Han, S. & C. Rizos , 1995a. Standardization of the variance-covariance matrix for GPS rapid static positioning. *Geomatics Research Australasia*, 62, 37-54.
Han, S. & C. Rizos, 1995b. A new method of constructing multi-satellite ambiguity combinations for improved ambiguity resolution. *Proc. ION GPS-95*, Palm Springs, California, 12-15 Sept., 1145-1153.
Han, S. & C. Rizos, 1996a. Improving the computational efficiency of the ambiguity function algorithm. *J. of Geodesy*, 70(6), 330-341.
Han, S. & C. Rizos, 1996b. Centimeter GPS kinematic or rapid static survey without ambiguity resolution. *Surveying and Land Information System*, Journal of the American Congress on Surveying & Mapping, 56(3), 143-148.
Han, S. & C. Rizos, 1996c. Progress and constraints of real-time carrier phase-based marine GPS positioning. *Gravity, Geoid and Marine Geodesy*, Springer-Verlag, 1997, 712-719.
Han, S. & C. Rizos, 1996d. Comparison of GPS ambiguity resolution techniques. Proc. *Int. Symp. on Global Positioning Systems, Digital Photogrammetry Systems, Remote Sensing & Geographical Systems (Geo-Informatics'96)*, Wuhan, P.R.China, 16-19 October, 136-146.
Han, S. & C. Rizos, 1997. An instantaneous ambiguity resolution techniques for medium-range GPS kinematic positioning. *Proc. ION GPS-97*, Kansas City, Missouri, 16-19 Sept., 1789-1800.
Han, S., K. Wong & C. Rizos, 1997. Instantaneous ambiguity resolution for real-time attitude determination. Proc. *Int. Symp. on Kinematic System in Geodesy, Geomatics & Navigation*, Banff, Canada, 3-6 June, 409-416.
Landau, H. & U. Vollath, 1996. Carrier phase ambiguity resolution using GPS and GLONASS signals. *Proc. ION GPS-96,* Kansas City, Missouri, 17-20 Sept., 917-923.
Lapucha, D., R. Barker & Z. Liu, 1995. High-rate precise real-time positioning using differential carrier phase. *Proc. ION GPS-95*, Palm Springs, California, 12-15 Sept., 1443-1449.
Leica, A., J. Li, J. Beser & G. Mader, 1995. Processing GLONASS carrier phase observations - theory and first experience. *Proc. ION GPS-95*, Palm Springs, California, 12-15 Sept., 1041-1047.
Raby, P. & P. Daly, 1993. Using the GLONASS system for geodetic survey. *Proc. ION GPS-93*, Salt Lake City, Utah, 22-24 Sept., 1129-1138.
Raquet, J. & G. Lachapelle, 1997. Long-distance kinematic carrier-phase ambiguity resolution using a reference receiver network. *Proc. ION GPS-97*, Kansas City, Missouri, 16-19 Sept., 1747-1756.
Rizos, C., S. Han & B. Hirsch, 1997. A high precision real-time GPS surveying system based on the implementation of a single-epoch ambiguity resolution algorithm. Proc. *38th Australian Surveyors Congress*, Newcastle, Australia, 12-18 April, 20.1-20.10.
Teunissen, P.J.G., 1994. A new method for fast carrier phase ambiguity estimation. Proc. *IEEE Position Location & Navigation Symp. PLANS'94*, Las Vegas, Nevada, 11-15 April, 562-573.
Townsend, B. & R. Fenton, 1994. A practical approach to the reduction of pseudo-range multipath errors in an L1 GPS receiver. *Proc. ION GPS-94*, Salt Lake City, Utah, 20-23 Sept., 143-148.
Wanninger, L., 1995. Improved ambiguity resolution by regional differential modelling of the ionosphere. *Proc. ION GPS-95*, Palm Springs, California, 12-15 Sept., 55-62.
Wübbena, G., A. Bagge, G. Seeber, V. Böder & P. Hankemeier, 1996. Reducing distance dependent errors for real-time precise DGPS applications by establishing reference station networks. *Proc. ION GPS-96*, Kansas City, Missouri, 17-20 Sept., 1845-1852.

KINEMATIC POSITIONING USING ADAPTIVE FILTERS AND MULTIPLE DGPS RECEIVER CONFIGURATIONS

A. M. Bruton, Department of Geomatics Engineering,
The University of Calgary.
2500 University Drive, NW,
Calgary, Alberta, Canada.

Abstract

The accuracy of trajectory determination by differential GPS (DGPS) is limited mainly by receiver noise, atmospheric errors, orbital errors and multipath. Independent estimates of a trajectory can be obtained if several reference stations are used to generate differential corrections. Using these estimates and an adaptive filtering system it is possible to estimate, characterize and remove some of the errors mentioned above.

The objective of this paper is to introduce the adaptive filter as a means of estimating and reducing the errors that influence the accuracy of a trajectory determined by DGPS. An analysis of trajectory estimation for an aircraft using a multiple DGPS receiver configuration and an adaptive filtering system is described. Data collected during an airborne survey carried out over the Canadian Rocky Mountains is used in the analysis. Four ground-based reference stations were spread out over approximately 100 km, providing independent estimates of the trajectory of the aircraft.

It is briefly reviewed how an adaptive filter can be used to estimate the correlated part of two discrete-time signals. Since the true trajectory of the aircraft is the underlying signal in each of the four estimates, the adaptive filtering system proves to be a useful tool for the determination of that trajectory and the effect of the errors that influence it. Examples from the analysis are used to demonstrate the effectiveness of the technique.

Introduction

An adaptive filtering system provides a means of estimating and reducing the errors in estimates of the trajectory of a vehicle in motion. Acting as an interference canceler, an adaptive system can isolate the correlated part of two sequences. This concept is very useful if the two sequences represent an underlying signal of interest and each is contaminated by different interference.

An airborne survey was carried out over the Rocky Mountains in September, 1996. Two east-west flight lines, each approximately 100 km in length, were used in this study. The data was collected using PC computers at four ground-based reference stations located as shown in Figure 1. Double-differenced differential processing was performed in post-mission using L1 data collected at each of the ground stations, providing four independent estimates of the trajectory of the aircraft. Taking a simplified approach, the accuracy of each trajectory is limited mainly by receiver noise, atmospheric errors and multipath.

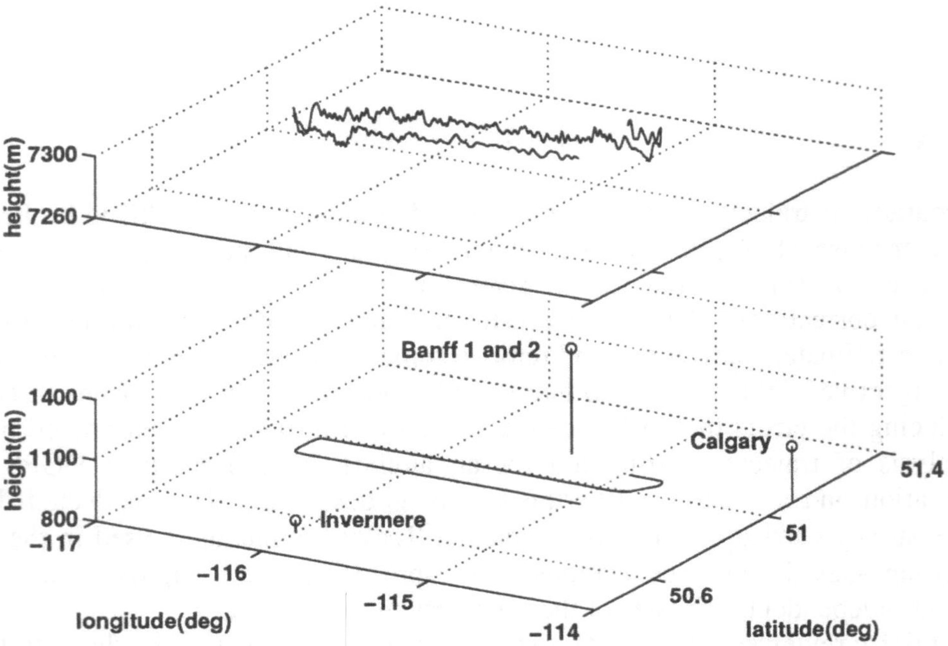

Figure 1: Flight Profile and Relative Positions of Ground Stations

An effective way to use an adaptive filter to estimate the errors present in multiple estimates of a trajectory is to form differences between the estimates and use these as input to the system. The difference between two estimates is a measure of the influence of the errors at both reference stations. If two such sequences are formed with one station in common, however, then the correlated part corresponds to the influence of the errors at that station. These can then be used to correct the estimate made using that station.

The basic adaptive system is briefly introduced below. It is shown that when used as an interference canceler, the output is the correlated part of two input sequences. This concept was implemented and examples are given showing that estimates of the trajectory of the aircraft are significantly improved. An adaptive system implemented in this way can also be used to isolate individual sources of error at a reference station. By taking advantage of the fact that multipath errors are different at each station, it is demonstrated that station multipath can be isolated by careful selection of the location of those stations.

Using the Basic Adaptive System as an Interference Canceler

Figure 2 shows how the basic adaptive system is configured, where x_k and y_k are the input and output of the system respectively, d_k is the desired sequence and $\varepsilon_k = d_k - y_k$ is the error sequence. An adaptive filter is a time-variant system that is designed to work for cases of non-stationary input signals. With noisy data as input, the transfer function $H_k(z)$ will adapt to minimize the effects of noise at the output at time t_k.

The applications of an adaptive system vary depending on the source of the desired sequence. When the input and desired sequences are derived independently, the system acts as an interference canceler. Conceptually this means that noise (or more generally, interference) present in both signals is estimated and removed.

Consider the following general form of the input and desired sequences:

$$d_k = s_k + n_{1\,k} \quad \text{and} \quad x_k = s_k' + n_{2k}, \tag{1}$$

where s and s' are correlated with each other and n_1 and n_2 are not correlated with each other or with s and s'. In this case, the output sequence y will be an estimate of s, the correlated part of x and d. The error sequence will therefore be an estimate of n_1, the part of d not correlated to any part of x. In summary:

$$y_k = \hat{s}_k \quad \text{and} \quad \varepsilon_k = \hat{n}_{1\,k}. \tag{2}$$

For further details concerning adaptive filters, the reader is referred to HAYES (1996).

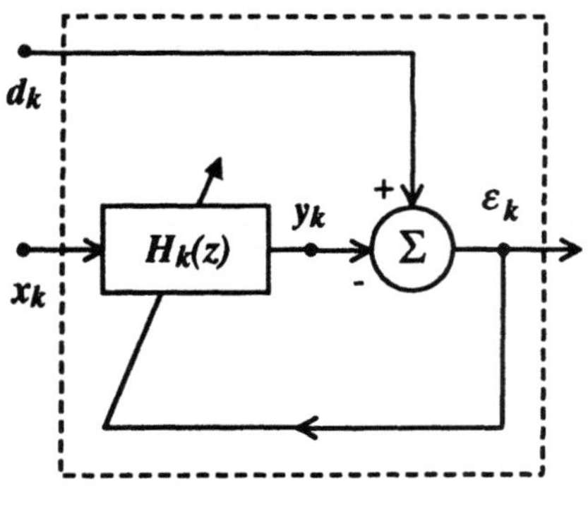

If $\quad d_k = s_k + n_{1k}$ and $x_k = s_k' + n_{2k}$,

where
$E[s_n\, s'_{n-j}] = p_k \quad \forall j, n$
$E[s_n\, n_{1\,n-j}] = 0 \quad \forall j, n$
$E[s_n\, n_{2\,n-j}] = 0 \quad \forall j, n$
$E[n_{1\,n}\, n_{2\,n-j}] = 0 \quad \forall j, n$

and p_k is generally an unknown cross-correlation function, then

$y_k = \hat{s}_k$ and $\varepsilon_k = \hat{n}_{1\,k}$.

Figure 2: Using the Basic Adaptive System as an Interference Canceler

Implementation of the Adaptive System

Differences formed between the four independent estimates of the trajectory are used as input to the adaptive filter. Table 1 summarizes the components of each estimate and the errors present in some of the differences that were formed, ignoring the effect of GPS receiver noise.

By using the differences (labeled 5-10) as input to the adaptive filter, it is possible to isolate the influence of the errors at individual stations. These can then be used to correct the estimates (labeled 1-4). Table 2 shows the input and output of the adaptive filter in several cases according to equation (2).

Table 1: Summary of the components of estimates of the trajectory and some differences. (Station abbreviations: B-Banff, I-Invermere, C-Calgary, P-plane)

		motion	unmodeled atmospheric errors between plane and:			multipath at station:				
Trajectory Estimates:			B	I	C	B1	B2	I	C	P
1.	B1	+	+			+				+
2.	B2	+	+				+			+
3.	I	+		+				+		+
4.	C	+			+				+	+
Differences:										
5.	B1-I		+	-		+		-		
6.	B1-C		+		-	+			-	
7.	B1-B2					+	-			
8.	I-C			+	-			+	-	
9.	I-B2		-	+			-	+		
10.	B2-C		+		-		+		-	

Table 2: Summary of the input and output of the adaptive filter in several cases.

	d	x	y							
11.	5.	6.	+			+				
12.	8.	9.		+				+		
13.	5.	7.				+				
14.	5.	10.	+							

The differences are typically of very low frequency. For this reason and to facilitate the adaptive process by removing unnecessary parts of the signal, the differences were band-limited to 0.055 Hz before being used as input to the filter. The normalized Least-Mean-Squares (NLMS) algorithm was implemented, as described in STEARNS and DAVID (1996).

Examples of Correcting Estimates of Trajectories

Consider the output y corresponding to row 11 in Table 2. This sequence is an approximation of the influence that the errors at station B1 have on the estimate of the trajectory and can be used to improve the estimate made from station B1. In the same way, the output y in row 12 can be used to correct the estimate made from station I.

Figure 3a shows the difference between the estimates of aircraft height made using stations B1 and I before and after correction using the adaptively estimated errors. The dotted line represents the differences before correction and the dotted line represents them after correction. It is clear that there is considerable improvement. The standard deviation of the disagreement is reduced from 9.3 cm to 3.2 cm over the whole time period. Notice that the level of improvement increases steadily in the period from 2:00 to 2:15 during which the filters were converging. Also notice the small jump in the agreement at around 2:30. This time corresponds roughly to the start of the second flight profile and might be due to large changes in the signal statistics. Figure 3b shows the frequency spectrum of the differences before and after correction. It is clear by inspection that the band where the improvement is greatest is below 0.005 Hz. This corresponds to long-term multipath and unmodeled atmospheric effects.

An Example of Separating the Sources of Error at a Reference Station

Because stations B1 and B2 in the Banff area were separated by several metres, the effects of atmospheric errors are the same at both. Because multipath errors are different at each ground-based reference station, their influence can be recognized and partially removed. Using an adaptive filter it is possible to estimate the effects of multipath at each of B1 and B2 and the effects of unmodeled atmospheric errors at the Banff stations. Consider row 13 of Table 2, which is a direct estimate of the effects of multipath at B1. The difference B1-B2 in row 7 can then be used to obtain an estimate of the multipath at B2. The influence of the unmodeled atmospheric error at the Banff stations is given by row 14 of Table 2. The solid line in Figure 4a shows the estimate of the influence of multipath at station B1. The corresponding frequency spectrum is shown in Figure 4b. The multipath at station B1 can also be estimated indirectly using the difference between the output in rows 11 and 14 of Table 2. It is clear by inspection of Table 1 that these two ways of obtaining the estimate of multipath are not independent and should yield the same result. The dotted line in Figure 4a shows this second estimate and demonstrates a high level of consistency of the method and that the scale of the solutions is correct.

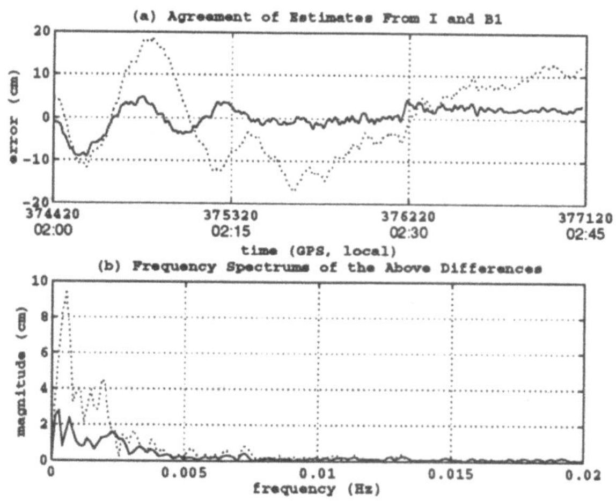

Figure 3: Agreement of Two Estimates Before and After Correction

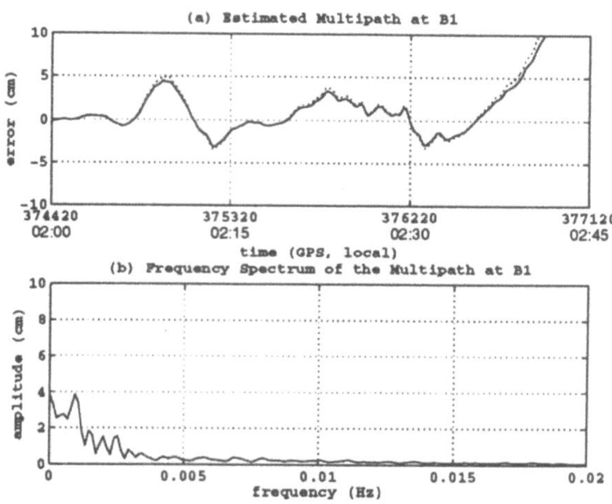

Figure 4: The Estimates of Multipath at Station B1

Notes on Accuracy

By comparing the output of the filter in a given case to the output if the input and desired sequences are reversed, a measure of the absolute accuracy of the technique is obtained. Obviously, if the algorithm is successfully estimating the correlated part of these signals, then the two outputs should agree closely. In all cases considered in this analysis, this agreement was at the 2-3 cm level (1-σ). This measure of the absolute accuracy agrees with the 3.2 cm level agreement (1-σ) between the corrected trajectories in Figure 3a.

Several simplifications about the sources of error in an estimate of the aircraft trajectory were made in this study. For example, possible errors due to the float ambiguity solution and orbital errors were ignored. It is important to understand that an adaptive filter will not correct for any error present in all estimates. With this in mind, it has been demonstrated that the absolute accuracy of the corrections made to a trajectory is at the 3 cm level, while the accuracy of the corrected trajectory is still subject to the limitations of the DGPS estimation.

Conclusions

The adaptive filter was introduced as a means of estimating the correlated part of two discrete-time signals. It was discussed how it can be used as a tool to estimate, analyze and reduce the errors in a trajectory estimated using DGPS. Examples were drawn from an analysis, showing that the errors are significantly reduced by the technique. Using multipath as an example, it was also shown that individual sources of error can be isolated. It was seen that including the convergence period, the absolute accuracy with which the influence of a single error source can be estimated is at the 3 cm level. The accuracy of corrected trajectories is therefore at this level and subject to the limitations of the DGPS estimation. It was seen that for the data set used in this analysis, the errors influencing the trajectory are of very low frequency, corresponding to atmospheric errors and long-term multipath. Future research goals include evaluating the performance of different adaptive filter algorithms and combining corrected estimates to improve accuracy. Overall accuracy might be further improved by averaging the corrected trajectories or by using them as input to a second level of adaptive filters.

Acknowledgments

The author expresses his thanks to Dr. K.P. Schwarz for his support through each stage of the work presented in this paper.

References

HAYES, M. [1996] Statistical Digital Signal Processing and Modeling. John Wiley & Sons, Inc., pp. 493-553.

STEARNS, S.D. and R.A. DAVID [1996] Signal Processing Algorithms in Matlab. Prentice-Hall Signal Processing Series, pp. 290-315.

ESTIMATING THE RESIDUAL TROPOSPHERIC DELAY FOR AIRBORNE DIFFERENTIAL GPS POSITIONING (A SUMMARY)

J. Paul Collins and Richard B. Langley
Geodetic Research Laboratory
Department of Geodesy and Geomatics Engineering
University of New Brunswick, Fredericton, N.B. E3B 5A3 Canada

ABSTRACT

When post-processing dual frequency carrier phase data, the residual tropospheric delay can easily be the largest remaining error source. This error can contribute a bias in height of several centimetres even if simultaneously recorded meteorological data are used. This is primarily due to the poor representation of the water vapour profile in the tropospheric delay models. In addition, a lack of real-time meteorological data would force the scaling of either surface values or standard atmosphere values; these are also unlikely to accurately represent the ambient atmosphere.

To obtain the highest precision in kinematic GPS some advantage may be obtained by estimating this error source along with the position solution. The simple tests reported in this paper removed biases of several centimetres in height when estimating the residual tropospheric delay from GPS data recorded at an aircraft in flight. However, important limitations exist in the geometry of the satellite coverage which must be considered before the full reliability of the technique can be quantified.

INTRODUCTION

This paper provides a brief summary of our investigations into estimating the residual tropospheric propagation delay from GPS signals. This parameter is the remaining part of the tropospheric delay not predicted by empirical models. In post-processed dual frequency carrier phase data, it can easily be the largest remaining error source. Unlike most applications of the technique, we have used data recorded at an aircraft in flight. This idea was motivated by the fact that highly accurate aircraft positions are required for gravimetric, altimetric and photogrammetric surveying purposes. Increasingly, GPS is being used to provide the decimetre-level accuracy required for some of these techniques. This level of precision can be achieved using carrier phase observables, but we will show that unmodelled tropospheric effects could potentially contribute a bias of a similar magnitude.

When processing GPS observations, a value for the tropospheric delay is predicted using empirical models which must be provided with meteorological values of the ambient temperature, pressure and relative humidity. Unfortunately, even with accurate values, these models rarely predict the true delay with a high degree of accuracy. In theory the hydrostatic component of the delay can be predicted in the zenith to the millimetre level, however the highly variable nature of atmospheric water vapour means that the accuracy of the non-hydrostatic delay is at the centimetre, or even decimetre level.

In addition, when recording GPS data at an aircraft, it is often the case that no meteorological information is recorded at the same time. When processing this data, assumed meteorological values must be used, and in addition to the poorly modelled wet component, there could also be a bias contributed by the hydrostatic component.

The results presented here are a subset of those presented in *Collins and Langley* [1997b] in which results using a wider set of models are presented.

MODELLING CONSIDERATIONS AND DATA PROCESSING

A least-squares positioning model using double-differenced, dual-frequency, GPS carrier phase observations is implemented in the KARS processing software [*Mader*, 1996]. The code has been modified at UNB to allow for the estimation of the tropospheric delay as a scale factor at either the roving receiver or at both the rover and the reference receiver.

Unfortunately, there is a problem with using differenced data to estimate the residual tropospheric delay over short baselines. For this situation, there exists a strong mathematical correlation between the partial derivatives of the tropospheric delay. For baseline lengths up to several hundred kilometres the elevation angles to a particular satellite will be similar and hence so will the partial derivatives (but with opposing signs). Even if the meteorological conditions are drastically different at the ends of such a baseline, it is difficult for a least-squares model to separate the two contributions. The usual technique to overcome this problem is to fix the tropospheric delay at the reference station and to estimate the *relative* delay at the secondary station. We have used real-time meteorological data at the reference station to help minimise the error in the estimated residual delay.

The data set used in this paper consists of dual-frequency GPS data recorded at a two second interval at a reference station and an aircraft in flight. The flying time was approximately 103 minutes up to a maximum distance of 210 kilometres from the reference station at St. John's, Newfoundland (see Figure 1). A set of fixed integers for all satellites on both frequencies was derived. This was done by processing the flight data at various elevation cut-off angles while resolving the ambiguities "on-the-fly". Comparing the ambiguities from these solutions with ambiguities computed for the short static period before the flight, has enabled stable sets of integers to be selected. While confident that these are the correct values, without actually estimating these values in flight, we can only confirm this by examining the residuals of the positions solution to see if they diverge over time.

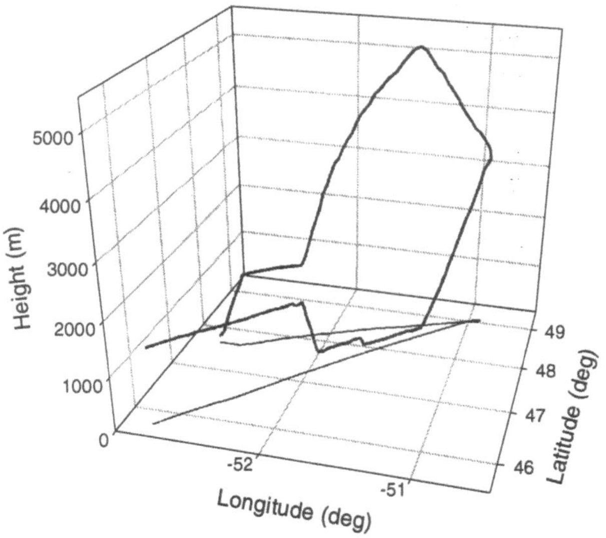

Figure 1. Flight path of aircraft for data used in this study.

Of the remaining error sources, the primary one is the satellite position error. To minimise it as much as possible, International GPS Service for Geodynamics (IGS) precise orbits were used. This leaves multipath and noise which should be of the order of centimetres or less for the carrier phase observable.

The solution was computed estimating the three-dimensional positions of the aircraft along with the residual tropospheric delay as a scale factor. No filtering was applied and no a-priori constraints were placed on any of the parameters. Each epoch provided an independent solution.

RESULTS

The results presented here used one tropospheric zenith delay and mapping function combination at the reference station and the aircraft. These were the *Saastamoinen* [1973] zenith delays using simultaneously recorded meteorological data and the mapping functions of *Niell* [1996] which only require position and day-of-year information. This model is denoted as SAANf in this paper.

Solution Residuals

Considering first of all the root-mean-square (rms) of the double-differenced carrier phase residuals after the least-squares adjustment, examination of Figure 2 shows the general improvement gained by estimating a residual tropospheric delay parameter. Overall a small improvement has been made, indicating that estimation of the residual tropospheric delay has reduced the impact of the errors in this model on the solution.

Residual Delay Estimates

Turning to the actual residual delays estimated, Figure 3 shows the residual delay estimated over the flight. The plot can be considered in two halves — before and after the 45 minute epoch. Consideration of the elevation angle trace shows that before this point in the flight there are no satellites at low elevation angles (< 10 degrees). As is well known in GPS, this limits the potential for adequately estimating the tropospheric delay. The wide variation in the first half of Figure 3, coupled with the large negative magnitude could mean that the residual estimates for this time span are unreliable.

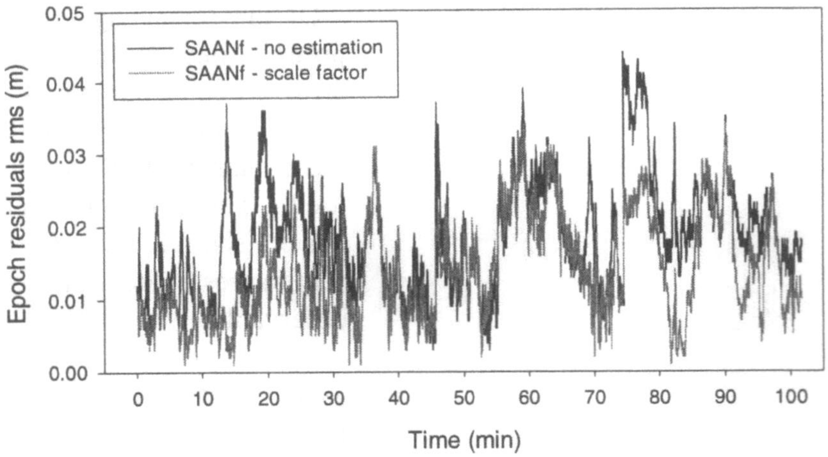

Figure 2. The rms double-difference carrier phase residuals with and without residual tropospheric delay estimation with real-time meteorological data.

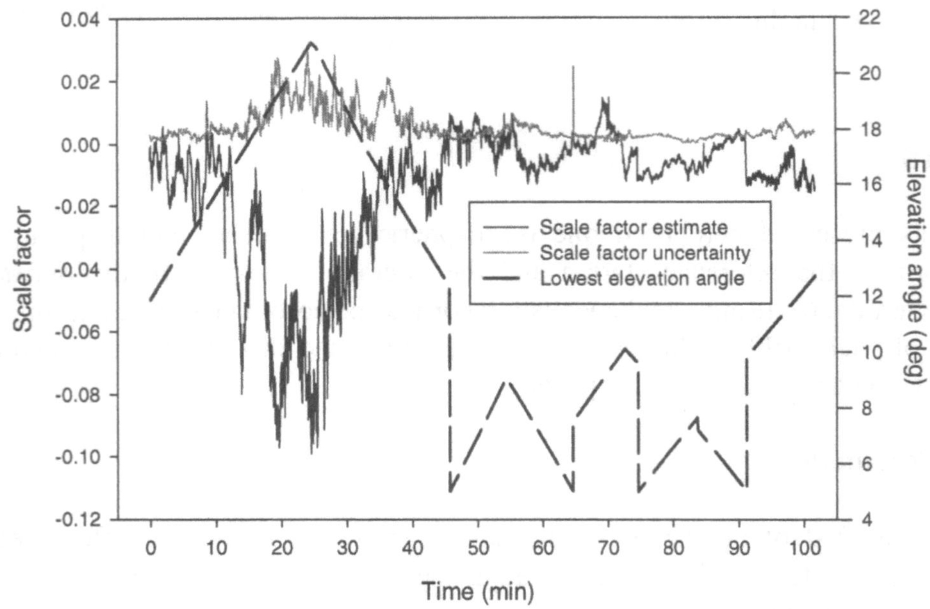

Figure 3. Comparison of residual tropospheric delay estimates and their uncertainty with the lowest elevation angle used in the solution.

Position Differences

Without residual delay estimation, we would consider the kinematic solution using SAANf to be the "best" obtainable because of its realistically-modelled zenith delays and mapping functions driven by real-time meteorological data. By estimating the residual delay we would hope to model any deviations from the average atmospheric structure implied by these models. Figure 4 shows the difference in the position components for solutions computed with and without residual tropospheric delay estimation. The difference in the height component is considerable: of the two sets of statistics for this data, even when considering only the "good" estimates after the 45 minute epoch, there is a mean bias in height of ~5 cm with an rms of ~9 cm.

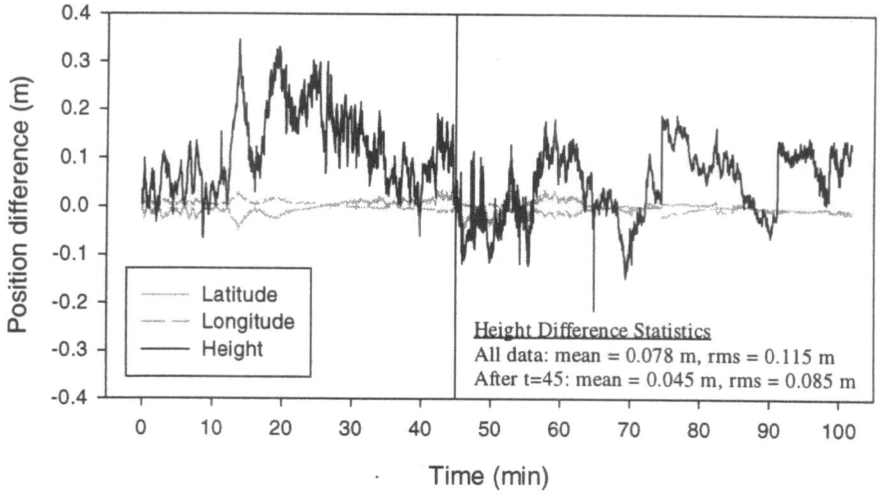

Figure 4. Difference in position solutions with and without residual delay estimation from predictions with real-time meteorological data.

It is possible under other conditions, for example using standard atmosphere meteorological values with the same zenith delay and mapping function combination, for the biases and differences shown in Figure 3 and Figure 4 to be much larger. However estimating the residual tropospheric delay appears to remove the impact of such biases. This is shown in Figure 5, where position differences between a solution computed with the "composite" tropospheric delay model UNB4 at the aircraft and the SAANf model solution are shown. (Note: UNB4 supplies meteorological data based on the 1966 U.S. Standard Atmosphere Supplements to the Saastamoinen and Niell algorithms. For more details see *Collins and Langley* [1997a]).

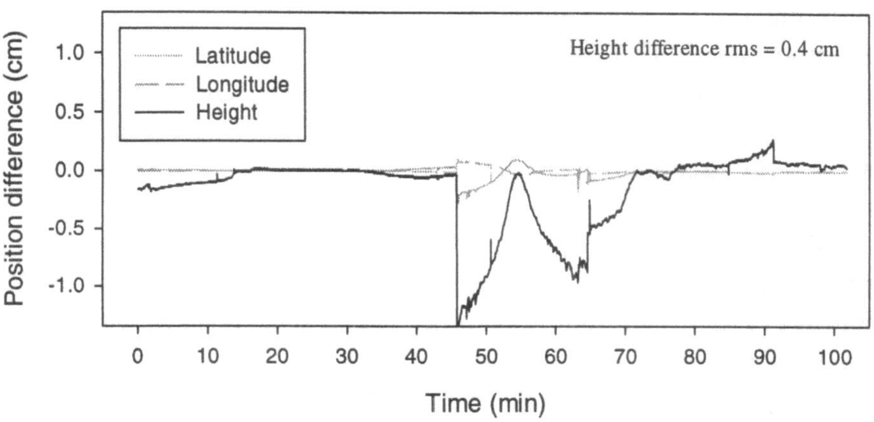

Figure 5. Position differences between the UNB4 solution and the SAANf solution. (Residual delay estimated in both solutions.)

CONCLUSIONS

We have presented in this paper a brief summary of our investigations into the effects of implementing residual tropospheric delay estimation from GPS data recorded at an aircraft in flight. The aim was to remove any unmodelled effects of the troposphere that cannot be predicted by empirical models, even when using meteorological measurements of the ambient atmosphere.

Estimating the residual delay appeared to almost wholly remove the impact of errors in the tropospheric delay model. However, the impact of the satellite geometry is important. It appears crucial that there exists data at low elevation angles (less than 10 degrees) for the tropospheric residual estimate to be meaningful. If the highest possible precision is required for aircraft positioning then estimation of a residual delay should be considered, otherwise biases of up to ten centimetres may be present in the solution.

This has been only a preliminary study and further work is required to study the condition of the normal equations of the least-squares adjustment and the reliability of the technique. New investigations could include the impact of antenna phase centre corrections, as the data is particularly sensitive to these at low elevation angles. Additionally, the implementation of a Kalman or other type of constraining least-squares filter could significantly enhance the technique by providing some a-priori constraints to the estimates.

REFERENCES

Collins, J.P. and R.B. Langley (1997a). *A Tropospheric Delay Model for the User of the Wide Area Augmentation System.* Final contract report prepared for Nav Canada, Department of Geodesy and Geomatics Engineering Technical Report No. 187, University of New Brunswick, Fredericton, N.B., Canada.

Collins, J.P. and R.B. Langley (1997b). "Estimating the residual tropospheric delay for airborne differential GPS positioning." *Proceedings of ION GPS-97*, Kansas City, Mo., 16-19 September, pp. 1197-1206.

Mader, G.L. (1996). "Kinematic and rapid static (KARS) GPS positioning: Techniques and recent experiences". *IAG Symposia No. 115*, Eds. G. Beutler, G.W. Hein, W.G. Melbourne and G. Seeber. IUGG/IAG, Boulder, Colo., 3-4 July. Springer-Verlag, Berlin, pp. 170-174.

Niell, A.E. (1996). "Global mapping functions for the atmosphere delay at radio wavelengths." *Journal of Geophysical Research*, Vol. 101, No. B2, pp. 3227-3246.

Saastamoinen, J. (1973). "Contributions to the theory of atmospheric refraction." In three parts. Bulletin Géodésique, No. 105, pp. 279-298; No. 106, pp. 383-397; No. 107, pp. 13-34.

HIGH-ACCURACY AIRBORNE INTEGRATED MAPPING SYSTEM

Dorota A. Grejner-Brzezinska

The Ohio State University Center for Mapping
1216 Kinnear Rd., Columbus, OH 43212
e-mail: dorota@cfm.ohio-state.edu

Abstract

A fully digital Airborne Integrated Mapping System (AIMS) for large-scale mapping and other precise positioning applications is currently under development at The Ohio State University Center for Mapping. AIMS is designed to be installed in an aerial platform and incorporates state-of-the-art positioning (differential GPS integrated with INS) and imaging (CCD) technologies. Preliminary test results show that platform position and orientation can be obtained with an estimated accuracy better than 7 cm and 10 arcsec, respectively, over long baselines.

Introduction

Two primary components of the AIMS system – the Positioning Module and an Imaging Module are combined and installed on an aerial platform. Tightly coupled Global Positioning System (GPS) and Inertial Navigation System (INS) comprise the Positioning Module. This module provides direct platform orientation with an accuracy of 4-7 cm in position (over long baselines) and better than 10 arcsec in orientation. Such accuracies support the extraction of geographically referenced information from the imaging component of AIMS without the need for ground control. The performance of the Positioning Module has been repeatedly tested, and has proven its high accuracy and reliability. This paper provides an overview of the Positioning Module of the integrated system, with special emphasis on a discussion of the airborne test results.

AIMS: Concept and Description

The architecture of AIMS, as presented in Figure 1, emphasizes flexibility that enables augmentation of a variety of sensors beyond the high-resolution CCD (Charge-Coupled Device) cameras, including Synthetic Aperture Radar (SAR) and radar or laser ranging devices. The GPS time is used to synchronize position information with measurements from the other sensors. Currently AIMS operates in a post-processing mode, however, the ultimate goal is to build a real- or near real-time system. It is anticipated that the system will require no ground control, except for the base station that enables differential GPS positioning. Although both technologies, GPS and inertial navigation, are well known and have been used for precise positioning for a number of years (Abdullah, 1997; Kerr, 1994; Lapucha, 1990; Schwarz , 1981, 1990, 1995), it should be pointed out that application of digital cameras in the aerial mapping field is a relatively new component in the mapping market.

The hardware components currently implemented in AIMS are two dual-frequency Trimble 4000SSI GPS receivers, a medium-accuracy and high-reliability strapdown Litton LN-100 inertial navigation system, and digital camera based on 4,096 by 4,096 CCD with a 60 mm by 60 mm imaging area (15-micron pixel size), manufactured by Lockheed Martin Fairchild Semiconductors (implementation under development). The LN-100 internal software was modified to provide raw IMU measurements, i.e., the velocity and angular rates measured in the IMU coordinate system. The main reason for this modification was the fact that the INS vertical channel needs to be externally stabilized, and any errors in external aiding result in a corruption of the inertial navigation solution. This problem is avoided by using an external strapdown solution based on the raw IMU data, and IMU calibration results estimated by the integrated filter. Updating INS with accurate GPS information at 1 Hz rate (or better) will assure platform orientation at the ~5 arcsec level. The tight integration scheme implemented in AIMS allows continuous precise positioning even if the number of tracked satellites drops below four.

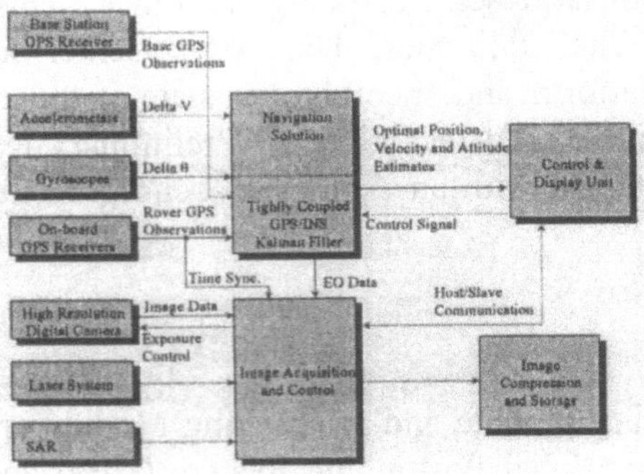

Figure 1. The structure of AIMS.

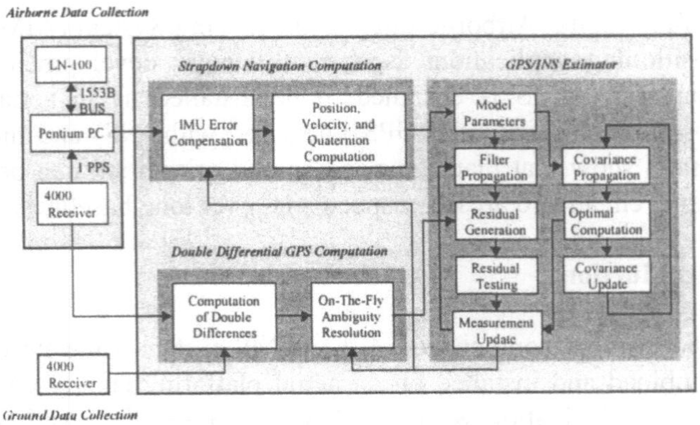

Figure 2. GPS/INS integration structure.

GPS/INS Integration and Kalman Filtering.

In AIMS a tight GPS/INS integration is implemented, as shown in Figure 2, where a single Kalman filter is used to process the GPS double-differenced phases, combined with the inertial solution, to optimally estimate the state unknowns (Da, 1997; Grejner-Brzezinska, 1997). Tight integration supports GPS cycle slip fixing and implementation of closed-loop INS error calibration. The inertial strapdown algorithm is implemented for processing the raw IMU data, providing the navigation solution. The state unknowns are errors in position, velocity, and orientation, three biases and three scale factors for the accelerometers, three gyro drifts, two deflections of the vertical and the gravity anomaly. In addition, GPS ionospheric delay is estimated for every satellite in the solution. Also, lever arm errors can be optionally included in the state vector. These are related to the unknown offsets of the GPS antenna from the IMU center. The INS error model adopted in our solution follows the approach specified by Bar-Itzhack and Berman (1988). The initial conditions for the navigation error parameters follow the recommendations of Litton Systems, Inc., for a high accuracy navigation instrument.

On-The-Fly (OTF) Ambiguity Resolution for the Integrated System.

A robust GPS/INS integration, as implemented in AIMS, provides fast and reliable correction of cycle slips and losses of lock affecting GPS measurements, and allows for fast OTF ambiguity recovery over long baselines. Initial L1 and L2 double-differenced phase ambiguities, and precise platform location, are provided by the real-time GPS software, as starting conditions for the integrated Kalman filter

(Grejner-Brzezinska, 1997). In addition, the ambiguities of the satellites whose tracking starts during the flight are resolved OTF, and cycle slips are detected and repaired based on the filter predictions. In the case of a total loss of lock, the four highest satellites are selected, and the search loop is activated. The INS error growth in AIMS is about 10 cm for the horizontal components, and about 20 cm for the vertical one, after a 50-second loss of GPS lock, as shown in Figures 3 and 4, which still enables instantaneous ambiguity recovery after the GPS signal is recovered. The OTF search loop was tested for speed and reliability of ambiguity recovery as a function of the quality of the INS prediction. Our tests showed that all cycle slips could be fixed instantly if positioning quality was not worse that 1 m per coordinate. In all of the tests performed, the system recovered the correct ambiguities.

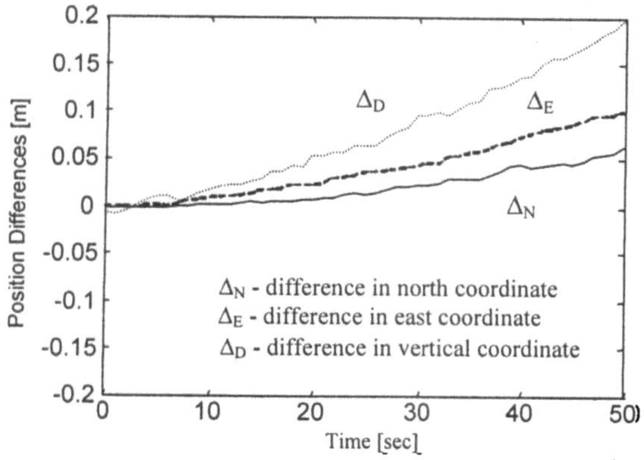

Figure 3. Differences between filter predictions and GPS positions; no GPS data in the filter.

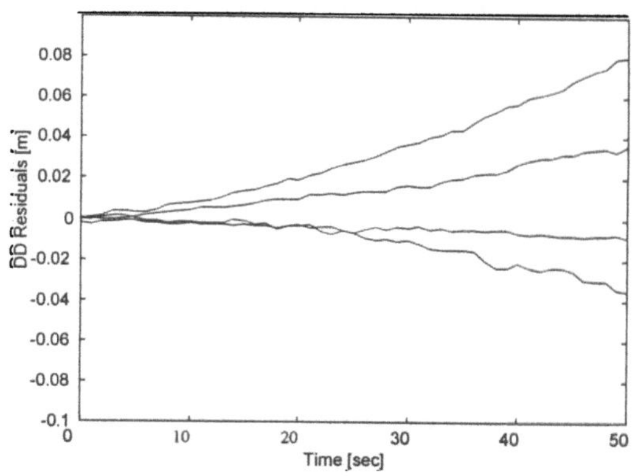

Figure 4. Differences between filter-predicted L1 DD and GPS DD; no GPS data in the filter.

Preliminary Test Results

The AIMS system performance has been tested under different conditions with varying base-rover separation. This paper presents the airborne test results of the flights performed on March 6, 1997 near St. Louis, Missouri, with the OMNI Solutions International Ltd. Beachcraft airplane, and on July 2, 1997 near Berkeley, California with the Hammon, Jensen, Wallen & Associates, Inc., two-engine Cessna 310P aircraft. In these tests the IMU data update rate was 256 Hz; however, a rate of 400 Hz was applied for data acquisition to guarantee no loss of data. Differential GPS observations were collected by a Trimble 4000 SSE for the St. Louis test and a Trimble 4000 SSI for the California test. A Zeiss RMK TOP aerial camera was used in the California test.

The level of double-differenced (DD) residuals and standard deviations (square roots of the diagonal terms from the covariance matrix) per coordinate for position, velocity and orientation angles represent the positioning quality in our tests. Since the standard deviations for the parameters behave very similarly for all the tests, the results from the 10-km flight test are plotted in Figures 5-7 as an example. Figures 8 and 9 present the double-differenced residuals for both flight tests. Figure 3 shows the differences between filter predictions without using GPS data in the filter, and GPS positions determined from an independent GPS positioning package. Figure 4 displays differences between filter-predicted double-differences and GPS double-differences, when no GPS data are applied in the filter. These results show that the average error growth rate is about 2 mm/s for the horizontal components, and about

4mm/s for the vertical, for the first 50 s of GPS total loss of lock, which indicates that most correlated IMU errors have been correctly estimated by the Kalman filter, and removed from the IMU measurements by the feedback update loop. Thus, the system is able to provide accurate velocity and position information even when GPS observations are not used for a short period of time. When GPS observations are used to update the filter at 1Hz rate, residuals range from 0 to 4 cm, after the removal of the outliers (Fig. 8 and 9). In Figure 8 only one satellite, PRN 7, has somewhat larger residuals, reaching 4 cm at the largest base-rover separation. PRN 7 also shows the largest discrepancy between the ionospheric estimates from the filter, and from the geometry-free combination (Figure 10). These indicate that both residuals and the ionospheric estimates are contaminated by an unmodeled effect, most probably differential troposphere that increases with the distance and altitude separation. Another example of the DD ionosphere estimates is presented in Figure 11.

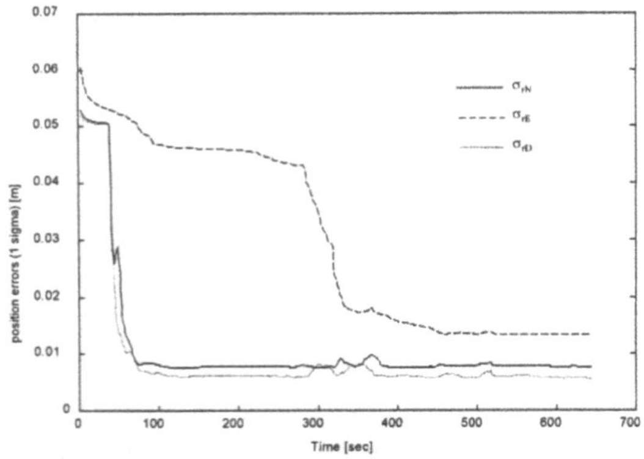

Figure 5. Position standard deviations.

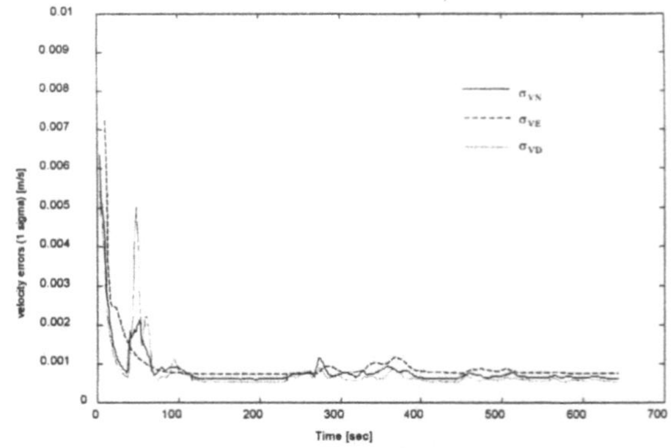

Figure 6. Velocity standard deviations.

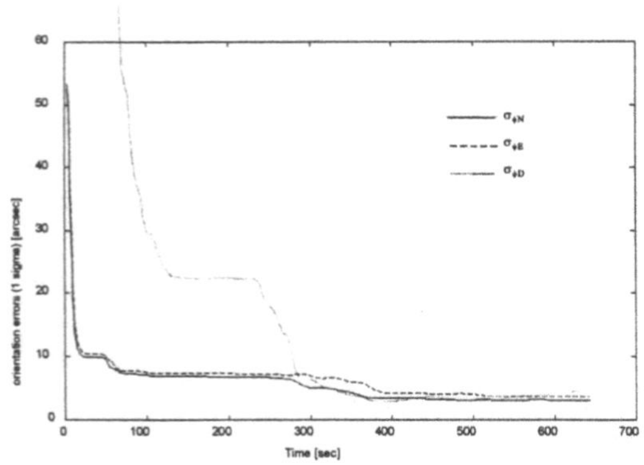

Figure 7. Orientation standard deviations.

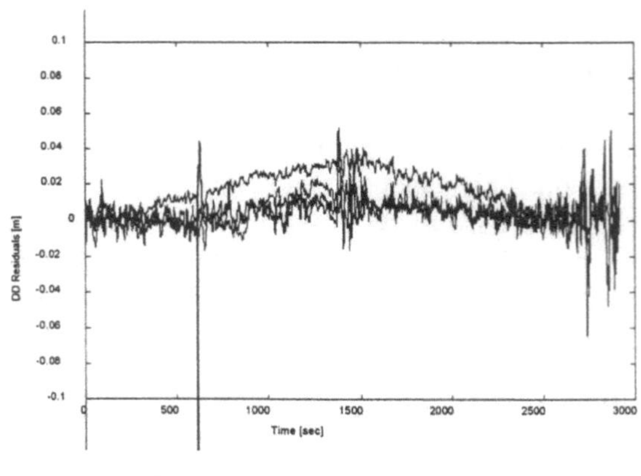

Figure 8. DD residuals, 180 km baseline, low level of GPS phase observation noise.

The comparison of the results obtained from GPS/INS, and independently from the aerotriangulation based on the California test, is currently being analyzed. Figure 12 presents the trajectory for the California flight test where the imaging data was collected. Even though we experienced some problems with the image time tags, we were able to compare horizontal coordinates of the four control points marked with an arrow in Figure 12. Preliminary results indicate that difference between GPS/INS and photogrammetric solutions ranges between 1-3 cm per coordinate component for three of the points, and reaches ~10 cm for the fourth one. More flight tests, where independent photogrammetric solution can be obtained, are needed to obtain a detailed performance assessment of our navigation system.

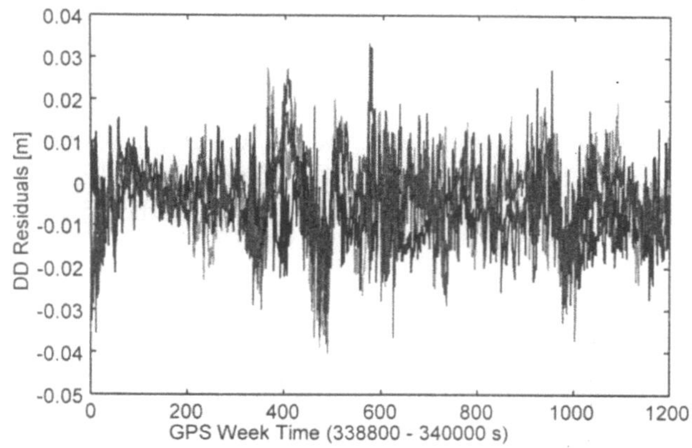

Figure 9. DD residuals, 25 km baseline, high level of GPS phase observation noise.

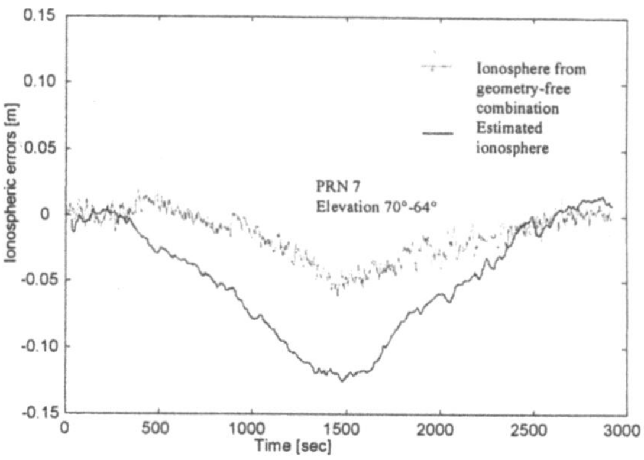

Figure 10. DD ionosphere estimates, 180 km baseline.

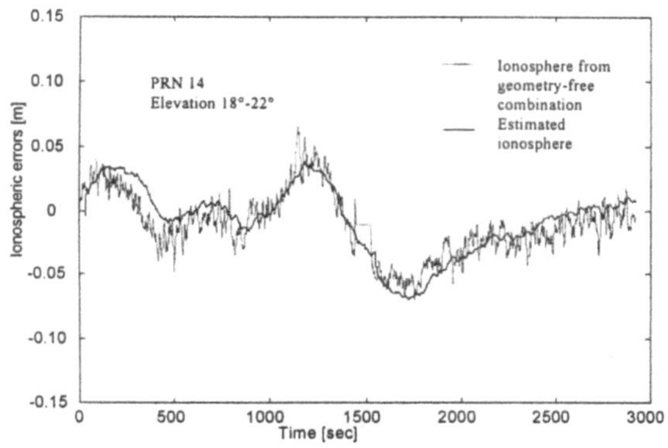

Figure 11. DD ionosphere estimates, 180 km baseline.

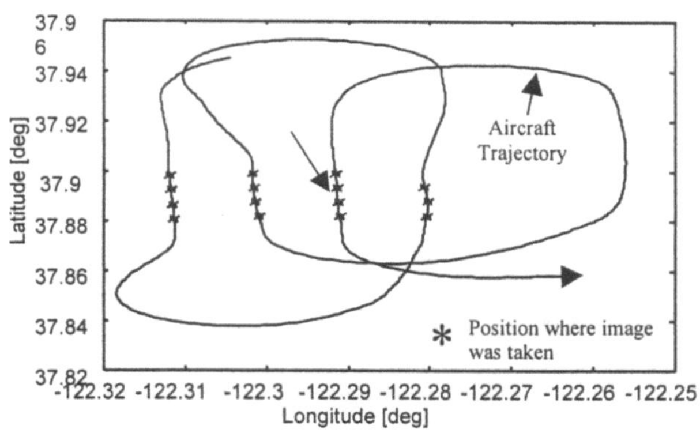

Figure 12. Aircraft trajectory, California flight test.

Conclusions

The results presented in this paper verify the high quality and reliability of the positioning component of the Airborne Integrated Mapping System that is currently under development at the Ohio State University Center for Mapping. The level of double-differenced residuals and the low rate of INS error growth when GPS data are not available demonstrate that the positioning system is well designed and calibrated, and most of the systematic effects are properly accommodated. However, there is still room for further improvements, especially in the modeling of differential atmospheric effects, and the attitude component of the system – a crucial factor that enables elimination of the ground control and traditional photogrammetric processing. Therefore, we plan to use 1'×1' (2'×2' optionally) grid values of deflections of the vertical in the strapdown navigation algorithm, and also to implement stochastic modeling of the tropospheric effect into the Kalman filter. Proper accommodation of the differential tropospheric effects might improve the estimation of the vertical positioning component of the platform.

One of the major goals of this system was to provide position and orientation of the aerial platform and the digital imagery with an accuracy that would permit the elimination of the ground control and the need for aerotriangulation. The results presented here look very promising, however, final conclusions about AIMS absolute accuracy can only be drawn following extensive testing against ground truth. Table 1 presents results of our preliminary simulations of the impact of the errors in the camera exterior orientation on the position accuracy of the ground points. Image size of 4k×4k, 10μ×10μ pixel size, focal length of 30 cm, and flight altitude of 1000 m were assumed; average errors in the initial position/orientation σ_{XoYoZo}, $\sigma_{\omega o\varphi o\kappa o}$ are given.

Table 1. Impact of the inaccurate exterior orientation on the ground positioning: simulation results.

σ_{XoYoZo}, $\sigma_{\omega o\varphi o\kappa o}$	Mean [m]	Standard Dev. [m]	Median [m]
5 cm, 5"	0.130	0.077	0.113
10 cm, 10"	0.260	0.153	0.226
20 cm, 20"	0.505	0.290	0.429
30 cm, 30"	0.768	0.481	0.651

Acknowledgments

This work is supported by the NASA Stennis Space Center, MS, Commercial Remote Sensing Program grant #NAG13-42, and a grant from Litton Systems, Inc. Trimble Navigation, OMNI International, Ltd. and Hammon, Jensen, Wallen & Associates, Inc. are acknowledged for providing GPS receivers and airplane time.

References

Abdullah, Q., (1997): Evaluation of GPS-Inertial Navigation System for Airborne Photogrammetry, Applanix, Applied Analytical Corporation.

Bar-Itzhak, I. Y., Berman, N. (1988): Control Theoretic Approach to Inertial Navigation Systems, AIAA *Journal of Guidance, Control, and Dynamics*, Vol. 11, No. 3, pp. 237-245.

Da, R., (1997): Investigation of Low-Cost and High-Accuracy GPS/IMU System, ION National Technical Meeting, Santa Monica, pp. 955-963.

Grejner-Brzezinska, D. A., (1997): Airborne Integrated Mapping System: Positioning Component, ION Annual Meeting, June 31 – July 2, Albuquerque, NM.

Kerr III, T. (1994): Use of GPS/INS in the Design of Airborne Multisensor Data Collection Mission (for Tuning NN-based ATR Algorithms), Proc. ION GPS, Salt Lake City, Utah, pp.1173-1188.

Lapucha, D. (1990): GPS/INS Trajectory Determination for Highway Surveying, *Kinematic Systems in Geodesy, Surveying and Remote Sensing,* Springer-Verlag, pp. 372-381.

Schwarz, K. P., (1981): A Comparison of Models in Inertial Surveying, pp. *Bulletin Geodesique*, Vol. 55, No. 4, pp. 300-314.

Schwarz, K. P. (1990): Kinematic Modelling – Progress and Problems, *Kinematic Systems in Geodesy, Surveying and Remote Sensing,* Springer-Verlag, pp. 3-16.

Schwarz, K. P. (1995): INS/GPS as a Georeferencing Tool for Multi-Sensor Systems, presented at the Mobile Mapping Symposium, May 24-26, Columbus, OH.

AIRBORNE GPS PERFORMANCE DURING A PHOTOGRAMMETRIC PROJECT

Paolo Baldi [*], Maria Marsella [**], Luca Vittuari [***]

[*] Dip. Fisica - Università di Bologna - E-mail: baldi@ibogfs.df.unibo.it
[**] DITS - Università. di Roma "La Sapienza"-E-mail:marsella@dits.ing.uniroma1.it
[***] DISTART-Università di Bologna - E-mail:vittuari@distart.ing.unibo.it

Abstract

Preliminary results obtained from the analysis of GPS data collected during an airborne photogrammetric project conducted in 1996 over the Island of Vulcano are described. The objective of this study is to determine an optimal processing procedure for obtaining high accuraacy projection center positions for large scale aerial photography applications. GPS measurements were collected using a multi antenna configuration both on the aircraft and on the ground. The performance of standard software for OTF differential kinematic processing was evaluated. In order to use redundant information both rigorous combination of independently computed solutions and simultaneous adjustment of GPS observation from multiple reference stations and mobile receiver were performed. GPS-derived camera projection centers are compared to the results from a aerotriangulation adjustment results. The results indicate that the GPS data can be helpful to preserve the accuracy level when there is a lack of ground control points.

Introduction

In order to acquire a detailed knowledge of the deformations related to the eruptive dynamics of the Vulcano Island, the use of a GPS-based aerial photogrammetry technique was investigated. Large scale photographic images were acquired in order to achieve sufficient accuracy of the mapping products and thus detect, by comparison with earlier products, large deformation patterns (Achilli et al., 1994). The collection of images and GPS data was carried out in September 1996 using three GPS reference ground stations operating in the project area plus an additional station located about 70 km away. On the aircraft a photogrammetric camera was interfaced with two GPS receivers connected through a splitter to the same antenna. The flight was performed under regular operational

International Association of Geodesy Symposia, Vol. 118
Brunner (ed.), Advances in Positioning and Reference Frames

conditions with no particular intention to avoid signal outages during aircraft turns. GPS kinematic solutions were obtained using both commercial and scientific software with OTF ambiguity resolution capabilities. Different data processing strategies were used in order to select the most appropriate and effective in satisfying the stringent accuracy requirements (< 20 cm). In particular, the performance of different On-The-Fly ambiguity resolution algorithms was studied. Performance of different GPS receiver models was also compared. In order to use all the redundant data, solutions obtained using the three reference stations were combined into a rigorous adjustment of kinematic vectors which connect the aircraft antenna to each reference station assumed as fiducial points with standard deviations of a few millimeters.
Software capable of multi-station adjustment of undifferenced GPS observable, was used in order to combine the correlated redundant data and perform a simultaneous estimation of ambiguities and relevant parameters, such as the coordinates of the rover station, atmospheric correction parameters, etc. GPS-derived position of the aircraft at the instant of image exposures was compared to those obtained independently by means of an aerotriangulation adjustment.

Airborne GPS Project

In September 1996 two GPS-photogrammetric flight mission were performed over the Vulcano Island. The GPS data were collected at the altitudes of 1000 and 1800 meters in order to acquire images at 1:5000 and 1:10000 scale. Distances between the aircraft and the reference stations on the island (VU00, VU04, VULC) did not exceed a few kilometers; but an additional reference station (RCFX) was operated at the airport located about 70 kilometers from the project area. The aircraft was equipped with a WILD RC20 camera connected to a dual GPS configuration consisting of a Ashtech ZXII and Trimble SSI receivers logging data respectively at 1 Hz and 0.2 Hz. Both receivers were connected through a splitter to the same antenna installed on the fuselage. The spatial offset between the GPS antenna and the camera projection center was about 1.5 meters in the vertical and a few centimeters in the horizontal components.
An accurate GPS control network was established in the project area and specially designed targets placed. The network included the three reference stations used for the differential kinematic solutions where GPS receivers were continuously operated during the flight mission, collecting observations every 1 second. The network was computed using the Bernese software and tied to ITRF through the Italian permanent station of Matera. Results of the network adjustment showed standard deviation of few millimeters both in the horizontal and vertical components. Table 1 summarizes the main characteristic of the GPS airborne missions and Figure 1 shows the horizontal aircraft trajectory on day 273.

Kinematic GPS Data Processing

GPS data collected on the aircraft and at the reference ground stations were processed by different software developed for handling kinematic GPS data, and witch included On-The-

Fly ambiguity resolution algorithms: PNAV (Precise Differential GPS NAVigation and Survey Program) (Ashtech, 1994), GEOTRACER GPS (TerraSat, 1994) and GEONAP-K (Wubbena, 1989).

Table 1 : Main characteristics of the Vulcano GPS airborne project

Airborne GPS Mission	
Period	**September 1996**
Location	**Vulcano Island, South Italy**
Aircraft	**Navajo**
Camera	**Wild RC 20**
Number of missions	**2 (Day 272/ Day 273)**
Hours flown	**2 (Day 272) / 3 (Day 273)**
Flight heights	**1000 m (1:5000) / 1800 m (1:10000)**
GPS receiver model	**Ashtech Z-XII / Trimble 4000SSI**
GPS acquisition rate	**1 sec / 5 sec**
Ground stations	**2 Ashtech Z-XII / 2 Trimble 4000SSI**
Cut-off angle	**10°**
Number of visible satellites	**6 / 7**

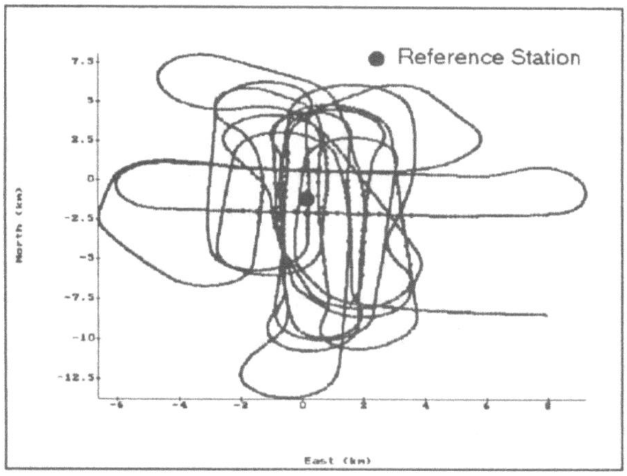

Fig. 1 Horizontal trajectory of aircraft on Day 273

PNAV, based on a Kalman filter and a smoother for combining backward and forward solution, can process all combination of GPS observables; and was used for a first analysis of the whole data set in order to access the overall quality and detect faulty observations. Results obtained from PNAV are summarized in Table 2, where the RMS of height differences are listed for each pair of reference stations. This level of repeatability can be achieved on most of the aircraft positions using standard processing procedures. During certain critical situations, due to unexpected disturbances on GPS signal acquired from one ground receiver (VU04), the standard procedure was not successful; and manual interaction to introduce waypoints, independently computed, to constrain the solution or to eliminate the affected observations was necessary.

A 1-hour dataset (from epoch 33000 to 37000), during which the photographic images were taken, was selected for further analysis with GEOTRACER GPS and GEONAP-K software. Comparing the results with PNAV, it was noted that the OTF algorithm implemented in GEOTRACER is more finely tuned resulting in a higher stability and slightly better RMSs of differences. It should be noted, though, that the estimated solutions resulted in a larger number of rejected positions. The main differences between the two software were observed during changes in the satellite configuration, as indicated in Figure 2. In order to correctly combine the redundant observations, the GEONAP-K software, capable of multi-station adjustment of undifferenced GPS observables, was used. Using the three reference stations on the island as fiducial points, and adopting a rigorous simultaneous dual frequency processing procedure (Seeber et al., 1995), high accuracy aircraft positions were estimated. Results from a single-station configuration showed differences with the multi-station solution at the 2-3 centimeter level RMS on the 3-D coordinates (Figure 3).

Table 2 RMS (cm) of the differences between trajectory solutions from two ground stations

stat.1-stat.2	RCFX-VU00	RCFX-VU04	RCFX-VULC
float/fixed solution	16.9	9.7	10.8
stat.1-stat.2	VU00-VU04	VU00-VULC	VU04-VULC
fixed solution	8.9	9.1	3.8
float solution	12.2	13.5	7.2

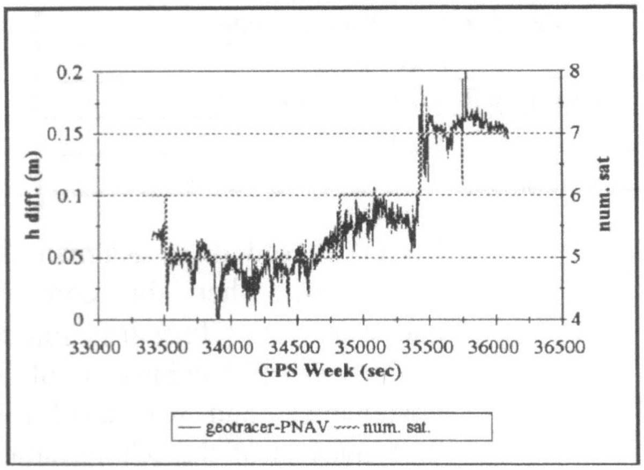

Fig. 2 Height differences between GEOTRACER and PNAV solutions

Comparison Between GPS and AT Projection Centers

As the GPS antenna and the camera system were spatially and temporarily connected, the coordinates of the camera projection center can be derived by interpolating the computed coordinates of GPS antenna position at the time of exposures and applying the measured spatial offset from the camera. GPS-derived projection centers can act as aerial control within the image processing procedure (Colomina, 1993), especially where the control network cannot be very dense due to operational constraints (Hothem, 1994).

In order to take advantage of the GPS aerial control and improve or at least maintain the quality of the relative mapping products, the GPS positions should fulfill specific accuracy requirements, that in case of 1:5000 photo scale are less than 20 cm. At the same time, when a conventional control network is established on the ground it is possible to estimate the complete set of image external parameters, that is position and orientation of the camera, by performing an aerotriangulation bundle adjustment. The accuracy of the estimated projection centers positions can be assessed from the standard deviation of the residuals on the control and photographically measured pass and tie points. For this test the AT adjustment was performed over a 5-strip photogrammetric block formed by 23 selected images, using 12 ground control points, from witch was obtained mean standard deviation values ranging between 5 and 25 centimeters (Table 3). It should be mentioned though, that the higher standard deviations for STP 2 and STP 5 can be explained by the lack of ground control points in the corresponding images.

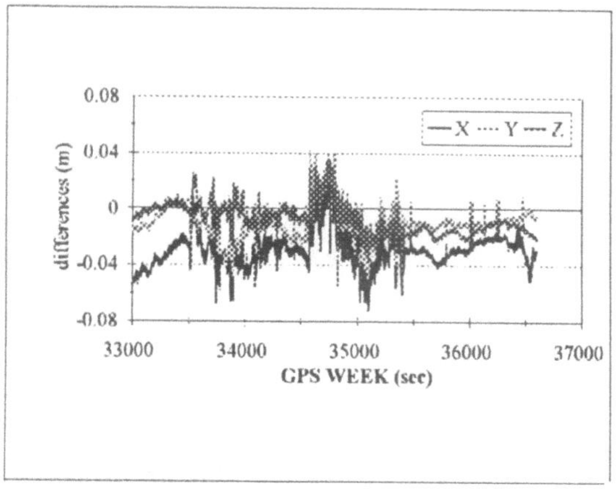

Fig. 3 Differences in XYZ components between single and multi-stations solutions (GEONAP-K)

Table 3 Mean value of standard deviations of the projection centers from AT adjustment

Standard deviations			
strip	**□E (m)**	**□N (m)**	**□h (m)**
STP2	0.207	0.186	0.067
STP3	0.104	0.149	0.054
STP4	0.104	0.119	0.051
STP5	0.248	0.225	0.151
STP8	0.111	0.128	0.046

The triangulated projection centers of the remaining strips can be used as an independent check of the GPS positions. The GPS projection centers to be compared with the AT results can be derived both by the rigorous adjustment of the three PNAV solutions (one for each reference station on the island) or by the multi-station GEONAP-K solution. The comparison between the two solutions shows the presence of small biases with different behaviour (Figure 4). The differences are about 5 centimeters for the horizontal components and 10 centimeters for the vertical component.

This indicates that the quality of airborne GPS data is high, so that we can obtain OTF position with higher accuracy both using a standard procedure and a more rigorous approach. The centimetric differences can be attributed to the different error models applied in the various software.

It should be noted that in case of a GPS-supported bundle block triangulation all the systematic discrepancies can be modeled and removed using the "shift and drift" approach.

An affine transformation was applied strip by strip between GPS and AT derived projection centers yielding residuals shown in Figure 5. The large residuals on strips 2 and 5 can be attributed to the lack of ground control points. In this case the use of high accuracy GPS control could be beneficial.

Conclusions

Different GPS kinematic processing were attempted in order to obtain high accuracy projection centers in the framework of a GPS-supported aerial photogrammetry project.

The use of multiple reference stations was advantageous both for controlling and constraining the kinematic solution. The use of accuracy GPS-derived camera centers can be helpful to preserve the accuracy level when there was lack of ground control points, as often occurs in remote and inaccessible areas.

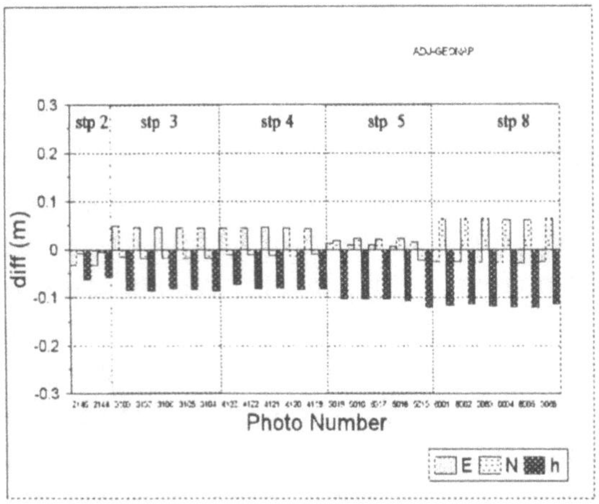

Figure 4 Comparison between adjusted PNAV and multi-station GEONAP-K solutions

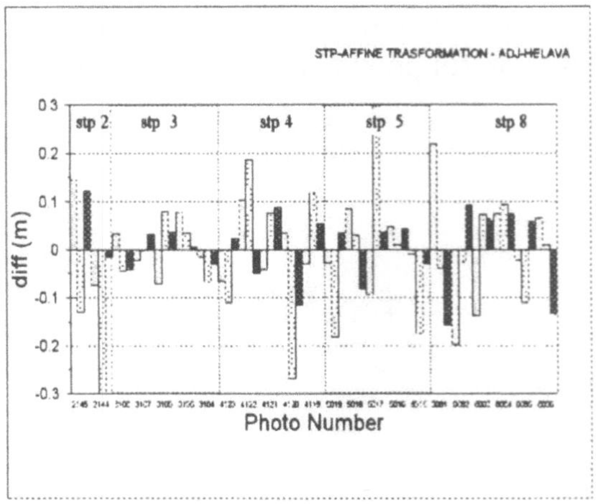

Figure 5 Comparison between GPS and AT derived projection centers.

Acknowledgment

This work was supported by CNR - Gruppo Nazionale di Vulcanologia (Italy).

References

Achilli, V., Baldi, P., Baratin, L., Mulargia, F. (1994). First results in applying aerophotogrammetry to Volcanology. Cahiers du Centre Europeen de Geodynamique et Seismologique, 8, 183-189.

Ashtech. (1994) - Precise Differential GPS Navigation and Surveying (PNAV) Software User's Guide, Pre-release version 2, Ashtech Inc., Sunnyvale, California, 316 p.

Colomina I. (1993) - A Note on the Analytic of Aerial Triangulation with GPS Aerial Control, Photogrammetric Engineering & Remote Sensing, Vol.59 , No.11, November 1993, 1619-1624.

Hothem, L., K.J. Craun, and M.A.Marsella (1994). "Operational Experience with Airborne GPS-Controlled Photography at the U.S. Geological Survey," Proc. of ASPRS and MAPPS Conference, Washington, D.C., pp. 51-63.

TerraSat (1994). "TOPAS Turbo 3.3 User's Manual," Hohenkirchen, Germany, 96p.

Seeber G., Boder W., Goldan H., Schmitz M., Wübbena G. (1995) Precise DGPS Positioning in Marine and Airborne Applications. In GPS Trends in Precise Terrestrial, Airborne and Spaceborne Applications, IAG Symposia 115, Springer, 202-21.

Wübbena G. (1989) The GPS adjustment software package GEONAP - Concepts and Models, Proceedings of 5th International Geod. Symp. On Satellite Positioning, Las Cruces, NM, 452-461.

SIGNAL DISTORTION IN HIGH PRECISION GPS SURVEYS

F.K. Brunner and H. Hartinger
Department of Engineering Surveying and Metrology
Technical University Graz
Engineering Surveying and Metrology
Steyrergasse 30
A-8010 Graz

Abstract

The aim of this paper is to show that GPS signal distortions can significantly influence the accuracy of the results of a GPS survey. Signal distortions may occur if the satellite signals propagate through a diffuse obstacle such as a bush and the distorted carrier phases arrive at the GPS antenna. As a result errors in the horizontal position of up to 2 cm and in the vertical position of up to 5 cm can occur. The results of several experiments will be shown. Different GPS choke ring antennas and receivers as well as different GPS processing software packages were used in order to prove that this effect is hardware and software independent. The signal distortion was evident in all experiments. The detailed investigation of the double difference phase data allowed the removal of the affected data which subsequently yielded the correct GPS results.

Kinematic GPS result of the railway track survey result in mm against arc length [m] of the track :

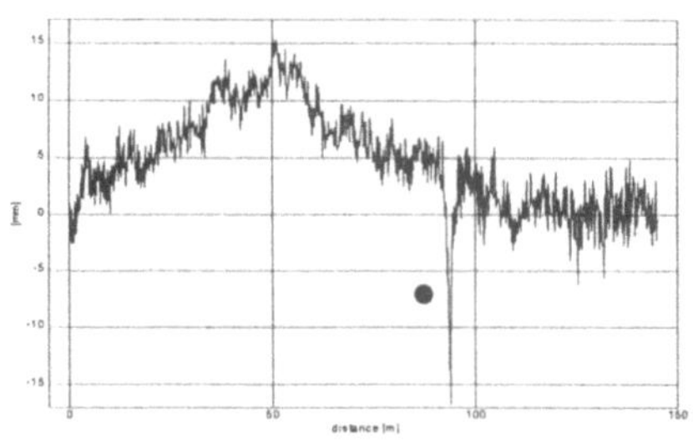

GPS result of railway track survey **with** signal distortion ●

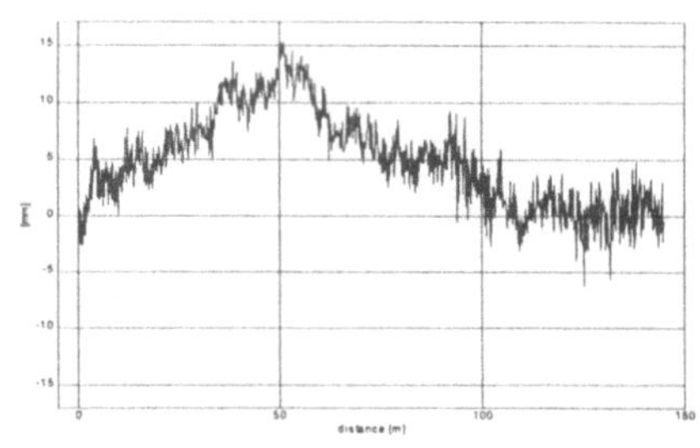

GPS result of railway track survey **after** signal distortion has been removed

Submitted to Survey Review

TESTING GPS APPROACHES FOR CIVIL AVIATION IN THE SWISS ALPS

A. Geiger, M. Scaramuzza, M. Cocard, H.-G. Kahle (Geodesy and Geodynamics Lab, Swiss Federal Institute of Technology, geiger@geod.ethz.ch)
H. Lang, R. Aebersold (Swiss Federal Office of Civil Aviation)
B. Meier, A. Dosé (Crossair, Switzerland)

Abstract

To analyse and test the feasibility of GPS-aided approaches for civil aviation in the Swiss Alps, an internationally co-operating group, leaded by the Swiss Federal Office of Civil Aviation (FOCA) has been set up. During several test flights and dedicated ground missions various problems, such as satellite visibility, multipathing, GPS signal interference and the influence of the topography on the navigation and the GPS system were addressed. The tests showed the potential of satellite based systems for approaches, in particular for use in rugged terrain.

The Project

The Swiss Federal Office of Civil Aviation (FOCA) has initiated a test program to gain insight into the use of GPS for approach and landing in rugged terrain, to gather data for reliable statistical analysis, and to draw conclusions on certification aspects. To reach this goal a group consisting of FOCA, Crossair, Swisscontrol, and the Institute of Geodesy and Photogrammetry of the Swiss Federal Institute of Technology (ETH) has been set up. The project is structured as a joint investigation with the co-operation of additional parties, such as the Technical University of Braunschweig, Telematica (Germany) and EUROCONTROL. The airport of Lugano-Agno in the southern part of the Swiss Alps was chosen as the test area because of its rugged terrain surrounding, thereby representing a demanding flight technical environment. The flights are primarily analyzed in view of the *Required Navigation Performance* (RNP) parameters, the accuracy, integrity, availability and continuity (Geiger et al., 1996).

Test Setup

In order to analyze the flights for the *Required Navigation Performance* (RNP) parameters, the accuracy, integrity, availability, and continuity a special data acquisition unit has been designed by the Swiss Aircraft Factory. This Airborne Data Acquisition and Recording System (ADARS, Fig. 1), which collects more than 100 different flight parameters (about 80 different ARINC labels, including inertial data, FMS, avionics-GPS data) and GPS data from an additional GPS-measurement unit has been installed on a SAAB 2000 aircraft (Fig. 2). At the same time data is acquired at the ground reference

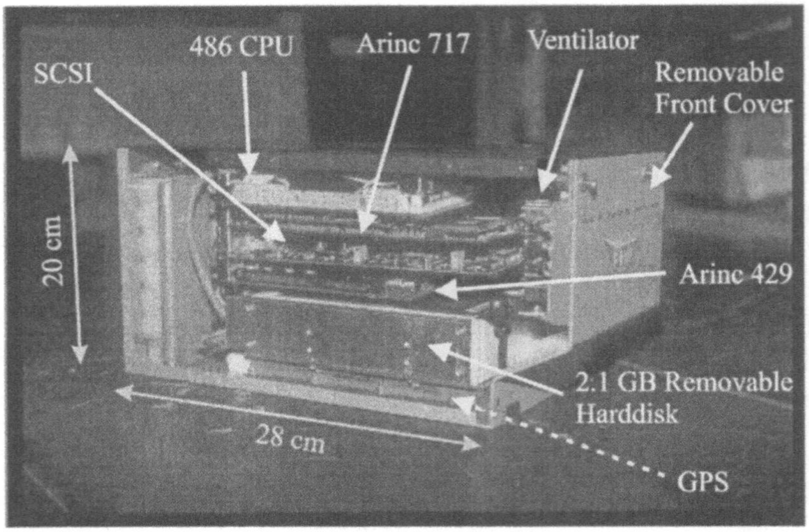

Fig.1: The data acquisition unit has been especially designed by the Swiss Aircraft Factory for the use in civil air carriers.

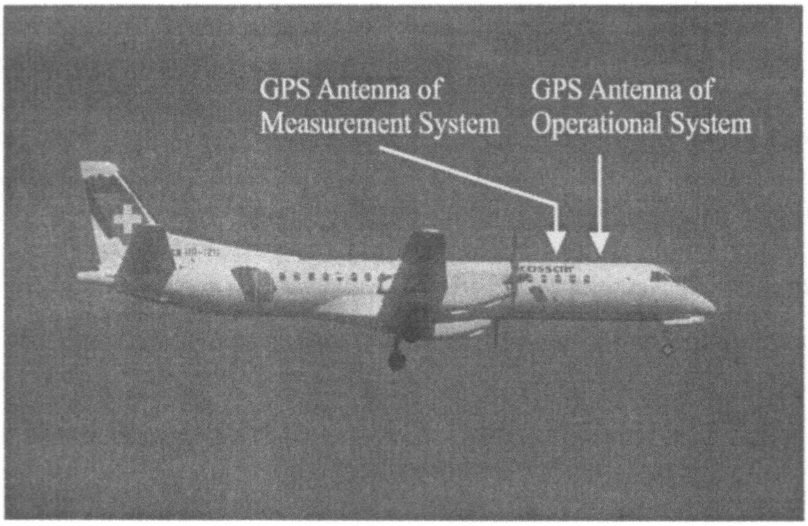

Fig. 2: Crossair's SAAB 2000 test carrier on the approach to the airport.

station. The concept of collecting range and phase data from an additional GPS equipment allows the comparison of different positioning methods. The additional GPS data is treated in a post-processing mode reaching an accuracy below the 10 cm level, and is, therefore, considered to provide the true path of the aircraft. By comparing the *true* path with the *desired course* the *Total System Error* (TSE) can be estimated. Since also the operational navigation solution is acquired the *Navigation System Error* (NSE) and the *Flight Technical Error* (FTE) can be determined (Fig. 3). These errors are put in relation to tunnel width (area around the desired course in which the aircraft path has to stay) , which defines the accuracy requested for specific segment of the approach. To certify the navigation system the NSE will be of most interest while TSE is of interest for the qualification of the total approach.

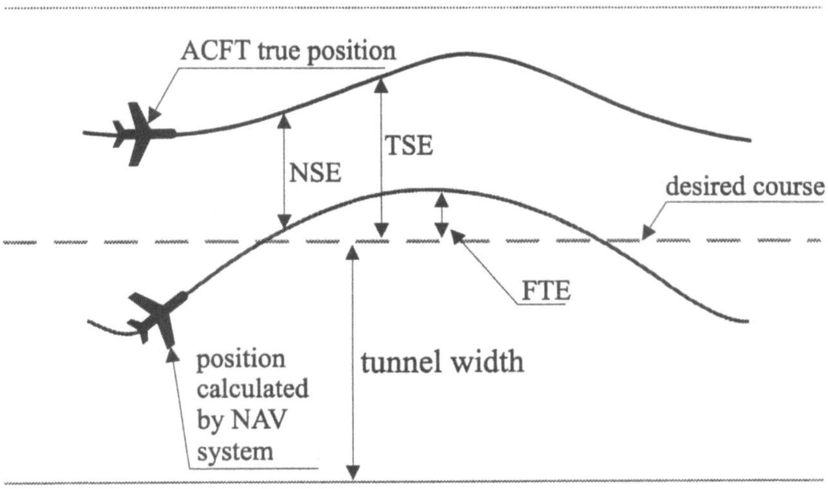

Fig.3: Definition of the nomenclature of errors (NSE = Navigation System Error; FTE = Flight Technical Error; TSE = Total System Error).

Special investigations

During the tests several problems have been encountered. Operational constraints and flight technical aspects have to be considered as well as electrical and software engineering solutions have to be found. Three of the geodetically interesting special investigations are described in the following.

Terrain effects

Terrain can cause problems to avionics even if the planned track follows a save altitude and is feasible for (D)GPS guidance. The ground proximity warning is controlled not only by the height above terrain (clearance) but also by the clearance change rate. Figure 4 shows a track with enough clearance but too high rates. This approach will therefore not be feasible for operational use. Detailed pre-calculations can be done by using high resolution DTM (digital terrain model), such as the DTM of the Swiss Topographic office (25m).

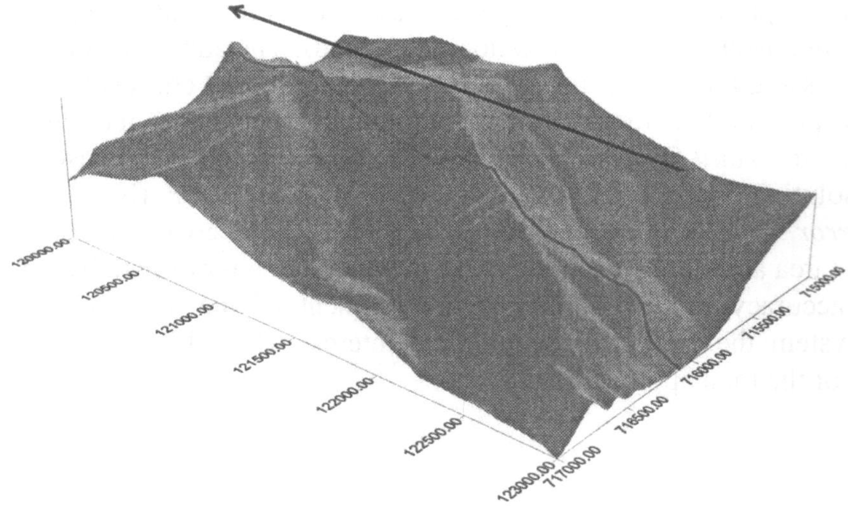

Fig. 4: Approach path over rough terrain causing problems to avionics because of steep slopes.

Multipathing

Multipath causes erroneous pseudodistance measurements. The error, if not technically mitigated, may reach several meters (code measurement). The advantage of using multifrequency systems could also be studied (Cocard and Geiger, 1995). Since DGPS will not help in reducing the effect, multipathing may cause major errors in position.

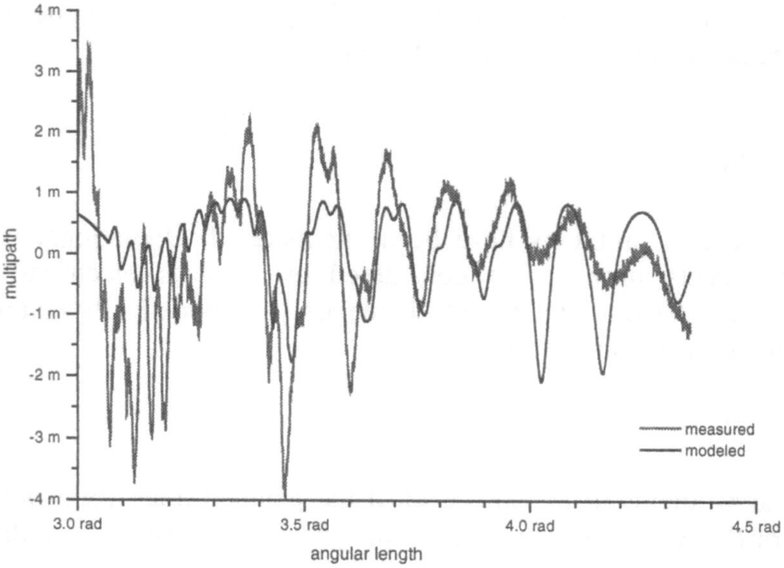

Fig. 5: Residuals of GPS C/A-Code measurements (SV 21) affected by multipath. The smooth curve represents the result of the modeling with multiple reflectors. (X-axis: topocentric along track angle of satellite)

First attempts to model the impact of multipath on pseudodistance measurements have been made. Multipath residuals are calculated by combining carrier and code phases. By modeling of the correlation process and by assuming a given reflector configuration it is possible to approximately reconstruct the multipath residuals.

Interference

GPS is vulnerable to disturbing and interfering radio frequencies, which can considerably decrease the signal to noise ratio of the GPS receiver. In extreme cases, to the GPS satellitesignals may completely be lost. Regions of interference (low signal to noise ratio) have been detected by measurement flights carrying spectrum analyzer and GPS-receivers (Scaramuzza and Geiger, 1995). In figure 6 the measured signal to noise strength is depicted. The value is represented in dependence of the elevation and the azimuth with respect to the location of the interfering source. By these measurements two distinct jamming radiation beams have clearly been detected (regions of low s/n, azimuth 25 degrees and 100 degrees, respectively).

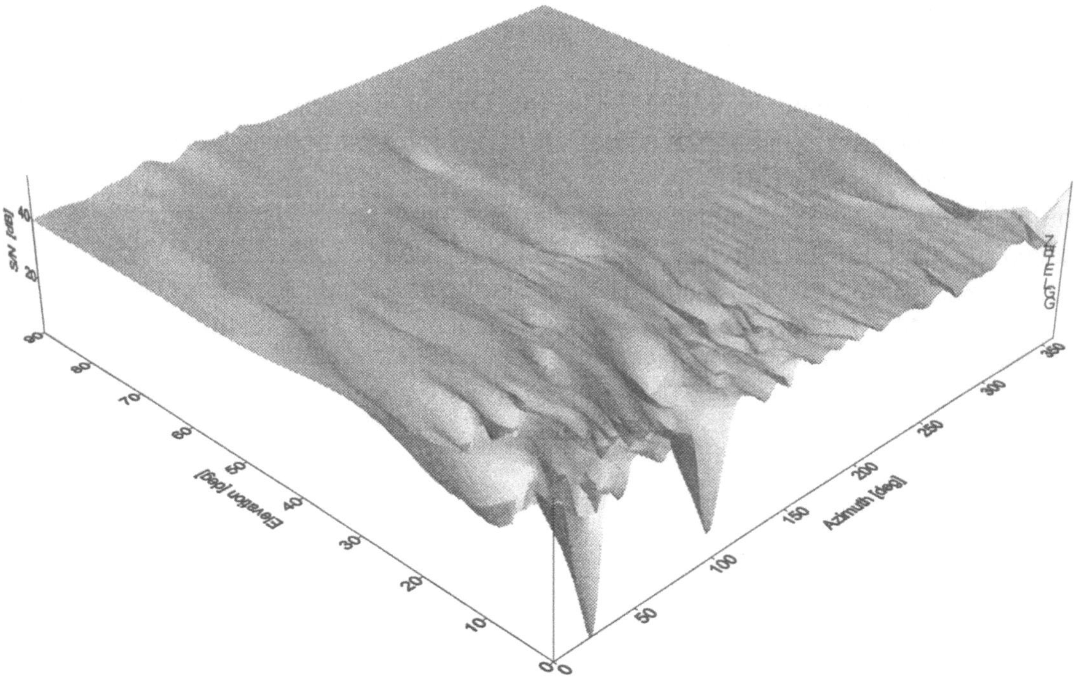

Fig. 6: GPS signal to noise ratio around an interfering source. The data has been acquired by Swisscontrol measurement flights in southern Switzerland (Schulte, 1995).

Conclusions

The described project shows the possibilities to implement geodetic methods into non-classical fields of geodesy. Precise kinematic GPS positioning, parameter estimation and modeling techniques are applied to flight technical problems of operational relevance. The first tests showed in principle the feasibility of the methods, however, different problematic areas have been detected, such as operational tasks, interference or topographic effects. These problems have to be solved before the definitive feasibility can be shown. The tests are now on redesign status and shall be continued for further statistical investigations in the near future.

References

Cocard, M., A. Geiger (1995): The importance of multifrequency systems for future GNSS. ISPA95, International symposium on precision approach and automatic landing. Deutsche Gesellschaft für Ortung und Navigation. p. 475-481.

Geiger, A., Scaramuzza, M. and Cocard, M., 1996. GPS guides civil aviation in Swiss Alps. Royal Institute of Navigation, Navigation News November/December 1996: 13 - 14, London.

Scaramuzza M., Geiger A. (1995): Use of Digital Terrain Models for Detection of Potentially Interfering Zones, Proceedings to 'GPS Interference - is it a Problem', London, 12.-13.10.1995, p. 145 - 150.

Schulte M. (1995): GPS Interference Search around the Swiss Lugano/Agno Airport, Proceedings to 'GPS Interference - is it a Problem', London, 12.-13.10.1995, p. 151-165.

Scaramuzza, M., Geiger, A., Lang, H., Aebersold, R., Dosé, A., Meier, B., Kummer, H. and Huwiler, M., 1997: Lugano-Trials: Experiences and First Results. EUGIN, DGON, GNSS 97, Proceedings.

www.geod.ethz.ch/GGL

STATIC AND KINEMATIC POSITIONING WITH GPS FOR THE CONSTRUCTION AND MAINTENANCE OF HIGH SPEED RAILWAY LINES

Heribert Kahmen
Dept. of Engineering Geodesy
Technische Universität Wien
Gußhausstraße 27 - 29
1040 Vienna, Austria
Fax: +43 1 504 27 21 E-mail: hkahmen@pop.tuwien.ac.at

Abstract

In Europe the densification and improvement of the high speed railway network has been made a priority, as it is generally accepted that railroads are most efficient for transporting people. In the near future high speed trains will connect all major towns with a speed of about 200 km/h. The Dept. of Engineering Geodesy, Technical University in Vienna, was involved in the development of fundamental concepts concerning geodetic networks and rail track alignment methodes.

The investigations have shown that the fundamental technical geodetic networks should be established in four steps. For a most economic organization of the measurement sessions simulation calculations have to be performed in advance. Approximate coordinates taken from a map and IGS orbit data were used.

For the railway track alignment a carriage was constructed carrying the rover of a DGPS system and other electronic sensors. The alignment calculations can be performed on-line, using algorithms based on Wiener filtering procedures.

Design of the networks

The fundamental concept of the network was applied to the design of a special 45 km long section of the new high speed railway line Cologne - Frankfurt in Germany (Kahmen, et al, 1998).

A maximum speed of 300 km/h shall be possible. That means that the networks have to satisfy the highest requirements concerning accuracy and reliability. In summary, the following were the major points of the network design:

- A fundamental network should provide points in the proximity of the railway line, whereby the spacing between them should be about 1 km.

- The relative accuracy should not exceed 1 cm.
- The fundamental network should be densified by traverses implemented on both sides of the railway line, whereby the spacing of the points should be about 200 m and the relative accuracy 1 cm.
- The network of tunnels and bridges should be integrated into the fundamental network.

It was decided to implement the network in four steps. The fundamental network should be established by GPS surveys in two steps, the densification with traverse nets by terrestrial surveys in a third step and finally by integration of the nets for tunnels and bridges in a fourth step. The main reasons for the hierarchical concept were:

- The connection with the national datum should be performed using a homogeneous network.
- The transformation parameters between the GPS network and the national network should be derived from a network of the highest accuracy.
- The points along the railway line should be of a high relative accuracy.

The fundamental network (step 1) consists of 12 points (Fig. 1).

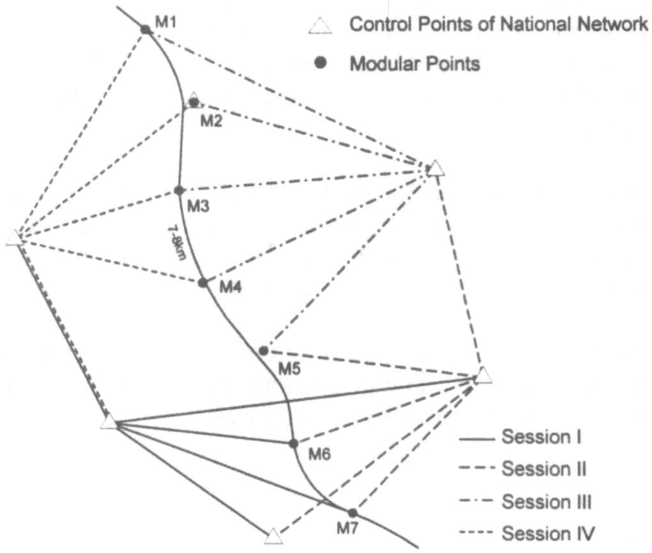

Fig. 1: Fundamental network (1st order)

Six points belong to the national network. The other points were established - referred to here as modular points - along the railway line with an average spacing of 7.5 km. These points could then be considered control points for the 2nd order network.

For the GPS surveys 6 receivers were used. The planning of the surveys was based on simulation calculations (see section 3). Four sessions were required each lasting 5 hours. For the least squares adjustment of the network (step 1) the vectors between the modular points, being measured during step 2, were used in addition. Thus they could be considered linearly independent.

The linear networks of step 2 were tied to the modular points of step 1. Every of those networks consists of 8 additional points with a spacing of 1 km (Fig. 2). For economical reasons the planning of the surveys was also based on simulation calculations (see section 3).

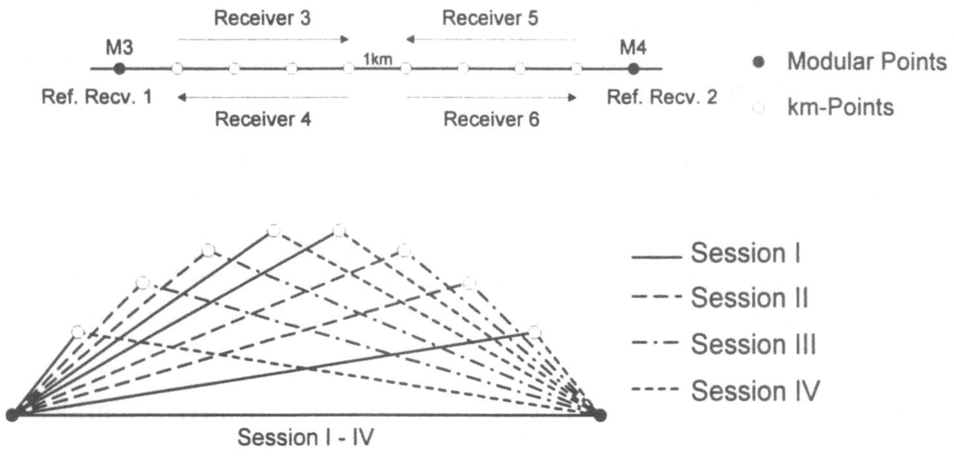

Fig. 2: Network (2nd order)

Six receivers were used, two of them occupied modular points (temporary reference stations) and the rest of them visited the unknown points. Four sessions were required each lasting 30 minutes.

The design of the densification network (Fig. 3) was also based on simulation calculations. Two traverses connect the 1-km-points of the fundamental network. Simulation calculations showed that it was necessary to add diagonal connections to increase the reliability of the network.

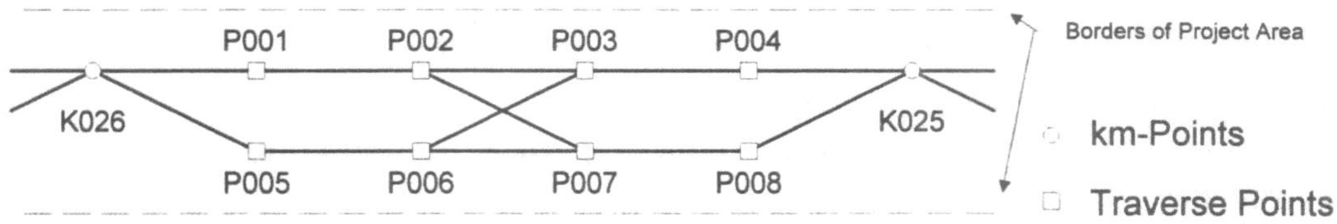

Fig. 3: Densification network (3rd order)

Simulation calculations

The accuracy and reliability analysis of the network, which had to be performed in advance, to organize the sessions in a most economical way, was based on a simulated least squares adjustment. For these calculations the geometry of the network and the stochastic parameters were needed. One part of the geometry was given by coordinates of fixed points of the national network. The rest of the information was taken from topographical maps showing the new points. While the a-priori standard deviations of the measurements were known from the operating characteristics of the GPS receivers the covariances had to be derived from simulated observations.

At first the coordinates of the network (step 1) available from the national network and from a 1 : 200,000 map were transformed into the "ITRF-System". Then, for the simulation of the observations, the GPS-Software Bernese 4.0 was used. Ephemeries and earth rotation parameters were provided from the CODE-Centre Bern. The reference time was chosen in such a way that the configuration of the satellites was comparable with the configuration during the measurements.

The simulations showed, that after an observation time of 5 hours, baselines up to 23 km in length long would have an accuracy of ± 3.7 mm (1 σ). Within an observation time of only 3 hours the accuracy decreased towards 6 mm. This accuracy was considered too small, as for the calculations based on real measurements, unresolved ambiguities and cycle slips had to be taken into account. Observation times longer than 5 hours did not really improve the results. Consequently an observation time of 5 hours was prescribed for the sessions.

The relative accuracy of the modular points is described by relative coinfidence ellipsoids; their horizontal components did not exceeding 3 mm. Consequently we could see hat enough clearance was available for unforeseen difficulties during the measurement procedures.

The simulated least squares adjustment of the network (step 2) was performed in a similar way. In that case, however, the observation time of one session was estimated a-priori with 30 minutes, based on past GPS experience in the Department.

The results obtained after the practical measurements confirmed, that GPS surveys can be planned with simulated least squares adjustments.

Integration of the fundamental network with nets for tunnels and bridges

For high speed railway lines a large number of tunnels and bridges have to be built in mountainous areas.

A tunnel net normally consists of at least 10 control points (Fig. 4a):

- one control point at each portal end section
- three control points about 1 km away from the portal points, to define the orientation of the tunnel traverse
- two control points on both sides of the tunnel.

The only condition for the design of the network was that the standard deviation of the cut through error should not exceed 1 cm.

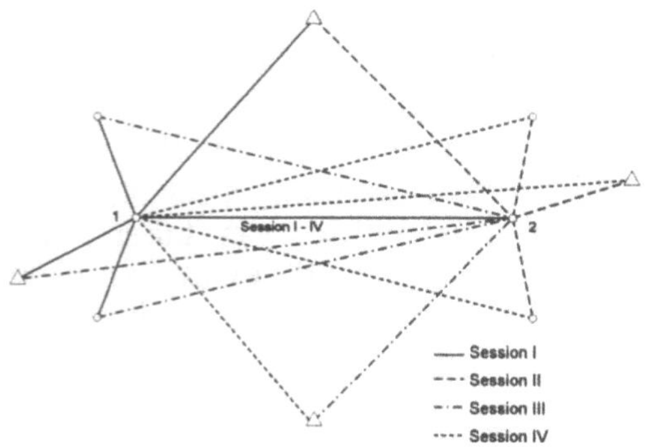

Fig. 4a: Design of a tunnel net and the planning of the measurement

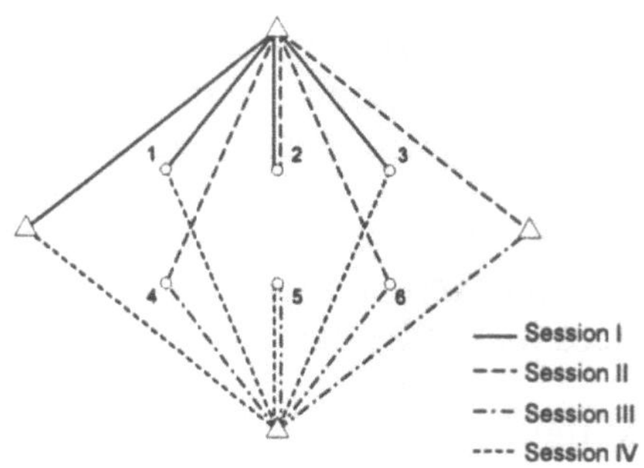

Fig. 4b: Design of a bridge net and the planning of the measurement

The control points at each portal end section and on both sides of the tunnel are needed to transform the local tunnel net into the national network. For the points at the portal, modular points or 1-km-points can be chosen. The two points on both sides of the tunnel have to be measured together with the GPS surveys of the network (step 2).

For the construction sites of bridges networks of high accuracy are also needed. The relative accuracy of the network (Fig. 4b) should not exceed 0.5 mm. The number of points depends on the length of the bridges and the topography. The network normally consists of squares with diagonals. Four control points are needed to transform the local network into the national network. Two of them can be modular points or 1-km-points, the other two are positioned on the left and right hand side of the bridge. A six parameter transformation is chosen, the scale factor is normally kept fixed.

The GPS-surveys of the tunnel and bridge-network follow in a fourth step. Examples of the observation plans are depicted in Fig. 4a and b.

Mathematical models for the maintenance of high speed railway tracks

For track-surveying special railway vehicles will be used in the near future, carrying a multi sensor system for data acquisition. A research project of the Department of Engineering Surveys of the University of Technology in Vienna showed, that GPS surveys can be used for the positioning of the measurement vehicle if Wiener filters are applied for the data evaluation procedures.

The measured data and the railway track can be described by signals, whereby three signal components have to be distinguished (Fig. 5). Mathematically the signals can be

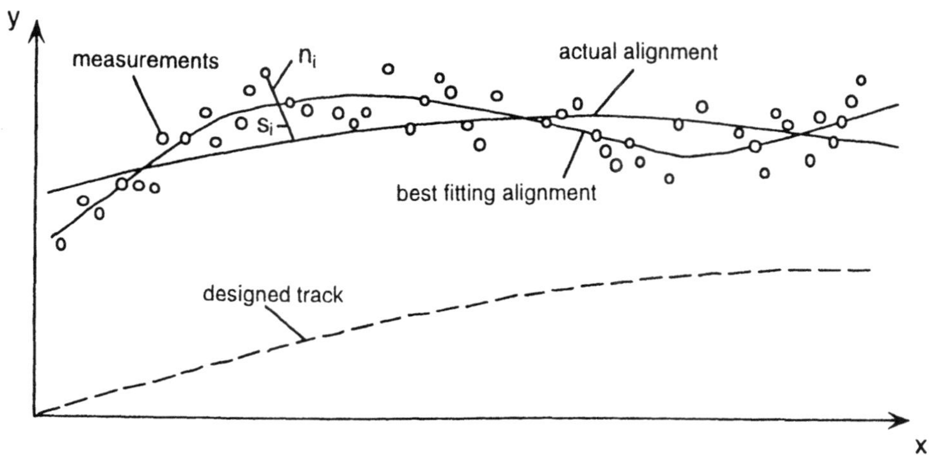

Fig. 5: Model of the Wiener filter evaluation procedure

described by the fundamental equation of the Wiener filter:

$$\underline{l} = \underline{Ax} + \underline{s} + \underline{n} \qquad (1)$$

where $\underline{l}$ is the observation vector, $\underline{Ax}$ is the function of the best fitting alignment, $\underline{s}$ describes the differences between the actual alignement and the best fitting alignment and $\underline{n}$ is a noise vector.

The function comprises elements, normally used for the alignment, such as straight lines, circles or clothoids. It must not conform to the trackdesign, however maximum or minimum alignment parameters and certain shifts of the rails per pendicular to the railway axis are not allowed to exceed specified limits. There should not be any shifts at the beginning and end of the surveyed track. This is met by increasing the weighting of the measurements at the beginning and the end of the track section in the Wiener filter evaluation procedure.

The evaluation process consists of 3 steps (Retscher, 1996).

In step 1 single track elements are computed. We get parameter vector $\underline{x}$ from:

$$\underline{x} = (\underline{A}^T \underline{Qu}^{-1} \underline{A})^{-1} \underline{A}^T \underline{Qu}^{-1} \underline{l} \tag{2}$$

where $\underline{Qu} = \underline{Qu} + \underline{Qs}$ is the cofactor matrix. The cofactor matrix $\underline{Qu}$ of the measurements and $\underline{Qs}$ of the signal can be estimated using the correlation function:

$$C(\Delta l) = Coe^{-\left(\frac{\Delta l}{a}\right)^2} \tag{3}$$

a can be estimated from $a \approx (2/5)\lambda$, where λ is a mean value of the wavelength of $\underline{s}$ and $\underline{l}$ the length of the elements.

The elements are connected in a 2 step procedure. Then certain conditions must be met

$$\begin{aligned} x_i\,(l) &= \underline{x_{i+1}}\,(l) = x \\ y_i\,(l) &= \underline{y_{i+1}}\,(l) = y \\ z_i\,(l) &= \underline{z_{i+1}}\,(l) = z \\ \tau_i(l) &= \tau_{i+1}(l) = \tau \\ \kappa\,(l) &= \kappa_{i+1}(l) = \kappa \end{aligned} \tag{4}$$

where τ is the azimuth of the tangents, κ the curvature.

In step 3 finally the signal $\underline{s}$ is computed:

$$\underline{s} = \underline{Q_s}\,\underline{Qu}^{-1}\,(\underline{l} - \underline{Ax}) \tag{5}$$

It contains the corrections (shifts) for the rail straightening machine. Practical results confirm that the high relative accuracy of 1...2 mm/10 m for the construction and maintenance of high speed railway tracks can be met (Retscher 1996).

References

Kahmen, H. and Wunderlich Th., Retscher, G., Kuhn, M., Plach, H., Teferle, F. N., Wieser, A.: Ein modernes Konzept zur Absteckung von Hochgeschwindigkeitstrassen. Allgemeine Vermessungs-Nachrichten 1998.

Retscher, G.; 3D-Gleiserfassung mit einem Multisensorsystem und linearen Filterverfahren. Geowissenschaftliche Mitteilungen. Heft 44. Institut für Landesvermessung und Ingenieurgeodäsie, TU Wien, 1996.

KINEMATIC GPS FOR ICE SHEET SURVEYS IN GREENLAND

K. Keller, R. Forsberg and C.S. Nielsen
National Survey and Cadastre (KMS)
Rentemestervej 8; DK-2400 Copenhagen NV; Denmark
(krk@kms.min.dk)

Abstract

This article presents results of repeated GPS surveys on the Greenland Ice Sheet for studying ice elevation changes, and for providing ground truth data for evaluation of remote sensing methods (e.g. satellite altimetry and SAR interferometry). The surveys have been carried out at several central sites of the Ice Sheet, as well as on local marginal ice caps, which are more sensitive to coastal climate changes. Kinematic GPS techniques have been used for surface profiling using snowmobile traverses and airborne laser altimetry, providing good agreements at the sub-metre level. Static GPS repeated surveys yield sinking and strain rates which are in good agreement with glacilogical mass balance models, indicating no major shrinking or growth of the Ice Sheet.

Introduction

This paper presents some of the data from the first year of the Danish project called ECOGIS (Elevation Changes Of the Greenland Ice Sheet), undertaken by:
1) National Survey and Cadastre – Denmark.
2) University of Copenhagen, Geophysical Dept.
3) Technical University of Denmark, Danish Center for Remote Sensing (DCRS).

In this project we will study the elevations and their changes over the Greenland Ice Sheet on a multi-disciplinary basis, employing geodetic field measurements, satellite radar altimetry, and synthetic aperture radar interferometry (InSAR) (Nielsen et al., 1997). The project focus will be on evaluating the accuracy of the emerging remote sensing height measurement techniques, and to provide base measurements of the Greenland Ice Sheet topography for future early detection of height changes. We will continue and verify earlier change measurements at the Greenland Ice Core Project (GRIP) drilling site at Summit. Comparisons of previous and current satellite altimetry results will also be done.

The field work during 1996 was carried out at 5 sites in Greenland (figure 1):

Site 1): Airborne laser altimetry and ground truth ski-doo traverses were done at the camp of the North Greenland Ice Core Project (NGRIP) placed on the North Central part of the Ice Sheet (75N;42W; elev. 2920 m a.s.l.).

At the following two ice cap stations airborne laser altimetry were carried out:

Site 2): Sukkertoppen Ice Cap, a small (25 km by 25 km) ice cap located at West Coast (66N;52W) some 150 km west of Kangerlussuaq (Søndre Strømfjord).

Site 3): The Geikie Plateau, also a small (25 km by 75 km) irregular ice cap with a very undulated topography located at the East Coast (70N; 26W).

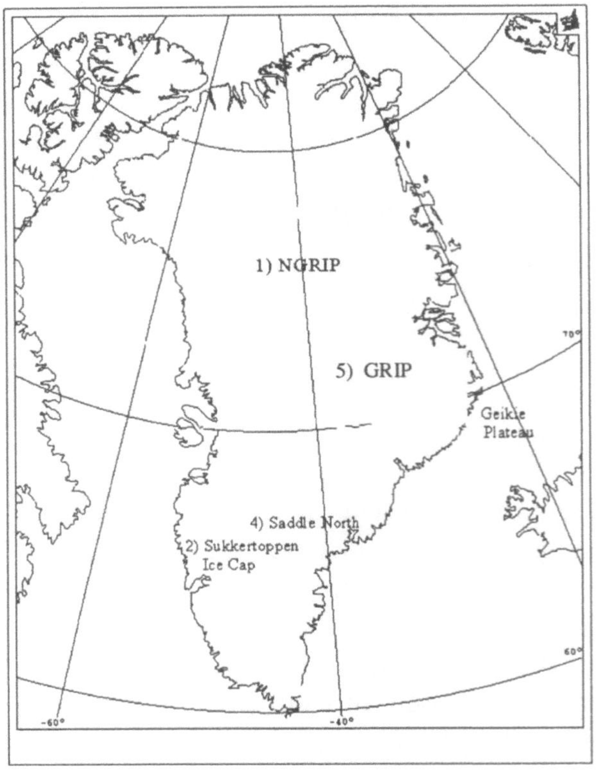

Figure 1. Map showing the 5 sites of fieldwork.

At Site 4), Saddle North, which is the topographic saddle point between the main Ice Sheet and the south ice cap kinematic GPS data were collected on short traverses. These data will not be treated here.

Site 5): The deep ice core drilling site GRIP at the Summit of the Greenland Ice Sheet (72.6N;37.6W).

Survey methodology

The primary goal of the geodetic surveys was to provide precise GPS ties between reference poles and coastal benchmarks, in order to monitor overall ice sheet changes. In addition local surface elevations and strain rates were measured to monitor ice volume changes and flow, and to provide the necessary height information within ERS-1 satellite altimetry footprints to gain an understanding of the return waveforms.

At the GRIP site at the center of the Ice Sheet a fiberglass pole has been frozen into the ice sheet at 80 m depth, and monitored by GPS on a yearly basis since 1992. Table 1 shows the height of the top of this pole, and the corresponding height above the snow surface. The pole on the average sinks by 22.4 cm/year, but this sinking is offset by the accumulation of new snow, so that the overall ice sheet at this point is in balance. This is also confirmed by glaciological modelling of the ice flow based on the GPS strain net established in the area of the GRIP site (Hvidberg et al; 1997), cf. Fig. 2.

Table 1. Movement of the reference marker at Summit (1992-1993-1994-1996).

	Annual change in position in cm.		
Year	Latitude	Longitude	Height
92-93	+ 7.4 cm N	+ 24.0 cm W	- 28.7 cm
93-94	+ 12.1 cm N	+ 16.7 cm W	- 22.0 cm
94-96	+ 9.2 cm N	+ 25.7 cm W	- 19.5 cm
Mean	+ 9.5 cm N	+ 23.0 cm W	- 22.4 cm

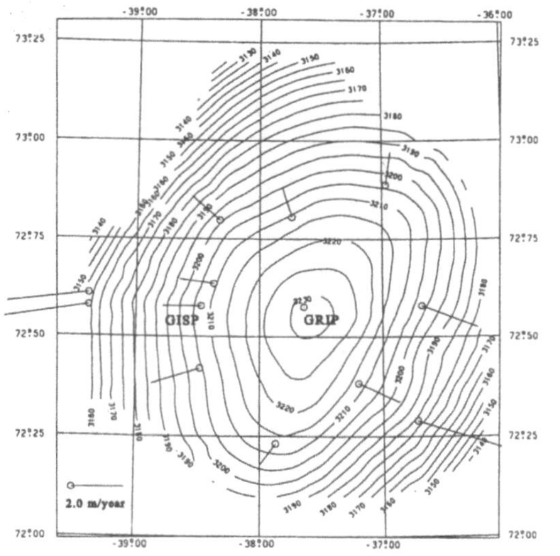

Figure 2. Strain rates and surface elevation contours at the GRIP site, summit of the Greenland Ice Sheet, as determined by repeated GPS surveys and surface surveys. Contour interval 5 m, arrows show annual ice movements.

The kinematic survey of the ice elevations were carried out by ski-doo traverses, where a GPS receiver is mounted on a sled, with a typical traverse speed around 25-30 km/h. Surveys were done relative to a central reference receiver using Trimble 4000 SSI receivers, processed by the GPSurvey or Geotracer software, and using IGS orbits. Results are tied to ITRF94 through the coastal GPS links.

To augment the surface surveys, additional airborne surveys were carried out by laser altimetry. A Greenlandair Twin-Otter was mounted with an Optech laser altimeter and a Trimble 4000 SSI unit, and kinematic GPS surveys were carried out over different areas with baselines up to 150 km. Since we did not have a proper INS at our disposal, a rough first order attitude correction of the laser was carried out using a dual-axis inclinometer (HP Planartechnik NS-45). By combining the horizontal accelerations measured by the inclinometer with the GPS-derived horizontal accelerations, a first-order estimate of roll and pitch can be obtained, and the GPS co-ordinates can be transferred to the footprint of the laser, provided an assumption is made on the heading (assuming zero yaw). The inclinometer corrections work reasonably well when flying low, but with larger elevations over rough ice these are not sufficient, as outlined below. We are therefore currently preparing a new custom-made INS unit based on Litef fibre-optic gyros and Schaewitz accelerometers. All data are logged on a laptop PC, using an in-house logging program designed by Jon Olsen of KMS.

The 1996 survey at NGRIP

Numerous geodetic surveys were carried out at the NGRIP camp during the 1996 field season. For ice modelling and mass balance studies a double 8 pole strain net was established at distances of 10 to 50 km from the camp. This was put up partly using Twin-Otter support, and partly by daily ski-doo traverses having distances from 50 km to 170 km. A second purpose of the traverses was to collect topographic ground truth data for Ice

Figure 3. Ski-doo GPS survey setup ("strap-down" GPS).

Sheet monitoring. Furthermore gravity measurements were made each 10 km along the traverses. These height data were collected using kinematic GPS mounted on a sledge carrying all kinds of equipment, including a complete polar survival kit (figure 3.)

The GPS was processed using the software Geotracer which gave quite good results. The overall accuracy of the GPS solutions is estimated to be approx. 5 cm, but due to small bumps and the antennas mounting on top of a sledge no more than a precision of 10 cm can be expected in the final heights measurements. Cross-over analysis shows that the 10 cm level has been achieved at the ski-doo traverses. The data from airborne laser altimetry is consistent with these measurements and all together the data are combined in a height model of the area around the NorthGRIP camp, as indicated in figure 4.

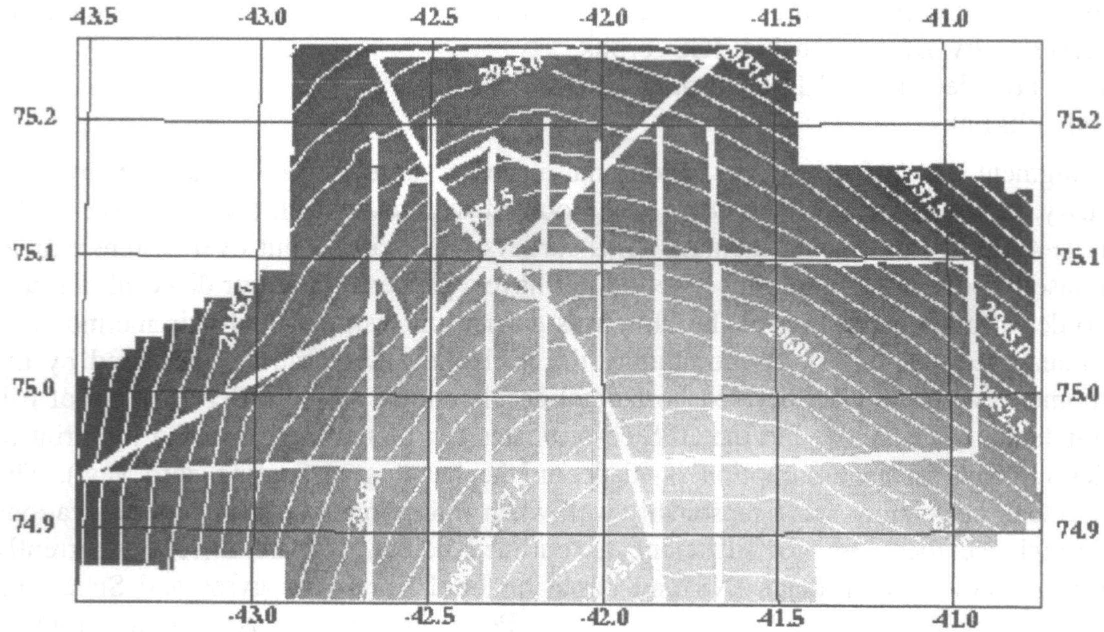

Figure 4. NGRIP height contours with data tracks, c.i. 2.5 m.

The surveys at Geikie Plateau

On July 16th and 17th 1996 two surveys over the Geikie Plateau were flown using Constable Pynt as base. Both survey tracks are shown on the map (Figure 5). The first one was mainly a reconnaissance flight which also was used to put out a depot containing radar reflectors and fuel for the operation in August. The second flight was a survey flight only and lasted for almost four hours. As reference for the survey a permanent pole was put in the gable of one of the buildings at the airport and the point was tied to the ITRF94 system. When we returned in August we found that the pole had been adopted as a windbag for the airport and a new reference pole had to be put up!

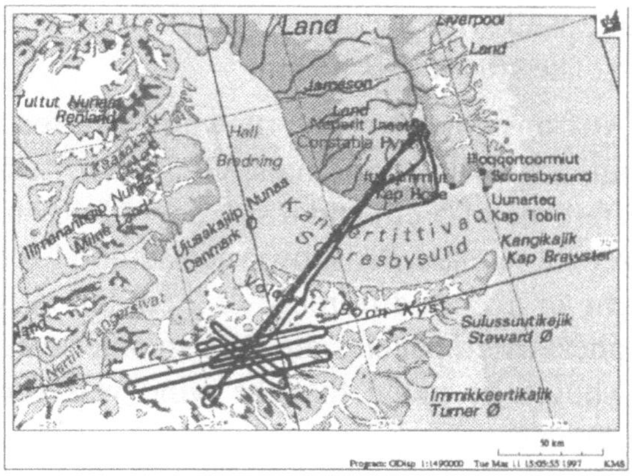

Figure 5. Flight tracks over the Geikie Plateau ice cap.

On August 18th and 19th an other flight was made to the Geikie Ice Cap for assembling and positioning radar reflectors, to provide ground truth data for an airborne synthetic aperture radar (SAR) campaign, originally planned for late August.

The SAR interferometry survey was going to provide digital elevation models of the region using the DCRS EMISAR system, mounted on a Royal Danish Air Force (RDAF) Gulfstream G-3 jet. Due to a tragic air crash in the Faroe Islands the RDAF postponed all scientific Greenland flights of the remaining G-3's, so the reflectors were never actually used as SAR targets, and may now be considered lost due to excessive snow accumulation in the area (estimated from the snow at the depot to be 4 m/year). Besides the depot and the four reflector sites three other sites were marked and positioned, among them a site which were considered as the top of the ice cap at an elevation of 2272 m located at 69° 56' 11" N; 25° 31' 47" W.

With the absence of airborne SAR, studies were done in the region on comparing ERS-1/2 tandem mission satellite SAR interferometry (Nielsen et al., 1997). The airborne SAR flight took place August 1997 as well as repeated laser measurements, and employed 4 new radar reflectors.

Comparisons of the surface and airborne GPS surveys

As an evaluation of the collected data a cross-over analysis was performed using different kinds of observations. The results of this analysis can be seen in table 2. At NGRIP the ski-doo measurements fit together very nicely. The overall fit is biased by 7 cm and the std. dev. is 10 cm which is very good considering that the data were collected in two weeks, where snowfall had occurred during the middle of the period. Further more was the sledge moving somewhat up and down due to small sastrugas on the surface, which also cause small irregularities in the data. These surface measurements fit

Table 2. Results of the cross-over analysis.

X-over analysis	Mean diff.	Std. Dev.
NGRIP GPS vs. GPS	0.07 m	0.10 m
NGRIP GPS vs. Laser	-0.38 m	0.33 m
Geikie laser vs. Laser	-0.02 m	1.95 m
Runway GPS vs. Laser	0.05 m	0.40 m

quite well with the airborne observations, even though there is an unexplained offset of 38 cm. The std. dev. is 33 cm.

At Constable Pynt the runway was used for validation of the laser altimeter heights and compared with a kinematic GPS survey using one of the airports trucks. It gave an overall difference of 5 cm and a std. dev. of 40 cm, which is satisfactory. The analysis of the data over the Geikie Pl. shows not as good results as on the runway test. The measurements have an overall difference of 2 cm but a std. dev. of 1.95 m which is not satisfactory. The highest discrepancies are seen in outer regions of the plateau where the slope is very large. It should be noted that the wind was very strong during the flight and caused a lot of turbulence.

Conclusions and future plans

On-The-Fly kinematic GPS processed using standard software is working very well both in an environment of low dynamics (ski-doo), and in a high dynamic environment such as an aircraft even on longer distances up to 200 km. The overall accuracy is believed to be within 10 cm. Results might be improved by using software capable of processing data from multiple reference stations. Integration of GPS and INS will also further increase precision, and this will be implemented in future surveys.

The airborne laser altimetry shows satisfying results over smooth areas such as the Greenland Ice Sheet, but the results from the Geikie Plateau indicate that INS data is required to resolve pitch, roll, and yaw over high undulating surfaces in order to determine the footprint of the laser measurements.

The purchased INS will also permit automatic GPS time-tagging of all the measurements, something which was not done in the current survey hence introducing another uncertainty, even though the laser altimeter data were correlated with the GPS data to check for any leap seconds.

The future plan is to repeat and extend the surveys of the sites shown at figure 1. Results from these surveys will then give input to a first estimate of the mass balance at the NorthGRIP area and other sites where e.g. height models will be improved due to further development in airborne GPS and InSAR height measurements.

References

C.S. Hvidberg, K. Keller, N.S. Gundestrup, C.C. Tscherning and R. Forsberg, 1997: *Mass Balance and Surface Movement of the Greenland Ice Sheet at Summit, Central Greenland*, Geophysical Research Letters, Vol. 24, No. 18, 2307-2310.

C.S. Nielsen, R. Forsberg, S. Ekholm, J.J. Mohr, 1997: *Merging of Elevations from SAR Interferometry, Satellite Altimetry, GPS and Laser Altimetry in Greenland*, ESA-SP-394, Proc. of 3rd ERS Scientific Symposium, in press.

A REWEIGHTED FILTERING ALGORITHM AND ITS APPLICATION TO OPEN PIT DEFORMATION MONITORING

Minghai Jia, Maria Tsakiri and Mike Stewart

School of Surveying and Land Information
Curtin University of Technology
GPO Box U1987, Perth 6845
Western Australia.

Abstract

The Global Positioning System (GPS) has been proved to be a capable surveying tool for monitoring the stability of steep walls in open pit cut mines. However, due to restrictions on satellite geometry and severe environment-dependent errors, as well as the uncertainty of the *a priori* knowledge of system noise variance, sub-centimetric accuracies using existing algorithms cannot always be achieved. This paper proposes a reweighted filtering algorithm which treats GPS deformation monitoring in an open pit environment as a kinematic surveying system. Estimating techniques of *a posteriori* weights of predicted states and observations are developed to match the weights appropriately and obtain reweighted filter solutions. Two different weight functions to reweight the predicted states and observations are established based on the state residuals and observation residuals. The proposed algorithm can automatically assign the larger weights to the more accurate predicted states and observations and the smaller weights to the less accurate predicted states and observations. Results from simulated GPS deformation monitoring data are shown using the proposed algorithm. Comparisons with the traditional Kalman filtering method indicate high levels of filter stability and accuracy for the proposed algorithm.

1. Introduction

In the open pit environment, GPS can offer several benefits over conventional surveying techniques. No 'line of sight' measurements between reference and objective points are required. In addition, the baseline lengths required in open pit deformation monitoring do not practically exceed 1km and the 'relative' data processing techniques ensure that GPS solutions remain unaffected by local atmospheric conditions for these baseline lengths.

The sub-centimetric level accuracy required to detect precursor movements to pit wall failures necessitates the use of carrier phase GPS processing techniques. Given the slow movement rates of open pit walls, the static nature of the deformation problem can be treated with an alternative approach, using GPS navigation techniques. In this way, repeat rapid GPS surveys of few seconds provide data which are integrated with data from previous surveys. This integration of all available data can be achieved by using optimal Kalman filtering estimation techniques (Tsakiri *et al* 1996) and the problem can be described by a *kinematic surveying system* (KSS). As open pit deformation networks are usually established by static control surveys, precise *a priori* knowledge of the coordinates of the points is not a problem. As a result, ambiguities can be resolved almost instantaneously in a process identical to the initialisation phase of a kinematic GPS carrier phase solution. A subsequent ambiguity fixed solution solving for coordinate changes produces the deformation solution. Provided sites are re-observed at a controlled time interval, based on real velocities of the monitoring points, motion and observation error should be less than one carrier phase half wavelength and will not affect ambiguity resolution. Using this approach, repeat surveys can produce ambiguity fixed solutions, theoretically good to better than 4mm (the noise of a raw double difference carrier phase observations).

Open pits remain one of the most challenging environments for the utilisation of GPS high precision technology. The main problem, affecting reliable and accurate solutions for deformation monitoring, is restricted satellite visibility, as monitoring points situated in pits can be up to several hundred metres deep and may be unable to observe the full available constellation. The small number of satellites and the resulting poor geometry can severely degrade an ambiguity fixed carrier phase solution. In addition, environment-dependent errors, such as multipath, can be the dominant noise source for monitored short baselines. Because of the small motions of the pit walls, attaining the optimum GPS solution is of fundamental importance.

This paper develops a new numerical-based method, namely, *one-step-reweighted filter* method, to deal with the residual errors in double difference carrier phase observations and access the *a posteriori* weights of the observations.

2. Kalman Filter for Kinematic Surveying Systems

The very slowly moving GPS deformation monitoring problem can be treated as a kinematic surveying system (KSS), which is composed of a state model with the stochastic disturbances and an observation model with the measurement noises. The description of the state and the observation model and their associated stochastic models at any epoch k is well documented (eg Bryson and Ho 1969). Kalman filter equations for KSS can be derived based on least squares principles (Bryson and Ho 1969). The Kalman filter optimal estimators (*ibid* 1969) are,

$$\hat{x} = \bar{x} + J(l - A\bar{x})$$
$$Q_{\hat{x}} = Q_{\bar{x}} - JDJ^T$$
$$J = Q_{\bar{x}}A^T D^{-1} \qquad (1)$$

$$d = l - A\bar{x}$$
$$D = AQ_{\bar{x}}A^T + Q_e$$

where, $\hat{x}$ is the optimal estimator of the state vector; $\bar{x}$ is the predicted state vector; l is the $n \times 1$ vector of observations; A is the $n \times t$ measurement matrix; J is the gain matrix; d and D are the innovation vector and its cofactor matrix respectively; $Q_{\bar{x}}, Q_e$ are the cofactor matrices for $\bar{x}$ and the vector e of observation noises respectively.

Based on equation 1 a reweighted filtered GPS carrier phase solutions can be derived.

3. One-step Reweighted Method

As mentioned previously, open pit environment-dependent errors and especially multipath are the major error sources in the monitoring data. These errors make the observation data from different satellites, with different directions and elevations, present significant heteroscedastic error properties. For example, the error effect is heavily dependent on the direction of the arriving signals from a specific satellite to the monitored point, which is usually very close to the open pit wall. The quantitative modelling of these type of errors is complex and not within the scope of this paper. However, the magnitude of these unmodelled errors in the observation data is sufficient to submerge the deformation information desired for sub-centimetric accuracy. These errors are reflected in the GPS double difference carrier phase filtered residuals and, based on these residuals, an appropriate weight can be given to observations obtained from satellites with different errors. A *one-step-reweighted* filter algorithm is developed in this section based on the evaluation of the filter residuals which substantially improve the solution at any epoch.

The GPS data processing technique, implemented in the software developed at Curtin University, comprises the use of two filter runs, as described in equations (1). The implementation of the first filter employs identical weights for all observations from different satellites due to lack of *a priori* quantitative knowledge of the environment-dependent errors, even though the observation errors are heteroscedastic. After the completion of the first filter, the filter residuals, which reflect the properties of the predicted states and the observation errors, are used to assess the *a posteriori* weights of the observations. Then, the filter is implemented again using the new weights to obtain the final results. In this filter the magnitude of the residuals define the assigned weight, in other words, observations with large residuals are assigned low weights and greater weights are given to the observations with small residuals. The *a posteriori* weights of the observations are expressed as a function of the their *a priori* weights and filter residuals (Yang 1994, Jia *et al* 1997),

$$\bar{p}_{ij}(v_i, v_j) = p_{ij}w_{ij}(v_i, v_j) \quad \text{for i, j=1, 2, ..., t, t+1, ..., t+n}$$

where, $\bar{p}_{ij}(v_i, v_j)$ is the *a posteriori* weight, $w_{ij}(v_i, v_j)$ is a weight factor function, and v_i, v_j include both measurement and state vector residuals. The weight factor functions can be directly defined based on robust estimation (eg, Hampel et al 1986, Yang 1994) and robust filter (Jia *et al* 1997) techniques. These should be continuous, even and non-negative functions of the filter residual.

Based on the above properties two known weight factor functions, Andrew's, and Tukey's (Andrew 1974, Beaton and Tukey 1974) are modified

(1) Modified Andrew's Weight Factor Function

$$w_{ij}(v_i, v_j) = \cos(v_i \,/\, c_0\sigma_0)\cos(v_j \,/\, c_0\sigma_0)$$

where σ_0 is a scale parameter and $c_0 = 0.5\pi$.

(2) Modified Tukey's Weight Factor Function

$$w_{ij}(v_i, v_j) = [1-|v_i v_j|/\sigma_0^2 c^2]^2$$

where $c_0 = 4.658$ or $c_0 = 6.0$ with known or unknown σ_0.

The adopted values for the so-called *tuning* constant c_0 are computed based on Hampel *et al* (1986). The chosen values for c_0 result in the 95% asymptotic efficiency on the standard normal distribution assuming the errors are Gaussian. This is important because in the one-step-reweighted filter, the values of constant c_0 should make the reweighted results on the homoscedatic observation error distribution to be as close to the unreweighted (ie Gauss distributed) results as possible.

The *a posteriori* weight matrices $\overline{P}_{\bar{x}}$ and $\overline{P}_l$ of the filtered states and the measurements respectively, based on the above weight functions, can be calculated as,

$$\overline{P}_{\bar{x}}(v_{\bar{x}_i}, v_{\bar{x}_j}) = P_{\bar{x}} * W_{v_{\bar{x}}} = (p_{\bar{x}_{ij}} w_{ij}(v_{\bar{x}_i}, v_{\bar{x}_j}))$$
$$\overline{P}_l(v_{l_i}, v_{l_j}) = P_l * W_{v_l} = (p_{l_{ij}} w_{ij}(v_{l_i}, v_{l_j}))$$

where, * denotes the Hadamard product of two matrices with the same dimensions (Rao 1973). Using the *a posteriori* weight matrices, and based on equation (1), the reweighted filter solutions are obtained.

4. Numerical Example

The effectiveness of the proposed one-step-reweighted filter algorithm is illustrated with results from simulated GPS monitoring examples in a situation of poor satellite geometry. In-house software for GPS simulation was used to generate data for a 30-second data span with 4 satellites and introduced error on the data from one satellite. The error was introduced after the 5 seconds to allow the filter to become stable.

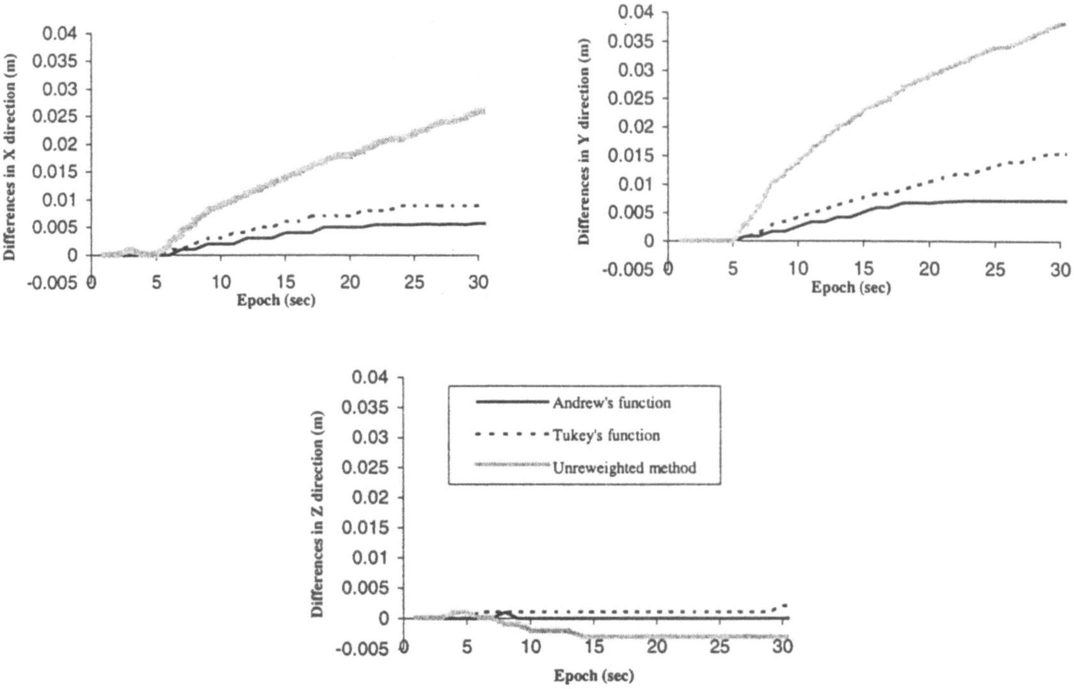

Figure 1. Comparison between reweighted methods and standard Kalman filter in X, Y, Z directions for observations from 4 GPS satellites

The GPS data are simulating a pit wall environment with an introduced masking of elevation angle from 0-55 degrees and azimuth from 90-135 degrees. This masking results in a data set of observations with only four satellites and hence, three double difference observations and a DOP (dilution of precision) value of 5.8. An environment-dependent error of 0.4 of a cycle (approximately 7.6 cm) is introduced to a random satellite (number 27) in every epoch. The differences between the 'truth' and the obtained results using a standard Kalman filter and the proposed reweighted method are shown in the three graphs of figure 1. All results refer to solutions in which ambiguities have been fixed. In all three graphs, the unreweighted standard Kalman filter produces the greatest differences of all, which reach almost 3.5 cm. The solution from the reweighted method is superior to the standard Kalman filter (unreweighted) solution for both functions. For the reweighted solutions the greatest errors are in the Y direction, which reach up to 1.5 cm, and the smallest errors are in the Z direction which are in the order of 0.1 cm. The differences from the 'truth' in X and Y direction are the greatest, reflecting a satellite direction effect resulted by the introduced error at the satellite. A comparison between the two reweighted functions indicate some differences with the Andrew's function tending to produce the smallest errors. In addition, there is an evident drifting pattern at the errors, mainly in Y direction, suggesting the influence of the simulated error contaminating the data from one satellite when observations from only 4 satellites are available. This means that although there are 3 double difference observations, in fact only 2 are unaffected and can be used reliably. Overall, however, there is an improvement in all three directions using the reweighted method, two times with Tukey's function and three times with Andrew's function. When more than 4 satellites are available and an error to the data of one satellite has been introduced, the reweighted method still offers improved solutions despite the adequate number of double difference observations (more than 3).

It must be mentioned here, that when no environment-dependent errors are included in the data set, as expected, the reweighted method produces almost identical results to the unreweighted standard Kalman filter.

5. Concluding Remarks

A filter method, namely a one step reweighted filter, has been developed to improve the accuracy of the GPS carrier phase solutions for deformation monitoring in open pits, where environment-dependent error contamination of the observations is likely to occur. The reweighted method reduces the influence of large errors by assigning appropriate weights depending on the GPS double difference residuals and implementing known weight functions which have been modified appropriately. The use of simulated tests have shown that the proposed method is capable of producing the required sub-centimetric accuracies with 4 satellites despite the existence of some large environment-dependent errors in the observations. In addition, the use of the reweighted filter suggests its applicability in restricted satellite geometry with results at the required level. A comparison between the weighted functions indicates an improvement using the modified Andrew's function which provides position errors of less than 5 millimetres. Further testing with real data will enable verification of the obtained results.

6. References

Andrew D F (1974). A Robust Method for Multiple Linear Regression. *Technometrics*, Vol. 16, 4:523-531.

Beaton A and J W Tukey (1974). The Fitting of Power Series, Meaning Polynomials, Illustrated on Band-Spectroscopic Data. *Technometrics*, Vol. 16, 2:147-157.

Bryson A E and Y H Ho (1969). Applied Optimal Control. Wiley, Inc., New York. ISBN 0 470 11481 9.

Hampel F R, E Ronchetti, E M Rousseeuvu and W Stahel (1986). Robust Statistics. WileySeries in Probability and Mathematical Statistics, ISBN 0-471-82921-8.

Jia M, M Tsakiri and X Ding and (1997). A new robust algorithm for kinematic surveying systems. International Symposium on Kinematic Systems in Geodesy, KIS97, June 3-6, Banff, Canada.

Rao CR (1973). Linear Statistical Inference and Its Applications, Second Edition. John Wiley & Sonc, New York.

Tsakiri, M, M Stewart and X Ding (1996). Robust Kalman Filter Architecture for GPS Deformation Monitoring. Proc. of the 8th FIG International Symposium on Deformation Measurements, June 25-28, Hong Kong, pp 275-285.

Yang Y (1994). Robust estimation for dependent observations, Manuscripta Geodaetica, 19: 10-17.

PERMANENT AUTOMATIC GPS DEFORMATION MONITORING SYSTEMS: A REVIEW OF SYSTEM ARCHITECTURE AND DATA PROCESSING STRATEGIES

Craig Roberts
Chris Rizos

School of Geomatic Engineering, The University of New South Wales
Sydney NSW 2052, Australia

ABSTRACT

Ground deformation due to volcanic magma intrusion, crustal motion, ground subsidence, etc., are phenomenon ideally suited for study using GPS. The *change* in length, height difference and orientation of baselines connecting GPS receivers in a carefully monumented ground network can be monitored. This is done by repeatedly measuring the same baseline components to an accuracy commensurate to, but preferably much higher than, the expected baseline component changes. Such GPS techniques are based on the "campaign" principle: the periodic (often annual) re-survey of a network of control points. However, over the last half decade or so, there has been a growing interest in the deployment of permanent, continuous GPS monitoring networks. The factors responsible for this trend include the declaration a few years ago of Full Operational Capability of the GPS system, and the steady decrease in price, size and power requirements of GPS receivers. Important additional factors have been the high cost of annual GPS surveys (manpower, travel, logistics, etc.), as well as the fact that geo-scientific research can be furthered because of the *continuous* measurement of a deformation phenomenon, rather than its *periodic* measurement.

The GSI network in Japan, the SCIGN network in California, and the SWEPOS network in Sweden are examples of large scale, continuous GPS networks for near-real-time crustal motion monitoring. However, smaller scale GPS arrays such as those on the Augustine volcano (Alaska), the Popacatepetl volcano (Mexico), the Kilauea volcano (Hawaii), and the Rabaul volcano (Papua New Guinea) reflect a growing interest in *local* continuous GPS volcano monitoring systems. GPS is also increasingly used to monitor engineering structures such as dams, bridges, offshore drill platforms, etc.

Developing an automatic GPS array system for such small scale monitoring applications is an engineering and software challenge. A network of permanent GPS receivers need to be deployed, often in an inhospitable environment in which they must operate reliably on a continuous basis. The GPS observations must be telemetered to a central computing facility where data processing occurs with minimum delay. Analysis of the time series of baseline results then takes place in order to detect any baseline component change between successive solutions which may be a precursor to failure or eruption. What are the *bases* of small scale monitoring systems? Often the solution has been simply to purchase commercial "off-the-shelf" real-time-kinematic (RTK) GPS systems. This is the high cost option, yet there are several hundred active volcanoes in the world, many located in the less developed countries, and the cost of GPS monitoring systems must be significantly reduced if the technology is to contribute to volcano hazard mitigation. An alternative approach is to develop a system based on single-frequency GPS receivers, integrated with communications and in-field computer sub-systems.

This paper discusses the characteristics of GPS monitoring networks and considers such issues as monument design, network design, GPS hardware, communication links, power supply, and data processing strategies.

INTRODUCTION -- GPS-BASED DEFORMATION MONITORING SYSTEMS

The first applications of "GPS geodesy" have been to measure the relative motion of tectonic plates and the deformation of active continental margins. These applications have grown to such an extent that there are now very few parts of the world which have not been studied (or are not at present under study) using the GPS technology. However, there is an increasing interest in GPS-based ground deformation measurement on a continuous basis, operating over small regional extents, using a variety of system configurations, which are optimised for specific monitoring applications. There are therefore several *bases* for GPS-based deformation monitoring systems.

Repeat GPS Surveys

In essence, GPS field observation campaigns are mounted to determine the relative position of groundmarks *at a certain epoch*. The data collection typically occurs over a period of several days or weeks, as teams of GPS technicians occupy the groundmarks for up to several days at a time, before moving on to other locations. The data is then collected, archived and prepared for processing back at the office of the organisation performing the survey. By repeating the survey campaign a number of times, any *change* in the baseline components between successive field campaigns is interpreted as a measure of the deformation. In the context of geodynamical studies, GPS receivers are relatively inexpensive and highly portable (compared to traditional space geodetic techniques), and this campaign-mode measurement strategy has been used to good effect for over a decade.

Many other deformation applications are also being addressed using repeat GPS survey methods. These include the measurement of ground subsidence due to fluid extraction or underground mining, and the monitoring of engineering structures or volcanos.

Permanent Geodetic Arrays

The large number of geodetic GPS receivers being purchased by organisations around the world, together with the escalating labour costs of GPS campaigns, has led to an increasing number of receivers being simply deployed at fixed sites on a permanent basis. Although many of these stations have been identified with the International GPS Service for Geodynamics (IGS), many permanent GPS receiver stations are being established in the form of sub-nets to support deformation monitoring. Hence patterns of deformation occurring over very short time scales, from a matter of a few minutes or hours, to several weeks, can be detected and studied. The Southern California Integrated GPS Network (SCIGN) in southern California (Bock et al., 1997) and the permanent Geographical Survey Institute (GSI) network in Japan (Tsuji et al., 1995) fall into this category. In addition to establishing the hardware arrays, data processing software has been implemented to ensure that the continuous stream of GPS data can be "digested". The software system consists largely of standard scientific software modified to run in "batch" mode with automatic data file handling and pre-processing capabilities. Further, these arrays are established in areas where the ground infrastructure is well developed, for example a situation that is generally atypical of applications such as the monitoring of volcano flanks, micro-faults, ground subsidence due to underground mining or fluid extraction, slope stability, and engineering structures such as dams, bridges, etc.

The authors contend that permanent GPS-based monitoring systems will be increasingly deployed across small areas, for a variety of deformation applications.

Deployment of Systems in Emergencies

In the case of volcano monitoring applications, there may be little or no activity over many years and the permanent deployment of a GPS array may not be justified. However, as soon as volcanic activity is detected (by whatever means), a range of instrumentation (including GPS) may be rapidly deployed by the responsible agencies. These systems must be highly portable and robust in order to function anywhere in the world. On the other hand, given that their deployment normally takes place *after* the commencement of inflation or ground uplift on a volcano's flanks, the sensitivity of the monitoring system need not be at the sub-centimetre level as the deformations are likely to be significantly greater than this.

Commercial GPS Surveying Systems

An example of an "off-the-shelf" system are the commercial "real-time kinematic" (RTK) products developed by several GPS manufacturers. Although marketed for tasks such as engineering surveys, etc., they can also be adapted for deformation applications. The following comments can be made:

- The integration of hardware, software, and telemetry sub-systems is well engineered.
- Due to limitations of the telemetry sub-system (generally UHF/VHF), and the "standard" double-differenced observation model that is used in such systems, the baseline length constraint is of the order of a few tens of kilometres at best.
- The coordinates are derived by the "rover" receiver(s), and hence must be transmitted back to a central monitoring facility if they are to be used in (near-)real-time.
- Typically there is no software that checks the baseline solutions to determine whether the observed variability is really a deformation signal or just "noise".
- Many deformational phenomena cannot be characterised as "kinematic" (except under extreme conditions just before a failure or eruption), but could be considered to be "static" over a period of one hour or so.

Certainly for many deformation monitoring applications that span relatively small areas, the RTK solution is the one of "first choice" as there are so few other options. However, because of its high cost it is likely to be only implemented in a small number of instances.

CHARACTERISTICS OF PERMANENT GPS-BASED SYSTEMS

We will restrict ourselves to permanent systems with the following characteristics:

- A network of well-monumented benchmarks upon each of which a GPS receiver and various auxiliary instrumentation is installed.
- One station within the network will be designated the Master Control Station (MCS), operating as the node of the network and where the computations are made.
- The areal extent of the network is of the order of several tens of kilometres.
- Capable of continuous autonomous operation.

Such permanent GPS-based monitoring systems can be categorised according to the following characteristics: (1) monument design; (2) network design; (3) GPS hardware; (4) communications links; (5) power supply; and (6) data processing strategy.

The characteristics of *individual* systems will tend to reflect, on the one hand, the nature of the deformational phenomenon being monitored, and on the other hand, the capabilities of the organisations responsible for installing and operating the systems, hence the following issues have also to be considered:

- Whether the deformation is mostly characterised by continuous, relatively small changes in the network geometry, or long periods of no deformation followed by sudden, violent changes in the position of the network benchmarks.
- Whether the deformational signal is mostly horizontal, or in the vertical.
- The physical environment in which the system will be deployed.
- The level of performance required of the monitoring system, e.g. the accuracy, the tolerable delay in production of the results, and so on.
- The level of GPS expertise within the organisation.
- The budget for equipment, and network construction, maintenance and operation.

Monument Design

The monuments must be stable. However, monuments designed to have stabilities at the millimetre level are very expensive and restricted to geodynamic networks such as SCIGN's. For many other applications the primary criteria may be that they are easy to construct (given the difficult environmental conditions that may be encountered).

Network Design

The siting of receivers is a critical issue, requiring careful reconnaissance and an understanding of the deformational phenomenon being monitored. Apart from the need to have stable, well defined benchmark monuments, the following concerns and constraints must also be addressed:

- The locations must have a clear view of the sky and at the same time facilitate line-of-sight communications as well as power generation.
- As the system will have to be deployed for long periods, security issues are significant and the involvement of local organisations is desirable.
- Multipath disturbance from nearby structures should be a minimum.
- If there is already instrumentation installed at the station, advantage may be taken of these facilities as they may have power, security and communications infrastructure.
- In order to "check" the deformation monitoring system, regular calibration may have to be conducted by means of periodic precise GPS surveys of the monuments.

GPS Hardware

To date deformation networks have consisted of "top-of-the-line" dual-frequency GPS receivers. These are also used in commercial RTK systems adapted for deformation use (Lowry & MacLeod, 1997). However, to significantly lower the hardware costs of future deformation systems (e.g. by a factor of ten) simpler GPS hardware must be considered.

Communication Links

Data telemetry is possibly the weakest "link" of any system. Without telemetry the data from the various network stations cannot be transferred to the MCS for processing. Where the infrastructure is available, telephone lines are, of course, the best data link option. However, in remote area locations such an option may not be available. The use of satellite communications technology would be preferable because of its versatility (e.g. the GPS array would not need to be deployed in such a way as to ensure line-of-sight communications), however it is generally too expensive. What is therefore often used is VHF/UHF radio technology.

Power Supply

Power is needed to operate the GPS receiver, the computer and the telemetry sub-systems. However, if the sites are not able to be serviced by mains power, a configuration based on batteries and a solar recharger will typically be used. Power management will therefore always be a significant challenge.

Data Processing Strategy

The heart of any GPS system is its software. In terms of carrier phase observation modelling we may distinguish between "GPS geodesy" and "GPS surveying" type processing. In the former many additional parameters are estimated, including satellite orbital elements, tropospheric scale parameters, etc. Such a processing strategy distinguishes the SCIGN and GSI networks that stretch across hundreds of kilometres. In the case of small scale systems, such as those based on "RTK-GPS surveying" modelling, the data processing software generally has the following characteristics:

- "On-the-fly" ambiguity resolution (AR) techniques as developed for use with dual-frequency receivers.
- Epoch-by-epoch "kinematic" data processing mode.
- The broadcast ephemerides contained within the Navigation Message is used.

However, there are several aspects of the deformation monitoring application that deserve comment as they do impact on the design of *optimised* data processing software:

- Because the receivers are more or less stationary, then the baseline components are already known (certainly to within 10cm), making AR more reliable.
- Several quality control algorithms will need to be implemented to detect bad data.
- Multi-baseline capability is desirable, so that the system can be easily scaled up to cater for a large number of GPS receivers.

Some Options for Permanent GPS-Based Deformation Monitoring Systems

Some of the defining characteristics and general features of three GPS-based deformation monitoring systems are summarised in Table 1. The deformation monitoring system may be custom designed, or based on "off-the-shelf" hardware and software components. If they are custom designed, then the following features may be incorporated:

(a) True continuous monitoring, or sampled monitoring every hour, or every day, etc.
(b) Single baseline mode, or based on a large network of receivers.
(c) Real-time processing at the base, or "near-real-time" (e.g. on a daily basis).
(d) Employ a kinematic baseline model, or a static baseline model, or selectable.
(e) Full duplex, half duplex or single duplex communications link between stations.

In Table 1, "Type 1" systems refers to the geodetic arrays such as SCIGN's and GSI's; "Type 2" refers to the commercial RTK-based system such as described in Lowry & MacLeod (1997); and "Type 3" refers to such custom designed systems as described in Hein & Riedl (1995), Han & Rizos (1996), Rizos et al. (1997).

Different hardware configurations, operational procedures as well as data processing strategies will be necessary to address the various deformation monitoring applications. However, the ideal system must be of relatively low cost, sensitive to the expected deformational signal and able to accommodate a large number of GPS receivers. All other design issues are minor in comparison to these. *There is a role for GPS-based deformation monitoring systems, but it will only be significant if the major constraints identified in this paper can be overcome.*

ACKNOWLEDGMENTS

The Australian Research Council is supporting a project to "Develop, Test and Deploy a GPS Array System for Continuous, Automatic Monitoring of Earth Deformations Arising From Volcanic Activity" (1996-1998). A follow-on project (for 1998-2000) has recently been approved. The first author is supported by a Trimble Navigation NZ scholarship.

REFERENCES

BOCK, Y., and others, 1997. Southern California Permanent GPS Geodetic Array: continuous measurements of regional crustal deformation between the 1992 Landers and 1994 Northridge earthquakes. *J.Geophys.Res.*, Vol.102, No.B8, 18013-18033.

HAN, S. & C. RIZOS, 1996. GPS network design and error mitigation for real-time continuous array monitoring systems. *9th Int. Tech. Meeting of the Sat. Div. of the U.S. Inst. of Navigation*, Kansas City, Missouri, Sept. 17-20, 1827-1836.

HEIN, G.W. & B. RIEDL, 1995. First results using the new DGPS real-time deformation monitoring system "DREAMS". *8th Int. Tech. Meeting of the Sat. Div. of the U.S. Inst. of Navigation*, Palm Springs, California, Sept. 12-15, 1467-1475.

LOWRY, A. & R. MacLEOD, 1997. PMoS - a real time precise DGPS continuous deformation monitoring system. *10th Int. Tech. Meeting of the Sat. Div. of the U.S. Inst. of Navigation*, Kansas City, Missouri, Sept. 16-19, 923-927.

RIZOS, C., S. HAN & C. ROBERTS, 1997. Permanent automatic low-cost GPS deformation monitoring systems: error mitigation strategies and system architecture. *10th Int. Tech. Meeting of the Sat. Div. of the U.S. Inst. of Navigation*, Kansas City, Missouri, Sept. 16-19, 909-917.

TSUJI, H., Y. HATANAKA, T. SAGIYA, & M. HASHIMOTO, 1995. Coseismic crustal deformation from the 1994 Hokkaido-Toho-Oki earthquake monitored by a nationwide continuous GPS array in Japan. *Geophys. Res. Let.*, Vol..22, No.13, 1669 -1672.

Table 1: General features of three GPS-based deformation monitoring systems.

CHARACTERISTIC	TYPE 1	TYPE 2	TYPE 3
Relative accuracy	10^{-8}	10^{-6}	10^{-6}
Dual-frequency	D-F	D-F	mixed
Coms link	telephone	UHF/VHF	various
Modelling sophistication	geodesy	surveying (baseline)	surveying (multi-baseline)
Kinematic or static model	static	kinematic	variable
Solution strategy	daily solutions	real-time (epoch-by-epoch)	near-real-time (session-by-session)
Centralised processing	yes	no, solutions sent back to base	yes
Limit to receivers	no	unknown	no
Operational expertise required	high	low to moderate	low
Areal extent	> 100 km	10 - 20 km	10 - 100 km
Portable	no	partly	yes
Cost/site	> US$40K	>US$20K	US$2-3K

MODIFIED GPS-OTF ALGORITHMS FOR BRIDGE MONITORING: APPLICATION TO THE PIERRE-LAPORTE SUSPENSION BRIDGE IN QUEBEC CITY

Rock Santerre and Luc Lamoureux
Centre de Recherche en Géomatique
Université Laval
Québec, Canada G1K 7P4

Abstract

Algorithms for On-The-Fly ambiguity resolution have been modified for the deformation monitoring of a suspension bridge. Instantaneous relative positioning of a deformation network at a precision of about ±5 mm horizontally and ±10 mm vertically has been achieved using GPS L1 phase observations. Particular attention has been paid to the modeling of relative tropospheric delay and to the phase center calibration between antennas of different types. Moving averages on station coordinates have also been applied to reduce multipath effects.

The Pierre-Laporte bridge is a 6-lane, 1040-m-long suspension bridge which crosses the St. Lawrence river in Québec City. Three 48-hour GPS sessions have been conducted during the months of July and October 1996 and February 1997. For each session, 5 GPS receivers were observing at a data rate of 2 seconds.

Daily variation in the vertical position of the antenna located at the deck center shows clear correlation with respect to temperature and vehicle loading. Transverse movement of the deck center has been monitored and correlated with (transverse) wind speed. Seasonal variation in temperature caused a vertical displacement of the deck center, a contraction of the towers as well as a displacement of the towers towards the river banks.

Modified OTF algorithms

In this study, a suspension bridge is seen as a "kinematic" deforming structure but with limited movement amplitude. The OTF (On-The-Fly) algorithms used to resolve GPS phase ambiguity, even if the receiver is in motion, have been modified for the particularity of a suspension bridge. The algorithms have been designed to work with single frequency receivers. The steps of the evaluation and the validation of L1 phase ambiguities and least-squares solutions are schematically presented in Figure 1 and described in the next paragraphs.

To reduce the number of ambiguity combinations, good a priori coordinates must be available. Unlike dam deformation GPS surveys, one cannot use the coordinates determined from a previous session because the deformation of a suspension bridge (even the towers) can easily exceed 10 cm (half of a L1 wavelength) between 2 sessions. The coordinates obtained from filtered code solutions or from double difference float solutions, as used in pure kinematic mode, give too many ambiguity sets to be tested. A solution is to start with a static-like processing of a short data span. The time span must be long enough to provide a positioning precision of ±5 cm (one quarter of a L1 wavelength), but within this time period the amplitude of the movement of the bridge should not exceed 5 cm. For bridge towers this criteria is always respected, but attention must be paid to the deck center movement when there are rapid fluctuations of temperature, wind speed or traffic jam during this "static" session. For the other epochs, the coordinates of the previous epoch are used as a priori coordinates. With the a priori coordinates, double difference ambiguity is calculated, for each pair of satellites, starting from the first epoch of the observation session, with the following equation:

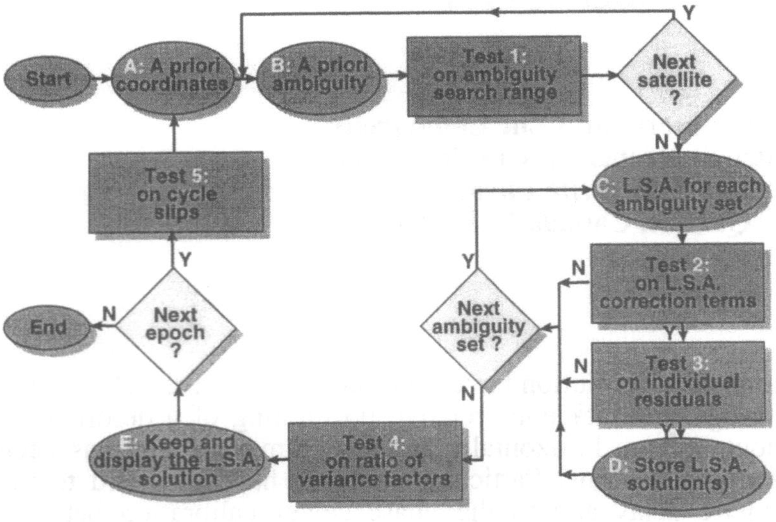

Figure 1: Data flow for ambiguity resolution and LSA solution.

$$\nabla \Delta \mathrm{N} = \frac{\nabla \Delta \Phi - \nabla \Delta \rho - \nabla \Delta \mathrm{dtrop}}{\lambda} \tag{1}$$

If the decimal part of equation (1) departs by more than 0.25 cycle from the closest integer value (this can be due to multipath or observation noise) two integer ambiguity values associated to this pair of satellites will be tested. If not, only the closest integer value will be retained.

Note that the ambiguities are not parameters to be estimated in the least-squares adjustment (LSA). In fact, different potential ambiguity combinations are tried and tested to isolate the correct ambiguity set among all possible candidates. Least-squares adjustment is performed with all ambiguity combinations and 3 criteria are used to identify the correct ambiguity set:

- the correction to the a priori coordinates must be smaller than 10 cm (in absolute value);
- the residuals must be smaller than 10 cm (in absolute value);
- the ratio of the second smallest and the smallest a posteriori variance factors must be larger than 2.

A test is also done to detect potential cycle slips between epochs. In this step, Doppler data at epochs i and i+1 are used to predict phase variation between epochs i and i+1, as follows:

$$\left| \left(\Phi_{i+1} - \Phi_i\right) - \left(\frac{D_i + D_{i+1}}{2}\right) \left(T_{i+1} - T_i\right) \right| \geq \lambda \tag{2}$$

If the difference (in absolute value) is larger than one wavelength, the observation at epoch i+1 is flagged, a new ambiguity value is calculated with equation (1), and the validation steps are again checked. Equation (2) is very efficient when the data rate is high and the receiver is static or moving slowly between the observation epochs.

To avoid mathematical correlation in double difference observations, the Hatch approach [Hatch, 1990] has been used. With this approach the observation weight matrix is kept diagonal and the residuals are associated to each satellite, instead of each pair of satellites. This last feature is very important for a detailed analysis of remaining unmodeled errors. At every epoch (at each 2 seconds in our case), four unknown parameters are solved in a least-squares adjustment.

GPS error sources and modeling

The main GPS errors to deal with are relative tropospheric delay, multipath and antenna phase center variation. The effect of other errors such as ionospheric refraction, ephemerides and Selective Availability (SA) can be assumed negligible for baselines as short as 1 km. The readers are referred to [Santerre, 1991] for a detailed analysis of GPS error propagation in GPS network.

Tropospheric refraction: Because some baselines of the monitoring network have height differences of about 60 m, attention has been paid to the modeling of relative tropospheric delay. Meteorological data collected at the level of the bridge deck were extrapolated in altitude for the GPS stations located on the 2 bridge towers. The height extrapolation algorithms for temperature, pressure and relative humidity are described in [Rothacher et al., 1986] and used as input for the Hopfield's tropospheric model.

Multipath: To avoid multipath as much as possible, antennas have been setup above any reflective objects at the reference stations and on the top of the 2 towers. At the deck center, the antenna was installed on a 4-m-high beam to avoid multipath (and obstructions) from the vehicles. However, the towers and suspension cables could be a source of multipath. A method to filter (short period) multipath effect is the use of the moving average of instantaneous coordinate results. The moving average window must be carefully selected, i.e., wide enough to reduce multipath effect but not too wide to avoid the loss of information about real bridge movement. Unfortunately, no chokering antennas were available to us during the observation sessions.

Antenna phase center variation: Three types of antennas have been used, namely Ashtech Z-XII and LD-XII and NovAtel 501. The relative phase centers of the antennas have been calibrated using a 1-m calibration beam. This aluminum beam, with precisely known length, was setup and leveled on a geodetic point and oriented (with an optical device) towards the direction of another geodetic point [Bourassa, 1994]. The comparison of the known baseline components of the calibration beam with the baseline components obtained from GPS observation sessions allows to determine relative phase center variation between pairs of GPS antennas. Four 24-hour observation sessions have been conducted for the relative phase center calibration of the 5 GPS antennas. The relative phase center offset never exceeded 2 mm in the 3 baseline components. These measured offsets are within the manufacturers technical specifications and are not significant, considering the precision associated to the calibration process. Accordingly, no further correction has been applied for the phase center variation.

Before surveying the Pierre-Laporte bridge, the methodology has been tested on known baselines with a device which mimics realistic bridge displacements. The length of the baselines and the height differences were typical of the bridge monitoring network. For independent and precise comparisons, the baseline height differences have been determined by geometrical leveling (relative geoid undulations were also taken into account). This analysis showed that instantaneous GPS relative positioning has typically a precision of about ±5 mm, horizontally and about ±1 cm, vertically, when PDOP factor does not exceed a value of 6.

Deformation monitoring network

Figure 2 shows the stations of the deformation monitoring network. The baseline length (D) and the height difference (Δh) between the stations are given in Table 1. Two reference stations (RIN1 and RIN2) have been setup in bedrock, on the north river bank close to the bridge. Stations TON and TOS are located on the top of the North and South towers, respectively. Station TACE is located on the deck center of the bridge.

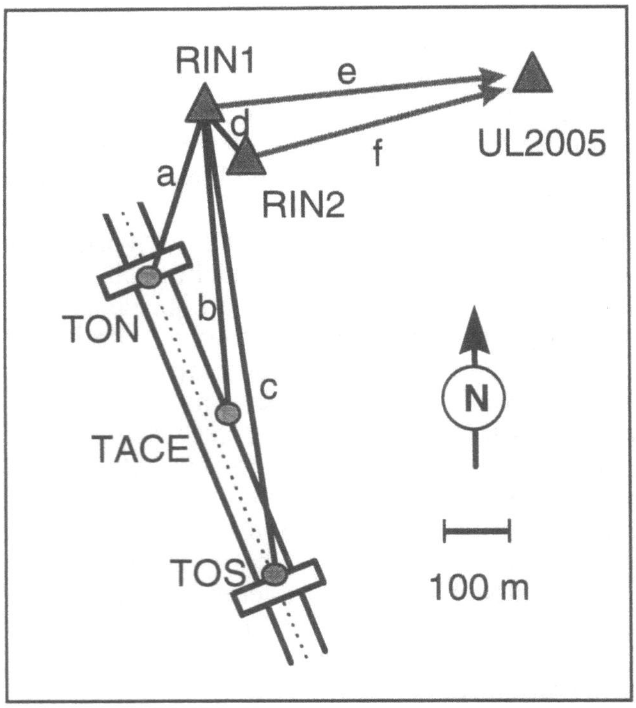

Figure 2: Deformation monitoring network of the Pierre-Laporte bridge.

Table 1: Baseline length and height difference.

Baseline	a	b	c	d	e	f
D (km)	0.3	0.7	1.0	0.1	3.5	3.5
Δh (m)	59	-4	59	-2	17	19

To verify the stability of the 2 reference stations, they have been connected, by GPS observations and precise geometric leveling, to station UL2005 located on the Laval University campus. This concrete pillar, located 4 km away from the bridge, is part of the GPS calibration network in Quebec City area, established by the Geodetic Surveys of Canada. Between the 3 sessions, the coordinates of stations RIN1 and RIN2, relative to UL2005, did not change significantly. So, the same set of coordinates for the reference stations was used for all 3 sessions.

The Pierre-Laporte bridge is a 6-lane, 1040-m-long suspension bridge which crosses the St. Lawrence river in Québec City. The center span is 670-m-long supported by 2 110-m-high towers. The bridge structure, of a total weight of 18,000 tons, is suspended at 2 main cables of 62 cm in diameter. Such an engineering structure is affected by strong constraints caused by winds, traffic loading and temperature variations.

Analysis of the movement of station TACE (deck center)

Figure 3 illustrates the vertical displacement of the deck center (station TACE) and the temperature as a function of time for the July 1996 session. The epochs where the temperature is the highest roughly correspond to the lowest vertical positions of the deck center. A variation of temperature of 8°C has caused a vertical displacement of about 12 cm. Thermal dilatation of the suspension cables explains this movement of the deck center. The vibrations caused by vehicle traffic on the bridge are more apparent in Figure 4 which is an enlargement of Figure 3, between 13:00 and 14:30 (EDT). Moreover, the influence of vehicle loading is clearly detected in Figure 4. In fact, one can see the impact of a car accident leading to the entire blockage of 3 lanes of the bridge, for about 30 minutes. The center deck of the bridge, half-full with stopped vehicles, has been bent by about 40 cm during this time period.

Figure 5 shows the transverse displacement of station TACE caused by the transverse wind speed. The wind speed was recorded on the bridge deck with an anemometer. The high correlation between these 2 quantities are clearly visible. For example, a transverse wind speed of 20 km/h (from the East direction) has caused a transverse displacement of the deck center of about 5 cm, in the West direction.

In Figures 3 to 5, the gray line, superimposed on the raw displacement results, is the moving average applied onto the coordinates to reduce the impact of observation noise and multipath. The length of the moving average window is 60 seconds, in Figures 3 and 4, and 120 seconds in Figure 5.

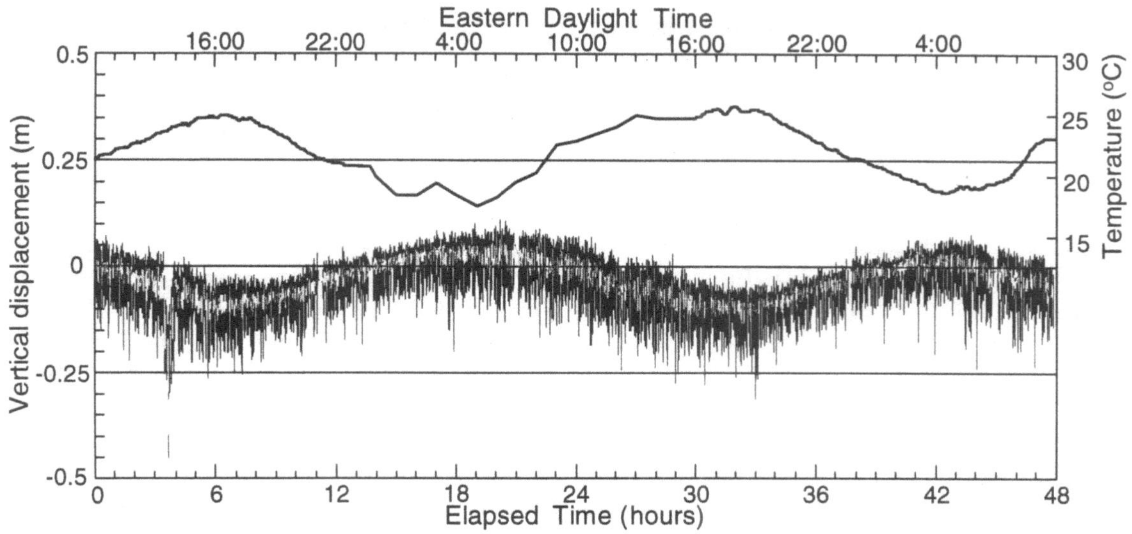

Figure 3: Vertical displacement of station TACE as a function of temperature (July 1996).

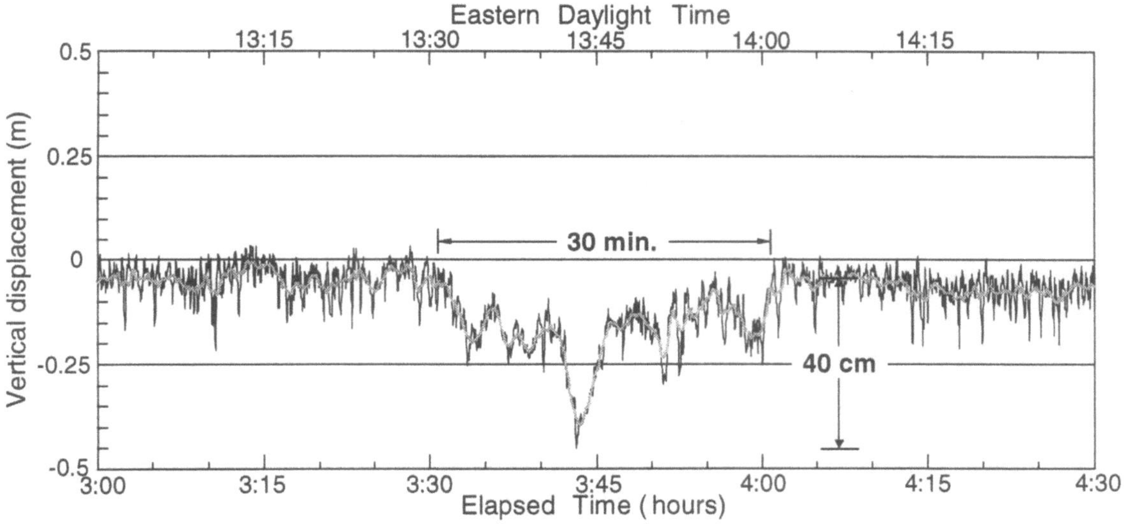

Figure 4: Vertical displacement of station TACE due to traffic loading during a car accident.

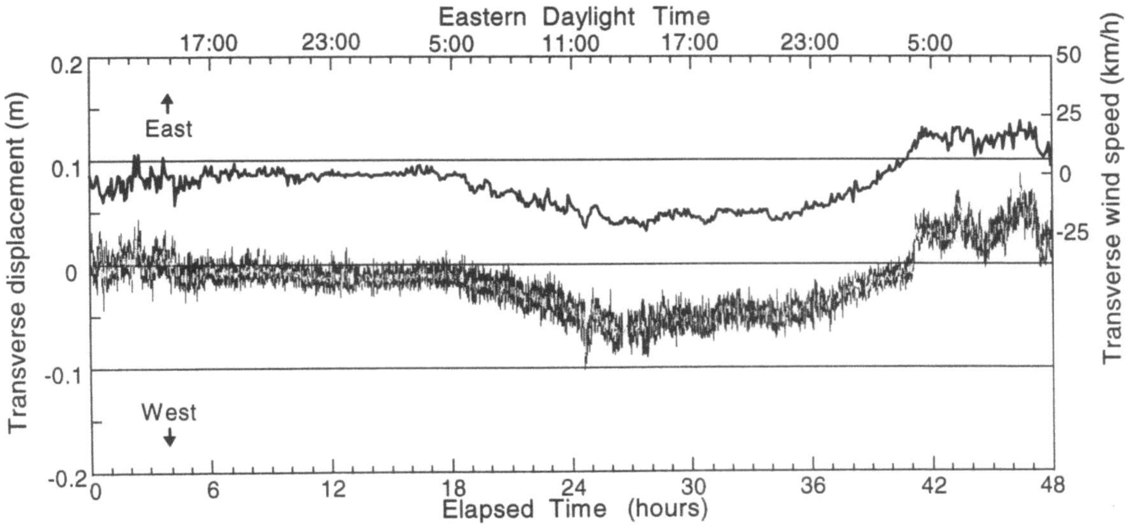

Figure 5: Transverse displacement of station TACE as a function of wind speed (October 1996).

Analysis of the seasonal displacements of the bridge (towers and deck center)

Table 2: Seasonal variations of the average coordinates of the bridge stations.

	variations between July and February sessions
station TON	L: 4.0 cm T: -1.0 cm V: -5.5 cm
station TOS	L: -3.8 cm T: -0.6 cm V: -6.1 cm
station TACE	L: 0.2 cm T: -4.5 cm V: 45.5 cm
temperature	-25°C
transverse wind speed	21 km/h

L: longitudinal, T: transverse, V: vertical

Table 2 presents the seasonal variations (between July 1996 and February 1997) of the average coordinates for each station. The analysis of these results shows that the 2 towers moved away from each other of 8 cm along the longitudinal axis of the bridge. The vertical variation of -6 cm at stations TON and TOS is a consequence of the thermal contraction of the 110-m-high towers. In fact, between the 2 sessions the average temperatures dropped by 25°C. The deck center (station TACE) rose of 46 cm between the 2 seasons. The thermal contraction of the suspension cables explains this movement. Finally, the average transverse displacement of station TACE (5 cm towards the West direction) is caused by the variation of the average transverse wind speed of 21 km/h between the 2 sessions.

Conclusions and further research

A methodology, algorithms and software have been developed for bridge deformation monitoring in post-processing. The precision of instantaneous relative positioning is about ±5 mm horizontally and ±10 mm vertically, when PDOP values are smaller than 6. The analysis of the displacements shows: i) a transverse movement of the deck center due to a transverse wind speed of about: -2 mm / 1 km/h; ii) a vertical displacement of the deck center with temperature variation of about: -2 cm / 1°C; iii) a vertical variation of the deck center correlated with traffic flow; iv) an elongation of the tower with respect to temperature variation of about: 2 mm / 1°C; and v) a displacement of the towers towards the river banks of about: -1.5 mm / 1°C.

Other GPS sessions (with a higher GPS data rate) for studying the high frequency movements of the bridge are planned. Provisions are made to add more receivers with chokering antennas. The methodology will also be implemented for real-time deformation monitoring and applied to other engineering structures such as towers, skyscrapers and dams.

Acknowledgments: We would like to thank the Quebec Transportation Department for their helps during the GPS campaigns, and the Natural Sciences and Engineering Research Council of Canada and the Laval University Faculty of Forestry and Geomatics for their financial supports.

References

Bourassa, M. (1994). "Étude des effets de la variation des centres de phase des antennes GPS." Mémoire de maîtrise, Département des sciences géomatiques, Université Laval, Québec, 109 p.

Hatch, R. (1990). "Instantaneous Ambiguity Resolution." IAG Symposium No. 107, Kinematic Systems in Geodesy, Surveying, and Remote Sensing, Banff, Alberta, Canada, pp. 299-308.

Rothacher, M. et al. (1986). "The Swiss 1985 GPS campaign." Proceedings of the Fourth International Geodetic Symposium on Satellite Positioning, Austin, Texas, pp. 979-991.

Santerre, R. (1991). "Impact of GPS satellite sky distribution." Manuscripta Geodaetica, 16, pp. 28-53.

THE GPS COMPONENT OF THE PROJECT FOR DIGITAL MAPPING OF THE KARST AQUIFER SYSTEM NEAR CURITIBA, BRAZIL†

Marcelo C. Santos, Antonio J. B. Vieira & Henrique Firkowski

Universidade Federal do Paraná
Curso de Pós-Graduação em Ciências Geodésicas
Caixa Postal 19011
81531-990 Curitiba PR Brazil
mcsantos@geoc.ufpr.br, berutti@geoc.ufpr.br, firk@geoc.ufpr.br

Abstract

The correct management of water supply, both surface and ground water, has tremendous environmental implications, directly affecting the quality of life of the population in general, with economical and political consequences. Very recently, part of the Karst aquifer system, which passes under part of the Brazilian Southern Region, has become under scrutiny for solving part of the problem of water supply for Curitiba and surroundings. The project for its digital mapping aims to correctly understand the processes governing this aquifer allowing for its use without depleting it.

This paper describes the GPS component of this project, applied to a test area of around 40 km^2. This component is basically composed of the surveying of the artesian wells and the definition of a local geoid. For the latter, the strategy adopted was to reposition two diferent geoidal maps, one given by the program MAPGEO from IBGE, and the other presented by Sá & Molina with a far greater resolution. Bench marks of the fundamental vertical network were used as constraints in the adjustments carried out for both geoidal maps and as control points ("ground truth"). Results indicate that a better local geoid was obtained based on the second geoidal map.

This project presents an application of Geodesy to sustainable development.

---

† A more detailed description of the work summarized here will be published in *GPS World* magazine.

KINEMATIC GPS POSITIONING WITH ADAPTIVE KALMAN FILTERING TECHNIQUES

Jinling Wang, Mike Stewart and Maria Tsakiri
School of Surveying and Land Information
Curtin University of Technology
GPO Box, U 1987, WA 6845
Perth, Australia

Abstract

In a kinematic GPS positioning system, state parameters are usually estimated using the Kalman filtering method. It has been noted that reliable ambiguity resolution, and estimation of other state parameters are highly dependent on the correct stochastic models for the GPS double difference measurements. In this paper, a real-time statistical procedure to estimate the covariance matrix of GPS double differenced measurements is presented. Test results indicate that with the proposed procedure, the reliability of ambiguity resolution and kinematic positioning results can be improved.

Introduction

In a kinematic GPS positioning system, the determination of the state parameters, which may include the position, velocity and acceleration of a moving platform and other parameters of interest, such as unknown carrier phase ambiguities, is achieved by continuously tracking GPS satellites in view and taking a series of measurements (code pseudo-ranges and carrier phases) to the satellites at constant intervals or epochs. The mathematical models involve the *dynamic model*, *measurement model* and *stochastic models*. The real-time estimates of the state parameters can be performed using the *Kalman filtering* technique (eg. Gao et al, 1996; Landau and Euler, 1992; Qin et al; 1992).

With the Kalman filtering technique, the state parameters can be not only estimated recursively in real-time mode, but also have the statistically defined optimal properties. It is well known, however, that the optimum filtering results are highly dependent on the correctness of the adopted stochastic models.

In GPS kinematic positioning, the precise carrier phase measurements with code pseudo-ranges can generate accurate positioning results. In order to avoid the effect of the errors in process noise covariance matrix on the state estimation, the filter can be set up to operate only on the measurement noise. However, stochastic modelling for GPS double differenced measurements still is a difficult task in real-time data positioning.

The widely used approach to construct the covariance matrices of GPS double differenced measurements is based on the variance-covariance propagation law, and it is always assumed that all the one way code or carrier phase measurements are independent and they have the same accuracy. Unfortunately, these assumptions are not realistic. The accuracy of the GPS code and carrier phase measurements may change with different measuring conditions. Adopting incorrect stochastic models for GPS measurements in data processing will inevitably result in unreliable statistics for ambiguity resolutions and biased positioning results (eg. Hatch and Euler, 1994).

In the literature, the procedures for the on-line estimation of the process and measurement noise matrices are termed *Adaptive filtering techniques* (e.g. Mehra, 1972, Chin, 1979). In this paper, one of the commonly used adaptive kalman filtering techniques will be reviewed. Based on the filtering residuals, a new adaptive filtering procedure will be presented. The performance of the new procedure will be tested with a GPS data set.

Kalman Filtering as Least Squares

The functional models used in GPS kinematic positioning can be expressed as

the dynamic model : $$x_k = \Phi_{k,k-1} x_{k-1} + \tau_k \ , \qquad (1)$$

and the measurement model : $$z_k = H_k x_k + \varepsilon_k \ , \qquad (2)$$

The corresponding stochastic models are further assumed as:

$$E(\tau_k \tau_i^T) = \begin{cases} Q_k & i = k \\ 0 & i \neq k \end{cases}, \quad E(\varepsilon_k \varepsilon_i^T) = \begin{cases} R_k & i = k \\ 0 & i \neq k \end{cases}, \text{ and} \quad E(\varepsilon_k \tau_i^T) = 0, \qquad (3)$$

where x_k is the state parameter vector which may contain the time-dependent position and velocity parameters and time-independent carrier phase ambiguity parameters, if they have not been resolved; $\Phi_{k,k-1}$ is the state transition; τ_k is the random error vector; z_k is the measurement vector; H_k is the corresponding design matrix; and ε_k is the measurement noises; Q_k is the so-called process noise covariance matrix; and R_k is the measurement noise matrix; k and i are the time indices.

For convenience in the following discussion, the Kalman filtering estimation formulae are derived using least squares method. The predicted state values $\bar{x}_k$ can be obtained by:

$$\bar{x}_k = \Phi_{k,k-1} \hat{x}_{k-1} \ , \qquad (4)$$

where $\hat{x}_{k-1}$ is the optimal estimator of the state parameters at the previous epoch $(k-1)$. The covariance matrix of $\bar{x}_k$ is expressed as:

$$Q_{\bar{x}_k} = \Phi_{k,k-1} Q_{\hat{x}_{k-1}} \Phi_{k,k-1}^T + Q_k \,, \tag{5}$$

By integrating the measurements y_k and the predicted values of state parameters $\bar{x}_k$, the optimal estimators of the state parameters x_k can be obtained using the least squares technique (e.g. Pelzer, 1985; Cross 1983). The Gauss-Markov models are :

$$E(l_k) = A_k x_k \quad \text{or} \quad l_k + v_k = A_k x_k \,, \tag{6}$$

$$C_{l_k} = P^{-1} = \begin{bmatrix} R_k & 0 \\ 0 & Q_{\bar{x}_k} \end{bmatrix}, \tag{7}$$

where $l_k = \begin{bmatrix} z_k \\ \bar{x}_k \end{bmatrix}$, $v_k = \begin{bmatrix} v_{z_k} \\ v_{\bar{x}_k} \end{bmatrix}$, $A_k = \begin{bmatrix} H_k \\ E \end{bmatrix}$.

In equation (6), v_{z_k} is the residual vector of the measurements z_k; $v_{\bar{x}_k}$ is the residual vector of the pseudo-measurements $\bar{x}_k$; E is the identity matrix; C_{l_k} is the covariance matrix of the measurement vector l_k; P is the weight matrix. With the least squares principle, the optimal estimators of the state parameters and their covariance matrix are formulated as (e.g. Pelzer, 1985):

$$\hat{x}_k = (A_k^T P_k A_k)^{-1} A_k^T P_k l_k = \bar{x}_k + G_k d_k \,, \tag{8}$$

$$Q_{\hat{x}_k} = (A_k^T P_k A_k)^{-1} = Q_{\bar{x}_k} - G_k Q_{d_k} G_k^T \,, \tag{9}$$

where G_k is the *gain matrix*; d_k is the *innovation vector* and Q_{d_k} is its covariance matrix, which are described as:

$$G_k = Q_{\bar{x}_k} H_k^T Q_{d_k}^{-1} \,, \tag{10}$$

$$d_k = z_k - H_k \bar{x}_k \,, \tag{11}$$

$$Q_{d_k} = R_k + H_k Q_{\bar{x}_k} H_k^T \,, \tag{12}$$

respectively. The innovation vector and its covariance matrix are important statistics used in the existing adaptive filtering techniques.

Adaptive Kalman Filtering Techniques

Various adaptive filtering techniques are divided into three categories: Bayesian, maximum likelihood (ML) and innovation approaches (Mehra, 1972). The Bayesian and

ML methods are computationally intensive and cannot be realistically used for real-time data processing (e.g. Chin, 1979). One of the well-known innovation approaches is the so-called *covariance-matching*. For simplicity, the process noise matrix is assumed known, and the identification of the measurement noise covariance matrix R_k only, is discussed.

Covariance Matching Procedure Based on Innovations

The basic concept of this procedure is to make the actual innovation covariance matrix consistent with their theoretical values (Mehra, 1972). The actual covariance matrix of the innovations sequence d_k is calculated approximately by its sample covariance:

$$\hat{Q}_{d_k} = \frac{1}{m}\sum_{i=1}^{m} d_{k-i} d_{k-i}^T , \tag{13}$$

where m is chosen empirically to give some statistical smoothing. By matching the above estimated covariance matrix with its theoretical form presented by (12), the measurement noise covariance matrix is estimated as:

$$\hat{R}_k = \hat{Q}_{d_k} - HQ_{\bar{x}_k} H^T = \frac{1}{m}\sum_{i=1}^{m} d_{k-i} d_{k-i}^T - HQ_{\bar{x}_k} H^T . \tag{14}$$

Many numerical tests indicate, however, that the innovations sequences are very sensitive to the approximate values used in the linearization of the GPS measurements equations. More importantly, with this procedure, it cannot be guaranteed that the resulting matrix $\hat{R}_k$ is positive definite.

A New Procedure Based on Filtering Residuals

In order to obtain a realistic estimator of the measurement noise covariance matrix, instead of the innovations sequence, the more precise estimator of the measurement noise level should be used. From equation (6), we obtain the measurement filtering residuals:

$$v_{z_k} = -R_k Q_{d_k}^{-1} d_k = H\hat{x}_k - z_k , \tag{15}$$

which, obviously, is the best estimator of the measurement noise level because the estimated values $\hat{x}_k$ (not the predicted values $\bar{x}_k$) of the state parameters are used in their computations.

Similarly to equation (13), an estimator of the measurement noise covariance matrix is derived from the filtering residuals as:

$$\hat{R}_k = \frac{1}{m}\sum_{j=1}^{m} v_{z_{k-j}} v_{z_{k-j}}^T , \tag{16}$$

After epoch m, $\hat{R}_k$ is updated with the incoming filtering residuals. m is called the width of the moving window. It is noted that, compared with the computation of the innovation vector, which is generated by the standard Kalman filtering process, the computation of the filtering residuals will involve some extra computations. Fortunately, the number of added calculations is very small.

Test Results and Analysis

A GPS kinematic data set was used to test the performance of the adaptive filtering technique. The data were collected on July 18, 1997, at the Fremantle port in Perth, Australia, using two Trimble 4000SSE dual frequency receivers. The rover receiver antenna was mounted on a boat, and was moving around the offshore test area (about 5.5km away from the reference station). During the 10 minutes of the experiment, 7 satellites were tracked. The data collection rate was 1Hz.

In the data processing, carrier phase and pseudo-range measurements from both L1 and L2 frequencies were used and double differenced measurements were formed. The true ambiguity values were recovered with the whole data set. The first epoch solution with fixed ambiguities was used as the initial values for the Kalman filtering. The initial standard deviations for L1 and L2 carrier phase and pseudo-range data are 1.0m, 1.0m, 0.010m, 0,012m, respectively. The width of the moving window in the adaptive filtering procedure was set to 200 epochs.

After 200 epochs, ambiguity resolutions were performed on-the-fly for each period of 10 epochs. A total of 40 batches were formed. In each batch, the solutions with and without using the new adaptive filtering (AF) procedure expressed by equation (16) have been obtained (in this test, the Covariance Matching Procedure did not generate a positive definite matrix $\hat{R}_k$). Some of the results are shown in Fig.1 and Fig. 2.

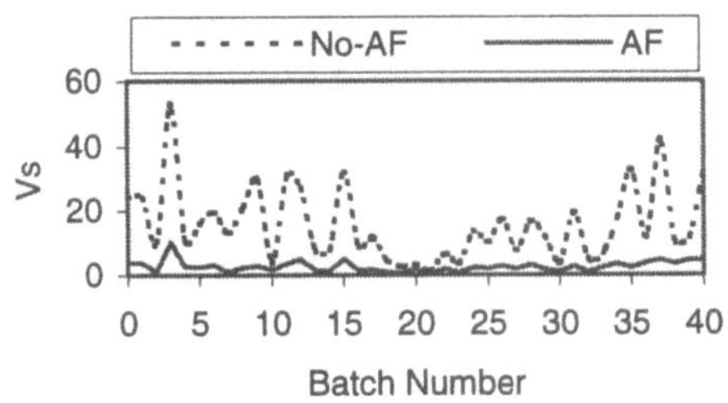

Fig. 1 : Ambiguity search space volume

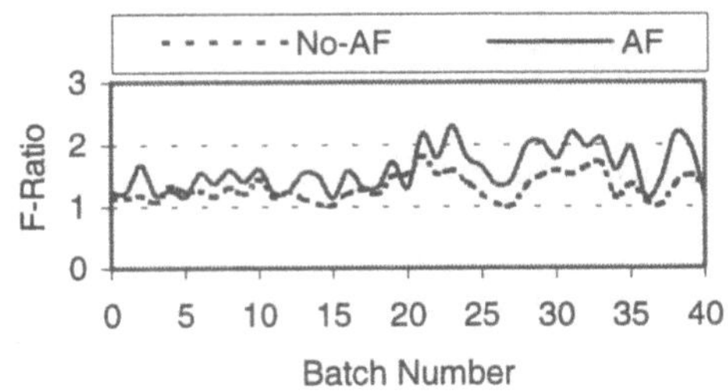

Fig. 2 : *F*-ratio values

Fig.1 shows the changes of the ambiguity search space volume Vs, which is defined by the determinant (times 10^8) of the ambiguity states covariance matrix. In all the cases, the ambiguity search spaces with the adaptive filtering procedure are smaller than those without using it.

Fig. 2 compares the *F*-ratio values used in the ambiguity resolutions (eg. Landau and Euler, 1992). In almost all the cases, the *F*-ratio values in the solutions with the adaptive filtering procedure are bigger than those without using it. More importantly, without using the adaptive filtering procedure, in up to 10% cases, the wrong ambiguities have been identified as the best ambiguity combinations. Therefore, the reliability of the ambiguity resolution is improved by using the adaptive filtering procedure.

Test results have also demonstrated the accuracy of the kinematic positioning results can be improved using the adaptive filtering procedure. For example, at epoch 200 when the filtering process was using the preset standard deviations for measurements, the estimated standard deviations of (x, y, z) coordinates were 0.009m, 0.021m and 0.010m, respectively. At epoch 201 when the filtering process began to use the estimated measurement noise covariance matrix with the adaptive procedure, the standard deviations of (x, y, z) coordinates are down to 0.002m, 0.006m and 0.002m, respectively.

Concluding Remarks

In GPS kinematic positioning, an incorrect measurement noise covariance matrix can result in the wrong ambiguity resolution and thus biased positioning results. The currently used method to construct such a covariance matrix is not reliable in some practical applications.

Based on the filtering residuals, a new adaptive filtering procedure has been proposed. With such an adaptive procedure, the measurement noise covariance matrix can be estimated in real time data processing. Test results have shown that with the new adaptive procedure, the reliability of the ambiguity resolutions and the accuracy of kinematic positioning can be significantly improved.

References

Chin L. (1979) Advances in Adaptive Filtering, In Leondes C.T. (ed.): *Advances in Control Systems Theory and Applications*, Academic Press, No. 15, pp.277-356.

Cross P. A. (1983) Advanced Least Squares Applied to Positioning Fixing, *working paper* No.6, Department of Land Surveying, Polytechnic of East London.

Gao Y., J. F. McLellan and J.B. Schleppe (1996) An Optimized Fast Ambiguity Search Method for Ambiguity Resolution on the fly, *Proceedings of IEEE PLANS*, Atlanta, Georgeia, pp.246-253.

Hatch R. and H.-J. Euler (1994) Comparison of several AROF kinematic techniques. *Proceedings of ION-GPS-94*, Salt Lake City, Utah, pp.363-370.

Landau H. and H-J Euler (1992) On-the-Fly Ambiguity Resolution for Precise Differential Positioning, *Proceedings of ION GPS-92*, Albuquerque, New Mexico, pp. 607-613.

Mehra R. K. (1972) Approaches to Adaptive Filtering, *IEEE Transactions on Automatic Control*, AC-17, Oct., pp. 693-698.

Pelzer, H (1985) Statische, Kinematische und Dynamische Punktfelder, In Pelzer (Ed.): *Geodätische Netze in Landes- und Ingenieurvermessung* II, Stuttgart, Wittwer.

Qin X., Gourevitch S., and M. Kuhl (1992) Very Precise Differential GPS: Developemnt status and results, *Proceedings of ION GPS-92*, Albuquerque, New Mexico, pp. 615-624.

AUTHOR INDEX

Abu, S. 49
Adam, J. 27
Aebersold, R. 351
Altamini, Z. 57
Aquino, M.H.O. 43, 101
Arciniegas, S. 168
Ashkenazi, V. 43, 101
Augath, W. 35

Baldi, P. 343
Baran, L.W. 252
Barbaste, J. 125
Barbato, F. 217
Beattie, D. 187
Becker, M. 49, 66
Beutler, G. 93
Blewitt, G. 8
Blick, G. 107
Blitzkow, D. 73, 205
Boonphakdee, C. 49
Boucher, C. 8, 57
Brouwer, F. 35
Brunini, C. 199, 277
Brunner, F.K. 349
Bruton, A.M. 325
Bruyninx, C. 18

Chang, C.C. 114
Chang, R.G. 114
Chao, J. 265
Chen, R. 259
Chen, W. 43
Cocard, M. 351
Collins, J.P. 331
Costa, S.M.A. 73, 187, 217

Da Fonseca Junior, E.S. 149
Dare, P. 120
Davies, P.B.H. 8
De Jonge, P.J. 223, 313
Ding, X. 155, 265
Dominici, D. 271
Dosé, A. 351
Dousa, J. 18
Drewes, H. 168, 174, 180, 193, 199, 211

Edwards, K. 125
Ehrnsperger, W. 18

Fachbach, N. 18
Fagard, H. 168
Ferraro, L. 18
Figurski, M. 18
Firkowski, H. 387
Forsberg, R. 363
Fortes, L.P.S. 73, 167
Franke, P. 66

Gaudet, R.J. 59
Geiger, A. 351
Gelo, S. 161
Gende, M. 277
Grant, D. 107
Grejner-Brzezinska, D.A. 337
Gurtner, W. 27

Han, S. 283, 307, 319
Harsson, B.G. 27
Hartinger, H. 349
Heflin, M.B. 8
Hernández, J.N. 193
Herring, T.A. 8
Houghton, H. 155
Hoyer, M. 168

Ihde, J. 27

Jaworski, L. 161
Jia, M. 369
Johansson, J.M. 18, 229
Jokela, J. 131

Kahle, H.-G. 351
Kahmen, H. 357
Kakkuri, J. 259
Kaniuth, K. 180, 199, 289

Keller, K. 363
Kleusberg, A. 277
Koivula, H. 137
Kouba, J. 8, 245
Kumar, M. 168

Lamoureux, L. 381
Lang, H. 35, 351
Langley, R.B. 331
Lin, L.-S. 295, 307
Luz, R.T. 73
Lyszkowicz, A. 161

Malys, S. 1
Marsella, M. 343
Marti, U. 143
Maturana, R. 168
Meier, B. 351
Menge, F. 237
Mertikas, S. 295, 307
Mingsamon, S. 49
Mireault, Y. 245
Moirano, J. 180, 199
Monico, J.F.G. 149, 205
Moore, T. 43, 101

Nardi, A. 18
Nielsen, C.S. 363

Odijk, D. 313
Oliveira, L.C. 205
Ollikainen, M. 137, 301

Pereira, K.D. 73, 168, 187
Pérez, R.R. 217
Poutanen, M. 137

Radicioni, F. 271
Reinhart, E. 49
Rizos, C. 283, 295, 307, 319, 375
Roberts, C. 375
Rothacher, M. 93

Sacher, M. 35
Saleh, H. 120
Saleh, J. 125
Santerre, R. 381
Santos, M.C. 205, 387
Scaramuzza, M. 351
Schaer, S. 93
Schlüter, W. 27, 66
Schmitz, M. 237
Schneider, D. 143
Seeber, G. 168, 237
Seeger, H. 49, 66
Seemüller, W. 211
Selli, S. 271
Shagimuratov, I.I. 252
Sillard, P. 57
Slater, J.A. 1
Springer, T.A. 18, 93
Stewart, M.P. 155, 265, 369, 389
Stoppini, A. 271
Stuber, K. 180
Subiza, W.H. 217
Swiatek, A. 161
Symons, L.J. 43

Teunissen, P.J.G. 313
Tiberius, C.C.J.M. 313
Torchetti, R. 168
Tremel, H. 180, 193
Troyer, L. 307
Tsakiri, M. 369, 389

Van Mierlo, J. 35
Vespe, F. 18
Vieira, A.J.B. 387
Vittuari, L. 343
Völksen, C. 237

Wang, J. 389
Wanninger, L. 86
Weber, G. 18, 66
Weber, R. 79
Wiget, A. 18
Woeppelmann, G. 27
Wübbena, G. 237

Zdunek, R. 161
Zielinski, J.B.

Springer and the environment

At Springer we firmly believe that an international science publisher has a special obligation to the environment, and our corporate policies consistently reflect this conviction.
We also expect our business partners – paper mills, printers, packaging manufacturers, etc. – to commit themselves to using materials and production processes that do not harm the environment. The paper in this book is made from low- or no-chlorine pulp and is acid free, in conformance with international standards for paper permanency.

Springer

MIX
Papier aus verantwortungsvollen Quellen
Paper from responsible sources
FSC® C105338

If you have any concerns about our products,
you can contact us on
ProductSafety@springernature.com

In case Publisher is established outside the EU,
the EU authorized representative is:
Springer Nature Customer Service Center GmbH
Europaplatz 3, 69115 Heidelberg, Germany

Printed by Libri Plureos GmbH
in Hamburg, Germany